Springer Texts in Business and Economics

Springer Texts in Business and Economics (STBE) delivers high-quality instructional content for undergraduates and graduates in all areas of Business/Management Science and Economics. The series is comprised of self-contained books with a broad and comprehensive coverage that are suitable for class as well as for individual self-study. All texts are authored by established experts in their fields and offer a solid methodological background, often accompanied by problems and exercises.

H. A. Eiselt • Carl-Louis Sandblom

Operations Research
A Model-Based Approach

Third Edition

 Springer

H. A. Eiselt
Faculty of Business Administration
University of New Brunswick
Fredericton, NB, Canada

Carl-Louis Sandblom
Department of Industrial Engineering
Dalhousie University
Halifax, NS, Canada

ISSN 2192-4333 ISSN 2192-4341 (electronic)
Springer Texts in Business and Economics
ISBN 978-3-030-97161-8 ISBN 978-3-030-97162-5 (eBook)
https://doi.org/10.1007/978-3-030-97162-5

© The Editor(s) (if applicable) and The Author(s), under exclusive license to Springer Nature Switzerland AG 2012, 2022

This work is subject to copyright. All rights are solely and exclusively licensed by the Publisher, whether the whole or part of the material is concerned, specifically the rights of translation, reprinting, reuse of illustrations, recitation, broadcasting, reproduction on microfilms or in any other physical way, and transmission or information storage and retrieval, electronic adaptation, computer software, or by similar or dissimilar methodology now known or hereafter developed.

The use of general descriptive names, registered names, trademarks, service marks, etc. in this publication does not imply, even in the absence of a specific statement, that such names are exempt from the relevant protective laws and regulations and therefore free for general use.

The publisher, the authors and the editors are safe to assume that the advice and information in this book are believed to be true and accurate at the date of publication. Neither the publisher nor the authors or the editors give a warranty, expressed or implied, with respect to the material contained herein or for any errors or omissions that may have been made. The publisher remains neutral with regard to jurisdictional claims in published maps and institutional affiliations.

This Springer imprint is published by the registered company Springer Nature Switzerland AG
The registered company address is: Gewerbestrasse 11, 6330 Cham, Switzerland

Once upon a time, there was a little boy who liked nothing better than to sit in his yard and look at the moon after it rose. One day, he asked his dad: "Daddy, I have been looking at the moon for many weeks now, and it always looks the same. Is this all there is to it then?" "Hmm," replied his father and pondered the situation. The next day, his father returned to his son, who was again looking at the moon. He handed him a book. "Here son," he said. "This is a book with pictures that show you the moon from many different angles. This may explain what the moon is all about." The little boy smiled as he read the book. It was a good book.
Jesús Peregrino

Preface

Since the 1960s, operations research (or, alternatively, management science) has become an indispensable tool in scientific management. In simple words, its goal on the strategic and tactical levels is to aid in the decision-making process and, on the operational level, automate decision making. Its tools are algorithms, procedures that create and improve solutions to a point at which optimal or, at least, satisfactory solutions have been found.

While many texts on the subject emphasize methods, the focus of this book is on the applications of operations research in practice. Typically, a topic is introduced by means of a description of its applications, a model is formulated, and its solution is presented. Then the solution is discussed and its implications for decision making are outlined. We have attempted to maximize the understanding of the topics by using intuitive reasoning while keeping mathematical notation and the description of techniques to a minimum. The exercises are designed to fully explore the material covered in the chapters, without resorting to mind-numbing repetitions and trivialization.

The book is designed for (typically second year) students of business management and industrial engineering. With the appropriate deletions, the material can be used for a one-semester course in the subject, while the complete material will be sufficient for a full-year course. Parts of the book require some, albeit rudimentary, knowledge of differential calculus. The reasoning and explanations are intuitive throughout. Each algorithm is followed by a numerical example that shows in detail how the method progresses. After presenting the applications and the techniques, each chapter ends with a number of fully solved examples that review the concepts covered in the chapter. Some more technical material and additional practice problems have been taken out and are available at the file-sharing site https://drive.google.com/drive/folders/1yCvgkZ4lrF_EruXQqJsZO5kWkcoM5xvE & the mirror site drive.google.com/drive/folders/17iILwG52uqwWxCMQa74csG4lEWtS4KTA. In addition to clarifications and modifications as well as corrections of some typos, the third edition includes new chapters on nonlinear programming and on reliability, material that is typically covered in courses on industrial engineering. Furthermore, some of the previously existing material has been streamlined.

It is our pleasure to thank all the people who have made this volume possible. Special thanks are due to Mrs. Jialin Yan and Mrs. Ramya Prakash. Their encouragement and assistance are much appreciated. Thanks also to the Buddha Man for his meticulous typing and to Mr. Logan Geraghty for his help with the figures. Without the help of all of these individuals, this book would not have seen the light of day. We like to thank all of them.

Fredericton, NB, Canada H. A. Eiselt
Halifax, NS, Canada C.-L. Sandblom

Symbols and Definitions

The list below includes the special symbols used in the book. While we have made every possible attempt to keep notation to a minimum, some symbolism will be required.

Notation

∈ :   Element of
⊆ :   Subset
⊂ :   Proper subset
∪ :   Union of sets
∩ :   Intersection of sets
∅ :   Empty set
|S|:   Cardinality of the set S

$x \in [a, b]: a \leq x \leq b$
$x \in [a, b[: a \leq x < b$
$x \in]a, b]: a < x \leq b$
$x \in]a, b[: a < x < b$

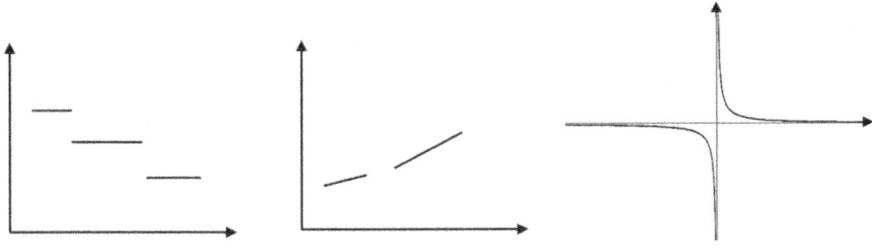

Fig. 1 Non-continuous functions

ix

$\lceil x \rceil$	:	Ceiling of x, the smallest integer greater than or equal to x
$\lfloor x \rfloor$	:	Floor of x, the largest integer smaller than or equal to x
$|x|$	:	Absolute value of x
$a := a + b$:		Valuation, a is replaced by $a + b$

Definition 1: In nonmathematical terms, we will call a function *continuous*, if it can be drawn without lifting the pen from the paper. In other words, it does not contain holes, jump discontinuities, or vertical asymptotes, as shown in Fig. 1.

Definition 2: A function is *differentiable*, if its derivative exists at all points (thus excluding noncontinuous functions, V-shaped functions or those with cusps, and functions with vertical tangent lines).

Contents

1   Introduction to Operations Research..................... 1
  1.1   The Nature and History of Operations Research........... 1
  1.2   The Main Elements of Operations Research............... 4
  1.3   The Modeling Process................................. 10
  Reference... 13

2   Linear Programming..................................... 15
  2.1   Introduction to Linear Programming..................... 15
  2.2   Applications of Linear Programming..................... 20
      2.2.1   Production Planning............................ 20
      2.2.2   Diet Problems.................................. 22
      2.2.3   Allocation Problems............................ 28
      2.2.4   Blending Problems.............................. 34
      2.2.5   Transportation and Assignment Problems.......... 38
      2.2.6   Dynamic Production-Inventory Models............. 45
      2.2.7   Employee Scheduling............................ 49
      2.2.8   Cutting Stock Problems......................... 52
  2.3   Graphical Representation and Solution.................. 70
      2.3.1   The Graphical Solution Method.................. 71
      2.3.2   Special Cases.................................. 80
  2.4   Postoptimality Analyses............................... 90
      2.4.1   Graphical Sensitivity Analyses................. 90
      2.4.2   Optimal Solution and Sensitivity Analyses
              on a Printout................................. 101
  2.5   Duality.. 114
  References.. 123

3   Multiobjective Programming.............................. 125
  3.1   Vector Optimization.................................. 126
  3.2   Solution Approaches to Vector Optimization Problems.... 131
  3.3   Goal Programming..................................... 133

4	**Nonlinear Programming**		143
	4.1	Introduction	143
	4.2	Applications of Nonlinear Programming	146
		4.2.1 Forestry Logging	146
		4.2.2 Maximizing Tax Revenues	147
		4.2.3 Optimizing Inventory and Queuing Models	148
	4.3	Properties of Nonlinear Programming Problems	149
	4.4	Solution Methods for Nonlinear Optimization	152
		4.4.1 Newton's Method for Unconstrained Optimization	153
		4.4.2 Constrained Optimization	154
	References		159
5	**Integer Linear Programming**		161
	5.1	Definitions and Basic Concepts	162
	5.2	Applications of Integer Programming	170
		5.2.1 Diet Problems Revisited	172
		5.2.2 Land Use	174
		5.2.3 Modeling Fixed Charges	175
		5.2.4 Workload Balancing	179
		5.2.5 Political Districting	181
	5.3	Solution Methods for Integer Programming Problems	183
		5.3.1 Cutting Plane Methods	183
		5.3.2 Branch-and-Bound Methods	188
		5.3.3 Heuristic Methods	194
6	**Network Models**		215
	6.1	Definitions and Conventions	215
	6.2	Social Network Analysis	218
	6.3	Network Flow Problems	220
	6.4	Shortest Path Problems	230
	6.5	Spanning Tree Problems	238
	6.6	Routing Problems	239
		6.6.1 Arc Routing Problems	241
		6.6.2 Node Routing Problems	249
	References		266
7	**Location Models**		267
	7.1	The Major Elements of Location Problems	267
	7.2	Covering Problems	270
		7.2.1 The Location Set Covering Problem	271
		7.2.2 The Maximal Covering Location Problem	274
	7.3	Center Problems	277
		7.3.1 1-Center Problems	278
		7.3.2 p-Center Problems	280
	7.4	Median Problems	281
		7.4.1 Minisum Problems in the Plane	281

		7.4.2 Minisum Problems in Networks	285
	7.5	Other Location Problems	290
	References	302

8	**Project Networks**	303
	8.1 The Critical Path Method	304
	8.2 Project Acceleration	312
	8.3 Project Planning with Resources	317
	8.4 The *PERT* Method	320
	Reference	332

9	**Machine Scheduling**	333
	9.1 Basic Concepts of Machine Scheduling	334
	9.2 Single-Machine Scheduling Models	336
	9.3 Parallel Machine Scheduling Models	340
	9.4 Dedicated Machine Scheduling Models	345
	Reference	352

10	**Decision Analysis**	353
	10.1 Introduction to Decision Analysis	353
	10.2 Visualizations of Decision Problems	355
	10.3 Decision Rules Under Uncertainty and Risk	357
	10.4 Sensitivity Analyses	363
	10.5 Decision Trees and the Value of Information	366
	10.6 Utility Theory	373

11	**Multiattribute Decision Making**	385
	11.1 The General Model and a Generic Solution Method	385
	11.2 TOPSIS	390
	11.3 The Analytic Hierarchy Process	391
	References	399

12	**Inventory Models**	401
	12.1 Basic Concepts in Inventory Planning	401
	12.2 The Economic Order Quantity (*EOQ*) Model	404
	12.3 The Economic Order Quantity with Positive Lead Time	407
	12.4 The Economic Order Quantity with Backorders	410
	12.5 The Economic Order Quantity with Quantity Discounts	412
	12.6 The Production Lot Size Model	414
	12.7 The Economic Order Quantity with Stochastic Lead Time Demand	417
	12.7.1 A Model that Optimizes the Reorder Point	418
	12.7.2 A Stochastic Model with Simultaneous Computation of Order Quantity and Reorder Point	419
	12.8 Extensions of the Basic Inventory Models	420
	Reference	424

13	**Stochastic Processes and Markov Chains**	425
	13.1 Basic Ideas and Concepts	425
	13.2 Steady-State Solutions	430
	13.3 Decision Making with Markov Chains	431
14	**Reliability Models**	439
	14.1 Fundamentals of Reliability Systems	440
	14.2 Time Aspects of Reliability	445
	14.3 Failure Rates and the Hazard Function	447
	14.4 Redundancy and Standby Systems	450
	14.5 Estimating Reliability	451
	Reference	455
15	**Waiting Line Models**	457
	15.1 Basic Queuing Models	458
	15.2 Optimization in Queuing	466
16	**Simulation**	473
	16.1 Introduction to Simulation	474
	16.2 Random Numbers and Their Generation	475
	16.3 Examples of Simulations	479
	16.3.1 Simulation of a Waiting Line System	480
	16.3.2 Simulation of an Inventory System	483
	References	494
17	**Heuristic Methods**	495
	Reference	504
	Appendices	505
	Appendix A: Vectors and Matrices	505
	Appendix B: Systems of Simultaneous Linear Equations	506
	Appendix C: Probability and Statistics	509
	References	515
	Index	517

Introduction to Operations Research

In its first section, this introductory chapter first introduces operations research as a discipline. It defines its function and then traces its roots to its beginnings. The second section highlights some of the main elements of operations research and discusses a number of potential difficulties and pitfalls. Finally, the third section of this chapter suggests an eight-step procedure for the modeling process.

1.1 The Nature and History of Operations Research

The subject matter, *operations research* or *management science* (even though there may be philosophical differences, we use the two terms interchangeably), has been defined by many researchers in the field. Definitions range from "a scientific approach to decision making" to "the use of quantitative tools for systems that originate from real life," "scientific decision making," and others. In the mid-1970s, the Operations Research Society of America (then one of the two large professional societies in the field) defined the subject matter as follows:

> Operations Research is concerned with scientifically deciding how to best design and operate man-machine systems usually under conditions requiring the allocation of scarce resources.

Today, the Institute for Operations Research and Management Science (INFORMS) markets operations research as the "science of better." What all of this essentially means is that the science uses indeed quantitative techniques to make and prepare decisions, by determining the most efficient way to act under given circumstances. In other words, rather than throwing large amounts of resources (such as money) at a problem, operations research will determine ways to do things more efficiently.

Rather than being restricted to being a toolkit for quantitative planners, operations research is much more: it is a way of thinking that does not just "do things," but, during each step of the way, attempts to do them more efficiently: the waitress, who provides coffee refills along the way rather than making special trips; the personnel manager, who (re-) assigns employees so as to either minimize the number of employees needed or schedule employees to shifts, so as to make them more pleasant; the municipal planner, who incorporates the typically widely diverging goals and objectives of multiple constituents or stakeholders when locating a new sewage treatment plant; and the project manager, who has to coordinate many different and independent activities. All of these individuals can benefit from the large variety of tools provided by operations research.

There are a number of obstacles that stand in the way of the extensive use of operations research in practice. One of those obstacles is awareness. If managers were to be able to realize that a problem may possibly benefit from the use of a quantitative analysis, it does not matter at all, whether or not they can perform the analyses themselves: there are plenty of specialists out there to do the job. The first step, though, requires someone to simply realize that operations research could be applied to a problem that presently requires a solution. This is the reason why we have written this book with a strong focus on applications.

When trying to find out where you are and where you are going, it is always a good idea to determine where you come from. The next few paragraphs will highlight some of the main milestones to operations research. Clearly, space limitations require us to cut many corners. We would like to refer to the eminently readable history of operations research by Assad and Gass (2011).

What are usually considered to be early contributions are usually advances in mathematics or statistics: Diophantine's discourses on integer solutions to linear equations in the third century AD are related to integer programming; Euler's work on the Königsberg bridge problem in 1736 is the first occurrence of graph theory; Pascal, Bernoulli, and Bayes have made major advances in statistics. All results found by these and many other scientists have put down a mathematical and statistical foundation, on which operations research (and many other disciplines) can rest comfortably.

While many authors credit the advances in the military in World War II to the birth of operations research, we believe that the groundwork was laid considerably earlier. F.W. Taylor is often called the "father of scientific management," when he performed his time studies in 1881. His main question was "what is the best way to do a job," which could very well be the motto of operations research. Henry L. Gantt introduced bar charts, "Gantt charts" in today's parlance, for scheduling problems, and Agner Krarup Erlang introduced the discipline of queuing in 1909 when working at the Copenhagen Telephone Exchange. Another contribution in the early days was made by F.W. Harris in 1913, when he developed the "economic order quantity" for inventory management, a result that is so robust that it is, in one way or another, used to this day. All of these individuals would today be referred to as industrial engineers, as their main concern was the smooth functioning of industrial processes.

It is hardly surprising that these early contributions occurred at a time that saw more complex industrial processes (the assembly line is but one example), a

1.1 The Nature and History of Operations Research

tremendous increase in the division of labor, and with it the need for coordination of activities.

Later notable work was performed by John von Neumann in the 1920s, when he introduced the theory of games to the world. In this context, the theory of games must be understood as the study of competitive situations. As such, it is little surprise that economists are among the main users of game theory. Leontief's input–output models and Kantorovich's mathematical planning models for the Soviet economy were main contributions in the 1930s. The 1940s saw Hitchcock's transportation problem, Stigler's diet planning, and the aforementioned advances based on military applications.

However, the main event occurred in August of 1947 when George Bernard Dantzig developed what is now called the simplex method for linear programming. Arguably, no other event has influenced the science of operations research more than this development. Other main developments are due to John F Nash, who extended von Neumann's results in game theory and proved some main theorems, Bellman's dynamic programming principle in 1950, and Kuhn and Tucker's optimality conditions for nonlinear optimization problems in 1951 (which was later discovered to be a reinvention of work by Karush in 1939). The year 1951 saw not only the first full publication of Dantzig's simplex method in the open literature, but also the first computer-based simplex method. As a matter of fact, the advances in computing hardware and software had a tremendous impact on the advances of operations research. Without the progress made in computer sciences, operations research would not have been able to gain the status it has today.

New results keep pouring in. Starting in 1950, the *Operational Research Quarterly* (later renamed the *Journal of the Operational Research Society*) was the first journal in the field published in the UK. The first American journal followed in 1952 with the *Journal of the Operations Research Society of America*. Today, many countries have their own operations research journals. To name a few, there are the *European Journal of Operational Research* (the largest operations research journal by size, close to 10,000 pages per year), *INFOR* (Canada), the *OR Spectrum* (Germany), *TOP* (Spain), the *Central European Journal of Operations Research* (Austria), the *Yugoslav Journal of Operations Research* (even though the country of Yugoslavia no longer exists), *OPSEARCH* (India), *Pesquisa Operacional* (Brazil), and many more. In addition, there are many specialist journals, such as *Computers & Operations Research*, *Mathematical Programming*, *Management Science*, *Naval Research Logistics*, and many others.

And wherever there is a national journal, more often than not there is a national society behind it. Each of these societies has an annual meeting, where researchers present their latest findings. Again, in addition to these national conferences, there are meetings devoted to special topics such as optimization, logistics, supply chains, location, transportation, scheduling, and many more. It should be apparent by now that the number of contributions can only be described as vast.

1.2 The Main Elements of Operations Research

This section will briefly explain the main elements of operations research. Essentially, operations research is concerned with quantitative (mathematical) models and their solution. This is actually where some people make a distinction: they claim that while management science is mostly concerned with models, operations research deals mostly with solution methods. The model that we build for a given scenario is a (hopefully close) picture of reality and consists of mathematical expressions and relations intended to describe the problem at hand. A model will never include all components a real situation does: features such as some of the choice criteria used by consumers are not fully observable and will have to be ignored; decision makers' risk aversion, while it may be included to some degree in a decision-making model, is not completely explainable and will have to be estimated or ignored. It is very important to distinguish between the original real-life problem and the mathematical model that is built. The step from the problem to the model entails gains and losses: while we are losing some information, we gain solvability and gain insights into the structure of the problem. As one of the founders of the science once wrote "the purpose of mathematical programming [a subtopic of operations research, eds.] is insight, not numbers."

The next issue to be decided upon is the level of aggregation. In conjunction with the decision maker the modeler will have to decide what to include in the model and what to leave out. Comprehensive models are nice and avoid problems with suboptimal solutions, but they are large, labor-intensive, and thus expensive. It depends on the specific situation what level is appropriate. The aforementioned suboptimal solutions may occur, if we were to optimize for only one department of a firm. For instance, the manager of a shipping department of a large firm that manufactured construction equipment consistently hired employees through a temp agency, even though this was much more expensive than hiring people directly. However, temporary employees were paid by headquarters, while the pay for regular employees came directly out of the manager's budget. Thus, hiring temporary employees was optimal for the manager, even though it cost the firm much more money.

The usual way to describe situations for operations research models is to first list everything we *want* to do, and everything that we *have to* do and respect. What we want to achieve (high profit, high levels of customer satisfaction, a large market share, low production costs, or similar) is summarized in our *objective(s)*. On the other hand, all requirements (such as budget limitations, limitations of a firm's capabilities with respect to manpower, knowledge base, available machinery, and others) are summarized in the *constraints*. While in most operations research tools, objectives and constraints are clearly separated, this is not always true in reality. Consider, for instance, a simple budget constraint. It will state that the amount of money we can spend in, say, a month cannot exceed the amount of money that we have. We can write this as a formal constraint, as long as we are aware of the fact that a constraint in operations research is considered absolute. This means that whatever the solution technique is, it will not consider any solution that violates this constraint. In other words, it will consider the constraint as hard. In practice, however, the

1.2 The Main Elements of Operations Research

constraint may not be that hard. We could, for instance, take out a loan and temporarily spend more than we have. While this is necessarily a temporary measure and we would like to avoid it, it is possible in practice. Written as a constraint, however, it is not possible. There are techniques that address this problem by formulating the models differently, allowing short-term deficits, but trying to minimize this. Notice that doing so transforms what we used to clearly consider a constraint into part of the objective function.

Another important issue to deal with is the issue of measurement. While many of the features of a model are easily measurable (profits, travel time, processing times, the length and width of a stretch of road to be paved are typical examples), others may not be. Particularly in models involving human psychological and sociological factors regarding the public sector, it is not always clear how to measure entities in the model. For instance, when locating a desirable public facility, how do we measure accessibility of the location? We may express this feature in terms of average distance to the potential clients of this facility, but other issues such as opening hours or parking may be just as relevant. A particular difficulty is presented when dealing with models that include such nebulous concepts as "fairness." Take the issue of a simple speeding ticket. The penalty is supposed to hurt the speeder and thus make an impression. An obvious constraint is that the law must treat everybody equally. So whoever speeds will receive the same penalty, say $100. Such a penalty will, of course, hurt people with a small annual income a lot more than somebody who makes several hundred thousand dollars a year. Thus, while the penalty is equal, the pain is not. Alternatively, one could assess the penalty as, say, 1% of the monthly net income, thus trying to distribute the pain equally (it still does not, as a $10 penalty for somebody with a $1000 monthly net income hurts more than a $1000 penalty for somebody with a $100,000 monthly income). This would lead to ridiculous penalties, such as $50,000 for speeding and, in the final analysis, is nonsensical as it negates any incentive to earn more, as prices, fines, and other expenses are adjusted accordingly. We leave decisions regarding this issue to the decision makers; the purpose of our discussion was to highlight the difficulty expressing some features of a model quantitatively.

Typically, in many problems, public and private, the objective is ill-defined or fuzzy. This will require the decision maker to formulate a surrogate or proxy expression instead. For instance, the measurable criterion "profit" may be a proxy for the "well-being of the company." Again, moving from the true objective to a proxy involves gains and losses: we lose as the proxy expression does not do exactly what we want our objective to do, but we gain by obtaining something that is measurable and thus can be included in a quantitative model. The choice of a suitable proxy expression is crucial for the applicability of the model.

Each operations research model will consist of mathematical expressions involving *parameters* and *variables*. While parameters are fixed and given numbers that we know (or can determine), but that are not under our control, variables are numbers that we do not know (but would like to know), and which are under our direct jurisdiction. Consider a few examples. The estimated demand for a product is a parameter, while the number of units that we make is a variable. The amount of beef we put in each can of chili is under our jurisdiction and thus a variable, while the nutritional content of beans in the can is not under our control and thus a parameter. The type of truck we use for a shipment and the route the truck takes are variables, while the location of our customer's warehouse is a parameter. The purpose of the solution method is then to determine the actual values of the variables, i.e., determine the production level, the quantity of beef in a can, and the type of truck and route that a truck takes.

As far as solution techniques are concerned, we distinguish between exact and heuristic solution techniques. An exact technique will find a solution that respects all constraints included in the model and optimizes the objective specified by the decision maker. Note that such a solution is actually optimal for the model but not necessarily for the problem: if the model is only a very rough approximation of the problem, the solution that is labeled "optimal" will not be very useful to the decision maker, as it is optimal only for the model, not the problem. Solutions obtained by heuristic or approximation methods are typically much easier to find, but, as the name implies, may not necessarily be the best (or even very good for that matter). For more details, readers are referred to Chap. 17. Solution techniques come in two versions: they are either closed-form solutions or iterative algorithms. A closed-form solution is essentially a formula or a set of formulas: we input the parameters and use the formula(s) to obtain values for our variables. Few models exist for which closed-form solutions exist. Most models require the use of *iterative algorithms*. (As an aside, the name algorithm derives from the Persian mathematician Al-Khwarizmi, who worked and published in algebra around 820 AD.) An iterative algorithm starts with an iteration counter, initially set to 0, and a solution x^0, possibly a guess, the solution that is presently employed, or some other solution, and then performs a test that checks whether or not the solution satisfies certain criteria (such as feasibility or optimality). If the solution passes the test, it is accepted and the algorithm terminates; otherwise, the present solution is modified by the algorithm, a new solution x^k (typically an improved guess) is generated, the iteration counter is increased by one, and the new solution returns to the test. At this point, we are in a loop, which is repeated, until an acceptable solution has been found. Each loop that involves the test and an improvement constitutes an iteration of the algorithm. Many modern large-scale problems require thousands, if not millions, of iterations to find a solution. This explains why the use of high-speed digital computers is crucial for the solution of nowadays models. The flow chart in Fig. 1.1 may explain the main ideas.

The above discussion has made frequent mention of the concept of solution. It is important to realize that a solution is a *set of instructions*. In production planning, it will tell the decision maker what quantities to produce, in diet planning it will tell the chefs what meals to prepare (e.g., in seniors' residences), and in transportation planning it will tell the dispatcher which trucks and which drivers to dispatch where, and with what loads. Associated with each solution is a value of the objective function, a value that tells the planner how much money will be made if a certain production plan is adopted, how much it will cost, if a certain meal schedule is followed, and what are the consequences if we schedule trucks and drivers in a

1.2 The Main Elements of Operations Research

Fig. 1.1 Steps of an iterative algorithm

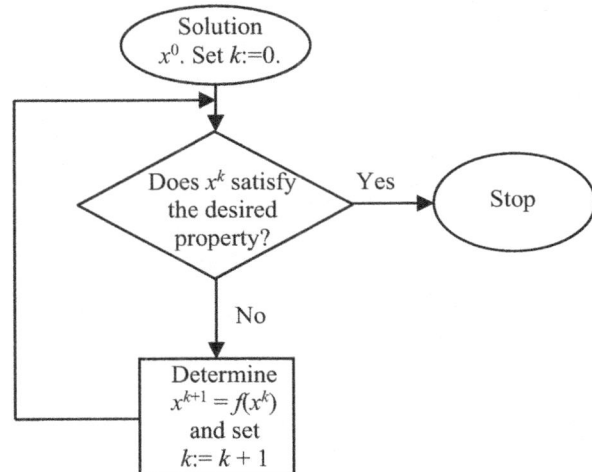

certain way. As it has no part in reconstructing a specific course of action, we will consider the value of the objective function the *consequence* of the solution, but not part of it.

There are four major concerns or phases when applying any operations research model. They are

1. *Feasibility* (can we do this?)
2. *Optimality* (is this the best we can do with what we have?)
3. *Sensitivity* (what happens if some of the input parameters or conditions beyond our control change)
4. *Implementability* (is the solution that we have obtained something that we can actually do?)

We will explain the first three phases in a very simple numerical

Example A company faces a demand for its product. The magnitude q of the demand is a function of the price p, both of which are to be set by the company. It is known that the price-quantity relation is $p = 16 - 2q$, meaning that starting at \$16 per unit, each unit increase of the demand decreases the price all customers are willing to pay for the product by \$2. Clearly, the function is defined only for quantities no larger than 8, as for quantities higher than 8, the price would be negative. In addition to the general quantity-independent fixed costs of \$20, the cost function is $C(q) = \frac{1}{2}q^2$. The company wants to maximize its profit.

We assume that the company will set the price p and the quantity q at such values that the market is cleared, i.e., the given price p will cause customers to purchase exactly the produced amount q. In view of the price-quantity relation, this means that fixing the value of one of the two variables will determine the value of the other one;

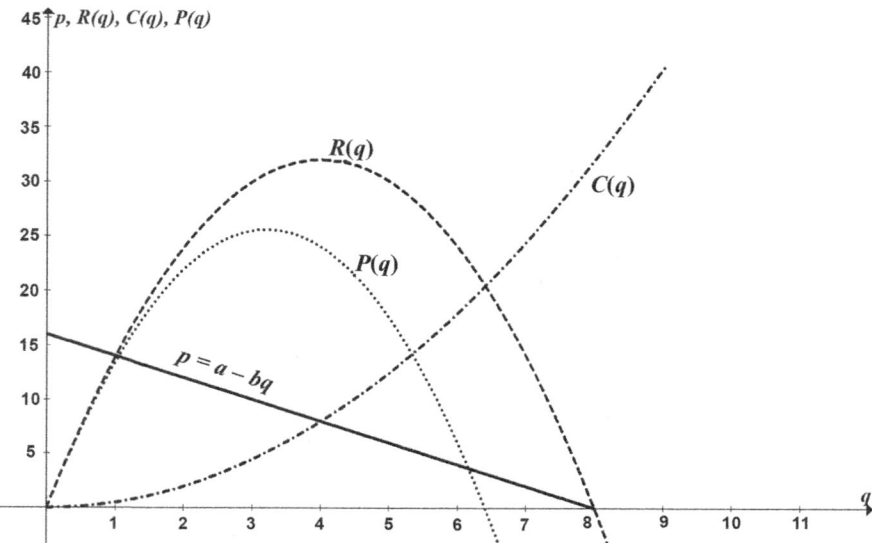

Fig. 1.2 Revenue, cost, and profit functions of example

in this sense there is only one variable whose value the company can freely set. In our discussion below, we select this to be the quantity q. In order to formulate the problem, we will employ a very simple version of what is known as the *decomposition principle*. Starting with a large component of the problem that we cannot model per se, it subdivides this component into smaller and smaller pieces, until we are able to find expressions for it. In this example, "profit" is such a component. We first decompose profit by using its definition as revenue minus costs. Now we have to deal with these two components separately. First take revenue. Again, we decompose the entity by using the definition "revenue equals price times quantity," which leaves us with these two expressions. At this point, we are able to deal with them directly, as we know that the unknown quantity has been defined as q and the price is $p = 16 - 2q$. Now consider costs. Decomposing costs, we obtain fixed and variable costs as its two components. The variable costs are $½q^2$, and, for simplicity, we will assume fixed cost of 0 for now. We can then put together the profit function (the composition phase) as

$$P(q) = (16 - 2q)q - ½q^2 = -2.5q^2 + 16q.$$

To facilitate our discussion, Fig. 1.2 provides a plot of the price-quantity relation (the solid line), the revenue function $R(q) = (16 - 2q)q$ (the dashed line), the cost function $C(q) = ½q^2$ (the dash-dotted line), and the profit function $P(q)$ (the dotted line).

We can now explain the first three major concerns by using this example. First consider feasibility. Let us assume that the planner considers it mandatory that we

1.2 The Main Elements of Operations Research

are not making losses so that any solution that provides a loss is not feasible. (This is not generally true: losses obviously occur and are generally unrelated to conditions of feasibility.) This leaves us with a *break-even analysis*, i.e., an analysis that determines the types of production plans that generate a nonnegative profit This condition is $P(q) = 0$, and in Fig. 1.2, we are looking for the quantities at which the profit function intersects the abscissa. In our example, this occurs at the quantity $q = 6.4$ (the highest quantity that generates a nonnegative profit). The reason for this somewhat counterintuitive result is that higher quantities result in (1) lower prices and (2) progressively higher costs due to the quadratic cost function. Introducing now fixed costs of, say, 20, the break-even relation is then $P(q) = 16q - 2.5q^2 - 20 = 0$, or $q \approx 1.7$ or 4.7. At this point, a positive profit can only be made for quantities between 1.7 and 4.7. Increasing the fixed costs further to, say, 30, no quantity will result in a positive profit.

Next consider the issue of optimality. The quantity that generates the maximal profit is the highest point of the profit curve and it occurs at the quantity $\bar{q} = 3.2$. The unit price of our product at that point equals $p = \$9.6$ and the resulting profit is $\$25.6$.

Finally, let us examine what happens if some of the input parameters change. We note that the optimal solution will be at $\bar{q} = 3.2$, regardless of the fixed cost. In Fig. 1.2, the fixed costs are merely a line parallel to the abscissa that moves up if the costs increase. In doing so, the area of positive profits shrinks. While the optimal quantity does not change, the profit at optimum does: with 0 fixed costs in our example, the profit at optimum is $P(\bar{q}) = 25.6$ while it shrinks to 5.6 if the fixed costs are \$20.

Another sensitivity analysis could ask what happens if the variable cost function in the original model (we typically do not consider compounded changes) were to change to the linear function $C(q) = 4q$. The break-even analysis determines a zero profit for $q = 6$, i.e., we will have positive profits if we make at most 6 units (recall that there is an upper bound of $q \leq 8$ in order to ensure positive prices). The optimal quantity is now $\bar{q} = 3$, with a price at optimum of $p = 10$, a revenue of $R(\bar{q}) = 30$, costs (with zero fixed costs) of $C(\bar{q}) = 12$, and thus a profit of $P(\bar{q}) = 18$. Introducing again fixed costs of, say, \$20 will eliminate positive profits altogether.

The last sensitivity analysis—again starting from the original situation with zero fixed costs—asks what happens if the price-quantity relation changes to $p = 16 - 1q$. Again, another inverted parabola is the profit function, and the break-even point, here again the maximal quantity that guarantees positive profits, is at $q = 10.6667$. Optimality is achieved for $\bar{q} = 5.3333$ with an associated profit of $P(\bar{q}) = 42.6667$. In general, in this modification, there is a larger range of quantities that guarantee a positive profit (the reason is that the price drops more slowly than in the original case). Since the cost function has not changed, the profit will also be higher.

In general, it is always a good idea to compare the results provided by a mathematical solver with those the decision maker or analyst comes up with intuitively. If optimized solution and intuition do not match, it is important to carefully check the model, as one of them will likely be wrong. Which one depends on the experience of the person involved and the care with which the model was formulated.

1.3 The Modeling Process

This section outlines some of the major steps that will be followed when formulating, solving, and implementing an operations research model. Clearly, each situation has its own idiosyncrasies and difficulties, but some general ideas are common to all of them. We will present the main eight steps below.

Step 1: Problem recognition In order to build a successful model, the first step is for someone to realize that it is not "business as usual," and that it is simply no longer good enough to follow the old "we have always done it like that" and, its sister expression in crime, "we have never done it like that." Problem recognition does not only include the realization that things are not what they were thought to be, but also what the potential for improvement actually is. The decision maker has to keep in mind that building a model is a lengthy and expensive process that is only worth undertaking if there appears to be significant potential for improvement. This step takes a manager who is fully familiar with the actual situation in the firm and, at the very least, somewhat familiar with what operations research can do.

Step 2: Authorization to model This step will require the analyst, who is in charge of model development, to convince management of the need to produce a model for (a part of) the operation. This "sale" of potential benefits to those who eventually have to pay for it is obviously crucial. It requires very good communication skills on the part of the analyst. Often, analysts make the mistake to get lost in their technical lingo, which not only annoys decision makers, but greatly reduces the chances of obtaining permission to model. Avoiding this pitfall requires also that the analyst clearly understands the mindset and way of thinking of the decision makers.

Step 3: Model building and data collection This is a step, in which this book can help with the model building. The first step of the modeler has to decide on the scope and the level of aggregation. Is it necessary to get into small details of the operations, or it is sufficient to adopt a macro view? This decision will have to be made in conjunction with the decision maker(s) and maybe has already been done in the previous step before the model building idea was developed and presented. Finding out what is actually important may take quite a while, but spending some time on it is definitely no waste. Once the level of aggregation has been decided upon, the modeler will determine who the relevant stakeholders are and what their objectives are. This is the time to decide on appropriate surrogate criteria. In addition to find out about from management what these objectives are, it will be necessary to learn about the constraints from the shop floor. This type of information is typically unknown in the corner offices, and it will be of tremendous benefit to the modeler and his model if such information is collected directly at the source. Some analysts went to great lengths to accomplish this. Gene Woolsey wrote about his modeling and data collection adventures in the practice journal *Interfaces* with stories that involved him getting a job as a worker in the unit he was supposed to model and seeing firsthand what the actual problems were. In addition to obtaining much better data than those typically available in offices, he also got to know the workers and their concerns. This would come as a significant benefit later on when the results of his modeling efforts would be implemented: the suggested solution would take these concerns into consideration and greatly enhance chances of worker acceptance of the changes, rather than resistance and passive boycotts of any suggested change. Once the data have been collected, the formal model will be built. This typically starts with a listing of the assumptions and simplifications that are made. This is important for the decision maker later on, when the time comes to accept or reject the recommendations made by the model. It is a lot better for the decision maker to see right away that a model is not really applicable and not implement it, rather than to implement it and having to live with the (dire) consequences later, just because some assumptions did not apply in the given situation. Since circumstances may change over time, the assumptions should be checked periodically with the decision maker, so as to avoid wasting time. Then the decision variables are defined based on what the decision maker(s) would like to know.

Step 4: Solution of the model This is a step in which the modeler's technical knowledge is required. Here, we use the appropriate computer software, and document the model properly so as to allow future users to more or less easily take over and use the model again without having to go through the entire process again. There are a few pitfalls. One of them is to use some software "just because we already have it." If the software does what we need, then it is perfectly all right to use it, but changing the model to suit existing software is a highly questionable procedure. Analysts have to keep in mind that a few thousand dollars for needed software are very little compared to the costs incurred by the model building team and the potential benefit. Another issue to keep in mind is that modeling and model solution nowadays—in contrast to the situation some decades ago—is an interactive process. In "the olden days," computational power was severely limited and expensive. As a result, modelers tried to develop the entire model as well as possible, then had it solved, and, when it came back with errors (which it always did), fixed the errors and resubmitted it. This required a lot of foresight, a lot of thought, and long and arduous searches for the errors. Nowadays, computing power is ubiquitous and cheap. A result is that modeling is typically done in an interactive fashion. A fairly small part of the model is developed first, then solved, and if there are any errors, they will be easy to find, as the model is still small. Once the analyst is happy with the modeled part, additional parts are added. The revised model is solved, and again, error detection is easy as anything wrong must be related to the new part. This process goes back and forth, until the entire model is developed.

Step 5: Model validation In this step, the analyst will determine whether or not the solution to the model that was obtained in the previous step does make sense in the context of the real problem. If this is not the case, the model is not (yet) a faithful representation of reality and it must be changed. In such a case, the process will shuttle between Steps 3 and 5. It is always a good idea to determine if the solution leads to a major change as compared to the present solution (if any). If this is the case, it is rather unlikely that the new solution will be adopted by the decision makers. It is also useful at this stage to include a number of sensitivity analyses in the package prepared for the next step.

Step 6: Model presentation In this step, the analyst(s) will "sell" the solution to management. In trying to convince management to buy into the newly developed plan, it helps a great deal if the assumptions are clearly stated, the solution is clearly presented (and at least tentatively checked against reality in the previous step), and some alternative scenarios are also presented. As a matter of fact, in all cases other than the lowest operational level, on which solutions are more or less automatically implemented, there are decision makers, whose job is to make decisions, not to accept or reject decisions from analysts. This means that the main function of operations research is not so much decision making, but preparing the decision. Presenting some workable decision alternatives is usually a good idea. This is a crucial step, which does not only decide about the future of the model, but possibly the future of the modelers themselves: no firm will keep employees whom it does not see contributing to the benefit of the organization, and producing models that are not used provides no benefit. As a matter of fact, among the models that decision makers actually asked to be built, only a surprisingly small portion was ever implemented. This is highly detrimental to the firm and their employees in particular and the profession as a whole.

Step 7: Implementation Given the acceptance by the decision makers (most likely with some modifications of the solution), the task is now to translate the model recommendations into practice. At this point it will help a great deal if those who have to live with the recommendation and the changes—the employees and workers—will accept the solution rather than sabotage it. If the modeler created goodwill in Step 3 of this procedure, listened to the concerns of those involved, and included them in the modeling process as far as possible, chances of acceptance are much enhanced. In addition, as the solution of the model is implemented, it is crucial to monitor the implementation each step of the way, so that it is possible to adjust the solution in case some unexpected changes occur.

Step 8: Monitoring and Control This phase of the process is largely overlooked. It includes the timely comparison of the plan and reality. Reality changes, and the plan should be adjusted accordingly. As the famous saying in the United States Marine Corps goes, "improvise, adapt, overcome." Furthermore, the more frequently adjustments can and are being made, the less dramatic they will have to be. As an example, assume that an individual has planned his budget for the upcoming year. Suppose that $1200 have been set aside for the purpose of entertainment. If the individual monitors the situation monthly, he may find by the end of January that he has already spent $500 on entertainment purposes. This leaves $700 for the remainder of the year, or $63.64 for each of the remaining 11 months. This is considerably less than the $100 that were planned for each month, but is still not too dramatic. Monitoring only each second month, the individual has not noticed that he spends way too much money and behaves in February in the same way as in January by spending another $500. Checking the situation by the end of February, there are only $200 for each of the remaining months, or $20 for each month. This is a much more severe decrease from the originally planned $100 per month and will be much more difficult to adhere to.

Reference

Assad AA, Gass SI (2011) Profiles in operations research: pioneers and innovators. Springer, New York

Linear Programming 2

This chapter will introduce linear programming, one of the most powerful tools in operations research. We first provide a short account of the history of the field, followed by a discussion of the main assumptions and some features of linear programming. Thus equipped, we then venture into some of the many applications that can be modeled with linear programming. This is followed by a discussion of the underlying graphical concepts and a discussion of the interpretation of the solution with many examples of sensitivity analyses. Each of the sections in this chapter is really a chapter in its own right. We have kept them under the umbrella of the chapter "Linear Programming" so as to emphasize that they belong together rather than being separate entities.

2.1 Introduction to Linear Programming

The purpose of this section is to provide a short introduction to linear programming problems. We first present a short historical account of the topic, followed by the main assumptions of linear programming problems, and finally some details about the optimization problems under discussion.

As already discussed in Chap. 1, linear programming problems were originally developed by Leonid Vitaliyevich Kantorovich (later Nobel prize laureate) in the 1930s before being first described by George Bernard Dantzig in 1947. His findings that included his "simplex method for linear programming" were presented in 1949 at a conference for research in economics, and some of the papers, including Dantzig's work, were published in 1951 in a volume edited by (later Nobel prize laureate) Tjalling C. Koopmans. Much work was done in the early years. The first programmed solution code based on Dantzig's simplex method was already developed in 1951, and many applications were first discussed in those years, among them the blending of aviation gasolines and trim loss problems. The dual simplex method by Lemke and the Hungarian method for assignment problems were also described and published in the 1950s. In 1963, Dantzig's book "Linear Programming and

Extensions" was published, an "instant classic." The year 1979 saw the publication of a paper by Leonid Genrikhovich Khachiyan, whose "ellipsoid method" made quite a stir. However, while it has some remarkable theoretical properties, it failed to perform well in practice and is not in use nowadays. On the other hand, Narendra Karmarkar's interior point method, first described in 1984, has slowly made some inroads. To this day, though, Dantzig's simplex method is the method of choice for the vast majority of all linear programming problems that are solved.

In order to structure our discussion, first consider a general mathematical programming problem. It can be written as the problem

$$P : \text{Max } z = f(x)$$

$$\text{s.t. } g(x) R b,$$

where P stands for problem (sometimes we have multiple problems under consideration, in which case we will write P_1, P_2, and so forth), $f(x)$ is the objective function, which is to be maximized, and z is the objective function value (e.g., profit, market share, sales, or cost) associated with the present solution x. This objective is to be optimized "subject to" (or "s.t." for short) the constraints $g(x) R b$. We will come to them and their meaning again later. The use of objectives and constraints goes back to the fundamental economic principle: either maximize the outcome for a given input or minimize the needed input for a required output. A typical example is for students to either decide to study a certain number of hours for a course (the input) and then try to maximize the grade in that course (the output), or they determined what output (grade) they are happy with and then minimize the effort (hours) to get there.

Linear programming problems are special cases of the above formulation. In particular, we have three fundamental assumptions in linear programming. They are:

1. *Deterministic property*
2. *divisibility*, and
3. *linearity*

Consider first the deterministic property. Simply put, it requires that all parameters are assumed to be known with certainty. (By the way, the antonym of deterministic is *probabilistic* or *stochastic*.) While this assumption appears reasonable in some instances, it is not in others. Consider these examples. While we know exactly how many machines we have for the processing of semi-finished products, these machines may fail unexpectedly, making their actual capacities probabilistic. Similarly, we know the magnitude of contracted sales, but we have only rough estimates concerning next month's additional demand. Given that much of the world is probabilistic, how can we possibly apply linear programming without making simplifications that may be so major so as to render the results unusable? (Remember the old adage: "if a model does not fit the problem, don't use it.") In general, there are two ways to circumvent the problem. One would be to resort to techniques that do not make that assumption—such as stochastic programming, a very complex field for which the available software is much less powerful than for deterministic linear programming—or to circumvent the difficulty by using sensitivity analyses. This is the procedure of choice for many decision makers. Suppose we have estimated next month's demand to be 500 units, but we are not sure about it. One way to handle this is to solve the problem with a demand of 500 units. Once the optimal solution is known, we attempt to find out what happens if the demand were to change to values other than 500. (Notice the words "what—if" that always indicate sensitivity analyses.) For instance, we can change the demand to, say 490, resolve the problem and find out what happens to our solution. This process can be repeated to the relevant values of the demand. Once this has been accomplished, the decision maker will have a clear idea what the effects on different levels of demand are on the solution of the problem. The limitations of this approach are obvious: the method is valid only if a small number of uncertain parameters exist, but it will drown the decision maker in massive amounts of data, if much of the information in the model is uncertain.

Next consider the issue of divisibility. Divisibility is ensured, if we allow the variables in the model to be any real number (even though we will restrict ourselves to rational numbers most of the time). Most importantly, divisibility allows variables to be noninteger. While many instances do not allow it, this assumption is typically less of a problem that we may first think. Suppose that our model includes a variable that expresses the number of cans of corn that we make and sell. It is obvious that this number must be integer, as partial cans cannot be sold. The main question is, though: who cares if we make 1,345,827 or 1,345,826 cans? Many of the numbers in our model will be approximations anyway, so that it is completely irrelevant if the number of cans in the optimal solution is integer or not. Simple rounding will solve the problem. On the other hand, suppose that the model under consideration includes a variable that expresses the number of houses to be built in a subdivision. Simply rounding a noninteger solution up or down may result in solutions that may not be as good as possible (or, depending on the constraints, not even feasible). If integrality of a variable is essential, users are well-advised to resort to techniques from integer programming, discussed in Chap. 5 of this volume.

Finally, the issue of linearity. (For a discussion of the issue of linearity, see Appendix B of this book.) The assumption in linear programming is that all expressions, the objective function as well as all constraints, are linear. One of the underlying assumptions that leads to linearity is proportionality. As an example, if we purchase one pound of potatoes for, say, 50¢, then proportionality means that 2 lbs will cost $1, 3 lbs will cost $1.50, and so forth. So far, our costs are proportional to the quantity and the cost function is $0.5x$ with $0.5 being the per-unit price of the potatoes and x denoting the quantities of potatoes we purchase. This is part of a linear function. However, if the quantities we purchase become large, the merchant may offer us a rebate, which will make the function nonlinear (or piecewise linear).

Having stated all of the main assumptions, we can now focus on the individual components of linear programming problems. First consider the objective function. In all of mathematical programming, objective functions either maximize a function or they minimize it. Often, the objective expresses the wishes of the decision maker in monetary terms, but nonmonetary objectives are possible and common in applications in engineering and in the public sector. Note that if the objective is Max $z = f(x)$, it can alternatively and equivalently be written as Min $-z = -f(x)$. Suppose that originally, the idea was to maximize profit, then the alternative objective is to minimize losses. In other words, each maximization function can be rewritten as a minimization function and vice versa. This eliminates the need for separate methods or treatment of problems with minimization and maximization objective.

Consider now the constraints. Rather than writing them in the very general form $g(x) \, R \, b$ that is often used in mathematical programming, we will write all constraints as *LHS R RHS*. What this means is that the left-hand side *LHS* is in some desired relation, denoted by R, to the right-hand side *RHS*. First, linear programming accepts relations of the type $\leq$, $=$, and $\geq$. These relations compare a feature of the solution at hand with a proscribed standard. For instance, the number of pallets of corn shipped out of a warehouse (the reality, what "is") on the left-hand side is compared to the quantity that our customers actually wanted (the stipulated target, the "should") on the right-hand side. Similarly, we can compare our actual expenditures on the left-hand side with the targeted expenditures (i.e., our budget) on the right-hand side. Typically, constraints are formulated so as to have a linear function on the left-hand side and a single parameter on the right-hand side. This bears some resemblance to hypothesis testing, in which we compare reality (some statistic from a sample) with a standard from a theoretical distribution.

We will defer our discussion of the many facets of constraints to Sect. 2.1 of this book, where a variety of applications of linear programming are discussed. At this point, we will only mention a few general principles that relate to constraints. As already discussed in Chap. 1, constraints are "hard" in the sense that the linear programming solver will return a "there exists no feasible solution" message in case the set of constraints does not allow a solution to be found. This means that constraints are absolute and must be satisfied exactly. In other words, given that we have $100, any plan that includes expenditures of say $100.01 will not be considered as feasible. In reality, often constraints are much softer. This will have to be considered in the formulation, for instance, by formulating constraints more loosely, e.g., consider the possibility of taking out a loan to increase the budget, or by allowing solutions that do not fully satisfy a customer's demand. Another important advice for modeling is to avoid equations whenever possible, using inequalities instead. Linear programming is well equipped to handle equations from a technical point of view, but equations are very restrictive and often lead to infeasibilities. We will return to this subject when we discuss applications and the graphical solution method.

In order to more fully explain the structure and features of constraints, we consider a very small numerical example. Suppose that we can manufacture computer speakers for the domestic market and speakers for export. The assembly of a domestic speaker takes 10 sec each, while an export speaker takes 12 sec. The capacity of the machine on which both speakers are assembled is 20 hrs

2.1 Introduction to Linear Programming

(or 72,000 sec). Our distributor has placed orders for 2500 domestic speakers and 3000 export speakers but is prepared to take more. Each domestic speaker nets us $40, while each export speaker contributes $45 to our overall profit.

Defining x_1 and x_2 as the number of domestic and export speakers made and sold, respectively, we can formulate the linear programming problem as

$$\text{P : Max } z = 40x_1 + 45x_2$$
$$\text{s.t.} \quad 10x_1 + 12x_2 \leq 72,000$$
$$x_1 \quad\quad\quad \geq 2500$$
$$x_2 \geq 3000.$$

Whenever an inequality of either "$\leq$" or "$\geq$" type is inputted, the solver will automatically add an additional variable to create an equation. These variables have an important interpretation. In particular, whenever we have a constraint of type *LHS* $\leq$ *RHS*, the solver automatically adds a *slack variable S* to the left-hand side of the constraint in order to transform the inequality to an equation. It then uses the constraint *LHS* + *S* = *RHS* with $S \geq 0$. As an example of what a slack variable means and why this procedure is valid, consider a simple budget constraint. In such a constraint, the left-hand side expresses the amount of money we actually spend, while the right-hand side specifies the amount of money that is available, so that the constraint reads "the amount of money spent must be less or equal than the amount of money available." The slack variable is then the difference between left- and right-hand side. In our budget constraint, the rewritten constraint then states that the amount of money used (the original *LHS*) plus the amount of money unused (the slack *S*) equals the amount of money available (the original *RHS*).

A similar procedure is used for inequalities of the "$\geq$" type. Here, the original constraint *LHS* $\geq$ *RHS* is transformed into an equation by subtracting a *surplus or excess variable E* from the left-hand side so as to arrive at the reformulated constraint *LHS* − *E* = *RHS*. For instance, in a production requirement, where the left-hand side expresses the actual production, while the right-hand side value specifies the smallest quantity that must be made, the excess variable specifies the amount by which the present solution exceeds the requirement.

Consider now the above numerical example and suppose someone has suggested the solution $x_1 = 2800$ and $x_2 = 3200$. First of all, this indicates that we have decided to manufacture 2800 speakers for the domestic and 3200 speakers for the export market. By plugging these values into the objective function, we can determine the profit level for this solution as $z = 40(2800) + 45(3200) = \$256,000$. We can also check if this solution is feasible: the plan requires $10(2800) + 12(3200) = 66,400$ sec to make, which is less than the available assembly time. As a matter of fact, given that we have 72,000 sec on the assembly machine and this solution uses 66,400 sec, there is a slack capacity of $72,000 - 66,400 = 5600$ sec. The next two demand constraints are also satisfied: since we make 2800, this is more than we need to (actually, an excess of $2800 - 2500 = 300$ units); similarly, 3200 units for the export market are $3200 - 3000 = 200$ units in excess of the requirement.

As an aside, the solution discussed in the previous paragraph is not optimal. At optimum, we make 3600 units for the domestic market and 3000 units for the export market. This results in zero slack capacity for the assembly machine and a surplus of 1100 units for the domestic market and a zero surplus for the export market. The profit at optimum is $279,000.

2.2 Applications of Linear Programming

This section presents a variety of linear programming applications. Each of these applications is a prototype, in the sense that "real" applications in practice will build upon these formulations by adding lots of "bells and whistles." However, the major use of learning about these bare-bones formulations is to understand the type of formulations they present and the way the variables are defined, which is typical for the particular type of application.

The following subsections describe in detail how some scenarios can be modeled. They may be simplistic, but the models presented here include the most important features of the application under consideration.

2.2.1 Production Planning

The formulation that we present in this section is a very basic prototype of production planning models. It is probably the simplest possible problem to be formulated, and the purpose to present it at this point is to introduce some of the basic ideas of modeling. In this context, suppose that there are three products we can manufacture. For simplicity, we will simply refer to them as P_1, P_2, and P_3, respectively. The three products sell for $20, $15, and $17 per unit, respectively. These products are manufactured on two machines M_1 and M_2. Each of the two products has to be processed on both machines. At this level, the order in which to process the products on the machines is irrelevant. The two machines have capacities of 9 and 8 hrs, respectively, within the planning period. After that time, the machines will require regular maintenance, a task for which they have to be shut off and are no longer available. The processing times of the three products on the two machines (in minutes per quantity unit) are shown in Table 2.1.

It is important to understand that the machine capacities and their usage are meant in the following way. Suppose there is a clock attached to the machine that shows the amount of time that is still available on the machine before the next scheduled

Table 2.1 Processing times of the products on the machines

	P_1	P_2	P_3
M_1	3	5	4
M_2	6	1	3

2.2 Applications of Linear Programming

maintenance. Initially, the clock on the first machine shows 9 hrs or 540 min. Suppose now that four units of P_2 are processed on M_1. Given that the processing time per unit is 5 min, the available time decreases by 20 min from 540 to 520. This continues until one of the machines has no capacity left. The process is considered automatically in the formulation, we present it here so as to stress the way time is managed here. It is often misunderstood that a capacity of, say, 8 hrs means a workday and the objective is to plan what product is processed at what time on which machine. This type of micro planning is discussed in detail in Chap. 9 of this volume.

The production costs are \$120 per hour of machine time on M_1, and \$90 per hour of machine time on M_2. In other words, operating time costs \$2 and \$1.50 per minute on the two respective machines. Given the operating times, the unit processing costs of the three machines on M_1 are \$6, \$10, and \$8, respectively, while the unit operating costs on M_2 are \$9, \$1.50, and \$4.50, respectively. This results in overall unit processing costs of \$15, \$11.50, and \$12.50 for each of the three products, respectively. Considering the selling prices, we have profit contributions of \$5.00, \$3.50, and \$4.50 per unit of each of the three products.

The problem can then be formulated so as to maximize the overall profit, while respecting the capacity constraints of the two machines. In order to formulate the problem, we first need to define the variables. In general, variables are defined, so that they can provide "shop floor answers," i.e., numbers that tell the user of the solution what to do. In this example, we would like to know the number of products that we will manufacture. How much time we will use on the two machines in the process does not have to be defined as a variable, since it is a direct consequence of how many products we will make. The same holds for the profit, which is also a consequence of what we do.

Defining x_1, x_2, and x_3 as the number of units of the respective products P_1, P_2, and P_3 that we will manufacture and sell (note that we assume here that everything we make can also be sold!), we can then start to formulate the objective function. Knowing that the objective is to maximize the profit, which is the sum of profits that derive from the making and selling of products, we consider one such term at a time. For instance, making and selling one unit of P_1 nets \$5, and we have now decided to make x_1 units of it. This means that the profit that derives from the making and selling of P_1 will be $5x_1$. Similar expressions can be derived for the other two products, resulting in the objective function shown in the formulation below.

Next, consider the constraints. We need to formulate one constraint for each machine. Each of these constraints will state that the actual usage of the resource called processing time does not exceed the processing time that is available. The time available is known, so the right-hand side of the constraints is easy. In order to be more specific, we will deal with the first machine and formulate the actual time that is used on M_1. First, we note that the time will be consumed by the making of the three products. Consider one such product at a time, say P_1. We know that it takes 3 min to process one unit of P_1 on M_1, and we have decided to make x_1 units of P_1. This means that the processing time on M_1 that is used for the making of P_1 is $3x_1$ minutes. Similarly, we will use a total of $5x_2$ minutes on M_1 to make all the units of P_2 that we will manufacture, and we need $4x_3$ minutes on M_1 to make all of P_3. The sum of these processing times now must be less than or equal to the available time of 540 min. This is the capacity constraint for M_1.

The capacity constraint for the second machine is formulated in similar fashion. Finally, we have to ensure that we are making nonnegative numbers of the products—after all, we are in the business of manufacturing and selling the products, rather than buying them, which is what negative values of the variables would indicate in this context. The formulation can then be written as the problem

$$P : \text{Max } z = 5x_1 + 3.5x_2 + 4.5x_3$$
$$\text{s.t.} \quad 3x_1 + 5x_2 + 4x_3 \leq 540$$
$$6x_1 + 1x_2 + 3x_3 \leq 480$$
$$x_1, \quad x_2, \quad x_3 \geq 0.$$

Solving the problem, the optimal solution indicates that the decision maker should make and sell 20 units of P_1, no units of P_2, and 120 units of P_3. The associated total profit is \$640.

2.2.2 Diet Problems

The diet problem is not only among the first linear programming problems to be solved, but it is arguably also the most intuitive model. As a matter of fact, the problem was studied in 1939 by the (later Nobel laureate) George Stigler in the context of determining a cost-minimal nutritious food for "an active economist who lives in a large city." Incidentally, he included 77 different foods in his model. His solution, while not optimized, was found to be quite close to the optimal solution for his data. An interesting account of the history of the diet problem is found in Garner Garille and Gass (2001).

In general, two versions of the problem can be thought of. They are roughly equivalent to the fundamental economic principle, applied to the determination of a diet. The two criteria are cost and nutritional value, where cost is an input factor, while nutritional value is an output. Thus, we can either try to minimize the cost (the input), while guaranteeing at least a certain and predetermined nutritional value (the output), or we can attempt to maximize the nutritional value (the output), while using no more than a prespecified dollar value (the input). In order to determine which of the two versions is more suitable, consider this. In the latter approach, the constraint is a simple resource constraint that states that the total amount of money spent on food cannot exceed the decision maker's budget. However, the objective function is more complex. It is stated to maximize the nutritional value. The problem with this is that the measure "nutritional value" is not a simple number, it is what is usually referred to as multidimensional measure, as it comprises a large number of measures: protein, carbohydrates, fats, vitamins, minerals, etc. And it is obviously not meaningful to simply add those different units together so as to obtain a single one-dimensional measure.

2.2 Applications of Linear Programming

Table 2.2 Input data for the sample problem

	Hamburger 3½ oz	Medium fries 4 oz	Cheesecake 2.8 oz	Nutritional requirement
Calories	250	380	257	[1800; 2200]
Fat	13%	31%	28%	≤100%
Cost per serving	$1.59	$2.19	$2.99	

On the other hand, the former version of the problem is much easier to handle. The objective is then a simple (one-dimensional) cost minimization, and the "nutritional content" is relegated to the constraints. And this is where the requirements can be handled quite easily: for each nutritional component, we can formulate one or two constraints, an upper and/or a lower bound. As a simple numerical example, consider three foodstuffs, e.g., hamburger, fries, and cheesecake (indicating that we are using the term "diet" in the widest possible sense). As far as nutrients are concerned, we consider only calories and fat in our example. Table 2.2 shows the nutritional contents of the foodstuffs, the nutritional requirements, and the prices per serving of the foods.

Note that our numerical example expresses the calories in terms of the usual calories (kCal) and we have an upper and a lower bound for it. (Note that in practice, the recommended caloric intake depends on gender and age of an individual, as well as other factors.) On the other hand, the fat content is expressed in terms of the recommended daily requirements, and it is an upper bound.

The problem in the cost minimization version can then be formulated as follows. First, we define the variables. As usual, the variables are defined depending on what the decision maker would like to know but does not at the moment. In this model, the decision maker's goal is to determine the quantities of the individual foods that are to be included in the daily diet, so that we will define variables x_1, x_2, and x_3 as the quantities of hamburgers, medium fries, and cheesecakes in the diet. As long as we use well-defined "servings" as units and use the same units for each food in the objective and all constraints, there is no problem inadvertently adding apples and oranges, or hamburgers and fries for that matter. What we are adding are their nutritional contents, as we show in this example.

First consider the objective function. The general idea is to minimize the costs of all foodstuffs in the diet. Consider one of these foods at a time. As shown in Table 2.2, each hamburger costs $1.59. Since there will be x_1 hamburgers in the diet, the total costs incurred by hamburgers in the diet are $1.59x_1$. Similarly, the costs that can be attributed to fries and cheesecake are $2.19x_2$ and $2.99x_3$, respectively, so that the objective function is the sum of these three terms.

Let us now formulate the constraints. In doing so, we consider one requirement at a time. First, we focus on the lower bound of the constraints. What we would like to express is that the caloric content of the diet should be at least 1800 calories. The calories in the diet derive from the three foods. The number of calories in the diet that derive from hamburgers are $250x_1$, the number of calories from fries are $380x_2$, and the calories from cheesecake are $257x_3$. Adding up these three expressions results in the total number of calories that we actually have in the diet. This number should not fall short of 1800 (the first constraint), and it should also not exceed 2200 (the second constraint). Note that both of these constraints have the same left-hand side.

The constraint that regulates the fat content of the diet is constructed similarly. Overall, the constraint should state that the content of fat in the diet should not exceed 100% of the recommended daily value. Where does the fat in the diet come from? In this example, it derives from hamburgers (for a total of $13x_1$), fries (for a total of $31x_2$), and cheesecake (for a total of $28x_3$). This total fat content should then not exceed 100%.

In summary, the problem can then be formulated as follows.

$$P : \text{Min } z = 1.59x_1 + 2.19x_2 + 2.99x_3$$

$$\begin{aligned} \text{s.t.} \quad 250x_1 + 380x_2 + 257x_3 &\geq 1800 \quad &\text{(calories, lower bound)} \\ 250x_1 + 380x_2 + 257x_3 &\leq 2200 \quad &\text{(calories, upper bound)} \\ 13x_1 + 31x_2 + 28x_3 &\leq 100 \quad &\text{(fat, upper bound)} \\ x_1, \quad x_2, \quad x_3 &\geq 0. \quad &\text{(nonnegativity constraints)} \end{aligned}$$

In addition to the lower and upper bounds on the caloric intake and the upper bound on the fat content of the diet, we include the nonnegativity constraints on the quantities of foods included in the diet, as the idea is to eat rather than regurgitate. Incidentally, the optimal solution to this formulation specifies that the planner eats $6\frac{1}{3}$ hamburgers, ½ serving of fries, and no cheesecake. Doing so will provide 100% of the allowable fat intake, generate 1800 calories, and cost $11.32.

While this solution may be optimal with respect to the problem as specified above, it is clearly riddled with a number of problems. The number of hamburgers in the diet is obviously much too high for all but the most determined junk food junkies. It will require additional constraints or the introduction of other foodstuffs which are added in the loop that attempts to reconcile the solution of the model with the real problem of finding a cost-effective diet. In order to illustrate the modeling process, we use a somewhat larger problem with real data. The data of the eight foods included and the 11 nutrients is found in Table 2.3.

As far as the daily nutritional requirements are concerned, we have identified the following parameters. The diet should include

- Between 1800 and 2200 calories
- No more than 78 g of fat
- No more than 300 mg of cholesterol
- No more than 2300 mg of sodium
- Between 225 and 325 g of carbohydrates
- At least 28 g of fiber
- At least 50 g or protein, and
- At least 1300 mg of calcium
- At least 18 mg of iron

2.2 Applications of Linear Programming

Table 2.3 Nutritional contents and prices of foods in the model

	Pasta	Tomato juice	Clam chowder	Beef chuck	Cheddar cheese	Orange juice	Bologna	Potato chips	Required quantity
Weight	16 oz	8 fl. oz	16 fl. oz	16 oz	16 oz	8 fl. oz	16 oz	16 oz	
Calories	563	41	316	1396	1824	112	1392	2400	$\in [1800, 2200]$
Fat	2.6 g	0.1 g	9 g	90 g	150.4 g	0.6 g	112	153.6 g	≤ 78
Cholesterol	0 mg	0 mg	18 mg	372 mg	480 mg	0 mg	272	32 mg	≤ 300
Sodium	12.8 mg	653 mg	3353 mg	220 mg	2816 mg	1 mg	3344	2832 mg	≤ 2300
Carbs	119 g	10.3 g	39.2 g	0 g	6.4 g	25.4 g	25.6	233.6 g	$\in [225, 325]$
Fiber	12.8 g	1 g	2.7 g	0 g	0 g	0.6 g	0	24 g	≥ 28
Protein	24.3 g	1.8 g	18 g	136 g	113.6 g	1.7 g	68.8	36.8 g	≥ 50
Calcium	67 mg	24 mg	149 mg	72 mg	3264 mg	26 mg	384	320 mg	≥ 1300
Iron	6.4 mg	1 mg	5.4 mg	12 mg	5 mg	1.4 mg	5 mg	8 mg	≥ 18
Potassium	198 mg	556 mg	410 mg	1088 mg	448 mg	496 mg	1424	6032 mg	≥ 4700
Vitamin C	0 mg	44 mg	9 mg	0 mg	0 mg	123 mg	5 mg	176 mg	≥ 90 mg
Price per serving	$2.30 for 16 oz	61¢ per 8 fl. oz	$2.24 for 16 fl. oz	$8.87 per lb	$3.36 per 16 oz	46¢ for 8 fl. oz	$1 for 16 oz	$4.84 for 16 oz	

- At least 4700 mg of potassium, and
- At least 90 mg of Vitamin C

The nutritional values of eight foods are shown in Table 2.3. Note that prices and nutritional contents of all drinks have been normalized to 8 fl. oz, while the servings of all foods are shown in 16 oz. This is not necessary but has been done in order to allow easy comparisons between foods.

We first define variables $x_1, \ldots, x_8$ for the number of servings of the eight foods outlined in Table 2.3. Following the ideas developed above for the small problem, we can now formulate the diet problem. The formulation is as follows.

P : Min $z = 2.3x_1 + .61x_2 + 2.24x_3 + 8.87x_4 + 3.36x_5 + .46x_6 + 1x_7 + 4.84x_8;$

s.t.

$563x_1 + 41x_2 + 316x_3 + 1396x_4 + 1824x_5 + 112x_6 + 1392x_7 + 2400x_8 \geq 1800$
(Calories, lower bound)

$563x_1 + 41x_2 + 316x_3 + 1396x_4 + 1824x_5 + 112x_6 + 1392x_7 + 2400x_8 \leq 2200$
(Calories, upper bound)

$2.6x_1 + 0.1x_2 + 9x_3 + 90x_4 + 150.4x_5 + 0.6x_6 + 112x_7 + 153.6x_8 \leq 78$
(Fat)

$18x_3 + 372x_4 + 480x_5 + 272x_7 + 32x_8 \leq 300$
(Cholesterol)

$12.8x_1 + 653x_2 + 3,353x_3 + 220x_4 + 2816x_5 + 1x_6 + 3344x_7 + 2832x_8 \leq 2300$
(Sodium)

$119x_1 + 10.3x_2 + 39.2x_3 + 6.4x_5 + 25.4x_6 + 25.6x_7 + 233.6x_8 \geq 225$
(Carbohydrates, lower bound)

$119x_1 + 10.3x_2 + 39.2x_3 + 6.4x_5 + 25.4x_6 + 25.6x_7 + 233.6x_8 \leq 325$
(Carbohydrates, upper bound)

$12.8x_1 + 1x_2 + 2.7x_3 + 0.6x_6 + 24x_8 \geq 28$
(Fiber)

$24.3x_1 + 1.8x_2 + 18x_3 + 136x_4 + 113.6x_5 + 1.7x_6 + 68.8x_7 + 36.8x_8 \geq 50$
(Protein)

2.2 Applications of Linear Programming

Table 2.4 Nutritional content of additional foods

Criterion	Milk	Apples
Weight	8 fl. oz	16 oz
Calories	146	213
Fat	8 g	0.7 g
Cholesterol	32 mg	0 mg
Sodium	96 mg	5 mg
Carbs	12.8 g	56.5 g
Fiber	0 g	9.8 g
Protein	8 g	1.1 g
Calcium	272 mg	25 mg
Iron	8 mg	1 mg
Potassium	349 mg	439 mg
Vitamin C	0 mg	18 mg
Price per serving	22 ¢ for 8 fl. oz	87¢ per lb

$67x_1 + 24x_2 + 149x_3 + 72x_4 + 3264x_5 + 26x_6 + 384x_7 + 320x_8 \geq 1300$
(Calcium)

$6.4x_1 + 1x_2 + 5.4x_3 + 12x_4 + 5x_5 + 1.4x_6 + 5x_7 + 8x_8 \geq 18$
(Iron)

$198x_1 + 556x_2 + 410x_3 + 1088x_4 + 448x_5 + 496x_6 + 1424x_7 + 6032x_8 \geq 4700$
(Potassium)

$44x_2 + 9x_3 + 123x_6 + 5x_7 + 176x_8 \geq 90$
(Vitamin C)

$x_1, x_2, \ldots, x_8 \geq 0.$
(Nonnegativity constraints)

Solving the model, we find that there is no feasible solution. Debugging the model shows that the sodium, fiber, calcium, and potassium constraints are incompatible. An obvious course of action is then to change the model to include a food that is low in sodium, and high in fiber, calcium, and potassium. In this example, we have chosen whole milk (3.25% fat). The nutritional values for milk are shown in the second column of Table 2.4. The number of servings of milk in the diet is defined as x_9, and the appropriate terms are added to the constraints, i.e., adding $0.22x_9$ to the objective function, and $146x_9$ to the left-hand sides of the first two constraints, $8x_9$ to the fat constraint, and so forth.

After the addition of milk, there still exists no feasible solution. The incompatible constraints are now *sodium*, carbs upper bound, *fiber, calcium,* and *potassium* (the nutrients that were incompatible before are shown in italics). As a potential remedy, we add apples with skin as the tenth food; see the third column of Table 2.4. The quantity of apples in the diet is denoted by the new variable x_{10}, and the appropriate additions to the model are made: add $0.87x_{10}$ to the objective function, add $213x_{10}$ to the left-hand sides of the first two calorie constraints, add $0.7x_{10}$ to the left-hand side of the fat constraint, etc.

This time we do obtain a solution to the problem, which is shown in Table 2.5 as "Solution 0."

By any reasonable standard, this is a weird diet with lots of milk. Limiting the milk intake to no more than two servings (that's still one pint per day), we add the constraint $x_9 \leq 2$, and obtain solution 1, again shown in Table 2.5.

This diet looks a little better, but two pounds of apples a day seems like an overkill. It is also apparent that the restriction on milk has resulted in a 16% increase of costs. This is, of course, not surprising, as each new nonredundant constraint further restricts the feasible set, so that the new solution cannot possibly be better (cheaper) and is typically worse (more expensive) than its predecessor.

Restricting the quantity of apples in the diet to one pound, i.e., one serving, we obtain Solution 2, shown again in Table 2.5. Note that the costs have increased by another 22%. However, for the first time, this combination of foods resembles something that could be called a realistic diet, even though the large quantity of orange juice is less than ideal. In this solution, fat, the upper bound on carbohydrates, fiber, calcium, and potassium as well as the bounds on milk and apples are tight, i.e., the actual intake in terms of nutrients equals the respective bounds in the constraints. As a matter of fact, whenever left- and right-hand sides equal each other in a solution, the constraint is referred to as a *bottleneck*. The identification of bottlenecks is crucial in optimization, as a bottleneck indicates that a resource is at its limit and as such deserves particular attention. If, for instance, successive solutions to a problem such as those above for a diet problem show that, say, the sodium upper limit constraint is a bottleneck in many of the solutions, then it would probably be a good idea to search for additional foodstuffs that are low in sodium.

Back to the problem at hand, it would probably be a good idea at this point to include foods that are low on carbs and fat, and rich in fiber, calcium, and potassium. It may also be a good idea to limit the liquids in the diet. Example: If we were to limit the total liquid intake to four servings (that's one quart, as each serving is 8 oz), we would write $x_2 + x_3 + x_6 + x_9 \leq 4$. If this constraint were to be introduced at this point, we would again be in a situation, in which no feasible solution exists.

We terminate our discussion at this point, which is not to suggest that the present diet is reasonable: the purpose of this section was to introduce the reader to the iterative nature of the modeling process that repeatedly revises the model based on the present solution.

2.2.3 Allocation Problems

Allocation problems are one of the most prominent areas of application in linear programming. All models in this class have in common that they deal with the allocation of scarce resources to (economic) activities. At times, more than one scarce resource exists, in which case the modeler can choose any one of them as the

Table 2.5 Solutions after model modifications

Solution #	Pasta	Tomato juice	Clam chowder	Beef chuck	Cheddar cheese	Orange juice	Bologna	Potato chips	Milk	Apples	Cost
0	0	0	0	0	0	0.31	0	0.0057	9.37	2.82	$4.69
1	0	0	0	0	0.167	3.4	0	0.22	2	2.12	$5.46
2	0.81	0.46	0	0	0.15	3.53	0	0.22	2	1	$6.66

Table 2.6 Types of investments and their features

Investment type	Expected annual interest/dividend	Expected annual increase in value	Average risk per dollar
Real estate	0%	18%	20
Silver	0%	10%	12
Savings account	2%	0%	1
Blue chip stocks	3%	6%	7
Bonds	4%	0%	3
Hi-tech stocks	0%	20%	30

basis for the mathematical formulation. This will be further elaborated upon below. Due to the many different types of allocation problems, we will present two such scenarios below.

First consider an investment allocation problem. Problems of this nature were first formulated by Markowitz in the early 1950s as nonlinear optimization problems. Here, we will discuss a linear version of the problem. In this problem, the decision maker has to decide how much of the scarce resource (money) to allocate to different types of investments. As we will see below, this observation already indicates how to define the variables.

In our numerical example, the investor has $300,000 that can be invested. In addition to the money at hand, it is possible to borrow up to $100,000 at 12% interest. This money can be used for leveraging (borrowing to invest). The investor has narrowed down the choices to six alternatives, shown in Table 2.6. The table also shows the expected annual interest or dividend for the investment alternatives, the expected annual increase of the value of the investment, and an indication of the risk of the investment (per dollar).

We will consider two versions of the problem: the first version attempts to maximize the expected value of the assets at the end of the planning period (1 year), while the second version minimizes the total risk of the investment.

First consider version 1. The value of the assets after 1 year equals today's value of the investment plus the expected interest or dividend plus the expected change in value within a year minus the amount of money that was borrowed (principal and interest). In addition to the restricted availability of money already mentioned above, the decision maker faces the following constraints:

- The expected value of assets (exclusive interest) at the end of the planning period should be at least 7% higher than at the beginning.
- Invest at least 50% of all the money actually invested in stocks and bonds combined.
- Invest no more than 20% of total amount available (excluding the amount borrowed) in real estate and silver combined, and.
- The average risk of the portfolio should not exceed 10.

2.2 Applications of Linear Programming

In order to formulate the appropriate model, we first must define the problem variables. The scarce resource in this application is money, so that we define x_j as the dollar amount invested in the j-th alternative. Here, x_1 will denote the money invested in real estate, x_2 the money invested in silver, and so forth.

We can then formulate the objective function. Each dollar invested in real estate will produce no interest or dividend, but will gain an expected 18%, so that the investment of $1x_1$ will have appreciated to $1.18x_1$. A dollar invested in silver and a savings account will appreciate to \$1.10 and \$1.02, respectively. A \$1 investment in blue chip stocks is expected to be worth \$1.06 after a year plus a dividend of 3¢, making it worth \$1.09. The remaining investments are dealt with in a similar fashion. From our revenue, we must deduct the amount borrowed (x_7) plus the interest to be paid on the borrowed amount ($0.12x_7$), making it $1.12x_7$.

Consider now the constraints. An obvious restriction is that the investor cannot invest more money than is available. The amount invested is nothing but the sum of all individual investments, i.e., $x_1 + x_2 + x_3 + x_4 + x_5 + x_6$, while the amount that is available is the sum of the \$300,000 at hand plus the amount that is borrowed, viz., x_7. This is the budget constraint (2.1). Another straightforward restriction limits the amount that can be borrowed to \$100,000. This constraint is shown in (2.2).

The constraint that requires the invested money to show a growth of at least 7% can now be formulated as follows. The actual value of the assets at the end of the planning period uses the expected annual gains from Table 2.6, resulting in $1.18x_1$, $1.10x_2$, etc. The required increase of the invested money by at least 7% is the product of 1.07 (the principal plus the required increase) and the invested amount, which is the sum of the first six variables. This is shown in Constraint (2.3).

The constraints (2.4) and (2.5) model the requirements that at least 50% of the money invested must be invested in stocks and bonds combined, and that no more than 20% of the money available can be invested in real estate and silver combined. Note the difference between the two constraints: while in (2.4), we are dealing with a portion of the amount actually invested (the sum of the first six variables), constraint (2.5) refers to the total amount available to the investor (exclusive the amount that may be borrowed), which equals \$300,000.

The average risk of the portfolio equals the total risk divided by the amount invested, i.e., $\frac{20x_1+12x_2+x_3+7x_4+3x_5+30x_6}{x_1+x_2+x_3+x_4+x_5+x_6}$, which is not to exceed a value of 10. Multiplying the inequality by the (nonzero) denominator results in relation (2.6). The nonnegativity constraints (2.7) conclude the formulation. Version 1 of the model can then be summarized as follows.

P : Max $z = 1.18x_1 + 1.10x_2 + 1.02x_3 + 1.09x_4 + 1.04x_5 + 1.20x_6 - 1.12x_7$

s.t. $x_1 + x_2 + x_3 + x_4 + x_5 + x_6 \leq 300,000 + x_7$ (2.1)

$x_7 \leq 100,000$ (2.2)

Table 2.7 Optimal solutions to the different versions of the investment allocation model

Investment type	Version 1 with 12% interest on borrowed money	Version 1 with 10% interest on borrowed money	Version 2
Real estate	60,000	60,000	0
Silver	0	0	0
Savings account	0	0	160,500
Blue chip stocks	234,782.61	321,739.13	0
Bonds	0	0	160,500
Hi-tech stocks	5,217.39	18,260.87	0
Amount borrowed	0	100,000	21,000

$$1.18x_1 + 1.10x_2 + 1.00x_3 + 1.06x_4 + 1.00x_5 + 1.20x_6$$
$$\geq 1.07\,(x_1 + x_2 + x_3 + x_4 + x_5 + x_6) \tag{2.3}$$

$$x_4 + x_5 + x_6 \geq 0.5\,(x_1 + x_2 + x_3 + x_4 + x_5 + x_6) \tag{2.4}$$

$$x_1 + x_2 \leq 0.2\,(300,000) \tag{2.5}$$

$$20x_1 + 12x_2 + x_3 + 7x_4 + 3x_5 + 30x_6 \leq 10\,(x_1 + x_2 + x_3 + x_4 + x_5 + x_6) \tag{2.6}$$

$$x_1, x_2, \ldots, x_7 \geq 0. \tag{2.7}$$

Table 2.7 shows the optimal solution of this model with a 12% interest on borrowed money (as formulated above), and a slight modification with 10% interest on the amount we borrow.

The resulting profits of the two versions that maximize the expected return turn out to be $332,973.9, and $333,408.7, respectively. The average risk is 10 in both cases.

Version 2 of the investment model is very similar. It deletes the risk constraint (2.6) from the formulation and uses its left-hand side (the actual risk of the portfolio) in a minimization objective. Furthermore, we require an appreciation of at least 7% on the available money (exclusive the borrowed amount), i.e., $321,000 at the end of the planning period. A summary of the optimal solutions is shown in Table 2.7. At 12% interest, Version 2 generates a return (after interest payment) of $328,110. The risk of this solution is only 2.

Note the jump that occurs in the amount of money borrowed: apparently, given an interest of 10% on borrowed money, it is still worthwhile to borrow and invest, while an increase to 12% is prohibitive, so that no money is borrowed, and no leveraging occurs. Also note the very different solution provided by Version 2 of the model.

Another allocation problem is found in the allocation of manpower—the scarce resource in this context—to tasks. In this specific application, we deal with allocating police officers to districts. Clearly, not all districts in a city are created

2.2 Applications of Linear Programming

Table 2.8 Protection provided by police officers and smallest allowable protection

	A	B	C	D	E
Additional marginal protection per officer	3	7	10	5	4
Lowest acceptable protection	40	50	70	60	40

equal, so that the impact a single policeman provides to a district will be different for different areas. In the numerical illustration below, a city has a total of 67 police officers to be allocated among five districts $A, B, \ldots$, and E. Table 2.8 shows the degree of protection offered to a district for each officer assigned to the district.

For example, if six police officers are assigned to district A, then the degree of protection is $6(3) = 18$. Suppose that due to the intricacies of the job, it is not possible to hire additional police officers in the short run. Actually, each police officer who is not assigned to one of the districts can be hired out to a private security company for a fee of $100 per officer and day. Other than the house taxes (a flat fee for our purposes that does not have to be included in the problem), we assume that hiring out police officers is the council's only income. Council wants to set up a linear programming problem that maximizes the department's income (due to hiring out police officers). The following restrictions have to be observed:

- Each district must have at least the minimum protection as specified in the table above.
- The average protection is at least 50, and.
- We have to allocate at least 50% more officers to districts A, B, and C combined than to districts D and E combined.

Formulating the problem, we have to define the variables first. In this context, the scarce resource consists of police officers, so that we can define x_j as the number of police officers assigned to district j. For simplicity, we use $x_A, x_B, x_C, x_D,$ and x_E for the five districts under consideration. For the same reason, we ignore the fact that the number of police offers assigned to a district will have to be an integer number. The objective function then maximizes the product of the revenue for a police officer offers hired out for a day ($100) and the number of police offers hired out. The number of police officers hired out for private security work equals the difference between the number of police officers available (67) and the number of officers assigned to the five districts (the sum of all five variables).

The first constraint simply requires that we cannot assign more police officers than we have. Constraints (2.9)–(2.13) ensure that each district receives at least the minimum protection required in Table 2.8. Constraint (2.14) requires that the average protection in the five districts is at least 50, and constraint (2.15) ensures that the first three districts have at least twice as many police officers allocated to them as have the last two. As usual, the nonnegativity constraints complete the formulation. The model can then be written as follows.

$$P : \text{Max } z = 100\left[67 - (x_A + x_B + x_C + x_D + x_E)\right]$$

s.t.
$$x_A + x_B + x_C + x_D + x_E \leq 67 \qquad (2.8)$$
$$3x_A \geq 40 \qquad (2.9)$$
$$7x_B \geq 50 \qquad (2.10)$$
$$10x_C \geq 70 \qquad (2.11)$$
$$5x_D \geq 60 \qquad (2.12)$$
$$4x_E \geq 40 \qquad (2.13)$$
$$(3x_A + 7x_B + 10x_C + 5x_D + 4x_E)/5 \geq 50 \qquad (2.14)$$
$$x_A + x_B + x_C \geq 1.5(x_D + x_E) \qquad (2.15)$$
$$x_A, \ x_B, \ x_C, \ x_D, \ x_E \geq 0. \qquad (2.16)$$

Solving the problem reveals that the number of police officers allocated to the five districts are $13\frac{1}{3}$, $12\frac{2}{3}$, 7, 12, and 10, respectively. This means that a total of 55 police officers are allocated, leaving 12 police officers to be hired out, so that city council's daily income from this allocation is \$1200. Furthermore, this allocation results in protection levels of 40, $88\frac{2}{3}$, 70, 60, and 40. In other words, all districts except the second receive the minimal protection required, while district B receives $(88\frac{2}{3})/50 \approx 1.77$ times the protection that is minimally needed.

2.2.4 Blending Problems

All blending problems have in common that they take a number of given ingredients or raw materials and blend them in certain proportions to the final products. In other words, in the process we are creating something new by mixing existing materials. Typical examples of blending are coffees, teas, whiskeys, tobaccos, perfumes, gasoline, and similar products. Clearly, there are some rules for the blending process. For instance, in order to ensure a specific taste, it may be required that a blend includes at least a certain proportion of a raw material.

The process is shown in Fig. 2.1. On the left, there are m buckets with given quantities of raw materials, while on the right, there are n empty buckets that are to be filled with known quantities of the blends. In the blending process, we take, one at a time, a certain number of scoops from each of the raw materials and transfer them into the buckets on the right. Once sufficient quantities have been transferred to the "Product" buckets on the right, all that is left to do is stir, package, and sell.

This figure not only demonstrates the actual process, but it also allows us to see how the variables in blending problems are to be defined. What we need to know in

2.2 Applications of Linear Programming

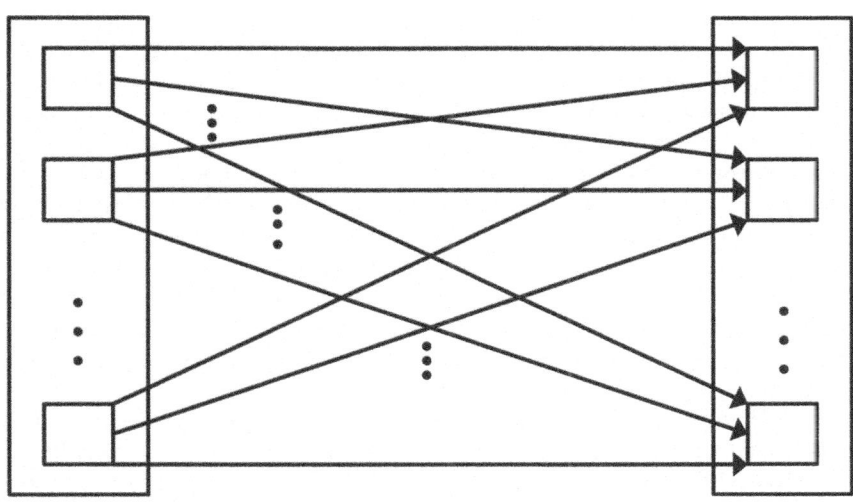

m raw materials   *n* products

Fig. 2.1 Blending ingredients

Table 2.9 Blending rules for the wines

Basic wines	Blends	
	Filzener Hexenhammer	*Leiwener Hosenscheisser*
Riesling	[.45; .55]	[.20; .50]
Müller-Thurgau	[.10; .15]	[.10; .60]
Silvaner	[.35; .35]	[.30; .40]

any specific blending problem is how many "scoops" of each raw material goes into each of the "Product" buckets. In more general terms, we define x_{ij} as the quantity of raw material i that goes into product j.

As a numerical example, suppose that we are to blend two table wines from the Moselle region in Germany. The two blends are the *Filzener Hexenhammer* and the *Leiwener Hosenscheisser*. Both products are blends of wines from three white grapes, viz., *Riesling*, *Müller-Thurgau*, and *Silvaner*. The original wines are available in quantities of 10,000, 5000, and 6000 gallons at a cost of $8, $6, and $5 per gallon. The estimated demands for the two blends are 7000 and 8000 gallons (our customers will purchase these or any higher amounts from us) and the estimated sales prices are $16 and $18 per gallon, respectively.

The rules that have to be followed when blending the two wines are summarized in Table 2.9. The meaning of these figures is best explained by some of the numbers. For instance, the interval [.45; .55] in the *Riesling* row and the *Filzener Hexenhammer* column of the table indicates that at least 45% and at most 55% of the *Filzener Hexenhammer* blend must be *Riesling*. Note that the *Silvaner* content of the *Hexenhammer* must be exactly 35%. It is also noteworthy that while the ranges for the *Hexenhammer* blend are quite tight (which usually indicates a well-controlled and high-quality product), the ranges for the *Hosenscheisser* are very wide, clearly indicating a cheap product that consists of, loosely speaking, more or less whatever happens to be cheap and available.

As discussed above, we need six variables in this example. They are x_{11} (the quantity of *Riesling* in the *Hexenhammer*), x_{12} (the quantity of *Riesling* in the *Hosenscheisser*), x_{21} (the quantity of *Müller-Thurgau* in the *Hexenhammer*), and x_{22}, x_{31}, and x_{32} which are defined analogously. Before formulating the objective, it is beneficial to first determine the quantities of the basic wines that are used in the blending process and the quantities of the blends that are made in the process. For the time being, let us assume that there are no losses in the process (e.g., spillage or thirsty employees).

Consider the basic wines and their uses first. Each of those wines is either put into the *Hexenhammer* or the *Hosenscheisser* blend—that is all we can do with them in this context. Thus, the quantity of *Riesling* that we use is $x_{11} + x_{12}$, the quantity of *Müller-Thurgau* used in the process is $x_{21} + x_{22}$, and the quantity of *Silvaner* that is used is $x_{31} + x_{32}$. Similarly, we determine the quantities of the blends that are produced. The quantity of each blend that is made in the process is nothing but the sum of its ingredients. In our example, the quantity of *Hexenhammer* that we blend equals $x_{11} + x_{21} + x_{31}$, and the quantity of *Hosenscheisser* we make is $x_{12} + x_{22} + x_{32}$. The objective function is then to maximize profit, which, in turn equals the difference between revenues from the two blends and the costs of the three basic wines. It is shown as relation (2.17) below.

As far as constraints go, we have three different types of constraints. First there are the supply constraints that require us not to use more of the basic wines than we can get, secondly there are the demand constraints that state that we have to make at least as many gallons of the blends as our customers are estimated to demand, and thirdly and finally, there are the blending constraints.

Since the quantities of the ingredients (the basic wines) and the products (the blends) have already been determined above, the first two types of constraints are easy to formulate. They are shown as constraint groups (2.18) and (2.19) in the formulation below. The blending constraints (group 2.20) are formulated as follows. Consider the content of *Riesling* in *Hexenhammer*. One of the blending constraints will state that the quantity of *Riesling* in *Hexenhammer* (which is x_{11}) must be at least 45% of the total quantity of *Hexenhammer*, which has already been determined as $x_{11} + x_{21} + x_{31}$. In other words, we can write $x_{11} \geq .45(x_{11} + x_{21} + x_{31})$. This is the lower bound of this combination of basic wine and blend. The upper bound is formulated in similar fashion as $x_{11} \leq .55(x_{11} + x_{21} + x_{31})$. This is repeated for all combinations of basic wines and blends. The complete formulation is then

$$P : \text{Max } z = [16(x_{11} + x_{21} + x_{31}) + 18(x_{12} + x_{22} + x_{32})]$$
$$- [8(x_{11} + x_{12}) + 6(x_{21} + x_{22}) + 5(x_{31} + x_{32})] \quad (2.17)$$

2.2 Applications of Linear Programming

$$\text{s.t.} \quad x_{11} + x_{12} \leq 10{,}000$$
$$x_{21} + x_{22} \leq 5000 \qquad (2.18)$$
$$x_{31} + x_{32} \leq 6000$$

$$x_{11} + x_{21} + x_{31} \geq 7000$$
$$x_{12} + x_{22} + x_{32} \geq 8000 \qquad (2.19)$$

$$x_{11} \geq .45(x_{11} + x_{21} + x_{31})$$
$$x_{21} \geq .1(x_{11} + x_{21} + x_{31})$$
$$x_{31} \geq .35(x_{11} + x_{21} + x_{31})$$

$$x_{12} \geq .2(x_{12} + x_{22} + x_{32})$$
$$x_{22} \geq .1(x_{12} + x_{22} + x_{32})$$
$$x_{32} \geq .3(x_{12} + x_{22} + x_{32})$$

$$x_{11} \leq .55(x_{11} + x_{21} + x_{31})$$
$$x_{21} \leq .15(x_{11} + x_{21} + x_{31})$$
$$x_{31} \leq .35(x_{11} + x_{21} + x_{31})$$
$$x_{12} \leq .5(x_{12} + x_{22} + x_{32}) \qquad (2.20)$$
$$x_{22} \leq .6(x_{12} + x_{22} + x_{32})$$
$$x_{32} \leq .4(x_{12} + x_{22} + x_{32})$$

$$x_{11}, x_{12}, x_{21}, x_{22}, x_{31}, x_{32} \geq 0.$$

The optimal solution of the problem uses 3850 gallons of *Riesling*, 700 gallons of *Müller-Thurgau*, and 2450 gallons of *Silvaner* in the *Hexenhammer*, so that a total of 7000 gallons of that blend are made. Note that this is exactly the quantity that is desired. Similarly, we use 3983.33 gallons of *Riesling*, 4300 gallons of *Müller-Thurgau*, and 3550 gallons of *Silvaner* in the *Hosenscheisser* blend, making it a total of 11,833.33 gallons of that blend. This is considerably more than the minimum of 8000 gallons that were required. Also note that in the process, we use a total of 7833.33 gallons of *Riesling* (with 2166.67 gallons left over), we use all of the 5000 gallons of *Müller-Thurgau* and all of the 6000 gallons of *Silvaner* that are available to us. The overall profit is $202,333.33.

It is easy to add other relevant constraints. Suppose that we need to control the alcoholic content of the blends as well. For that purpose, assume that the *Riesling* has an alcohol content of 8%, *Müller-Thurgau* has 7.5%, and *Silvaner* has 6%, and it is desired that the *Hexenhammer* has at least 7.3% alcohol content. Note that the present solution contains .08(3850) + .075(700) + .06(2450) = 507.5 gallons of alcohol. Given that we are making 7000 gallons of the blend, this means that the present alcohol content of *Hexenhammer* is 507.5/7000 = 7.25%, which is not sufficient. The constraint requiring at least 7.3% can be written as $.08x_{11} + .075x_{21} + .06x_{31} \geq .073(x_{11} + x_{21} + x_{31})$. The new optimal solution does not change the quantity or composition of *Hosenscheisser*, but dramatically changes the composition of *Hexenhammer*. It now contains 5916.67 gallons *Riesling*, 2366.67 gallons of *Müller-Thurgau*, and 3350 gallons of *Silvaner*. The profit has decreased to $198,466.7, roughly a 2% decrease.

2.2.5 Transportation and Assignment Problems

Transportation problems have been introduced into the discussion by Hitchcock in 1941, thus predating the advent of linear programming by half a dozen years. It must be understood that there is not a single model that encompasses all, or even most, transportation scenarios. What we call "the" transportation problem in this context is a very simple prototype, which exhibits some basic structures that are typically inherent in transportation problems. The structure has three essential components. On the one hand, there are *origins*, where certain quantities of a single homogeneous good are available (so that one unit of the good is exactly the same, regardless from which origin it is taken). We can think of the origins as warehouses at which goods are stored. For simplicity, we will use the terms warehouse and origin interchangeably. Given that there is a total of m origins, we assume that there is a known supply of s_i at origin i.

The second set of components concerns the *destinations*. This is where given quantities of the good are needed. We will refer to them as either destinations or customers. We assume that there are n destinations, and suppose that there is a known demand for our product at a magnitude of d_j at destination j. So far, we have existing goods on one side, and the need for goods (or, equivalently, existing demand), on the other. The task is now to ship the existing goods from the origins to the destinations, so as to respect the existing quantities at the origins, satisfy the demand at the destinations, and organize the shipments as efficiently as possible. As a matter of fact, we can visualize the structure of the problem by considering the graph used in our discussion of blending problems (Fig. 2.1). Rather than input factors on the left and blended products on the right, we have origins and destinations, and instead of taking a number of ladles full of raw materials and put them into a bucket for a blended product, we ship a number of units from a warehouse to a customer. The visuals and the story are very different, but the mathematical structure is the same.

When requiring the efficiency of the shipments, we will need a criterion by which to measure efficiency. As usual, we will employ a monetary criterion. Here, an obvious choice is the cost of transportation. In order to introduce cost components, we assume that there is a cost of c_{ij} to ship a single unit of the good from origin i to destination j. Fixed unit cost of transportation mean a linear transportation cost

2.2 Applications of Linear Programming

function, as one unit shipped from i to j costs c_{ij}, two units cost $2c_{ij}$, and so forth The assumption of linearity of the cost function is crucial to our model. Assume that we are shipping pallets of bottles from regional distribution centers to supermarkets. Suppose now that the capacity of a truck is 40 pallets. Then the transportation cost for one pallet is one trip, the costs for two pallets are one trip (except for the cost of loading and unloading the same as the cost for a single pallet), the costs for three pallets are still mainly the costs of a single trip, etc. In other words, it costs about the same to ship one pallet or up to 40 pallets. Shipping pallet number 41, though, requires a second trip, causing the costs to jump to the equivalent of two trips. The result is a step function with jumps at the truck capacities. One way to "restore" the linearity assumption in this model is to measure the number of units shipped in terms of truckloads. Disregarding the differences in loading and unloading costs for different numbers of pallets, the cost function now includes points on a linear function (those with an integer number of truckloads). It is very important to note that the transportation network in this basic problem allows only direct transportation from an origin to a destination. It is *not* permitted to first ship from one origin to another to consolidate the load, to use round trips (similar to multi-stop shopping), or similar non-direct routes.

Before formulating a small example, it is necessary to distinguish between balanced and unbalanced transportation problems. A transportation problem is called *balanced*, if the sum of all supplies equals the sum of all demands. In balanced transportation problems, it will be possible to empty all the warehouses and satisfy all demands of our customers. In *unbalanced* transportation problems, we either have more units than customers want (the total supply exceeds the total demand), or we need more units than we have (the total demand exceeds the total supply). In the former case, we will be able to satisfy all demand, but some units will remain unshipped at some of the warehouses, while in the latter case, we will empty all the warehouses, but some demand will remain unsatisfied. We will discuss the balanced case first and then offer modifications that deal with unbalanced problems.

Consider a transportation problem with two origins and three destinations. The supplies at the origins are 30 and 20 units, respectively, while at the destinations, there are demands of 15, 25, and 10, respectively. Clearly, the total supply and the total demand both equal 50, making the problem balanced. The unit transportation costs are shown in the matrix $\mathbf{C}$ below, where an element in row i and column j shows the value c_{ij}. For instance, shipping one unit from origin 1 to destination 3 costs \$4.

$$\mathbf{C} = \begin{bmatrix} 1 & 7 & 4 \\ 2 & 3 & 5 \end{bmatrix}.$$

The three types of parameters—the supplies, the demands, and the unit transportation costs—completely describe the transportation problem. The model will minimize the total transportation costs, while ensuring that supplies and demands are respected. In order to formulate the problem, we define decision variables x_{ij} that denote the quantity that is shipped (directly) from origin i to destination j. The objective function is then composed as follows. First of all, we compute the total transportation costs along each origin–destination connection. As an example, in our numerical illustration, the link between origin 2 and destination 1 carries unit transportation costs of $2 and the quantity we ship on that link is x_{21}, making the total transportation cost on that connection $2x_{21}$. Adding the expressions on all links results in the overall total cost, which is then to be minimized in the objective function of the model as shown below.

Next, consider the constraints of the problem. Generally speaking, there will be two sets of constraints: the first set ensures (for balanced problems such as this example) that the flow of goods out of each origin exactly equals the quantity that is available at that origin. The second set of constraints requires that the flow of goods that is received at a customer site equals the demand of that customer. As an example, for the first set of constraints, consider the second origin. From this origin, goods can be shipped (directly) to destination 1 (the quantity shipped on this link is x_{21}), to destination 2 (the quantity is x_{22}), and to destination 3 (with a quantity of x_{23}). Consequently, the total quantity shipped out of origin 2 is $x_{21} + x_{22} + x_{23}$, which is supposed to equal 20 units, the quantity available at origin 2. The remaining supply constraints are formulated similarly.

We are now able to formulate the demand constraints. As a numerical example, consider customer (or destination) 3. The goods that the customer receives either come from origin 1 (a quantity of x_{13}) and from origin 2 (a quantity of x_{23}). As a result, the total quantity received at destination 3 is $x_{13} + x_{23}$, which should equal the demand of 10 at that point. Again, the other demand constraints are formulated in similar fashion. The complete formulation is then as follows:

$$\begin{aligned}
P : \text{Min } z = {}& 1x_{11} + 7x_{12} + 4x_{13} + 2x_{21} + 3x_{22} + 5x_{23} \\
\text{s.t.} \quad & x_{11} + x_{12} + x_{13} = 30 \\
& x_{21} + x_{22} + x_{23} = 20 \\
& x_{11} \phantom{+ x_{12} + x_{13}} + x_{21} = 15 \\
& \phantom{x_{11} +} x_{12} \phantom{+ x_{13} + x_{21}} + x_{22} = 25 \\
& \phantom{x_{11} + x_{12} +} x_{13} \phantom{+ x_{21} + x_{22}} + x_{23} = 10 \\
& x_{11}, x_{12}, x_{13}, x_{21}, x_{22}, x_{23} \geq 0.
\end{aligned}$$

The problem above is written in a way that makes the underlying structure clearly visible. It is apparent that each variable appears exactly twice in the constraints, once in a supply constraint and once in a demand constraint. The structure is very special (the matrix of coefficients is called totally unimodular), which ensures that as long as all supplies and demands are integer, one can prove that there will exist at least one optimal solution to the problem that is also integer. The special structure of the problem has given rise to specialized solution techniques, such as the MODI (MOdified DIstribution) method. With increasing computational power, their importance has diminished, so that we have chosen not to include them in this book, but instead describe them on the file-sharing site associated with this book. Another feature of the problem should be mentioned, even though we will not exploit it here.

2.2 Applications of Linear Programming

Given m origins and n destinations, the problem will have mn variables and $(m + n)$ constraints. However, one constraint is redundant (more specifically: linearly dependent). This can be seen in the above formulation by adding all supply constraints and then subtracting the first two demand constraints. The result will be the third demand constraint. As a result, we will have only $(m+n-1)$ independent constraints to be considered. This number features prominently in the aforementioned special solution methods.

The optimal solution to our example can be shown in the following transportation plan $\mathbf{T}$. In row i and column j, it shows the optimal value of the variable x_{ij}.

$$\mathbf{T} = \begin{bmatrix} 15 & 5 & 10 \\ 0 & 20 & 0 \end{bmatrix}.$$

In other words, we ship 15 units from origin 1 to destination 1, 5 units from origin 1 to destination 2, 10 units from origin 1 to destination 3, and all 20 units that are available at origin 2 directly to destination 2. It is easy to ascertain (by multiplying the elements of $\mathbf{T}$ with the corresponding elements of $\mathbf{C}$ and adding them up) that the total transportation cost at optimality is $\bar{z} = 150$. It is also worth mentioning that each nondegenerate solution to a transportation problem has exactly $(m+n-1)$ variables at a strictly positive level. In case of degeneracy, there may be even fewer positive variables.

As far as extensions go, we will first look into unbalanced problems. First of all, given any unbalanced problem, we can no longer formulate the problem with all equation constraints, as total supply and demand are no longer equal. In case of the total supply exceeding the total demand, we will not distribute the entire supply to our customers, so that some supply will be left over. This means that we can formulate the demand constraints as equations (which pull the total quantity of demand from the supplies into the network) as we have in a balanced problem, while we write the supply constraints as less-or-equal-than constraints to ensure that we do not ship more than we have out of the origins. That way, the customer demand is satisfied everywhere, while some units are left over in one or more of the origins. In the above example, let the supplies be 30 and 23 rather than 30 and 20. The optimal solution will ship 27 units out of origin 1 and 23 out of origin 2, leaving three units unassigned at origin 1.

The case in which the total supply falls short of the total demand is dealt with similarly. Here, we will not be able to satisfy the entire demand, but in order to come as close as possible of doing so, we will use our entire supply. This means that we formulate the supply constraints as equations (thus pushing the entire supply into the network), while the demand constraints will be written as less-than-or-equal constraints. In that way, the total supply is shipped to customers, which will still leave some unsatisfied demand at one or more customers. In the above original example, suppose that the demands are now 15, 25, and 12, respectively, rather than 15, 25, and 10. The optimal solution ships 15 units to destination 1, 23 units to destination 2, and 12 units to destination 3, so that customer 2 will be left with an unsatisfied demand of 2 units. Other extensions of the basic model may include penalties for unsatisfied customers (loss of goodwill) and units left over in a warehouse (inventory costs).

Two interesting extensions are called *reshipments* and *overshipments*. Both modifications can be interpreted as some type of sensitivity analysis, in that we change some of the existing assumptions. In the case of reshipments, we use the same transportation network, but allow transportation on routes that are not direct. In other words, a reshipment would be given if we were not to send a unit from, say, origin 1 to destination 1 directly, but ship it from origin 1 to destination 3 (or some other destination) first, then back to, say, origin 2, and from there on to destination 1. Clearly, in order to actually use reshipments (or back-and-forth shipments), it must be advantageous to do so. Consider again our example above. At optimum, we ship five units from origin 1 to destination 2. To ship a single unit on that route, it costs $7. Instead, we could ship up to five units from origin 1 to destination 1 for $1, back to origin 2 for an additional $2, and from there to destination 2 for an additional $3. Hence, it costs $6 to ship a single unit on this somewhat circuitous route, $1 less than on the direct connection. Reshipping five units this way will save $5(1) = \$5$ for total transportation costs of $145. While reshipping may use routes that go back-and-forth multiple times, it is unlikely that such routes will exist in practice. The mathematical formulations of reshipments use absolute values of variables, as the value of a variable such as $x_{ij} = -5$ indicates that five units are shipped back from destination j to origin i. This is not really problematic, but it makes the formulation somewhat more unwieldy.

Overshipments are another way of improving on the optimal solution to the basic problem by changing the assumptions somewhat. The idea is to allow additional flow through the transportation network, which, paradoxically, may actually reduce costs. Again, in our example consider the possibility to add one unit of supply to origin 2 (for a total of 21), and one unit of demand to destination 1 (for a total of 16). The optimal solution to that problem will move 51, rather than 50, units through the transportation network at a cost of 147, a decrease of $3 from the original solution. Such a decrease is possible in this example, because we now ship one additional unit on the link from origin 1 to destination 1 (costing an additional $1), and on the connection from origin 2 to destination 2 (costing an extra $3), but we can now transport one less unit on the expensive link from origin 1 to destination 2 (which saves $7). This explains the net savings of $+1 + 3 - 7 = -\$3$. While reshipments were easy to implement (just modify the plan of what is shipped where), overshipments are considerably more difficult to apply. In order to benefit from overshipments, we need additional units at the right origin, and we have to convince at least one customer to accept more units than originally demanded, which is typically done by transferring some of the savings to the customer willing to take more than originally desired.

Other extensions of the basic problem have been discussed as well. One such extension includes capacities applied to the transportation links. Another, quite natural, modification includes not only direct shipments as used here, but allows transshipment points, at which the goods may be unloaded, temporarily stored, and reloaded onto other trucks. In some practical applications, such transshipment points

2.2 Applications of Linear Programming

do not just allow the consolidation of the loads, but also permit changing transportation modes, e.g., from truck to rail. Capacity constraints on the transportation points are a further natural feature to be included in a model. While some of these extensions may be incorporated in network models (see Chap. 6 of this volume), planners will generally resort to standard linear programming formulations to include all the desired features in the model.

A variety of other applications of "transportation problems" exist, some having absolutely nothing to do with shipping units from one place to another. One such example are the dynamic production–inventory models in Sect. 2.2.6. Here, the "origins" represent the periods of productions, while the "destinations" are the periods of consumption (or demand). A link from origin i to destination j exists, if $i \leq j$. The constraints then require that the outflows of the origins do not exceed the production capacities, while the inflows of the destinations must be at least as large as the known demand. Other applications assign groups of workers to shifts, or differently equipped military units to potential targets. Although the structure of the mathematical models may be identical, they still represent completely different physical situations.

While transportation problems have a specialized structure, *assignment problems* are even more specialized. Consider a set of n employees that are to be assigned to n tasks. We can use no more than 100% of an employee time in the allocation, and to each task, we must assign 100% of one or more employees' time. Each allocation bears a certain cost. As an example, consider typists who, quite naturally, have different abilities. Suppose that one of the tasks involves technical typing. In order to perform the task, some typists may already know how to do it and can perform the type of task quite efficiently (meaning that there will be low costs of assigning this employee to the task), while other typists may have to be trained, necessitating higher assignment costs. The problem is then to assign employees' time to tasks, so as to minimize the overall assignment costs.

In order to formulate the problem, we again first define the appropriate variables. Here, we define x_{ij} as the percentage of worker i's time that is assigned to task j. In order to explain the way the model is formulated, suppose that we have three employees and three tasks. The assignment/training costs are shown in the cost matrix

$$\mathbf{C} = \begin{bmatrix} 4 & 3 & 1 \\ 8 & 5 & 3 \\ 2 & 6 & 2 \end{bmatrix}.$$

If, for instance, we will use 20% of employee 1's time for task 2, then the costs are $.2(3) = .6$. The objective function will then minimize the sum of all assignment costs.

As far as the constraints of the model go, we have one set of constraints that specify that the sum of proportions of an employee's time must add up to 100%. Similarly, the sum of proportions of employees' time devoted to any one task must also add up to 100%. Given this structure, the problem can then be formulated as follows.

$$P : \text{Min } z = 4x_{11} + 3x_{12} + 1x_{13} + 8x_{21} + 5x_{22} + 3x_{23} + 2x_{31} + 6x_{32} + 2x_{33}$$

$$\text{s.t.} \quad \begin{aligned}
x_{11} + x_{12} + x_{13} & = 1 \\
x_{21} + x_{22} + x_{23} & = 1 \\
x_{31} + x_{32} + x_{33} & = 1 \\
x_{11} + x_{21} + x_{31} & = 1 \\
x_{12} + x_{22} + x_{32} & = 1 \\
x_{13} + x_{23} + x_{33} & = 1 \\
x_{11}, x_{12}, x_{13}, x_{21}, x_{22}, x_{23}, x_{31}, x_{32}, x_{33} & \geq 0.
\end{aligned}$$

It becomes apparent that the assignment is a very close relative of the transportation problem discussed above. More specifically, since the constraints and the objective are exactly the same, we can view assignment problems as transportation problems with all supplies and demands equal to one. Given that, there will be at least one optimal solution to the problem that has all variables integer. It follows that at least one optimal solution has all variables either equal to zero or equal to one. In many publications, the variables are assumed to be zero or one from the start, but this feature is really not an assumption of the general problem, but a consequence of the structure of the problem. It does, however, provide for an easier statement of the problem: the variables indicate whether or not an employee is assigned to a task (they equal one if he is and zero if not), and the constraints state that each employee is assigned to exactly one task, and each task is performed by exactly one employee.

Similar to the transportation problem, specialized algorithms were developed for the assignment problem. Of particular mention is the "Hungarian Method," a technique based upon a combinatorial theorem by the Hungarian mathematician Jenö Egerváry. Again, due to the generally available increased computational power, the problem can easily be solved as a general linear programming problem, and the importance of specialized methods has diminished, and we will not discuss it here, but relegate it to the website associated with this book.

Assignment problems have a number of applications, some not obviously related to assignments. The earliest story reported by G.B. Dantzig in his 1963 book on linear programming refers to it as the "marriage problem." The story is that a father has a number of daughters whom he wants to marry off. (It would work the same way with sons, in case this were desired.) There are a number of prospects for the matching, but each particular match requires a certain amount of dowry based on the (in-) compatibility of the couple. The thrifty father's overall objective is to minimize the total amount of dowry he will have to pay.

The marriage story in its original form does not appear to be among the prime applications of assignment problems, though. Instead, decision makers may attempt to match entities such as sports teams with the objective of maximizing the audience's appeal (and with it revenue, as appealing games—those with

2.2 Applications of Linear Programming   45

long-standing rivalries or those among teams with close standings—tend to attract larger crowds.) Assignment problems are also very important in the solution of problems such as the *traveling salesman problem*, see Chap. 6 of this volume.

There are some well-known and well-documented extensions to assignment problems, such as generalized assignment problems and quadratic assignment problems. Both types of extensions are not only very difficult from a computational point of view, but also beyond the scope of this volume.

2.2.6 Dynamic Production-Inventory Models

This section describes models, in which decision makers do not only have to answer the "how many" question as we have seen in many of the previous applications, but they also require an answer to the question "when" to produce. In the simplest case, assume we only consider a single product. Furthermore, suppose that the time frame of interest has been subdivided into small time units, in which production occurs. Throughout this section, we will refer to these units as "months." Within each month, production occurs, and customers take out products based on their demand.

Based on the production capacities, it may now not be possible to satisfy the demand in each month by the production in the same month. In order to avoid undersupplying our customers, we can produce more than the demand indicates during the earlier months of the planning period and keep the surplus in stock. This will, of course, cause inventory holding costs to be incurred. Given that the production costs may vary between the months, it may actually be preferable to manufacture goods earlier in the planning period rather than later, but the decision will depend on the relation between the production costs and the inventory holding costs. We will leave these decisions to the optimizer and our model.

Before presenting a numerical example, it is important to discuss the exact sequence of events within each month. At the beginning of each month, we take stock. Since nothing has happened to the inventory between this point in time and the end of the previous month (we assume that no theft, spoilage, spillage, etc. occurs; such events could be included, if desired), the inventory level at the beginning of the month will denote the number of units carried over from the previous month, which, in turn, will determine the inventory holding costs. Then production occurs. After the desired production quantity is made, customers take products out of our warehouse according to the estimated demand. Whatever is left after that will be carried over to the next month, and the process begins anew.

As a numerical illustration, consider the following scenario. The planning period ranges from the beginning of January of some year and ends at the end of April. The estimated demand, production capacity, and unit production costs are shown in Table 2.10.

In addition, it costs 5¢ to carry over one unit from the end of January to the beginning of February, 15¢ to carry over one unit from the end of February to the beginning of March, and another 15¢ to hold one unit in stock between the end of March and the beginning of April. The decision maker has an opening inventory of

Table 2.10 Parameters for the dynamic production—inventory model

	Month 1 (January)	Month 2 (February)	Month 3 (March)	Month 4 (April)
Estimated demand	80	70	130	150
Production capacity	120	140	150	140
Unit production cost	$1.00	$1.10	$1.20	$1.25

20 units in the beginning of the planning period and desires to have nothing left at the end of the planning period.

In order to formulate the model, we quite naturally need two types of variables, one for production and the other for inventory. Denote the production variables by x_1, x_2, x_3, and x_4, which are defined as the quantities to be manufactured in months 1, 2, 3, and 4, respectively. Similarly, we define the parameters d_1, d_2, d_3, and d_4 as the demand in periods 1, 2, 3, and 4, respectively. Before defining the inventory variables, we need to decide at which point to measure the inventory level. Given our problem description above, we may decide to count inventory at the beginning of each period. Alternatively, it is possible to determine the inventory level at the end of a period, which we leave as exercise in Problem 3 at the end of this section. Here, we denote by I_1, I_2, I_3, I_4, and I_5 the variables that denote the inventory levels at the beginning of the periods 1–5. Note that the inventory levels I_1 and I_5 are not variables, but parameters whose numbers we know: as outlined above, the opening inventory is $I_1 = 20$, and the closing inventory must be $I_5 = 0$.

The objective function is then a simple cost minimization function that consists of two main components, the production costs and the inventory costs. As far as constraints go, there are two types. First, there are the simple production capacity constraints, which specify that in no period can we produce more than our capacity allows. Secondly, there are the inventory balancing constraints. They state that the inventory level at the beginning of period t equals the inventory level at the beginning of the previous period $t-1$ plus our production in the previous period minus the demand in the previous period. Formally, we can write $I_t = I_{t-1} + x_{t-1} - d_{t-1}$ for $t = 2$–5. These constraints are the same as the usual balancing constraints in accounting, which state that what you have in your account today equals what you had yesterday plus the deposits yesterday minus yesterday's withdrawals. Note that the inventory balancing constraints implicitly include the requirement that the demand must be satisfied, so that separate constraints to that effect are not needed. Our model can then be formulated as follows.

2.2 Applications of Linear Programming

$$P: \quad \text{Min } z = 1x_1 + 1.1x_2 + 1.2x_3 + 1.25x_4 + .05I_2 + .15I_3 + .15I_4$$

$$\begin{aligned}
\text{s.t.} \quad x_1 &\leq 120 \\
x_2 &\leq 140 \\
x_3 &\leq 150 \\
x_4 &\leq 140 \\
I_2 &= I_1 + x_1 - 80 \,(\text{or, as } I_1 = 20, x_1 - I_2 = 60) \\
-I_3 + I_2 + x_2 &= 70 \\
-I_4 + I_3 + x_3 &= 130 \\
-I_5 + I_4 + x_4 &= 150 \,(\text{or, as } I_5 = 0, x_4 + I_4 = 150) \\
x_1, x_2, x_3, x_4, I_2, I_3, I_4 &\geq 0.
\end{aligned}$$

Notice that the model does not include constraints $I_t + x_t \geq d_t$ to ensure that sufficient products are available to satisfy the demand. The reason is that the nonnegativity constraints of the inventory variables guarantees that sufficient products are available. This is no longer the case if the demand occurs in a period before the production; if so, the additional constraints are needed.

The optimal production schedule has us manufacture 120, 10, 140 and 140 units of the product in the 4 months, and the inventories carried over between months 1 and 2, 2 and 3, and 3 and 4 are 60, 0, and 10 units, respectively. The sum of the production and inventory costs is $478.50.

The solution makes intuitive sense, as the low production level in February is a result of the lower production costs in January and the low inventory costs between January and February. On the other hand, the inventory carrying costs are significant after February, so that inventories only occur between March and April, and these are necessary as the April demand exceeds the production capacity in that month.

A potential extension of the model may consider warehouse capacities. In other words, we may impose limits on the number of units we can keep in stock. Such constraints are easily incorporated in this formulation. If in our numerical example the largest possible inventory levels between months 1 and 2, months 2 and 3, and months 3 and 4 are 40, 50, and 50 units, respectively, we add the constraints

$I_2 \leq 40$
$I_3 \leq 50$
$I_4 \leq 50$.

With these additional constraints, the production levels in the four periods are revised to 100, 30, 140, and 140 units, respectively, so that the inventory levels between the periods are 40, 0 and 10 units, respectively. The total costs for this system then (marginally) increase to $479.50. Note that this formulation assumes that warehouse capacity is only used "overnight." If production takes place in the morning and sales in the afternoon, then we need to store not only I_t units in period t, but $I_t + x_t$ for all periods under consideration.

Table 2.11 Optimal solution of the production-inventory problem

	Month 1	Month 2	Month 3	Month 4
Month 1	60	60	0	0
Month 2	–	10	0	0
Month 3	–	–	130	10
Month 4	–	–	–	140

This problem can also be formulated in an alternative fashion. This may not be the best way to formulate the problem, but it nicely demonstrates that the same scenario may be formulated in very different ways. Rather than using separate variables for production and inventory, we can define double-subscripted variables x_{ij} that indicate how many units of the product were manufactured in month i for use in month j. Such a formulation will require some additional preprocessing. For instance, in order to determine the objective function coefficient for the variable x_{14} in the numerical example in this section, we need to add the production costs in month 1 (when the product is made) and the inventory holding costs from month 1 to 2, those from month 2 to 3, and those from month 3 to 4 for a total of $\$1 = .05 + .15 + .15 = \$1.35$. The other coefficients in the objective function are determined similarly.

In addition, there will be two sets of constraints. The first are again the constraints that require the production capacities to be respected. For instance, the total production in Month 1 will be $x_{11} + x_{12} + x_{13} + x_{14}$, the production in Month 2 will be $x_{22} + x_{23} + x_{24}$, and similar for the remaining 2 months. The second set of constraints that are needed are then the demand constraints. The number of units available in, say, Month 3 is $x_{13} + x_{23} + x_{33}$ and this number must be at least as large as the demand in that month. The problem can then be formulated as follows.

$$P : \text{Min } z = 1x_{11} + 1.05x_{12} + 1.2x_{13} + 1.35x_{14} + 1.1x_{22} + 1.25x_{23} + 1.4x_{24}$$
$$+ 1.2x_{33} + 1.35x_{34} + 1.25x_{44}$$

$$\text{s.t.} \quad x_{11} + x_{12} + x_{13} + x_{14} \leq 120$$

$$x_{22} + x_{23} + x_{24} \leq 140$$

$$x_{33} + x_{34} \leq 150$$

$$x_{44} \leq 140$$

$$x_{11} \geq 60$$

(January's demand is reduced by the available opening inventory)

$$x_{12} + x_{22} \geq 70$$

$$x_{13} + x_{23} + x_{33} \geq 130$$

$$x_{14} + x_{24} + x_{34} + x_{44} \geq 150$$

$$x_{11}, x_{12}, x_{13}, x_{14}, x_{22}, x_{23}, x_{24}, x_{33}, x_{34}, x_{44} \geq 0.$$

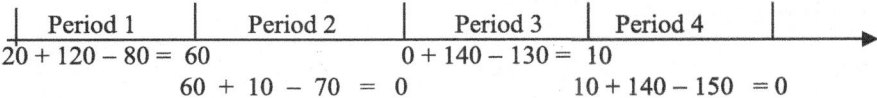

Fig. 2.2 Production and inventory levels

It is not difficult to see that the structure of this formulation is again the same as that of the transportation problem. However, it must be ensured that the "consumption month," i.e., the month in which the product is sold, is not before the "production month," i.e., the month, in which the product is made. In other words, all variables, in which the first subscript is smaller than the second subscript, must be forced to zero.

The optimal solution can best be summarized in a table such as that shown in Table 2.11.

The cost at optimum is $478.50, obviously the same as in the other formulation. The actual production levels in the 4 months can be determined by adding the values of the variables in the rows, while the sums in the columns result in the months' demand. It is also a good idea to plot the inventory changes on a timeline that lists the opening inventory of each month, adds the production within the month, and then subtracts the demand later that month. This is shown in Fig. 2.2.

Hence, the inventory levels at the beginning of the periods 2, 3, and 4 are 60, 0, and 10 units, respectively; again, reflecting the same result obtained earlier.

Incorporating limits on the warehouse capacity is also easy in this formulation. The number of units put in stock between January and February is $x_{12} + x_{13} + x_{14}$, the level of stock between February and March is $x_{23} + x_{24}$, and the level between March and April is x_{34}. All that needs to be done is to require these expressions not to exceed 40, 50, and 50, respectively, and we will again obtain the same result computed earlier for the formulation with explicit inventory variables.

Which of the formulations is used depends on the preferences of the user. The former model with the explicit inventory variables has the advantage of having $2n$ variables, given again n months within the planning period, while the latter model with the double-subscripted variables requires $½n^2$ variables. However, since the value of n is typically quite small and modern linear programming solvers can easily deal with formulations that have hundreds of thousands of variables, this should not be a concern.

2.2.7 Employee Scheduling

In contrast to the applications discussed up to this point, the models described in this, and the next section require that all variables are integer. However, since this feature is just a simple addition to the formulation, we have chosen to discuss these models in this chapter.

The employee scheduling model in this section is, in some sense, also an allocation problem. However, it has its own character, so that it justifies its own section. Consider a recurring situation in which employees have to be assigned to shifts. The number of employees required to be on the job varies throughout the day during a

Table 2.12 Personnel requirements during 4-hr time slots

Shift	0600–1000	1000–1400	1400–1800	1800–2200	2200–0200	0200–0600
Required number of employees	17	9	19	12	5	8

variety of time slots. For instance, a bus route will require significant service during the early morning and afternoon rush hours, while there will not be much service during lunch hour or late at night. Similar requirements exist for nurses, pilots, cashiers in grocery stores, and similar scenarios.

The difficulty with this problem is that we are typically not able to hire casual labor whenever needed, but we will have to use permanent employees. Consequently, the objective of the problem is to use the smallest number of employees (a proxy for costs) and still be able to staff the position(s) throughout the day.

In our numerical example, assume that a regular shift is 8 hrs and assume that there are 4-hr time segments during which personnel requirements have been observed. The personnel requirements during the 4-hr time slots are shown in Table 2.12 using a 24-hr clock.

Assume that shift work can start every 4 hrs at 6 AM, 10 AM, and so forth, and that an employee cannot work on more than one shift. Our decision is then how many employees to hire, who should start their respective shifts at each of these points in time. This means that we can define variables x_{06}, x_{10}, x_{14}, x_{18}, x_{22} and x_{02} as the number of employees who start their shift at 6 AM, 10 AM, 2 PM, and so forth. The total number of employees required is then the sum of all of these variables. As far as the constraints go, we have to require that a sufficient number of employees is present during each time slot. Consider, for instance, the time slot between 1400 and 1800 hrs, during which at least 19 employees are needed. The employees working during this time slot are those whose shift starts at 1000 hrs plus those who start working at 1400 hrs. This means that during this time slot $x_{10} + x_{14}$ employees will be working, a number that must be at least 19. Similar constraints have to be formulated for all six-time slots. The formulation can then be written as follows, where again, we ignore the necessary integrality requirements for reasons of simplicity.

2.2 Applications of Linear Programming

Table 2.13 Starting times of employees in optimal solution for 4-hr time slots

Start of shift	0600	1000	1400	1800	2200	0200
Number of employees	14	7	12	0	5	3

Table 2.14 Personnel requirements during 2-hr time slots

Shift	0600–0800	0800–1000	1000–1200	1200–1400	1400–1600	1600–1800
Required number of employees	17	11	9	7	13	19
Shift	1800–2000	2000–2200	2200–2400	2400–0200	0200–0400	0400–0600
Required number of employees	12	8	5	3	3	8

$$P : \text{Min } z = x_{06} + x_{10} + x_{14} + x_{18} + x_{22} + x_{02}$$

s.t.
$$x_{06} \qquad\qquad\qquad\qquad\qquad + x_{02} \geq 17$$
$$x_{06} + x_{10} \qquad\qquad\qquad\qquad\qquad \geq 9$$
$$x_{10} + x_{14} \qquad\qquad\qquad\qquad \geq 19$$
$$x_{14} + x_{18} \qquad\qquad\qquad \geq 12$$
$$x_{18} + x_{22} \qquad\qquad \geq 5$$
$$x_{22} + x_{02} \geq 8$$
$$x_{06}, \ x_{10}, \ x_{14}, \ x_{18}, \ x_{22}, \ x_{02} \geq 0.$$

Problems of this type typically have multiple optimal solutions. The problem as formulated has an optimal solution that requires a total of 41 employees. The starting times of their shifts are shown in Table 2.13.

Note that this solution has the exact number of required employees during all time slots, except for the time 1000–1400, where only 9 employees are needed, while 21 employees are available. In other words, there are 12 employees idle between 10 AM and 2 PM.

As an extension of the above model, assume that it is now possible to start the employees' shifts each 2 hrs rather than each 4 hrs. Similarly, the time requirements are known for 2-hr rather than 4-hr segments throughout the day. For instance, the 17 employees that were needed between 6 AM and 10 AM in the above problem will be required only between 6 AM and 8 AM, while between 8 AM and 10 AM only 11 employees are needed. The personnel requirements during the 2-hr time slots are shown in Table 2.14. Note that the larger requirement of each two adjacent time slots that correspond to a 4-hr time slot in the above example equals the requirement of that 4-hrs slot. In that sense, we are using the same example, just a finer grid.

The problem can then be formulated as follows.

Table 2.15 Starting times of employees in optimal solution for 2-hr time slots

Start of shift	06	08	10	12	14	16	18	20	22	24	02	04
Number of employees	0	0	7	4	3	5	0	0	0	3	0	14

$$\begin{aligned}
P: \operatorname{Min} z = x_{06} + x_{08} + x_{10} + x_{12} + x_{14} + x_{16} + x_{18} + x_{20} + x_{22} + x_{24} + x_{02} + x_{04} \\
\text{s.t.} \quad x_{06} &\geq 17 \\
x_{06} + x_{08} &\geq 11 \\
x_{06} + x_{08} + x_{10} &\geq 9 \\
x_{06} + x_{08} + x_{10} + x_{12} &\geq 7 \\
x_{08} + x_{10} + x_{12} + x_{14} &\geq 13 \\
x_{10} + x_{12} + x_{14} + x_{16} &\geq 19 \\
x_{12} + x_{14} + x_{16} + x_{18} &\geq 12 \\
x_{14} + x_{16} + x_{18} + x_{20} &\geq 8 \\
x_{16} + x_{18} + x_{20} + x_{22} &\geq 5 \\
x_{18} + x_{20} + x_{22} + x_{24} &\geq 3 \\
x_{20} + x_{22} + x_{24} + x_{02} &\geq 3 \\
x_{22} + x_{24} + x_{02} + x_{04} &\geq 8 \\
x_{24} + x_{02} + x_{04} &\geq 17 \\
x_{02} + x_{04} &\geq 11 \\
x_{04} &\geq 9
\end{aligned}$$

$x_{06}, x_{08}, x_{10}, x_{12}, x_{14}, x_{16}, x_{18}, x_{20}, x_{22}, x_{24}, x_{02}, x_{04} \geq 0.$

The optimal solution of this problem requires now only 36 employees, and the starting times are shown in Table 2.15.

Particularly noteworthy are the more than 12% savings in the number of employees that must be hired. In general, it is not surprising that the finer grid used here provides a solution that is at least as good as that with 4-hrs time slots. The reason is that the previous solution can still be implemented, and it would still provide a feasible solution. However, with the additional possible starting times, there are additional possibilities that may—and in this case do—allow us to find a better solution.

2.2.8 Cutting Stock Problems

In this model, the variables quite naturally must assume integer values. *Cutting stock problems* (or, alternatively, *stock cutting* or *trim loss problems*) are among the early applications of integer linear programming. The first studies concerned paper rolls, whose width is fixed, but which can be cut to desired lengths. The decision maker then has a number of larger rolls of paper of given length, which he has to cut down to smaller rolls that are in demand. This is what is called a one-dimensional problem, as only the length of the rolls is cut.

2.2 Applications of Linear Programming

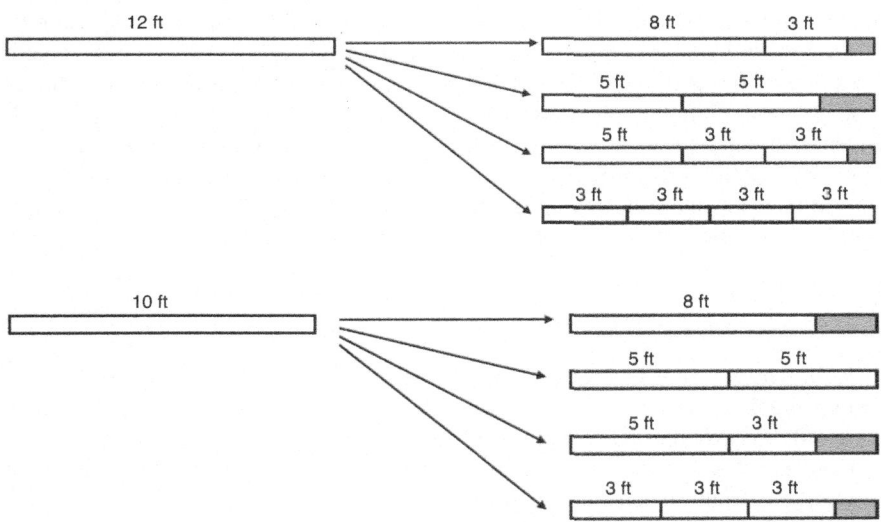

Fig. 2.3 Cutting patterns

In order to explain the formulation, suppose that a home improvement store carries wooden rods in a standard profile and width. They presently have two lengths, 12 ft and 10 ft. In particular, they have twenty 12 ft rods and twenty-five 10 ft rods in their warehouse. Management anticipates a need for sixty 8 ft rods, forty 5 ft rods, and seventy-five 3 ft rods. In order to obtain the desired lengths, we can either cut existing rods at a cost of 50¢ per cut, or purchase new rods at a cost of $2, $1.50, and $1.10 for the 8 ft, 5 ft, and 3 ft rods, respectively.

There are two common types of objectives. Management could either attempt to minimize the waste produced in the process or could minimize the costs incurred in the process. Minimizing waste is a popular option, yet it is nontrivial from a conceptual point of view. A small piece, say a 2 ft rod, cannot be used and is considered a complete waste. A larger piece, say a 4 ft rod, however, while twice the size of a 2 ft rod, but may be no waste at all, as it might be used to satisfy some future demand for 4 ft sizes, even if it is not used in this planning period. Since minimizing waste is really just a proxy for cost, we will simply minimize the total cost incurred in the process.

In order to formulate the problem, it is mandatory that we first devise a cutting plan. A cutting plan will include all meaningful cutting patterns. Patterns that are undesirable, either because they produce too much waste, are too difficult to cut, or for some other reason, are simply not included in the cutting plan.

The cutting plan for this example is shown in Fig. 2.3, where the meaningful cutting patters are numbered 1–8 from top to bottom.

We can now define the decision variable y_1 as the number of times that pattern 1 is cut, and similarly for the remaining seven patterns. This takes care of the cutting we have to do. In addition, we also require variables v_1, v_2, and v_3 that determine the number of 8 ft, 5 ft, and 3 ft rods that we purchase in addition to cutting longer rods to the required sizes. Consequently, the objective function consists of two major components, the cost of cutting and the cost of purchasing. Consider first the cutting costs. Pattern 1 requires two cuts, so that each time pattern 1 is cut, it will cost $1. Given that pattern 1 is cut y_1 times, the cost contribution of the first pattern is $1y_1$. Similarly, pattern 8 requires three cuts or $1.50 each time it is cut. Since pattern 8 is cut y_8 times, its cost contribution is $1.5y_8$. Adding the purchasing costs for the rods that are newly bought, the objective function can then be formulated as

$$\text{Min } z = (1y_1 + 1y_2 + 1.5y_3 + 1.5y_4 + 0.5y_5 + 0.5y_6 + 1y_7 + 1.5y_8) \\ + (2v_1 + 1.5v_2 + 1.1v_3),$$

with the terms in the first bracket being the cutting costs, while the terms in the second bracket are the purchasing costs. As far as the constraints are concerned, there will be two types, *viz.*, supply and demand constraints.

Consider first the supply constraints. In words, they state that the number of patterns cut from an existing length cannot exceed the number of rods that are available. Constraints of this type have to be formulated for each existing length. Consider first the 12 ft length. It is used in patterns 1, 2, 3, and 4, which, as we already know, are cut y_1, y_2, y_3, and y_4 times, respectively. Given that we have twenty 12 ft rods available, we can formulate the supply constraint for the 12 ft length as

$$y_1 + y_2 + y_3 + y_4 \leq 20. \tag{2.21}$$

Similarly, the supply constraint for the 10 ft rods is

$$y_5 + y_6 + y_7 + y_8 \leq 25. \tag{2.22}$$

The demand constraints are somewhat more difficult to formulate. As an example, consider the first required length of 8 ft. It is produced by patterns 1 and 5. Each time we cut pattern 1, we generate one 8 ft rod. Since we cut this pattern y_1 times, the number of 8 ft rods produced by cutting pattern 1 is $1y_1$. Similarly, since each time pattern 5 is cut, we generate a single 8 ft rod, we make a total of $1y_5$ 8 ft rods by cutting pattern 5. Since the only other way to obtain 8 ft rods is to purchase them (and we already have decided to buy v_1 of them), the total number of 8 ft rods that we will have is $1y_1 + 1y_5 + v_1$, a number that must be large enough to satisfy the demand of 60 units. The demand constraint for the 8 ft rods can thus be written as

$$y_1 + y_5 + v_1 \geq 60. \tag{2.23}$$

As far as 5 ft rods are concerned, the cutting plan reveals that they are generated by patterns 2, 3, 6 and 7. Since each time pattern 1 is cut, we produce two 5 ft rods, and as pattern 1 is cut y_1 times, we will produce a total of $2y_1$ 5 ft rods by cutting pattern 1. Applying similar arguments for the other three patterns that generate 5 ft rods, the constraint for these rods is

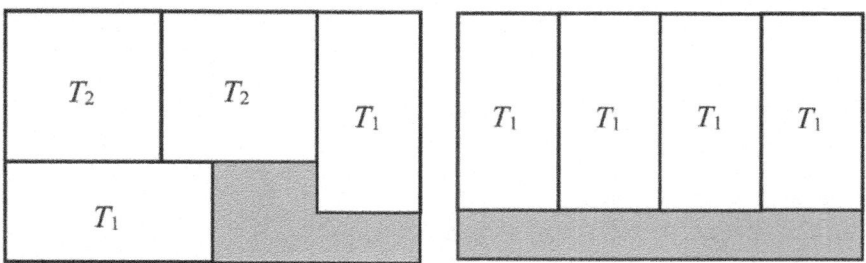

Fig. 2.4 Cutting patterns for boards of plywood

$$2y_2 + 1y_3 + 2y_6 + 1y_7 + v_2 \geq 40. \tag{2.24}$$

Finally, consider the 3 ft rods. They are generated by patterns 1, 3, 4, 7 and 8. Each time one of these patterns is cut, we generate 1, 2, 4, 1, and 3 of the required 3 ft rods. As a result, we can write the constraint for the 3 ft rods as

$$1y_1 + 2y_3 + 4y_4 + 1y_7 + 3y_8 + v_3 \geq 75. \tag{2.25}$$

The cutting stock problem can then be written as

Min $z = 1y_1 + 1y_2 + 1.5y_3 + 1.5y_4 + 0.5y_5 + 0.5y_6 + 1y_7 + 1.5y_8$
$+ 2v_1 + 1.5v_2 + 1.1v_3$
s.t. constraints (2.21) − (2.25)
$y_1, y_2, \ldots, y_8; v_1, v_2, v_3 \geq 0$ and integer.

Solving the problem results in $\bar{y}_1 = 2, \bar{y}_2 = 0, \bar{y}_3 = 0, \bar{y}_4 = 18, \bar{y}_5 = 5, \bar{y}_6 = 20, \bar{y}_7 = 0$, and $\bar{y}_8 = 0$, as well as $\bar{v}_1 = 53, \bar{v}_2 = 0$, and $\bar{v}_3 = 1$. This leaves none of the existing rods left over, and the demand is exactly satisfied.

Solving the same problems with demands of 20, 15, and 18 for the 8 ft, 5 ft, and 3 ft rods results in $\bar{y}_1 = 6, \bar{y}_2 = 0, \bar{y}_3 = 0, \bar{y}_4 = 3, \bar{y}_5 = 14, \bar{y}_6 = 8, \bar{y}_7 = 0$, and $\bar{y}_8 = 8$, as well as $\bar{v}_1 = 0, \bar{v}_2 = 0$, and $\bar{v}_3 = 0$. In this case, nothing is purchased, we have eleven 12 ft rods and three of the existing 10 ft rods left over, and the demand is exactly satisfied for the 8 ft and 3 ft rods, while one 5 ft rod is cut but not used.

As expected, things get more complicated when cutting is possible in two dimensions. However, it is not the formulation that becomes more difficult, but the cutting plan that may now include many patterns. Just imagine cutting an irregular piece of fabric that may be shifted and tilted by infinitesimally small amounts to any of its sides, resulting in infinitely many patterns. To simplify matters, assume that the patterns are regular. As a numerical example, consider a single board of plywood of size 5 ft by 8 ft, and assume that we need two types of boards, type T_1 is of size 2 ft × 4 ft, and type T_2, which measures 3 ft square. Two of the many possible patterns are then shown in Fig. 2.4, where the shaded parts indicate waste.

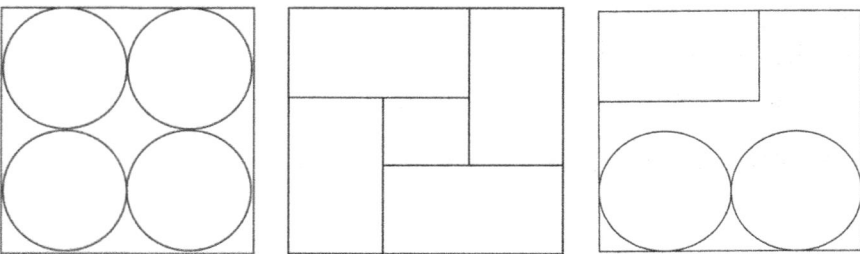

Fig. 2.5 Cutting patterns

Table 2.16 Hourly production capabilities of the two machines

	P_1	P_2	P_3
M_1	3	5	10
M_2	6	4	12

When evaluating these patterns, we could resort to the amount of waste that is generated. The first pattern uses 34 sq. ft out of the given 40 sq. ft for a usage rate of 85%. Pattern 2, in contrast, uses only 32 sq. ft for a usage rate of 80%, so it appears that pattern 1 is more efficient. This is, however, not necessarily the case. Pattern 2 is easy to cut: Adjust the saw once to a 1 ft width, cut off the shaded part at the bottom, readjust the machine to a 2 ft cutting width, and continue to cut the four desired T_1 pieces. Things are much more complicated when cutting pattern 1. First of all, we should point out that only so-called *guillotine cuts* are feasible in many applications. These are cuts that cut the existing piece all the way through. This restriction is important, as non-guillotine cuts will result in many operator errors unless we use automated methods such as laser cutting. Using only guillotine cuts, it appears best to cut pattern 1 by first adjusting the machine to a 2 ft width, cut the T_1 piece on the right with the piece of waste at the bottom, then turn the remaining board and cut the T_1 piece at the bottom (this way, we do not have to readjust the machine all the time), then readjust the machine to a 3 ft cutting width and cut the top left piece in the middle, generating the two T_2 pieces. And then we have to readjust the machine again to cut off the waste at the bottom of the two T_1 pieces. It should have become clear that while this pattern uses more of the existing board, it is much more complicated and thus costly to cut. A simple example, which demonstrates that not all patterns can be cut by guillotine cuts is shown in the middle figure in Fig. 2.5 in Problem 11 at the end of this chapter.

Exercises

Problem 1 (production planning): Solve a variant of the standard production planning problem. Three products P_1, P_2, and P_3 are manufactured on two machines M_1 and M_2. Each of the products must be processed on both machines in arbitrary order. The unit profits of the products are $18, $12, and $6, respectively, and the machine capacities are 24 and 16 hrs per planning period. Table 2.16 indicates how many units of the products can be made each hour.

2.2 Applications of Linear Programming

Table 2.17 Input data for Problem 2

	Marketing	Organizational behavior	Accounting	Operations research	Finance
Marginal improvement of mark	5	4.5	5.5	3.5	5.5
Marks required for passing the course	50	55	60	50	50

In addition, it is required that at least ten units of the second product are made. Formulate a profit-maximizing linear programming model for this problem.

Solution: Defining x_1, x_2, and x_3 as the quantities of the products to be made, the objective function below is formulated as usual. However, before we formulate the constraints, we have to adjust the units. The entries in Table 2.16 are expressed in terms of quantity units per hour. Multiplying them by the variables as usual would result in the meaningless units (quantity units)^2 per hour. Instead, we need to convert the entries in the table to hours per quantity units or, more conveniently, minutes per quantity unit. Multiplying by the variables (measured in quantity units), we obtain minutes used in the production, which can then be related to the capacities. The formulation can then be written as follows.

$$\begin{aligned} P: \text{Max } z = {}& 18x_1 + 12x_2 + 6x_3 \\ \text{s.t.} \quad & 20x_1 + 12x_2 + 6x_3 \leq 1440 \\ & 10x_1 + 15x_2 + 5x_3 \leq 960 \\ & \phantom{10x_1 + {}} x_2 \phantom{{} + 5x_3} \geq 10 \\ & x_1, \quad x_2, \quad x_3 \geq 0. \end{aligned}$$

Incidentally, the optimal solution prescribes that 43.5, 10, and 75 units of the respective products are made for a total profit of \$1353.

Problem 2 (allocation of time to courses): A student is planning the coming semester. In particular, he is attempting to allocate the weekly number of hours of study to the individual courses he is taking. Each hour of study will increase his mark by a certain quantity (starting at zero). Table 2.17 shows the marginal improvements of the marks given each hour of study (per week) as well as the marks required for passing the course.

For example, if our student were to allocate 15 hrs (per week) to marketing, then his final mark is expected to be 15(5) = 75, which means passing the course.

The student's objective is to minimize the total number of hours studied. In addition, the following constraints have been identified:

- A passing grade should be achieved in each course.
- Obtain an average grade of at least 64.
- Suppose that the student has the option to flip hamburgers at McDonalds in his spare time. This job pays \$10 per hour. Assuming that the student has a total of 80 hrs available for study and flipping, he must also make at least \$100 per week.

- The number of hours allocated to operations research should be at least 20% of the number of hours allocated to the other four subjects combined.

Solution: Given that time is the scarce resource, we define x_j as the number of hours allocated to studying the j-th subject, $j = 1, \ldots, 5$. The objective as well as the individual and overall passing requirements and the need to spend at least 20% of his studies on operations research are formulated in a straightforward fashion. The need to make at least \$100 in the hamburger shop is formulated by first determining the hours used for hamburger flipping, which is the number of hours available overall (here 80) minus the hours used for studying (the sum of variables). These hours are then multiplied by the hourly wage of \$10, which is then the amount of money made. This amount is then required to be at least \$100. Another important—and frequently forgotten—part of this type of problem is the inclusion of constraints that limit the grades to 100. As a matter of fact, if these constraints were omitted, our student would aim for passing grades in most courses and allocate a large number of hours to one course, in which good marks are easy to obtain, so that he receives in excess of 100 marks. Clearly, this is not possible, making the additional limitations necessary.

$$P : \text{Min } z = x_1 + x_2 + x_3 + x_4 + x_5$$

$$\begin{aligned}
\text{s.t. } 5x_1 &\geq 50 \\
4.5x_2 &\geq 55 \\
5.5x_3 &\geq 60 \\
3.5x_4 &\geq 50 \\
5.5x_5 &\geq 40 \\
5x_1 &\leq 100 \\
4.5x_2 &\leq 100 \\
5.5x_3 &\leq 100 \\
3.5x_4 &\leq 100 \\
5.5x_5 &\leq 100 \\
5x_1 + 4.5x_2 + 5.5x_3 + 3.5x_4 + 5.5x_5 &\geq 5(64) \\
10[80 - (x_1 + x_2 + x_3 + x_4 + x_5)] &\geq 100 \\
x_4 &\geq .2(x_1 + x_2 + x_3 + x_5) \\
x_1, x_2, x_3, x_4, x_5 &\geq 0.
\end{aligned}$$

If the problem were solved, we find that the student studies a total of about 66½ hrs: 10 hrs for Marketing, 12.22 hrs for Organizational Behavior, 18.18 hrs for Accounting, 14.29 hrs for Operations Research, and 11.82 hrs for Finance, resulting in marks of 50, 55, 100, 50, and 65 in the five areas, i.e., minimum passing grades in

2.2 Applications of Linear Programming

Table 2.18 Parameters for the sample problem

	Month 1		Month 2		Month 3	
	Product A	Product B	Product A	Product B	Product A	Product B
Production capacity	70	40	80	30	80	10
Unit production cost	$3.10	$10.50	$3.20	$10.80	$3.80	$12.00
Estimated demand d_A, d_B	50	30	60	10	100	40

Marketing, Organizational Behavior, and Operations Research, while the 100 marks in Accounting and the 65 marks in Finance are significantly better than the minimum requirements.

Problem 3 (reformulation of the dynamic production-inventory problem): Formulate the problem in Sect. 2.2.6 with inventory variables that are defined at the end of the period.

Solution: Define now I_0, I_1, I_2, I_3, and I_4 as the inventory levels at the end of periods 0 (the beginning of the planning period), 1, 2, 3, and 4. We can then use the same formulation provided in Sect. 2.2.6, except that we need to replace I_t by I_{t-1} for $t = 1, 2, 3,$ and 4. Doing so results in the formulation

$$P : \text{Min } z = 1x_1 + 1.1x_2 + 1.2x_3 + 1.25x_4 + .05I_1 + .15I_2 + .15I_3$$

$$\text{s.t.} \quad x_1 \leq 120$$
$$x_2 \leq 140$$
$$x_3 \leq 150$$
$$x_4 \leq 140$$
$$I_1 = I_0 + x_1 - 80 \text{ (or, as } I_0 = 20, x_1 - I_1 = 60)$$
$$-I_2 + I_1 + x_2 = 70$$
$$-I_3 + I_2 + x_3 = 130$$
$$-I_4 + I_3 + x_4 = 150 \text{ (or, as } I_4 = 0, x_4 + I_3 = 150)$$
$$x_1, x_2, x_3, x_4, I_1, I_2, I_3 \geq 0.$$

Using an interpretation that reflects the inventory variables as defined here, the solution is again the same as before.

Problem 4 (a two-product production–inventory model): A firm manufactures two products. Their production capacities for the two products, unit production costs, and estimated demands are shown in Table 2.18.

The opening inventories of the two products are 0 and 10 units, respectively. At the end of Month 3, we do not want any inventories left. The inventory carrying costs are 20¢ per unit of product A and 50¢ for each unit of product B. These costs are incurred whenever one unit of a product is carried over from 1 month to the next. The total inventory levels (for both products combined) from Month 1 to Month 2 should not exceed 40 units, while the total inventory level between Months 2 and 3 should not exceed 50 units. Formulate a linear programming model for finding a cost-minimizing production plan.

Solution: The decision variables are x_{A1}, x_{A2}, and x_{A3} as the production quantities of product A in the months 1, 2, and 3, and x_{B1}, x_{B2}, and x_{B3} as the production quantities of product B in the 3 months. In addition, the inventory levels at the beginning of periods 1, 2, 3, and 4 (where the inventory level at the beginning of period 4 equals the inventory level at the end of period 3) for the two products are defined as I_{A1}, I_{A2}, I_{A3}, and I_{A4}, and I_{B1}, I_{B2}, I_{B3}, and I_{B4}, respectively.

The formulation of the problem is then

$$P : \text{Min } z = 3.1x_{A1} + 3.2x_{A2} + 3.8x_{A3} + 10.5x_{B1} + 10.8x_{B2} + 12x_{B3}$$
$$+ .2I_{A2} + .2I_{A3} + .5I_{B2} + .5I_{B3}$$

$$\text{s.t. } x_{A1} \leq 70$$
$$x_{A2} \leq 80$$
$$x_{A3} \leq 80$$
$$x_{B1} \leq 40$$
$$x_{B2} \leq 30$$
$$x_{B3} \leq 10$$
$$I_{A1} = 0$$
$$I_{B1} = 10$$
$$I_{A2} = I_{A1} + x_{A1} - 50$$
$$I_{A3} = I_{A2} + x_{A2} - 60$$
$$I_{A4} = I_{A3} + x_{A3} - 100$$
$$I_{B2} = I_{B1} + x_{B1} - 30$$
$$I_{B3} = I_{B2} + x_{B2} - 10$$
$$I_{B4} = I_{B3} + x_{B3} - 40$$
$$I_{A4} = 0$$
$$I_{B4} = 0$$
$$I_{A2} + I_{B2} \leq 40$$

2.2 Applications of Linear Programming

Table 2.19 Optimal solution of Problem 4

	Month 1			Month 2			Month 3			
P_1	I_{A1}	x_{A1}	d_{A1}	I_{A2}	x_{A2}	d_{A2}	I_{A3}	x_{A3}	d_{A3}	I_{A4}
	0	+50	−50	=0			20	+80	−100	=0
				0	+80	−60	=20			
P_2	I_{B1}	x_{B1}	d_{B1}	I_{B2}	x_{B2}	d_{B2}	I_{B3}	x_{B3}	d_{B3}	I_{B4}
	10	+30	−30	=10			30	+10	−40	=0
				10	+30	−10	=30			

$$I_{A3} + I_{B3} \leq 50$$

$$x_{A1}, x_{A2}, x_{A3}, x_{B1}, x_{B2}, x_{B3}, I_{A1}, I_{A2}, I_{A3}, I_{A4}, I_{B1}, I_{B2}, I_{B3}, I_{B4} \geq 0.$$

The optimal solution is shown in Table 2.19. The associated costs are $1498.

Problem 5 (blending of tobaccos): A large tobacco manufacturer has the choice of buying four tobacco types *Virginia*, *Burley*, *Latakia*, and *Kentucky*. Once sufficient quantities have been purchased, they will make three blends of pipe tobacco, viz., *Sweet Smell*, *Brown Lung*, and *Black Death*. The following information is available.

- The four types of tobacco cost $3, $6, $5, and $2 per pound (in the order they were mentioned).
- The final blends are sold by the 4 oz pouch, i.e. there are four pouches per pound.
- The blends sell for $7, $9, and $12 per pouch (in the above order).
- *Sweet Smell* consists of 20% *Virginia*, 50% *Burley*, and 30% *Latakia*, *Brown Lung* is blended from 40% *Latakia* and equal proportions of the remaining tobaccos, and *Black Death* is 80% *Kentucky* and 20% *Latakia*.
- The four tobaccos are available in limited quantities. We may purchase up to 300 lbs of *Virginia*, 500 lbs of *Burley*, 100 lbs of *Latakia*, and 50 lbs of *Kentucky*.
- Our customers have placed orders for exactly 500 pouches of *Sweet Smell* and 400 pouches of *Brown Lung*. There are no firm orders for the expensive *Black Death*, but we are certain to be able to sell between 80 and 120 pouches.

(a) Formulate a linear programming problem for the above situation. Define the variables clearly.
(b) Assume that there is a 5% loss in the blending process. Explain the changes in the formulation.

Solution:

(a) As usual, the variables are denoted by x_{ij} and defined as the quantity of i-th raw tobacco in j-th blend. The problem is very similar to that in Sect. 2.2.6. The only major difference is that the raw materials are measured in pounds, while the products are sold by the pouch. As four pouches make a pound, we need to convert pouches to pounds by multiplying the quantities of the products by 4. The problem can then be formulated as:

$$P : \text{Max } z = 7(x_{11} + x_{21} + x_{31} + x_{41})4 + 9(x_{12} + x_{22} + x_{32} + x_{42})4$$
$$+ 12(x_{13} + x_{23} + x_{33} + x_{43})4 - 3(x_{11} + x_{12} + x_{13}) - 6(x_{21} + x_{22} + x_{23})$$
$$- 5(x_{31} + x_{32} + x_{33}) - 2(x_{41} + x_{42} + x_{43})$$

$$\text{s.t. } x_{11} + x_{12} + x_{13} \leq 300$$
$$x_{21} + x_{22} + x_{23} \leq 500$$
$$x_{31} + x_{32} + x_{33} \leq 100$$
$$x_{41} + x_{42} + x_{43} \leq 50$$

$$4(x_{11} + x_{21} + x_{31} + x_{41}) = 500$$
$$4(x_{12} + x_{22} + x_{32} + x_{42}) = 400$$
$$4(x_{13} + x_{23} + x_{33} + x_{43}) \leq 120$$
$$4(x_{13} + x_{23} + x_{33} + x_{43}) \geq 80$$

$$x_{11} = 0.2(x_{11} + x_{21} + x_{31} + x_{41})$$
$$x_{21} = 0.5(x_{11} + x_{21} + x_{31} + x_{41})$$
$$x_{31} = 0.3(x_{11} + x_{21} + x_{31} + x_{41})$$
$$x_{12} = 0.2(x_{12} + x_{22} + x_{32} + x_{42})$$
$$x_{22} = 0.2(x_{12} + x_{22} + x_{32} + x_{42})$$
$$x_{32} = 0.4(x_{12} + x_{22} + x_{32} + x_{42})$$
$$x_{42} = 0.2(x_{12} + x_{22} + x_{32} + x_{42})$$
$$x_{33} = 0.2(x_{13} + x_{23} + x_{33} + x_{43})$$
$$x_{43} = 0.8(x_{13} + x_{23} + x_{33} + x_{43})$$
$$x_{11}, x_{12}, x_{13}, x_{21}, x_{22}, x_{23}, x_{31}, x_{32}, x_{33}, x_{41}, x_{42}, x_{43} \geq 0.$$

The optimal solution purchases 45, 82.5, 83.3, and 44 lbs of the respective raw tobaccos, and produces 500, 400, and 120 pouches of the three tobacco blends. The profit is $7404.50.

(b) Suppose that there is a 5% loss in the blending process. This can easily be accounted for as follows. Everywhere the quantity of a product is referred to, it is replaced by the quantity multiplied by $1 - 5\% = 0.95$. In this formulation, we replace $(x_{11} + x_{21} + x_{31} + x_{41})$ by $(x_{11} + x_{21} + x_{31} + x_{41})(.95)$ and similar for $(x_{12}$

2.2 Applications of Linear Programming

Table 2.20 Input data for Problem 6

Component	Availability (in barrels)	Octane number	Vapor pressure	Cost per barrel
Naphta	30,000	85	10	$53
Hydrocrackate	45,000	79	4	$48
Reformate	20,000	101	9	$62
Alkylate	15,000	108	5	$69

$+ x_{22} + x_{32} + x_{42}$) and ($x_{13} + x_{23} + x_{33} + x_{43}$) in the objective function, and the second set of constraints. It would also be easy to include different losses for different products. The solution here purchases 47.37, 86.84, 87.89, and 46.31 lbs of the respective raw tobaccos, and sells again 500, 400, and 120 pouches of the three tobacco blends. The profit is now $7344.74, about a 1% decrease as compared to the case without losses.

Problem 6 (Blending of gasolines): Table 2.20 describes components in a petroleum refinery that can be used to blend gasoline:

The purpose is to blend two types of gasoline, *Regular* and *Premium*, so as to minimize the overall costs required to satisfy the demand. The *Regular* brand consists of *Naphta*, *Hydrocrackate*, and *Reformate*, while *Premium* consists of *Naphta*, *Hydrocrackate*, and *Alkylate*. The total contracted demand for gasoline is 80,000 barrels.

- *Regular*: The octane number must be at least 87 and the vapor pressure cannot exceed 7.2.
- *Premium*: The octane number must be at least 91 and the vapor pressure cannot exceed 6.8.

Assume that there are no losses in the blending process and that all quantities blend linearly by quantity, and formulate a cost-minimizing linear programming model for this situation.

Solution: Given the four "raw materials" *Naphta*, *Hydrocrackate*, *Reformate*, and *Alkylate* along with the two products *Regular* and *Premium*, we can define the variables x_{ij} as the quantity of raw material i in product j.

$$P: \text{Min } z = 53(x_{11} + x_{12}) + 48(x_{21} + x_{22}) + 62x_{31} + 69x_{42}$$

$$\text{s.t. } x_{11} + x_{12} \leq 30,000 \quad \text{(availability of } Naphta\text{)}$$
$$x_{21} + x_{22} \leq 45,000 \quad \text{(availability } Hydrocrackate\text{)}$$
$$x_{31} \leq 20,000 \quad \text{(availability of } Reformate\text{)}$$
$$x_{42} \leq 15,000 \quad \text{(availability of } Alkylate\text{)}$$

$$x_{11} + x_{12} + x_{21} + x_{22} + x_{31} + x_{42} = 80,000 \text{ (demand)}$$

$$85x_{11} + 79x_{21} + 101x_{31} \geq 87(x_{11} + x_{21} + x_{31}) \quad \text{(octane } Regular\text{)}$$
$$10x_{11} + 4x_{21} + 9x_{31} \leq 7.2(x_{11} + x_{21} + x_{31}) \quad \text{(vapor pressure } Regular\text{)}$$

$$85x_{12} + 79x_{22} + 108x_{42} \geq 91(x_{12} + x_{22} + x_{42}) \quad \text{(octane } Premium\text{)}$$
$$10x_{12} + 4x_{22} + 5x_{42} \leq 6.8(x_{12} + x_{22} + x_{42}) \quad \text{(vapor pressure } Premium\text{)}$$

$$x_{11}, x_{21}, x_{31}, x_{12}, x_{22}, x_{42} \geq 0.$$

The optimal solution produces a total of 80,000 barrels, 70,833.33 regular and 9166.67 barrels premium. Other than the resource Reformate, none of the resources is fully used. The octane constraints are both binding, so is the vapor pressure constraint for regular. The vapor pressure constraint for premium has some slack. The total profit is $4,305,211.

Problem 7 (blending with exact requirements): A fish processing plant makes two types of fish sticks, the *Scrumptious Skipper* and the *Delicious Sailor*. The *Skipper* consists of exactly 30% pollock, 40% haddock, and 30% sole, while the *Sailor* contains 30% pollock, 20% haddock, and 50% sole. A one-pound package of the *Skipper* sells for $2.50, while a one-pound pack of the *Sailor* retails for $3.50. There are 4000 lbs of pollock, 3000 lbs of haddock, and 3000 lbs of sole available in the plant. Formulate a profit-maximizing linear programming problem for this situation.

Solution: The usual thought would be to define the variables x_{ij} as the quantity of the i-th type of fish in the j-th type of fish sticks. The formulation is then

$$P: \text{Max } z = 2.5(x_{11} + x_{21} + x_{31}) + 3.5(x_{12} + x_{22} + x_{32})$$

$$\text{s.t.} \quad x_{11} + x_{12} \leq 4000$$

$$x_{21} + x_{22} \leq 3000$$

$$x_{31} + x_{32} \leq 3000$$

$$x_{11} = .3(x_{11} + x_{21} + x_{31})$$

$$x_{21} = .4(x_{11} + x_{21} + x_{31})$$

$$x_{31} = .3(x_{11} + x_{21} + x_{31})$$

$$x_{12} = .3(x_{12} + x_{22} + x_{32})$$

$$x_{22} = .2(x_{12} + x_{22} + x_{32})$$

$$x_{32} = .5(x_{12} + x_{22} + x_{32})$$

$$x_{11}, x_{12}, x_{21}, x_{22}, x_{31}, x_{32} \geq 0.$$

While this is a correct formulation, it is too large for what it does. Due to the fact that the blending requirements have to be satisfied exactly, we could simply formulate variables x_1 and x_2 as the quantities of the two fish stick packages that we make, and obtain the simpler formulation

$$P' : \text{Max } z = 2.5x_1 + 3.5x_2$$
$$\text{s.t.} \quad .3x_1 + .3x_2 \leq 4000$$
$$.4x_1 + .2x_2 \leq 3000$$
$$.3x_1 + .5x_2 \leq 3000$$
$$x_1, \quad x_2 \geq 0.$$

In both cases, we make 6428.57 packages of the *Skipper* and 2142.86 packs of the *Sailor* for a profit of $23,571.43. This shorter formulation is possible as there is a fixed relation between the number of packages of the two products and the quantity of the fish input (e.g., the quantity of pollock in the packages of *Skipper* is exactly 0.3 times the quantity of *Skipper* packages). This was not the case in Problems 5 and 6.

Problem 8 (a transportation problem): Consider a school district's problem to assign student from different villages to central schools. Typically, with the closing of small neighborhood schools of the "little red schoolhouse" type and the establishment of larger centralized schools, it has become necessary to bus the students to the schools (as these distances would even make Abe Lincoln take the bus). The objective is to ensure that all students must be able to take a bus, and school capacities cannot be violated.

Suppose there are three villages with 30, 50, and 20 students each. The two centralized schools have capacities of 70 and 60 students, respectively. The distances between the villages and the schools are shown in the matrix **C** below.

$$\mathbf{C} = \begin{bmatrix} 20 & 15 \\ 40 & 30 \\ 60 & 20 \end{bmatrix}.$$

(a) Formulate a transportation problem for minimizing the total mileage traveled by the students.
(b) Suppose that the buses available to the district each have a capacity of 35 students. Formulate constraints to ensure that there are no overfilled buses.
(c) In addition to the constraints under (b), it is now required that each school is filled to at least 75% capacity.

Solution:

(a) We first define variables, so that x_{ij} denotes the number of students bused from village i to school j. The model can then be formulated as follows.

$$P : \text{Min } z = 20x_{11} + 15x_{12} + 40x_{21} + 30x_{22} + 60x_{31} + 20x_{32}$$

$$\text{s.t.} \quad \begin{aligned}
x_{11} + x_{12} &= 30 \\
x_{21} + x_{22} &= 50 \\
x_{31} + x_{32} &= 20 \\
x_{11} + x_{21} + x_{31} &\leq 70 \\
x_{12} + x_{22} + x_{32} &\leq 60 \\
x_{11}, x_{12}, x_{21}, x_{22}, x_{31}, x_{32} &\geq 0.
\end{aligned}$$

Incidentally, the solution is summarized in the optimal transportation plan

$$\overline{\mathbf{T}} = \begin{bmatrix} 30 & 0 \\ 10 & 40 \\ 0 & 20 \end{bmatrix}.$$

The total mileage to bus all students to schools is 2600, and while the second school is filled to capacity, the first houses only 40 students, well shy of its capacity of 70.

(b) The most obvious way to formulate this constraint is to impose capacities on all routes, i.e., write six additional constraints $x_{11} \leq 35, x_{12} \leq 35, \ldots, x_{32} \leq 35$. This is, however, unnecessary, as only buses leading out of the second village could possibly have more students than the bus capacity allows. Hence it is sufficient to add the two constraints $x_{21} \leq 35$ and $x_{22} \leq 35$. The new solution has the transportation plan

$$\overline{\mathbf{T}} = \begin{bmatrix} 25 & 5 \\ 15 & 35 \\ 0 & 20 \end{bmatrix}$$

requiring a total mileage of 2625, a very minor increase from the original 2600.

2.2 Applications of Linear Programming

(c) Filling the schools to at least 75% of capacity requires the two schools to house at least $70(.75) = 52.5$ and $60(.75) = 45$ students. Since integrality is required and these are lower bounds on the number of students, we have to round up the first number to 53. We then add the constraints

$$x_{11} + x_{21} + x_{31} \geq 53$$

$$x_{12} + x_{22} + x_{32} \geq 45$$

to the problem. The optimal solution is then shown in the transportation plan

$$\overline{\mathbf{T}} = \begin{bmatrix} 30 & 0 \\ 23 & 27 \\ 0 & 20 \end{bmatrix}$$

which has an associated total mileage of 2730.

Problem 9 (an assignment problem): The manager of an experimental farm faces the task of trying out three fertilizers, each based on some of the main ingredients in commercial products: *Phosphorus*, *Biosolids* (mostly domestic septage), and *Nitrogen*. Each of the fields is to be planted with one of the three crops, viz., *Wheat*, *Rye*, and *Corn*. The crops react differently to the fertilizers, and small-scale lab tests have suggested the yields shown in the matrix **A**, in which a_{ij} indicates the expected yield of crop i given that only fertilizer j is used. The fertilizers and yields blend linearly, e.g., if the field of wheat were fertilized with 30% phosphorus and 70% biosolids, the yield would be $0.3(5) + 0.7(1) = 2.2$. The manager needs to ensure that no more than 100% of the available quantity of fertilizers is used and that each crop receives 100% fertilization.

$$\mathbf{A} = \begin{bmatrix} 5 & 1 & 7 \\ 3 & 4 & 5 \\ 6 & 3 & 2 \end{bmatrix}.$$

(a) Determine an optimal fertilization plan that maximizes the yield.

(b) Suppose that a new product *Potassium* became available, which results in yields of 4, 6, and 6 if applied to the three respective crops. Does that revelation have any effect on the fertilization plan?

Solution:

(a) Define x_{ij} as the proportion of fertilizer j that is applied to the field with crop i. The problem can then be written as

$$P: \text{Max } z = 5x_{11} + x_{12} + 7x_{13} + 3x_{21} + 4x_{22} + 5x_{23} + 6x_{31} + 3x_{32} + 2x_{33}$$

$$\text{s.t. } x_{11} + x_{12} + x_{13} = 1$$

$$x_{21} + x_{22} + x_{23} = 1$$

$$x_{31} + x_{32} + x_{33} = 1$$

$$x_{11} + x_{21} + x_{31} = 1$$

$$x_{12} + x_{22} + x_{32} = 1$$

$$x_{13} + x_{23} + x_{33} = 1$$

$$x_{ij} \geq 0 \text{ for all } i, j.$$

The optimal solution is $x_{13} = x_{22} = x_{31} = 1$, and $x_{ij} = 0$ otherwise. In other words, wheat is fertilized entirely with nitrogen, rye only with biosolids, and corn exclusively with phosphorus. The expected yield is 17.

(b) Adding the new fertilizer potassium changes the formulation as follows (the additions are shown in boldface):

$$P: \text{Max } z = 5x_{11} + x_{12} + 7x_{13} + \mathbf{4x_{14}} + 3x_{21} + 4x_{22} + 5x_{23} + \mathbf{6x_{24}} + 6x_{31} + 3x_{32} + 2x_{33} + \mathbf{6x_{34}}$$

$$\text{s.t. } x_{11} + x_{12} + x_{13} + \mathbf{x_{14}} = 1$$

$$x_{21} + x_{22} + x_{23} + \mathbf{x_{24}} = 1$$

$$x_{31} + x_{32} + x_{33} + \mathbf{x_{34}} = 1$$

$$x_{11} + x_{21} + x_{31} \leq 1$$

$$x_{12} + x_{22} + x_{32} \leq 1$$

$$x_{13} + x_{23} + x_{33} \leq 1$$

$$\mathbf{x_{14} + x_{24} + x_{34} \leq 1}$$

$$x_{ij} \geq 0 \text{ for all } i, j.$$

The new optimal solution is similar to the previous solution, except that it replaces the biosolids applied to rye by potassium. The expected yield increases by close to 12% to 19.

Problem 10 (a one-dimensional cutting stock problem): A home building store faces the following problem. Its customers demand half-inch plywood in the sizes 4 ft × 3 ft, 4 ft × 5 ft, and 4 ft × 6 ft. Customer demand for these three sizes is estimated to be at least 20, 50, and 40, respectively, and customers are prepared to pay $7, $9, and $10 for each of these sheets, respectively. The store must generate these sizes by cutting up standard 4 ft × 8 ft sheets, each of which costs them $6. Up to 100 such sheets are available. Furthermore, each cut costs the store $1.50. Formulate a model that indicates to the decision maker how to cut up the 4 ft × 8 ft sheets so as to maximize his profit. Clearly show the cutting patterns and define your variables properly.

Solution: Since the widths of the desired sheets are all 4 ft, we only have to consider a single dimension. The cutting plan includes patterns with two 3 ft × 4 ft boards (and the resulting 2 ft × 4 ft piece of waste), one 3 ft × 4 ft board and a 5 ft × 4 ft board without any waste, and a single 6 ft × 4 ft board with the resulting 2 ft × 4 ft piece of waste. The number of cuts performed according to these three patterns are denoted by y_1, y_2, and y_3, respectively. The problem can then be formulated as

$$P : \text{Max } z = [7(2y_1 + y_2) + 9(y_2) + 10(y_3)] - 6[y_1 + y_2 + y_3] - 1.5[2y_1 + y_2 + y_3]$$
$$= 5y_1 + 8.5y_2 + 2.5y_3$$

$$\begin{aligned}
\text{s.t. } 2y_1 + y_2 &\geq 20 &&\text{(generate at least twenty 4 ft} \times \text{3 ft sheets)} \\
y_2 &\geq 50 &&\text{(generate at least fifty 4 ft} \times \text{5 ft sheets)} \\
y_3 &\geq 40 &&\text{(generate at least forty 4 ft} \times \text{6 ft sheets)} \\
y_1 + y_2 + y_3 &\leq 100 &&\text{(up to 100 sheets of wood are available)} \\
y_1, \ y_2, \ y_3 &\geq 0 \text{ and integer}
\end{aligned}$$

The optimal solution does not cut the first pattern and uses the second and third pattern 60, and 40 times, respectively, resulting in a profit of $610. Changing the demands to at least 20 (as is) for the first size, exactly 50 for the second, and at most 40 for the third size, we only cut the second pattern 100 times.

Problem 11 (a two-dimensional cutting stock problem): A planner has 30 sheets of plywood of size 10 ft × 10 ft. They presently need 20 sheets in the shape of disks of diameter 5 ft as well as 15 sheets of plywood in the shape of 4 ft × 6 ft rectangles. The cutting patterns that are considered by the decision maker are shown in Fig. 2.5.

It costs $2.00 to cut a disk and $1.50 to cut a rectangle. This price includes all required cuts and is independent of the pattern the shape is cut from. Alternatively, we could purchase a disk at $4.50 and a rectangle for $3.00 each.

Formulate a linear programming problem that

- Minimizes the cost of obtaining the required shapes
- Does not use more 10 ft × 10 ft sheets than are available
- Produces the required numbers of disks and rectangles, and
- Ensures that the cutting results in no more than 30% of waste.

Define all variables clearly.

Solution: Define y_j, $j = 1, 2, 3$ as the number of times the j-th pattern is cut. In addition, define variables v_4 and v_5 as the number of disks and rectangles that are purchased. The problem can then be formulated as:

$$P : \text{Min } z = 8y_1 + 6y_2 + 5.5y_3 + 4.5v_4 + 3v_5$$

s.t.
$$y_1 + y_2 + y_3 \leq 30$$
$$4y_1 + 2y_3 + v_4 \geq 20$$
$$4y_2 + 1y_3 + v_5 \geq 15$$
$$.2146y_1 + .04y_2 + .3673y_3 \leq .3(y_1 + y_2 + y_3)$$
$$y_1, y_2, y_3, v_4, v_5 \geq 0 \text{ and integer,}$$

where the coefficients on the left-hand side of the last constraint are the proportions of waste given a 10 ft × 10 ft sheet.

2.3 Graphical Representation and Solution

Having described a variety of different applications of linear programming problems, this section will first demonstrate how linear programming problems can be represented graphically. We then discuss in some detail a graphical solution method. This is followed by a discussion of a number of special cases that may occur in the modeling and solution processes, and how to react to them as an analyst.

Throughout this chapter, we will restrict ourselves to the case of two variables in order to accommodate easy graphing. This does, of course, mean that the technique we describe in this section is not made for the solution of realistic problems that typically have tens of thousands of variables. Much rather, the purpose of this discussion is to create an understanding of what happens in the solution of linear programming problems and what the difficulties are, regardless of the size of the problem.

The first subsection will demonstrate how constraints and objective functions can be graphed and how the problem can be solved graphically. Based on the understanding of this material, Sect. 2.3.2 will then discuss a number of special situations

2.3.1 The Graphical Solution Method

As discussed in the introduction to linear programming, each model consists of an objective function and a number of constraints. This subsection will first demonstrate how to plot constraints, and then show how to deal with objective functions, and then put it all together in the graphical solution method.

Assuming that we have two variables x_1 and x_2, a constraint could be a linear function such as $3x_1 + 2x_2 \leq 6$. In order to plot this constraint, it is easiest to first consider the associated equation $3x_1 + 2x_2 = 6$. It is known that the line in two dimensions is uniquely determined by two points. In order to do so, we can simply set either of the variables to any value we like and solve for the other variable, resulting in one of the required points. Repeating this step with a different value will result in a second point. The straight line that leads through both of these points is then the set of all points that satisfy the equation.

In our example, setting $x_1 = 0$ leads to $2x_2 = 6$ or $x_2 = 3$, so that the first point is $(x_1, x_2) = (0, 3)$. The second point can be obtained by setting $x_2 = 0$, which leads directly to $3x_1 = 6$, or, equivalently, $x_1 = 2$. As a result, our second point is $(x_1, x_2) = (2, 0)$. The straight line in Fig. 2.6 is the set of points that satisfy $3x_1 + 2x_2 = 6$.

So far, we have determined that an equation is represented in two dimensions as a straight line. (Note that in a single dimension an equation is just a point.) In three dimensions, an equation would be represented by a plane, so that in general, we

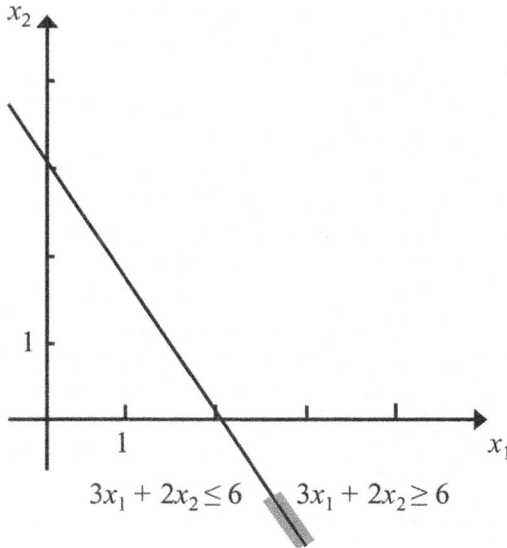

Fig. 2.6 Hyperplane and halfspaces

speak about an equation in any number of dimensions being represented by a *hyperplane*.

Back in our two-dimensional example, recall that the constraint in question is the inequality $3x_1 + 2x_2 \leq 6$, so that not only the set of points on the line are addresses. As a matter of fact, a $\leq$ or $\geq$ inequality will refer to the set of points on the line and all points in one of the two *halfplanes* generated by the line. The obvious question is then which of the two halfplanes is addressed by the constraint in question. Some people might believe that the halfplanes that belong to $\leq$ inequalities are below the line, while those of $\geq$ are above the line. This is not true, as each $\leq$ inequality can be rewritten as an equivalent $\geq$ inequality. In our example, the constraint $3x_1 + 2x_2 \leq 6$ is equivalent to its counterpart $-3x_1 - 2x_2 \geq -6$. Both constraints define exactly the same set of points.

A simple way to determine the proper halfplane is to choose any point that is not located on the line we have plotted and determine whether or not it satisfies the constraint in question. If so, then the point is located on the proper side of the line, otherwise the halfplane is on the other side. In our example, consider, for instance, the origin as a point. Its coordinates are (0, 0), so that the constraint $3x_1 + 2x_2 \leq 6$ reduces to $0 \leq 6$, which is correct. This means that the origin is on the "correct" side of the line, which allows us to determine the halfplane as being on the lower left side of the line. Had we chosen the point, say, (4, 2) instead, our constraint would have been $3(4) + 2(2) \leq 6$ or $16 \leq 6$, which is wrong, meaning that the point (4, 2) is located on the "wrong" side of the line. As a matter of fact, the set of points on the line and in the halfplane to the upper right of the line is determined by the constraint $3x_1 + 2x_2 \geq 6$. Figure 2.6 shows both the hyperplane and both halfplanes for all three types of constraints allowed in linear programming: $=$, $\leq$, and $\geq$. When graphing, the halfplane is always indicated by a hyperplane and a small flag at its end that indicates the appropriate halfplane.

Given that an equation is represented by a straight line in two and a plane in three dimensions, an inequality is represented by a halfplane in two dimensions and half the space in three dimensions (the separating line is given by the associated equation and the other half of the space is defined by the same inequality but with inverted inequality sign). This has led to the term *halfspace* that, in contrast to halfplane, which applies only to two dimensions, applies to the representation of an inequality of the $\leq$ or $\geq$ type in any number of dimensions.

At this point, we are able to plot the hyperplane or halfspace for each constraint in a given problem. In order to determine the *feasible set* (also called feasible region or set of feasible solutions), we first need to define a solution as feasible, if it satisfies *all* of the given constraints. This is not really a restriction, as if we do not want a constraint to be satisfied, why include it in the problem in the first place? Given that all constraints must be satisfied, the feasible set is then the intersection of the halfspaces and/or hyperplanes that correspond to all constraints of the problem.

As a numerical illustration, consider the following numerical example.

2.3 Graphical Representation and Solution

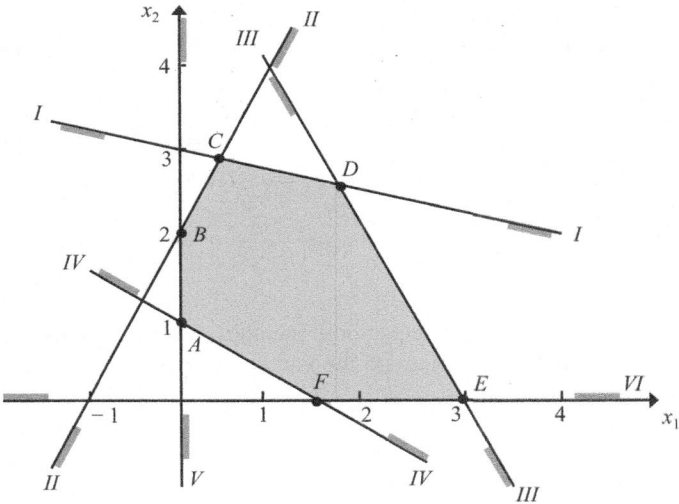

Fig. 2.7 Feasible region

$$P : \text{Max } z = 2x_1 + 3x_2$$

$$\text{s.t. } x_1 + 4x_2 \leq 12 \tag{I}$$

$$2x_1 - x_2 \geq -2 \tag{II}$$

$$5x_1 + 3x_2 \leq 15 \tag{III}$$

$$4x_1 + 6x_2 \geq 6 \tag{IV}$$

$$x_1 \geq 0 \tag{V}$$

$$x_2 \geq 0. \tag{VI}$$

The feasible set determined by the constraints of problem P, is shown as the shaded area in Fig. 2.7.

In our two-dimensional space, the feasible set is a linearly bounded polygon (in general, referred to as a *polytope*). It consists of the boundary with its linear segments and *corner points* (frequently referred to as *extreme points*) $A, B, \ldots, F$, as well as the interior. At each of the extreme points, at least two constraints are satisfied as equations. These constraints are usually referred to as *binding* (or *tight*) *at this point*. In our example, at point A, the constraints IV and V are satisfied as equations, at point B, constraints II and V are satisfied as equations, at point C, constraints I and II are satisfied as equations, and so forth.

If desired, we can then determine the exact coordinates of all extreme points by solving a system of simultaneous linear equations. For instance, for point A, the

Table 2.21 Coordinates of extreme points of sample problem

Point	Constraints binding at point	Coordinates (x_1, x_2)	z-value
A	IV, V	(0, 1)	3
B	II, V	(0, 2)	6
C	I, II	$\left(\frac{4}{9}, 2\frac{8}{9}\right) \approx (.4444, 2.8889)$	9.5556
D	I, III	$\left(1\frac{7}{17}, 2\frac{11}{17}\right) \approx (1.4118, 2.6471)$	10.7649
E	III, VI	(3, 0)	6
F	IV, VI	(1½, 0)	3

system of simultaneous linear equations includes all constraints satisfied as equations at this point. Here, these are the equations based on constraints *IV* and *V*, so that the system is

$$4x_1 + 6x_2 = 6 \quad \text{and}$$

$$x_1 = 0.$$

Replacing $x_1 = 0$ in the first equation and solving for x_2, we obtain $x_2 = 1$, so that the coordinates are $(x_1, x_2) = (0, 1)$.

Similarly, consider point *C*. At his point, the constraints *I* and *II* are binding, so that we have the set of simultaneous linear equations

$$x_1 + 4x_2 = 12$$

$$2x_1 - x_2 = -2.$$

A system like this can be solved by any of the pertinent methods, see Appendix B of this volume. One (albeit somewhat awkward) possibility is to use the *substitution technique*. Here, we solve the first equation for x_1, resulting in $x_1 = 12 - 4x_2$. We then replace x_1 by this expression in the second equation, so that we obtain $2(12 - 4x_2) - x_2 = -2$. Solving this equation for x_2 results in $x_2 = \frac{26}{9} \approx 2.8889$. Replacing x_2 by this value in $x_1 = 12 - 4x_2$ and solving for x_1 results in $x_1 = \frac{4}{9} \approx 0.4444$. Table 2.21 shows the points, the constraints that are satisfied as equations at that point, and their exact coordinates.

In *n* dimensions, each extreme point is determined by the intersection of at least *n* hyperplanes, so that we have to solve a system of at least *n* simultaneous linear equations in *n* variables to determine the coordinates for each of these extreme points.

Consider now the objective function. To simplify matters, we will at first ignore the constraints and deal exclusively with the objective function and its representation, before we combine objective function and constraints in the graphical solution method.

For now, consider the objective function Max $z = 2x_1 + 5x_2$. Ignoring the maximization for a moment, we have $2x_1 + 5x_2 = z$, which is nothing but a regular constraint with an unknown right-hand side value z. As discussed above, for any

2.3 Graphical Representation and Solution

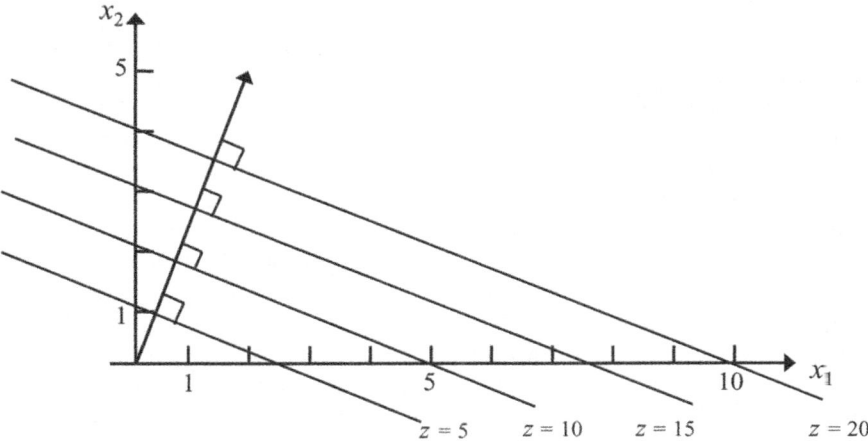

Fig. 2.8 Iso-profit lines

value of z, we have an equation that can be represented by a hyperplane in the space of variables, here the (x_1, x_2) space. Fig. 2.8 shows these lines for values of $z = 5$, 10, 15, and 20.

Depending on the type of objective function under consideration, these lines are usually referred to as *iso-profit lines*, *iso-cost lines*, or simply *contour lines*. Their name derives from the fact that all points on any one of these lines have the same value of the objective function. (You may be familiar with the contour lines on topographic maps. Those lines are really iso-altitude lines, or lines of equal altitude.) In other words, given the objective function under consideration, all points on the line labeled $z = 5$ are considered equally good by the decision maker. Similarly, a decision maker will consider all points on the line $z = 10$ as equally good—but better than those on the $z = 5$ line. In this example, the value of the objective function gets better in the northeasterly direction.

It is then possible to construct a vector that points into the direction, in which the objective function improves. This is the so-called *gradient of the objective function*, sometimes also referred to as the *direction of the objective function*. Formally, a gradient is the vector of partial derivatives, but here it is sufficient to think of it as the direction in which the solutions get better. The gradient is constructed as follows, where we use again our numerical example with the objective Max $z = 2x_1 + 5x_2$. Each term of the objective function can be thought of as the combination of the step direction and the step length. Here, x_1 means move to the right, and the coefficient 2 tells us to move 2 steps into that direction. The next term indicates that we should move 5 steps into the x_2 direction. Starting at an arbitrary point, we first move 2 steps into the x_1 direction, followed by 5 steps into the x_2 direction. The starting point is then connected to the end point, resulting in the gradient. Usually, we start these moves at the origin, but this is not necessary.

Observe that the gradient of the objective function is perpendicular to the iso-profit lines. Once we have the gradient, it is not necessary to explicitly plot any of the iso-profit lines. (In more than two dimensions, the gradient is a ray that is orthogonal—the generalization of perpendicular to n dimensions—to the iso-profit hyperplanes.) From a practical point of view, we can plot the gradient of the objective function and then push the perpendicular iso-profit lines as much into its direction as possible—the farther we push, the higher the profit.

Before putting it all together and describing the graphical solution technique, some properties of the objective function are worth mentioning. Suppose that in the above objective function each of the terms is measured in dollars. Assume now that we have decided to measure the profit in Indian rupees instead. Suppose that the present exchange rate is 75 rupees per dollar, so that the objective function is now Max $z' = 50x_1 + 375x_2$. Plotting this objective, we find that while the gradient is much longer, the direction of the objective function is exactly the same. As we will see later, such a change of currency will result in exactly the same solution as the original objective function, only the objective value changes: z' will be 75 times the value of z.

Another point of interest concerns minimization functions. What if the objective function minimizes some costs, e.g., Min $z = 3x_1 + 7x_2$? No special procedure is needed, as we can simply transform the minimization objective into an equivalent maximization objective by multiplying it by a negative number, e.g., (-1). This will result in the equivalent objective Max $-z = -3x_1 - 7x_2$. As far as the gradient of this function is concerned, it leads from the origin -3 steps into the x_1 direction (i.e., three steps to the left), followed by -7 steps into the x_2 direction (i.e., 7 steps down). Everything else remains exactly the same, the value of the objective function improves (i.e., gets smaller in case of a minimization function) as we shift the iso-cost lines more and more into the direction of the gradient. Figure 2.9 shows the gradients for the following objective functions:

(a) Max $z_1 = 4x_1 - 3x_2$
(b) Max $z_2 = -x_1 + 3x_2$
(c) Min $z_3 = -2x_1 - x_2$
(d) Min $z_4 = 2x_1 + 3x_2$

It is worthwhile to notice that if we have one function such as $z_1 = 4x_1 - 3x_2$ in the above example, maximizing the function leads to a gradient, which points into a southeasterly direction. Minimizing the same function leads into the northwest, diametrically opposed to the maximization of the same function.

We are now able to describe the complete graphical solution technique. After determining the feasible set, we plot the gradient of the objective function and move its iso-profit lines into the direction of the gradient, until we hit the last feasible point. While there are solutions with better objective function values beyond this point, none of them is feasible. Thus, the last feasible point into the direction of the gradient is the optimal point.

2.3 Graphical Representation and Solution

Fig. 2.9 Objective function gradients

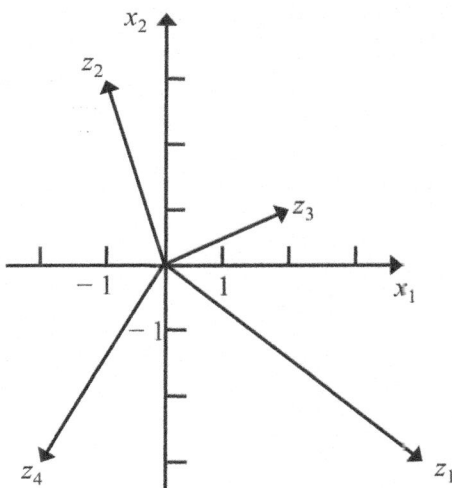

It is apparent that in this procedure, points in the interior of the feasible set cannot be optimal; any optimal solution will be on the boundary of the feasible set. In fact, Dantzig has proved his famous corner point theorem, which we will state here formally.

Theorem (Corner Point Theorem, Dantzig) At least one optimal solution is located at an extreme point of the feasible set.

The graphical solution method will identify such a corner point, whose exact coordinates we will have to determine next. As demonstrated earlier in this section, this is done by way of solving a system of simultaneous linear equations. Once the exact coordinates of the optimal solution point have been determined, all that is left to do is to determine the quality of the solution, as measured by the objective function. This is accomplished by replacing the variables in the objective function by their optimal values and thus computing the z-value.

We can summarize the procedure in the following steps:

Step 1: Graph the constraints and determine the set of feasible solutions.
Step 2: Plot the gradient of the objective function.
Step 3: Apply the graphical solution technique that pushes iso-profit lines into the direction of the gradient until the last feasible point is reached. This is the optimal solution $\bar{\mathbf{x}}$.
Step 4: Determine which constraints are satisfied as equations at $\bar{\mathbf{x}} = (\bar{x}_1, \bar{x}_2)$. Write them as equations and solve the resulting system of simultaneous linear equations for the exact coordinates of the optimal solution.

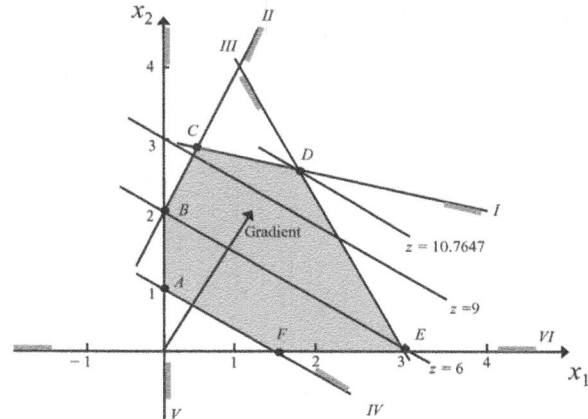

Fig. 2.10 Graphical solution technique

Step 5: Use the coordinates of the optimal point in the objective functions and compute the value of the objective function.

Applying the first two steps of this procedure to the problem stated in the beginning of this subsection, we obtain the graph in Fig. 2.10. Pushing now the iso-profit lines (some of which are shown) into the direction of the gradient, we find that the last feasible point on our way into a northeasterly direction is the extreme point *D*. This is the optimal point $\bar{\mathbf{x}}$. The constraints *I* and *III* are binding at this point, so that we solve the system of simultaneous linear equations that consists of relations *I* and *III* written as equations, i.e.,

$$x_1 + 4x_2 = 12$$

$$5x_1 + 3x_2 = 15.$$

The optimal solution is $(\bar{x}_1, \bar{x}_2) = \left(1\tfrac{7}{17}, 2\tfrac{11}{17}\right) \approx (1.4118, 2.6471)$, and the associated value of the objective function is $\bar{z} = 10\tfrac{13}{17} \approx 10.7647$. It can be shown that this solution is not only optimal but is the *unique optimal solution* to the problem. Sometimes, an optimal point is found, such that at least one of its neighboring extreme points has the same value of the objective function, and as such is also optimal. This would happen in our problem, if the same objective function were to be minimized rather than maximized. In this case, we would find the points *A* and *F* both as optimal solution points with $\bar{z} = 3$. More about this issue can be found in the next subsection on special cases.

While the graphical method as demonstrated above is an exact method (no approximations were made in the process), its use is to explain the main concepts and difficulties involved in solving linear programming problems. The reason is that practical problems have not two or three, but tens of thousands of variables, making graphing impossible. Since the graphical solution technique uses the exact pictorial knowledge of the feasible set, it will be necessary to find an algebraic technique that

2.3 Graphical Representation and Solution

is independent of the graphical image. Dantzig's *simplex method* is such a tool. Rather than moving through the feasible space directly to the optimal solution, the simplex method is an *incremental technique* that starts with a feasible solution (which can be determined by "some" technique), improves it, tests whether or not an optimal solution has been found, and if not, increases the solution further. It does so by moving on the boundary of the feasible set from one extreme point to an adjacent extreme point. The method also belongs to the class of *feasible* (and improving) *direction methods*. This means that a step from a (feasible) extreme point to an adjacent extreme point is only made if the solution remains feasible and the value of the objective function improves in the process. A feature of the feasible set, called *convexity*, guarantees that if a point is found none of whose neighbors has a better z-value than the one we are presently at, this is an overall (i.e., global) optimal solution.

To demonstrate a simplex path, i.e., the sequence of extreme points generated and examined by the simplex method, consider again the example of Fig. 2.10 and assume that we have "somehow" determined point A as a starting point. Point A has two neighboring extreme points F and B. Both are feasible, so that moves are possible. However, while the move from A to B improves the value of the objective function as B is on a higher iso-profit line, the move from A to F will leave the value of the objective function unchanged. (This is one of the special cases discussed in the next subsection.) Since we are looking for improvements, the simplex method will move to point B. At that point, we have again two neighboring extreme points, *viz.*, A and C. While a move from B to A retains feasibility of the solution, the z-value would decrease, disallowing such move. On the other hand, moving from B to C maintains feasibility and improves the value of the objective function, so that the simplex method makes this move. At point C, we have again two neighbors, which are B and D. Moving to B is not allowed, as this would decrease the z-value. On the other hand, a move to D not only maintains feasibility, but also increases the value of the objective function. The neighboring extreme points at point D are C and E. Moving either way will keep the solution feasible, but in both cases, the value of the objective function will decrease. At this point, the method terminates with the message that point D is an optimal solution.

While examples have been constructed in which the simplex algorithm performs very poorly, the average performance of the algorithm has been excellent. Given a problem with m constraints, there is consensus that on average, the simplex algorithm needs to examine only $1\frac{1}{2}m$ extreme points. In each step, we need to examine an extreme point, which means we must solve a system of simultaneous linear equations. Traditionally, computational details of this method, which is considered to be one of the ten top algorithms of the twentieth century, have been included in texts such as this. Given the abundance of software (some of it even free) and the fact that users do not need to know details about how the method functions, we will not discuss it in this book. For a full treatment, interested readers are referred to Eiselt and Sandblom (2007), the file-sharing site that accompanies this book, or one of the many books available on the subject.

Finally, we would like to address the question why, given the tremendous computing power of today's equipment, we do not simply enumerate all extreme points, determine their exact coordinates and their objective values, and then choose the one with the best objective value (meaning the highest value for maximization and lowest value for minimization problems), which then will be the optimal solution. Given Dantzig's corner point theorem, the procedure is certainly valid in that it will find an optimal solution. However, as the example below will clearly demonstrate, it is of no practical value.

As an example, consider a problem whose constraints are $0 \leq x_1 \leq 1, 0 \leq x_2 \leq 1$, and so forth for all n variables. With two variables, the feasible set is a square with the four corner points $(x_1, x_2) = (0, 0), (1, 0), (1, 1)$, and $(0, 1)$. With three variables, the set is a unit cube with eight corner points $(x_1, x_2, x_3) = (0, 0, 0), (0, 0, 1), (0, 1, 0), (1, 0, 0), (0, 1, 1), (1, 0, 1), (1, 1, 0)$, and $(1, 1, 1)$. Continuing in this fashion, we will have a feasible set in the shape of a hypercube with 2^n extreme points. Given a very small practical problem with only $n = 100$ variables, there are $2^{100} \approx 10^{30}$ corner points. Given the fastest machines today that can deal with more than 10^{15} floating point operations per second (called 1 petaflop), and assuming that one such operation can determine one extreme point (it cannot: we would need to solve a system of simultaneous linear equation with 100 variables and 100 equations), we would still need close to 670,000 years to solve the problem. Considering the fact that practical linear programming problems do not have 100 variables but maybe hundreds of thousands of variables, this shows the complete uselessness of enumeration techniques for linear programming problems.

2.3.2 Special Cases

This subsection discusses incidents that can occur when a linear programming problem has been formulated and submitted for solution. The first two cases are really error messages to the modeler. They require immediate intervention, as the solver will not be able to continue. The last three cases are of a more technical nature, it is good to know about them, but no user intervention is required. Below, we will discuss each of these issues graphically and illustrate it by means of a numerical example. In Sect. 2.4.2, we describe how these special cases appear on the computer printout of an optimal solution.

1. *There exists no feasible solution.*

The nonexistence of a feasible solution is related by the solver to the analyst. Whenever that happens, this must be taken as an error message. This is also the time for the user to intervene, as the solver is not able to act any further. The error message indicates that constraints are too tight, meaning that there is a contradiction among the constraints. This message is very often—but by no means exclusively—received by inexperienced analysts, who include too many constraints in their model, some of which refer to situations they *wish* to happen rather than those that *must* happen. It is

2.3 Graphical Representation and Solution

Fig. 2.11 Contradictory constraints

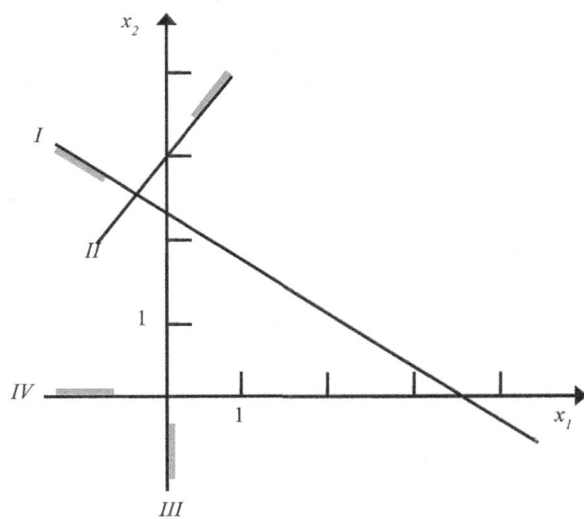

very important to understand that in a model, the constraints are absolute, meaning that they cannot be violated.

As an illustration, consider the following numerical example. Since this special case is caused exclusively by the constraints, we will not include an objective function in our description. Suppose that the feasible set is determined by the following set of constraints:

$$\text{P}: 2x_1 + 3x_2 \leq 7 \tag{I}$$

$$-x_1 + x_2 \geq 3 \tag{II}$$

$$x_1 \geq 0 \tag{III}$$

$$x_2 \geq 0. \tag{IV}$$

A graphical representation of the problem is shown in Fig. 2.11. Clearly, there is a contradiction between the constraints. To see this algebraically, rewrite constraint *II* as $x_2 \geq 3 + x_1$, which, as constraint *III* requires that $x_1 \geq 0$, implies that $x_2 \geq 3$. Similarly, constraint *I* can be rewritten as $3x_2 \leq 7 - 2x_1$. As $x_1 \geq 0$ per constraint *III*, this implies that $3x_2 \leq 7$ or, equivalently, $x_2 \leq 2⅓$. This is an obvious contradiction to the requirement that $x_2 \geq 3$.

The question often arises as to which of the constraints actually causes the nonexistence of feasible solutions. To investigate this question, consider again the example in Fig. 2.11. If constraint *I* were to be deleted, the feasible set would be the cone that is constructed by constraints *II* and *III* with the vertex at (0, 3). If constraint *II* were to be deleted, the feasible set is the triangle with vertices at (0, 0), (3½, 0), and (0, 2⅓). If constraint *III* were to be deleted, the feasible set would be the set that is determined by the halfspaces of constraints *I*, *II*, and *IV*, which has its vertices at the points $(-.4, 2.6)$ and $(3.5, 0)$. Finally, if constraint *IV* were to be deleted, there would still be no feasible solution.

In summary, we have seen that the deletion of any one of the constraints *I*, *II*, and *III* causes infeasibility to disappear. This means that the question "which constraint causes the infeasibility" is indeed the wrong question: it is not a single constraint that causes infeasibility, but the incompatibility of a number of constraints, here *I*, *II*, and *III*. Many commercial solvers will not only provide the decision maker with the "There exists no feasible solution" message, but also offer further help. Typically, this help comes in the form of an identification of the set of constraints that causes the infeasibility. If we were to add the constraints $x_1 \geq 2$ and $x_2 \geq 2$ to the region pictured in Fig. 2.7 with constraints *I–VI*, there would be no feasible solution, and the solver would notify the analyst that the constraints $5x_1 + 3x_2 \leq 15$, $x_1 \geq 2$, and $x_2 \geq 2$ *together* cause the infeasibility.

The next question is then how to deal with infeasibilities should they occur. The answer is that the planner has to decide which of the variables involved should be "loosened up." As an example of a budget constraint, loosening up such a "$\leq$" constraint would be accomplished by increasing the right-hand side value. In other words, if one of the constraints that causes the infeasibility is a budget constraint, then increasing the amount of cash available will make the problem less stringent. Similarly, if a customer requires at least, say, 100 units of a product and we must supply at least that many units to the customer, such a "$\geq$" constraint would be made looser by reducing this number by convincing the customer to accept a somewhat smaller quantity. Even equations can be relaxed somewhat. As an example, consider the equation $2x_1 + 5x_2 = 6$, in which the left-hand side indicates the quantity of certain critical nutrients or medicines that an individual consumes. Any equation can be expressed as two opposing inequality constraints, in our case as $2x_1 + 5x_2 \leq 6$ *and* $2x_1 + 5x_2 \geq 6$. Relaxing these two constraints means allowing a bit less than 6 and a bit more than 6 units of that nutrient in the diet, which can be achieved by changing the right-hand side values by some small value. For instance, instead of the original constraints, we could use $2x_1 + 5x_2 \leq 6.1$ and $2x_1 + 5x_2 \geq 5.8$, thus allowing a certain bandwidth within which there are feasible solutions. These are so-called *interval constraints* that marginally increase the size of the problem but are much easier to deal with from a computational and modeling point of view.

Often, however, it is not advisable to change the right-hand side value of a single constraint, as a very significant change may be required to achieve feasibility. It is often much easier to make a number of smaller changes on multiple right-hand side values. As an example, consider the example in Fig. 2.11. Changing the right-hand side of constraint *II* alone would require us to reduce the original value of 3 by more than 22% down to $2\frac{1}{3}$ before a feasible solution could be obtained. Similarly, the right-hand side value of constraint *I* must be increased from its original value of 7 to at least 9 before a feasible solution can be obtained; that is an increase of more than 28%. Alternatively, we could increase the right-hand side value of constraint *I* by 14% to 8 and simultaneously reduce the right-hand side value of constraint *II* by 11% down to 2.67 and obtain a feasible solution with these smaller changes that may be easier to implement.

2.3 Graphical Representation and Solution

Fig. 2.12 Unbounded "optimal" solutions

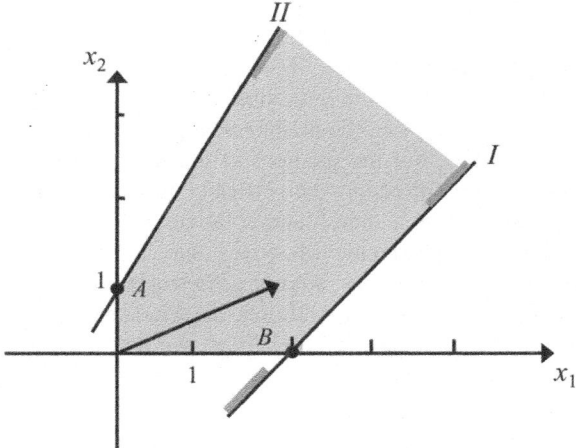

2. *Unbounded "optimal" solutions.*

The existence of unbounded "optimal" solutions is, in some sense, the opposite of nonexisting feasible solutions. This is true in the sense that in the previous case, the constraints were too tight and had to be loosened up, they are too loose here and require tightening. Most frequently, this case occurs if some constraints have been forgotten. Again, it is an error message that requires intervention from the analyst.

As an illustration of this case, consider the following numerical example.

$$P : \text{Max } z = 2x_1 + x_2$$

$$\text{s.t. } x_1 - x_2 \leq 2 \quad \text{(I)}$$

$$-2x_1 + x_2 \leq 1 \quad \text{(II)}$$

$$x_1, x_2 \geq 0.$$

The graphical representation of this problem is shown in Fig. 2.12. Using the graphical solution method, we notice that we can increase the value of the objective function to arbitrary levels by following the direction pointed out by the gradient of the objective function. (In case you should encounter arbitrarily high profits in practice, remember that you learned it from us first. We are happy with 10% of infinity.) Clearly, this cannot occur in practice, which is why we use the word "optimal" in quotes. In order to resolve the issue, it helps if the analyst asks how such arbitrarily good solutions can be obtained. For instance, in a production problem we would determine that we make money by making and selling products: given that machine capacities and demand are limited, profits are limited, too. So it would benefit the modeler to investigate whether or not the appropriate constraints have been included.

At first glance, it appears that the reason for the existence of unbounded "optimal" solutions is the fact that the feasible set is not bounded in one direction (in our example, the northeast). However, if we leave the feasible set in our example unchanged, but modified the objective function from its original Max $z = 2x_1 + x_2$ to Min $z = 2x_1 + x_2$, the graphical solution method would continue to use the same contour lines, but the gradient of the new objective function would point into a southwesterly direction, diametrically opposed to its previous direction. The (finite) optimal solution is then found at the origin.

This small example illustrates that a feasible set that is unbounded in one direction is a necessary, but not sufficient, condition for the existence of unbounded "optimal" solutions. In addition, the gradient of the objective function must also point "towards that opening." Here, we will leave this imprecise wording as it is; suffice it to mention that exact mathematical conditions for the existence of unbounded "optimal" solutions do exist.

3. *Dual Degeneracy*.

This is the first of those special cases that an analyst should know about, but that does not require intervention. Formally, dual degeneracy occurs if two adjacent extreme points of the feasible set have the same value of the objective function. An example is the problem shown in Fig. 2.10. As already discussed in the context of simplex paths, the point *A* has the same value of the objective function as its neighboring point *F*. That in itself is not really noteworthy. The situation, however, changes if dual degeneracy occurs at optimum. By definition, if we have an optimal solution and another extreme point with the same z-value exists, that point must also be optimal.

An example for this is again Fig. 2.10, but with the original objective function changed from Max $z = 2x_1 + 3x_2$ to Min $z = 2x_1 + 3x_2$. This changes the gradient of the objective function by 180°, leaving us with the points *A* and *F* as optimal solutions. Whenever two neighboring extreme points are optimal, all points between them are also optimal, i.e., dual degeneracy at optimum means the existence of *alternative optimal solutions*. In this example, the optimal coordinates of the points *A* and *F* are (0, 1) and (1½, 0), respectively, both having an objective value of $z = 3$. Some points on the line segment between these two points are $\bar{\mathbf{x}} = (¾, ½)$, $\left(\frac{9}{8}, \frac{1}{4}\right)$, and $\left(\frac{3}{8}, \frac{3}{4}\right)$. These points have an objective value of $z = 3$ and thus are optimal as well. However, none of them is an extreme point, and the simplex method will not generate them.

Note that the fact that points that are not corner points of the feasible set are optimal does not invalidate Dantzig's corner point theorem. All the theorem states is that *at least* one optimal solution is at an extreme point. This allows for the possibility of non-extreme points being optimal, but only if there is another optimal extreme point as well.

2.3 Graphical Representation and Solution

Fig. 2.13 Redundancy and primal degeneracy

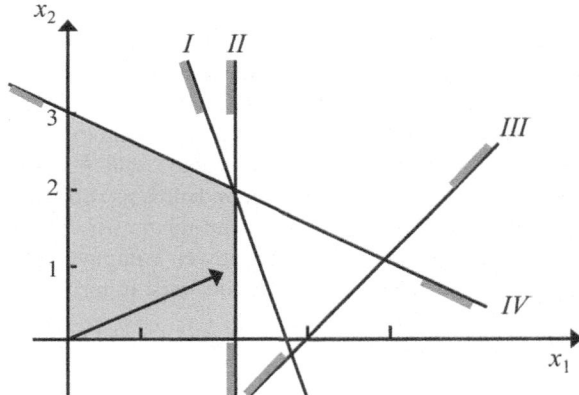

4. *Redundancy.*

Redundancy is an issue that relates to individual constraints. In particular, a constraint is said to be redundant, if it can be deleted from the set of constraints without changing the feasible set. A more detailed definition will distinguish between constraints that exhibit *strong redundancy* and those that are *weakly redundant*. While a weakly redundant constraint belongs to a constraint whose hyperplane shares at least one point with the feasible set, this is not the case for strongly redundant constraints.

As an illustration, consider the following numerical example (without an objective function, as it does not play a role in the discussion of redundancy).

$$P : 3x_1 + x_2 \leq 8 \qquad (I)$$

$$x_1 \leq 2 \qquad (II)$$

$$x_1 - x_2 \leq 3 \qquad (III)$$

$$x_1 + 2x_2 \leq 6 \qquad (IV)$$

$$x_1, x_2 \geq 0.$$

Figure 2.13 depicts the feasible set of this problem.

Constraint *III* is obviously redundant as its removal does not change the shaded area that symbolizes the feasible set. As a matter of fact, it is strongly redundant as its bordering hyperplane does not touch the feasible set at any point. In contrast, consider constraint *I*. It is also redundant (but just about so), so that its removal likewise does not change the feasible set. However, it is only weakly redundant, as it touches the feasible set at the point (2, 2). Had constraint *I* be instead $3x_1 + x_2 \leq 8.1$, it would have been strongly redundant, had it been $3x_1 + x_2 \leq 7.95$, it would not have been redundant at all, as it would have cut off the point (2, 2) and thus shaped the feasible set.

By their very definition, redundant constraints can be deleted from the problem formulation without changing anything. The problem is that there is no easy general way to recognize whether or not a constraint is redundant. Again, the graphical representation is deceiving as in it, redundancy can easily be detected. However, if all we have is a set of constraints that we cannot plot, detection of redundancy is a different, and much more complicated, matter. As a matter of fact, determining whether or not a constraint is redundant is as difficult as solving the existing linear programming in the first place. And this is why we can, and pretty much have to, ignore the issue and have the solver compute optimal solutions, regardless of whether or not redundant constraints exist in the formulation.

5. *Primal Degeneracy.*

While the issue of primal degeneracy is of very little, if any, practical concern, it is a very important theoretical matter. If primal degeneracy is not properly taken care of, the simplex method may "cycle," meaning that it generates the same points over and over again without ever reaching an optimal solution. In graphical terms, primal degeneracy occurs in two dimensions, if more than two planes of constraint intersect at a single point. In Fig. 2.13, the point (2, 2) exhibits degeneracy. As in two dimensions, the intersection of any two straight lines uniquely determines a point, we can think of points with primal degeneracy as "overdetermined." In n dimensions, primal degeneracy occurs if the hyperplanes of more than n constraints intersect at one point. Any modern code will include a module that deals with degeneracy, so that this is of no concern to users.

Exercises
Problem 1 (graphing constraints and objective, graphical solution method): Consider the following linear programming problem.

$$\text{Max } z = x_1 + x_2$$

$$\text{s.t. } 5x_1 + 2x_2 \leq 10 \quad \text{(I)}$$

$$x_2 \geq 1 \quad \text{(II)}$$

$$3x_1 + 5x_2 \leq 15 \quad \text{(III)}$$

$$x_1, x_2 \geq 0.$$

(a) Graph the constraints, determine the feasible set, and use the graphical solution method to determine the optimal solution point. Which constraints are binding at optimum? Compute the exact coordinates at optimum and calculate the value of the objective function at optimum.

Fig. 2.14 Graph for Problem 1

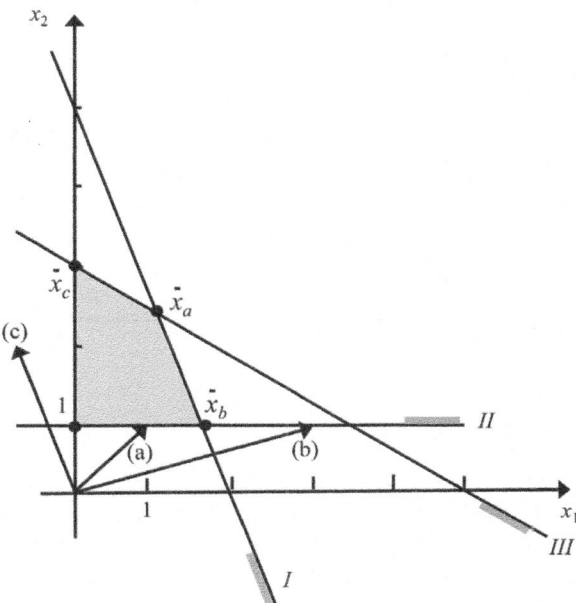

(b) What if the objective were Min $z = -3x_1 - x_2$? Plot the new objective, use the graphical solution method, determine the optimal solution point, its coordinates, and the value of the objective function at optimum.

(c) Repeat (b) with the objective function Min $z = x_1 - 2x_2$.

Solution:

The objective functions for (a), (b), and (c) are shown in Fig. 2.14. The optimal solutions are indicated by the points $\bar{\mathbf{x}}_a$, $\bar{\mathbf{x}}_b$, and $\bar{\mathbf{x}}_c$, respectively. The exact coordinates are $\bar{\mathbf{x}}_a = [1.0526, 2.3684]$ with $\bar{z}_a = 3.4211$, $\bar{\mathbf{x}}_b = [1.6, 1]$ with $\bar{z}_b = -5.8$, and $\bar{\mathbf{x}}_c = [0, 3]$ with $\bar{z}_c = -6$. The binding constraints at the three points are I and III, I and II, and III and the nonnegativity constraint $x_1 \geq 0$, respectively.

Problem 2 (graphing constraints and objective, graphical solution method): Consider the following linear programming problem.

$$\text{Max } z = 3x_1 + x_2$$

$$\text{s.t. } 6x_1 + 5x_2 \geq 30 \quad \text{(I)}$$

$$-x_1 + 2x_2 \geq 4 \quad \text{(II)}$$

$$x_2 \leq 5 \quad \text{(III)}$$

$$x_1, x_2 \geq 0.$$

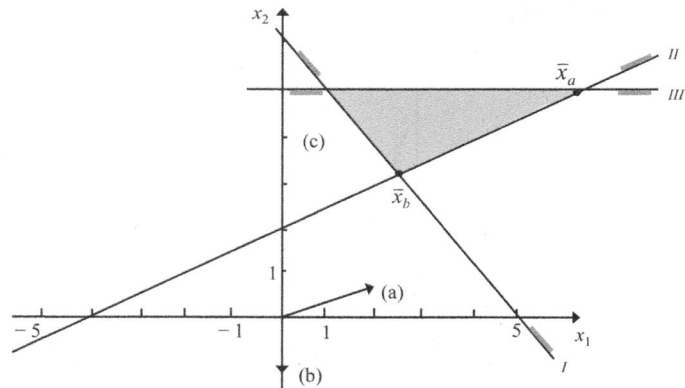

Fig. 2.15 Graph for Problem 2

(a) Graph the constraints, determine the feasible set, and use the graphical solution method to determine the optimal solution point. Which constraints are binding at optimum? Compute the exact coordinates at optimum and calculate the value of the objective function at optimum.
(b) Repeat (a) with the objective function Min $z = x_2$.
(c) What happens if constraint I were of the type "$\leq$" instead?

Solution:

(a) The feasible set is the shaded area in Fig. 2.15, and the optimal solution point is $\bar{\mathbf{x}}_a = [6, 5]$ with $\bar{z}_a = 23$. Constraints II and III are binding at optimum.
(b) The objective leads "straight down" as shown in the above figure. The optimal solution point is $\bar{\mathbf{x}}_b = [2.3529, 3.1765]$ with the objective value $\bar{z}_b = 3.1765$. At this point, the constraints I and II are binding.
(c) The feasible set would no longer be the shaded region, but the quadrilateral indicated by (c). Given the objective in (a), the optimal solution is again $\bar{\mathbf{x}}_b$.

Problem 3 (Dual degeneracy at optimum, alternative optimal solutions):
Consider the following linear programming problem:

$$P : \text{Min } z = x_1 + x_2$$

$$\text{s.t. } x_1 \leq 4 \quad (I)$$

$$x_1 + x_2 \geq 7 \quad (II)$$

$$x_1, x_2 \geq 0. \quad (III) \text{ and } (IV)$$

2.3 Graphical Representation and Solution

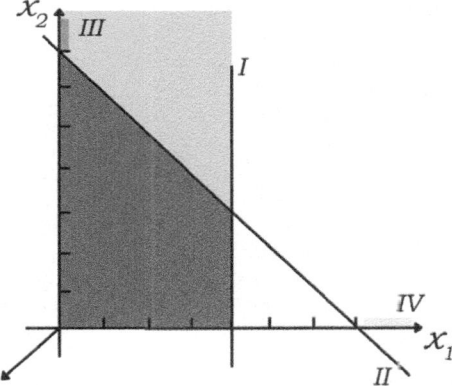

Fig. 2.16 Dual degeneracy and alternative optimal solutions

(a) Graph the feasible set and the gradient of the objective function. Which, if any, of the five special cases does apply?
(b) Consider the same problem but change constraint 2 from $x_1 + x_2 \geq 7$ to $x_1 + x_2 \leq 7$. What is the feasible set now? Which, if any, of the five special cases does apply?

Solution:

(a) The feasible set is shown as the lightly colored area in Fig. 2.16. The neighboring solutions $\mathbf{x} = (4, 3)$ and $\mathbf{x} = (0, 7)$ both have the objective value $z = 7$. The (along with all points on the line segment between them) are optimal. In other words, we have dual degeneracy at optimum.
(b) The feasible set is now the dark gray section in Fig. 2.16. We still have dual degeneracy, but no longer at optimum, which is now at $\bar{\mathbf{x}} = (0, 0)$.

Problem 4 (Primal degeneracy): Consider the following linear programming problem:

$$P : \text{Min } z = 3x_1 + x_2$$

$$\text{s.t. } x_1 \leq 5 \quad (I)$$

$$x_1 + x_2 \geq 5 \quad (II)$$

$$x_1, x_2 \geq 0. \quad (III) \text{ and } (IV)$$

Graph the feasible set and the gradient of the objective function. Which, if any, of the five special cases does apply?

Solution: The feasible set and the objective function of the problem are shown in Fig. 2.17. It is apparent that there is primal degeneracy at the point $\mathbf{x} = (5, 0)$, where

Fig. 2.17 Primal degeneracy

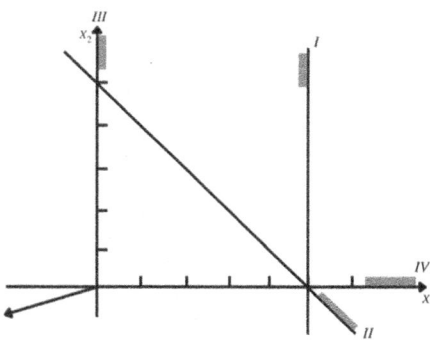

the three hyperplanes *I*, *II*, and *IV* intersect. Note that constraint *IV* is also weakly redundant.

2.4 Postoptimality Analyses

This section will investigate what can be thought of as the third phase in linear programming. In particular, it asks: "What happens, if...?" In simple words, we will examine what happens, if there is some change in some of the components of the linear programming problem that was formulated and solved earlier. The first subsection will explore what it means graphically when we make the proposed change, while the second subsection puts this knowledge to work and examines what managerial consequences the anticipated changes will have.

2.4.1 Graphical Sensitivity Analyses

Recall that one of the main assumptions of linear programming is the deterministic property. In other words, we assume that the structure of the problem and all of the parameters of the problem are assumed with certainty. In virtually all realistic cases, this is a troubling assumption: prices may or may not be known in advance, demand may be uncertain, machine capacities (due to unforeseen breakdowns), employees may call in sick, thus changing the availability of manpower, and so forth. How then can we justify using linear programming at all?

Using *postoptimality analyses* can be seen as a "trick" that allows us to get away with assuming that the deterministic property holds, while it actually may not. As an example, suppose that we are facing an uncertain demand, which, as past experience indicates, ranges from, say, 80 to 130 units. Furthermore, suppose that most of the time the demand is somewhere about 110 units. The idea is now to temporarily fix the demand at the level of 110 and solve the problem. Once this has been done, we perform sensitivity analyses by asking: "What happens, if the demand were to increase (from 110) by 10 (to 120)?" "What if it increases by 20 (to 130)?" "What

2.4 Postoptimality Analyses

if it decreases by 10, 20, or 30 units?" Information of this type can be obtained either by setting the demand to 120, 130, or any of the other values the decision maker is interested in and actually resolving the problem, or by gleaning the information from the printout. This information will then indicate how sensitive the solution is to changes in the input parameters. While some problems are not sensitive at all with respect to some changes, meaning that even significant changes in the original demand, prices, or other parameters do not result in major (or even any) changes of the solution, others are very sensitive, so that even minor changes in the input parameters change the solution a great deal. There is nothing we can do about it, but it is very important information to the decision maker. If it is known that a model is very sensitive to changes, the decision maker will have to be very cautious and monitor the process closely, obtaining additional and updated information at many steps along the way. This is not as necessary in problems that are rather insensitive to changes.

We distinguish between two different types of sensitivity analyses. The first type deals with *structural changes*, meaning the addition and deletion of variables and constraints. Changes of that nature are major and often dealt with by resolving the problem altogether. The second type of sensitivity analyses involves *parameter changes*. In these cases, only some of the numbers in the model change. Typically, we deal with *ceteris paribus* changes, i.e., we determine what happens if one number changes, while all other parameters remain unchanged. The advantage of such an analysis is that it separates the different changes and analyzes their effects. If we were to analyze simultaneous changes of a number of parameters, we would not be able to easily specify what actually causes the observed effect, e.g., the increase or decrease in the total costs.

We will first look at the changes that may occur when we either add or delete a variable or a constraint. The addition of variables and constraints is an issue during the modeling process when a model is built. Furthermore, it may occur after the problem has been solved when opportunities and requirements are added to the problem as time passes. Similarly, it may happen that over time some activities or possibilities no longer exist, or constraints have become obsolete. Whenever possible, we will deal with these cases on an intuitive level, which is pretty much all that can be done short of resolving the problem.

Consider the *addition of a variable*. In graphical terms, the addition of a variable corresponds to the increase of the dimensionality of the problem. It is useful to understand a variable as an opportunity, i.e., an additional activity that we may or may not undertake. Having the additional variable allows us to choose the level at which we engage in the activity (i.e., the value the variable assumes), while not including the variable in the model is the same as setting its activity level or value equal to zero. In this sense, adding a variable is the same as adding an opportunity. Doing so allows us to increase the level of the new possible activity from a zero level to any value within the constraints. The immediate conclusion is that the addition of a variable can never result in a deterioration of the objective function value (a decrease in maximization functions or an increase in minimization functions), but possibly an improvement. Formally, defining the original problem as P_{orig} with its optimal objective value $\bar{z}_{orig}$ and the problem with the added variables as P_{addvar} and its optimal objective value as $\bar{z}_{addvar}$, we know that $\bar{z}_{orig} \leq \bar{z}_{addvar}$ in problems with maximization objective and $\bar{z}_{orig} \geq \bar{z}_{addvar}$ in problems with minimization objective. Along similar lines, if P_{orig} has unbounded "optimal" solutions, then so does the modified problem P_{addvar}. On the other hand, if the original problem P_{orig} has no feasible solution, the added variable may or may not allow P_{addvar} to have feasible solutions.

There is little that can be said in general beyond this. Often, if variables are added in the modeling process, it may be useful not to start solving the new problem from scratch but to start with the previous optimal solution, which usually only requires a few additional steps to reach the optimal solution of the new problem. However, with the advances of optimization software, it often takes only seconds to resolve a problem, so that such considerations are no longer as important as they used to be.

The *deletion of a variable* can be discussed in analogous fashion. Again, let the original problem be P_{orig}, while the formulation without the now deleted variable is P_{delvar}. The objective function values of the two problems are defined as $\bar{z}_{orig}$ and $\bar{z}_{delvar}$, respectively. Deleting a variable is now equivalent to deleting an opportunity, or, more formally, forcing the value of a variable to zero. If the value of the variable that is to be deleted equaled zero in the optimal solution of P_{orig}, then the variable can be deleted without any change of the solution and $\bar{z}_{orig} = \bar{z}_{delvar}$ in both, maximization and minimization problems. The main reason for this result is that we are no longer allowing an activity that we did not engage in in the first place. On the other hand, if the value of the variable that is to be deleted was positive in the original problem P_{orig}, the deletion of this variable will deprive us from a worthwhile activity, which lets the objective value deteriorate. In other words, we have $\bar{z}_{orig} \geq \bar{z}_{delvar}$ for maximization and $\bar{z}_{orig} \leq \bar{z}_{delvar}$ for minimization problems. It is also straightforward that if P_{orig} is unbounded, then P_{delvar} may or may not be unbounded as well. Furthermore, if P_{orig} has no feasible solution, then neither does P_{delvar}.

The *addition of a constraint* is an issue that we can much more easily analyze, as it allows us to visualize the situation in a graph as the dimensionality of the problem does not change. Again, we restrict ourselves to small problems with only two variables, but the conclusions are valid for any general formulation. Consider again some arbitrary original maximization problem P_{orig} and assume that a constraint is added, resulting in P_{addcon}. By definition, adding a constraint means that the resulting problem P_{addcon} is more restricted, so that its objective value $\bar{z}_{addcon} \leq \bar{z}_{orig}$ in maximization problems and $\bar{z}_{addcon} \geq \bar{z}_{orig}$ in minimization problems. More specifically, if the new constraint is satisfied by the optimal solution of P_{orig}, then this solution will also be optimal for the new problem P_{addcon}. On the other hand, if the old optimal solution violates the new constraint, then the optimal solution of P_{addcon} is different from the optimal solution of P_{orig} and the objective value will be the same (in case of alternative optimal solutions) or be worse than before. Furthermore, if the original problem has unbounded "optimal" solutions, then the problem with the new constraint may or may not have bounded optimal solutions. If the original problem has no feasible solutions, then adding a constraint, which makes the problem even more constrained, will consequently not result in feasible solutions.

Finally, consider the *deletion of a constraint*. Again, assume that the problem P_{orig} has been solved, resulting in the objective value $\bar{z}_{orig}$. The problem without one or more of the constraints will be referred to as P_{delcon} and its optimal value of the objective function is $\bar{z}_{delcon}$. It is apparent that the new problem P_{delcon} is less restricted than the original problem P_{orig}, so that $\bar{z}_{delcon} \geq \bar{z}_{orig}$ holds for maximization problems, while $\bar{z}_{delcon} \leq \bar{z}_{orig}$ holds for minimization problems. As in the case of constraint additions, we can distinguish between two cases: either the constraint that is deleted was binding at optimum before it was deleted, or it was not. In case it was binding (and not weakly redundant), then it was, generally speaking, a constraint that held back the solution and with its removal, better solutions may exist. On the other hand, if the constraint was not binding, then it did not restrict the solution in the original problem, so that its removal cannot result in better solutions. Again, if unbounded "optimal" solutions existed in P_{orig}, then the removal of a constraint cannot change that regardless if it is binding or not. If P_{orig} did not have feasible solutions, the deletion of a constraint may or may not result in the problem P_{delcon} having feasible solutions.

Next consider parameter changes. In order to classify such changes, consider the following simple linear programming problem

$$\begin{align} \text{P}: \text{Max } z = &\ 5x_1 + 6x_2 \\ \text{s.t.} \quad &\ x_1 - 2x_2 \geq 2 \\ &\ 3x_1 + 4x_2 \leq 12 \\ &\ x_1, \ x_2 \geq 0. \end{align}$$

This model, as well as any other linear programming problem, includes three different types of parameters. The first are the *objective function coefficients* (typically denoted by $c_1, c_2, \ldots$ with the subscript indicating the variable the coefficient is associated with), which are the numbers found in the objective function (here the numbers $c_1 = 5$ and $c_2 = 6$). Depending on the application, they may be referred to as unit profits, cost coefficients, or similar names. The second type of parameter are the *right-hand side values* (which we usually denote by $b_1, b_2, \ldots$ with the subscript indicating the number of the constraint we are dealing with). In our example, these are the values $b_1 = 2$ and $b_2 = 12$. Again, depending on the specific applications, the right-hand side values may be referred to as resource availabilities, demands, inventory levels, or similar names. Finally, there are the *left-hand side coefficients*, which sometimes are called *technological coefficients* $a_{11}, a_{12}, \ldots, a_{21}, a_{22}, \ldots$ with the first subscript denoting the number of the row or constraint and the second subscript standing for the number of the column or variable. In our example, the technological coefficients are $a_{11} = 1, a_{12} = -2, a_{21} = 3$, and $a_{22} = 4$. Depending on the application, these values may symbolize the processing times of a product on a machine, the content of nutrients of a food item, the interest rate of a specific investment, or similar values. In this book, we will investigate changes of the

Fig. 2.18 Changing the coefficients of the objective function

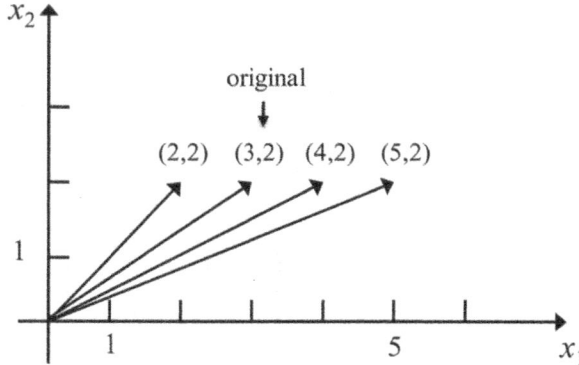

objective function coefficients and changes of the right-hand side values. For changes of the left-hand side parameters, we suggest to simply resolve the problem.

First, consider changes of the objective function coefficients. In order to explore what these parameter changes cause, we first look only at the objective function and ignore the constraints. To facilitate our arguments, consider the objective function Max $z = 3x_1 + 2x_2$. The gradient of the objective is shown in Fig. 2.18 and is labeled (3, 2). Suppose now that we want to examine changes of c_1, the value associated with the variable x_1. If this number, whose original value is "3," decreases to, say, "2," the gradient tilts in a counterclockwise direction to a position shown as (2, 2). If, on the other hand, c_1 were to increase to, say, "4," then the gradient will tilt in a clockwise direction to the position shown as (4, 2). If c_1 were to further increase to a value of "5," the gradient further tilts in a clockwise direction to the position shown as (5, 2).

We see that the increase of an objective function coefficient in a maximization function symbolizes the fact that the activity that corresponds to its associated variable has become more attractive or profitable. Thus, the gradient of the objective function is drawn more into that direction. In our example, we see that as c_1 increases from 3 to 4 and then to 5, the gradient is pulled further and further into the x_1 direction, i.e., to the right.

We note that since all the changes occur in the objective function, the feasible set will remain unaffected by these changes. This means that if there are no feasible solutions before the change, then there will be no feasible solutions after the change. On the other hand, the case of unbounded "optimal" solutions is different, as it does depend not only on the feasible set, but also on the gradient of the objective function. As a simple example, consider the following linear programming problem.

$$P : \text{Max } z = 2x_1 + 1x_2$$
$$\text{s.t.} \quad x_1 - x_2 \leq 2$$
$$x_1, \quad x_2 \geq 0.$$

This problem is shown in Fig. 2.19, in which the gradient of the objective function is labeled by its coefficient as (2, 1). Clearly, there are unbounded "optimal"

Fig. 2.19 Unbounded "optimal" solutions

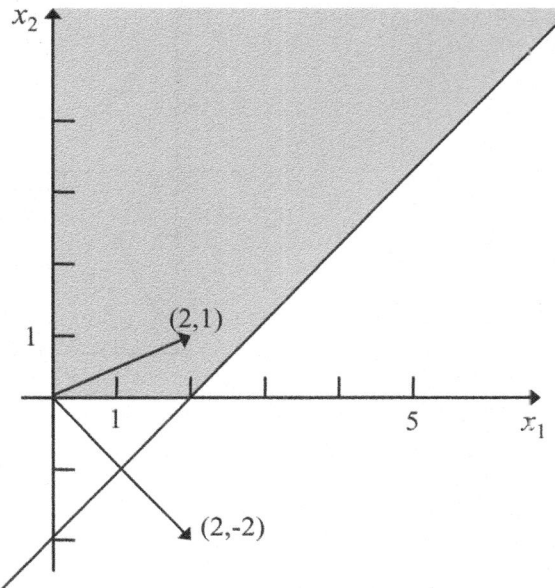

solutions to the problem. However, if c_2 decreases from its present value of "1" to less than "-2" [the gradient is shown as $(2, -2)$], the problem has a unique finite optimal solution at $\bar{x}_1 = 2$ and $\bar{x}_2 = 0$ with an objective function value of $\bar{z} = 4$, clearly a finite value.

We are now able to incorporate constraints in our discussion. In order to do so, consider the following linear programming problem:

$$
\begin{aligned}
\text{P}: \text{Max } z = {}& 1x_1 + 2x_2 \\
\text{s.t.} \quad & x_2 \le 3 \\
& 3x_1 + 2x_2 \le 11 \\
& x_1 - x_2 \le 2 \\
& x_1, \quad x_2 \ge 0.
\end{aligned}
$$

The problem can be visualized in the graph in Fig. 2.20.

Using the graphical solution technique (or any commercial solver), we determine that the point B is the unique optimal solution with coordinates $(\bar{x}_1, \bar{x}_2) = (1\tfrac{2}{3}, 3)$. Suppose now that we want to examine the sensitivity of the solution with respect to c_2, the objective function coefficient of x_2. If we were to increase the value that presently equals 2 to some higher value, our previous discussion indicates that the gradient of the objective function tilts in a counterclockwise direction. This does not, however, have any effect on the solution, which stays at point B. As a matter of fact, no finite value of c_2, regardless of how large, will change the solution, which remains at point B (meaning that the solution is very insensitive to increases of this

Fig. 2.20 Graphical solution technique

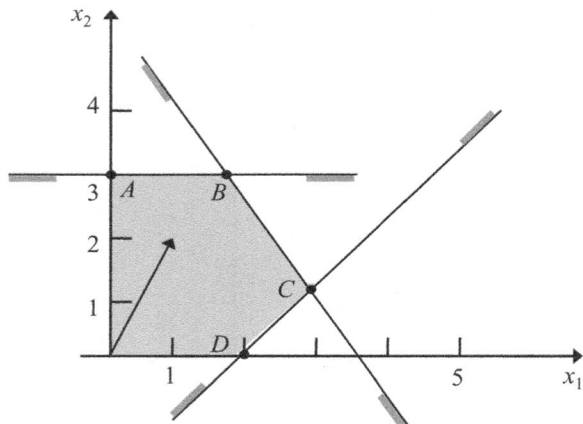

Table 2.22 Optimal solutions $(\bar{x}_1, \bar{x}_2)$ and objective values $\bar{z}$ for different values of c_2

Range of c_2	$]-\infty, -1[$	-1	$]-1, \frac{2}{3}[$	$\frac{2}{3}$	$]\frac{2}{3}, +\infty[$
Optimal solution point	D	D and C	C	C and B	B
Optimal coordinates $(\bar{x}_1, \bar{x}_2)$	(2, 0)	(2, 0) and (3, 1)	(3, 1)	(3, 1) and (1⅔, 3)	(1⅔, 3)
Optimal objective value $\bar{z}$	2	2	$3 + c_2$	3⅔	1⅔ + 3c_2

coefficient). Clearly, since $\bar{x}_2 = 3$ at optimum, the value of the objective function will change as the activity that the variable x_2 symbolizes becomes more and more valuable.

Consider now a decrease of c_2. Graphically, this means that the gradient of the objective function tilts in a clockwise direction. For small changes, the optimal solution remains at point B. However, once c_2 reaches the value of ⅔, point B is still optimal, but so is point C with coordinates $(x_1, x_2) = (3, 1)$ (and all non-extreme points on the line segment between these two points). This is clearly a case of dual degeneracy at optimum, i.e., alternative optimal solutions. Once c_2 decreases below the value of ⅔, point C is the unique optimal solution. Point C remains optimal until c_2 reaches the value of -1. At this point, points C and D are both optimal, again a case of alternative optimal solutions. If c_2 drops below -1, point D with coordinates $(x_1, x_2) = (2, 0)$ remains optimal, regardless how small the coefficient is.

We summarize the effects of the changes of c_2 in Table 2.22.

Similar analyses can be performed for each of the given objective function coefficients. Without further comments, Table 2.23 shows different ranges of c_1 and the resulting optimal solution points, their coordinates, and their values of the objective function.

So far, we have only looked at the effects of individual changes on the optimal solution. There is, however, an interesting result that considers the effects of

2.4 Postoptimality Analyses

Table 2.23 Optimal solutions $(\bar{x}_1, \bar{x}_2)$ and objective values $\bar{z}$ for different values of c_1

Range of c_1	$]-\infty, 0[$	0	$]0, 3[$	3	$]3, \infty[$
Optimal solution point	A	A and B	B	B and C	C
Optimal coordinates $(\bar{x}_1, \bar{x}_2)$	(0, 3)	(0, 3) and (1⅔, 3)	(1⅔, 3)	(1⅔, 3) and (3, 1)	(3, 1)
Optimal objective value $\bar{z}$	6	6	$6 + 1⅔c_1$	11	$2 + 3c_1$

simultaneous changes of several objective function coefficients. The rule is called the *100 percent rule* and it can be stated as follows.

100% Rule As long as the sum of the absolute values of the percentage increases or decreases of the objective function coefficients is no more than 100%, the optimal solution point remains optimal.

Formally, we denote the largest allowable increase of an objective function coefficient c_j by $\overline{\Delta}c_j$, while the largest allowable decrease of an objective function coefficient c_j is denoted by $\underline{\Delta}c_j$. In our example, the optimal solution for the original objective function Max $z = 1x_1 + 2x_2$ was $(\bar{x}_1, \bar{x}_2) = (1⅔, 3)$. As shown in Table 2.23, this solution remains optimal as long as c_1 (whose original value is $c_1 = 1$) does not increase by more than $\overline{\Delta}c_1 = 2$ to the upper limit of the range at $c_1 = 3$. Similarly, the solution remains optimal as long as c_1 does not decrease by more than $\underline{\Delta}c_1 = 1$ to the lower end of the range at $c_1 = 0$. Similarly, we obtain the values $\overline{\Delta}c_2 = +\infty$ and $\underline{\Delta}c_2 = 1⅓$. Suppose now that we want to investigate the effect of a simultaneous increase of the value of c_1 by Δc_1 and a decrease of the value of c_2 by Δc_2, the 100% rule then states that the optimal solution $(\bar{x}_1, \bar{x}_2) = (1⅔, 3)$ remains optimal, as long as the sum of actual increases in relation to their respective values $\overline{\Delta}c_j$ plus the sum of actual decreases in relation to their respective values $\underline{\Delta}c_j$ does not exceed $100\% = 1$. In our example, we obtain $\frac{|\Delta c_1|}{\overline{\Delta}c_1} + \frac{|\Delta c_2|}{\underline{\Delta}c_2} = \frac{|\Delta c_1|}{2} + \frac{|\Delta c_2|}{1⅓} \leq 1$. For instance, if we were to face a simultaneous increase of c_1 by ½ and a decrease of c_2 by ½, the condition tells us that $\frac{½}{2} + \frac{½}{1⅓} = ¼ + ⅜ = ⅝ < 1$, so that the optimal solution remains optimal. On the other hand, if the increase of c_1 were ¾ and the decrease of c_2 were 1, then we would have $\frac{3/4}{2} + \frac{1}{1⅓} = ⅜ + ¾ = 9/8 > 1$, so that the optimal solution will change.

A similar argument can be applied to a simultaneous decrease of c_1 and increase of c_2, or simultaneous increases or decreases of both cost coefficients. Increasing both cost coefficients simultaneously presents an interesting special case. As there is no finite upper bound on c_2, the 100% rule reduces to the regular limit on c_1 and no limit on c_2.

Consider now changes of a single right-hand side value b_i. Again, we will first examine the effects such a change on the constraint itself before investigating what

Fig. 2.21 Changing a right-hand side value

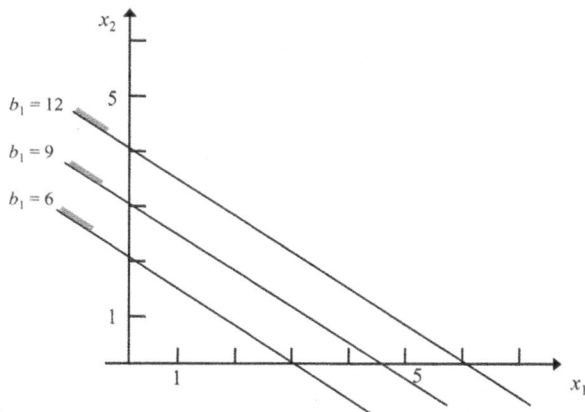

happens to optimal solutions. As a numerical example, consider the constraint $2x_1 + 3x_2 \leq 6$. The resulting hyperplane and halfspace are shown in Fig. 2.21, labeled as $b_1 = 6$.

If we were to modify the right-hand side value to, say, $b_1 = 9$, the hyperplane and halfspace would shift in parallel fashion to the position shown in Fig. 2.21 by $b_1 = 9$. A further increase to $b_1 = 12$ is shown in the figure as well. Decreases of the right-hand side value in our example would again result in a hyperplane and/or halfspace shifting in parallel fashion, but now in a southwesterly direction.

Given this result, we are now able to examine changes of right-hand side values in a linear programming problem. Before doing so, we note that such changes do in no way affect the objective function, but they may change the feasible set.

As an illustration, consider again the numerical example that was used to discuss changes of the objective function. For convenience, we restate the model here.

$$P : \text{Max } z = 1x_1 + 2x_2$$

$$\text{s.t. } x_2 \leq 3 \quad \text{(I)}$$

$$3x_1 + 2x_2 \leq 11 \quad \text{(II)}$$

$$x_1 - x_2 \leq 2 \quad \text{(III)}$$

$$x_1, x_2 \geq 0.$$

Figure 2.22 shows the feasible set (the shaded area) and the gradient of the objective function. The extreme points of the feasible set are 0, A, B, C, and D. The optimal solution is again at the point B with coordinates $(\overline{x}_1, \overline{x}_2) = (1\tfrac{2}{3}, 3)$ and value of the objective function $\overline{z} = 7\tfrac{2}{3}$.

Consider now changes of the second right-hand side value b_2. If b_2 increases from its present value of 11, then the hyperplane and halfspace of this constraint will shift into a northeasterly direction. For instance, if $b_2 = 15$, the feasible set is now

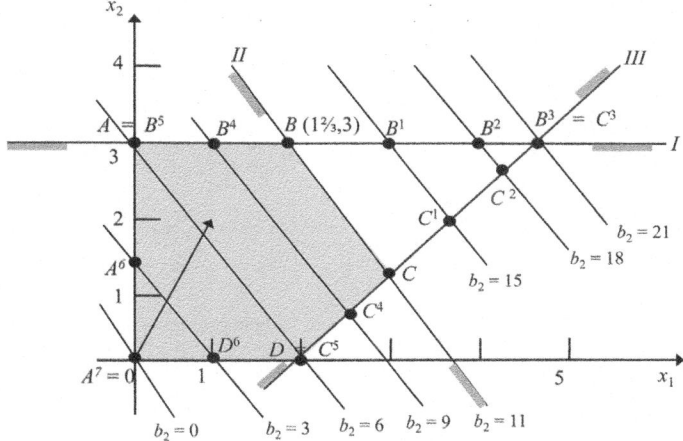

Fig. 2.22 Changing optimal points based on different right-hand side values

enlarged by the set with the extreme points B, C, B^1, and C^1. This has a direct effect on the optimal solution point, which has moved from its original position at point B to its new position at point B^1. Note that while the optimal coordinates of the optimal point have changed, one thing has not: the point is still determined by the intersection of the hyperplanes that belong to constraints I and II, the latter now with its new right-hand side value. Such a point is referred to as a *basis point* or simply a *basis*. We observe here that while the basis has not changed, the optimal solution has moved to a new location, albeit still at the intersection of hyperplanes I and II. A further increase of b_2 has the effect of changing the feasible set to $0, A, B^2, C^2$, and D, with an optimal solution at point B^2. Again, the basis has not changed (the point is still determined as the intersection of hyperplanes I and II), but its location has moved further to the right. A further increase to $b_2 = 21$ results in a feasible set with extreme points $0, A, B^3 = C^3$, and D. The optimal solution is now at points $B^3 = C^3$. We observe that at this point, hyperplanes I, II, and III now intersect at the same point, causing primal degeneracy. Any further increase of b_2 will not change the feasible set any further, constraint II is now redundant.

Return now to the original value of $b_2 = 11$ and slowly decrease this value. For instance, for $b_2 = 9$ the feasible set has shrunk to $0, A, B^4, C^4$, and D with an optimal solution at B^4. A further decrease to $b_2 = 6$ results in the feasible set with extreme points $0, A = B^5$, and $D = C^5$, indicating that primal degeneracy now occurs at $A = B^5$, and $D = C^5$. The optimal solution is now at $A = B^5$. A further decrease to $b_2 = 3$ results in a feasible set with extreme points $0, A^6$, and D^6 with an optimal solution at A^6. A further decrease to $b_2 = 0$ causes the feasible set to consist of only the origin, which, as it is the only feasible point, is now also optimal. Any further decrease of b_2 will cause the problem not to have any feasible solutions. Our results are summarized in Table 2.25.

Table 2.24 Optimal solution and objective values of sample problem for various values of b_1

Range of b_1	]−∞, 0[	0	]0, 1[	1	]1, 5½ [	5½	]5½, ∞[
Optimal coordinates	There exists no feasible solution	(2, 0)	$(2+b_1, b_1)$	(3, 1)	$(-\frac{2}{3}b_1 + 3\frac{2}{3}, b_1)$	(0, 5½)	(0, 5½)
Optimal objective value $\bar{z}$		2	$2 + 3b_1$	5	$7\frac{1}{3} - \frac{1}{3}b_1$	11	11

Table 2.25 Optimal solution and objective values of sample problem for various values of b_2

Range of b_2	]−∞, 0[	0	]0, 6[	6	]6, 21[	21	]21, ∞[
Optimal solution point	There exists no feasible solution	0	$A^7, \ldots, A$	$A = B^5$	$B^5, B^4, B, B^1, B^2, B^3$	$B^3 = C^3$	$B^3 = C^3$
Optimal coordinates $(\bar{x}_1, \bar{x}_2)$		(0, 0)	$(0, \frac{1}{2}b_2)$	(0, 3)	$(\frac{1}{3}b_2 - 2, 3)$	(5, 3)	(5, 3)
Optimal objective value $\bar{z}$		0	b_2	6	$4 + \frac{1}{3}b_2$	11	11

In general, our discussion has revealed that reducing a right-hand side value for a less-than-or-equal constraint from $+\infty$ to $-\infty$ causes the constraint to be redundant at first, then essential (shaping the feasible set, but possibly not changing the optimal solution), binding (shaping the feasible set and any change of the right-hand side value, regardless how small, changes the optimal solution), to so strong so as to cause the nonexistence of feasible solutions. Not each case has to go through all of these phases. For instance, changing the value of b_2 will result in the constraint first being redundant, then essential and binding, and then infeasible.

The ranges for changes of b_1 and b_3 can be similarly obtained and are shown in Tables 2.24 and 2.25, respectively. We must note, though, that the interpretation of these intervals is different from that in the case of changes in the objective function. While a range of c_j of, say, [3, 7] indicates that as long as c_j is anywhere between 3 and 7, the optimal solution will not change, the same interval for b_i will indicate that as long as b_i is within this range, the optimal basis will not change. In other words, for all changes within this range, the same constraints will be binding at optimum. The optimal solution point will however change. This makes the ranges for right-hand side values more difficult to interpret for managers.

Finally in this section, we want to apply the 100% rule to right-hand side changes. The rule itself is the same as that used for changes of objective function coefficients. Here, with right-hand side values of $b_1 = 3$, $b_2 = 11$, and $b_3 = 2$, we have obtained intervals of [1, 5½], [6, 21], and $[-1\frac{1}{3}, \infty[$, see Tables 2.24, 2.25, and 2.26. In other words, we have $\underline{\Delta}b_1 = 2, \overline{\Delta}b_1 = 2\frac{1}{2}, \underline{\Delta}b_2 = 5, \overline{\Delta}b_2 = 10, \underline{\Delta}b_3 = 3\frac{1}{3}$, and $\overline{\Delta}b_3 = \infty$

Table 2.26 Optimal solution and objective values of sample problem to various values of b_3

Range of b_3	$]-\infty, -3[$	-3	$]-3, -1\frac{2}{3}[$	$-1\frac{1}{3}$	$]-1\frac{1}{3}, \infty[$
Optimal coordinates $(\bar{x}_1, \bar{x}_2)$	There exists no feasible solution	$(0, 3)$	$(3+b_3, 3)$	$(1\frac{2}{3}, 3)$	$(1\frac{2}{3}, 3)$
Optimal objective value $\bar{z}$		6	$9 + b_3$	$7\frac{2}{3}$	$7\frac{2}{3}$

With anticipated changes of the right-hand side values from their present values to $b_1 = 4\frac{1}{2}$, $b_2 = 7$, and b_3 unchanged at 2, we obtain $\Delta b_1 = 1\frac{1}{2}$, $\Delta b_2 = -4$, and $\Delta b_3 = 0$. We can then compute $\frac{|\Delta b_1|}{\overline{\Delta b_1}} + \frac{|\Delta b_2|}{\overline{\Delta b_2}} = \frac{1\frac{1}{2}}{2\frac{1}{2}} + \frac{4}{5} = \frac{7}{5} > 1$, so that these changes violate the 100% rule and will not only cause the solution to change, but also the basis, i.e., different constraints will be binding after the change.

As a different example, assume that b_1 decreases from its present value of 3 to $2\frac{1}{2}$ (meaning that $\Delta b_1 = -\frac{1}{2}$), b_2 increases by $\Delta b_2 = 2$ to its new value of $b_2 = 13$, and the third right-hand side value decreases by $\Delta b_3 = -1$ from its present value of 2 to its new value of $b_3 = 1$. The condition is then $\frac{|\Delta b_1|}{\overline{\Delta b_1}} + \frac{|\Delta b_2|}{\overline{\Delta b_2}} + \frac{|\Delta b_3|}{\overline{\Delta b_3}} = \frac{\frac{1}{2}}{2} + \frac{2}{10} + \frac{1}{1\frac{1}{3}} = \frac{3}{4} < 1$, satisfying the 100% rule, so that after this change, the same constraints will be binding. Again, the solution may (and most likely will) change.

2.4.2 Optimal Solution and Sensitivity Analyses on a Printout

In this section, we will first provide a simulated printout that is typical for what analysts obtain from a computer upon solving a linear programming problem. We then explain the different features and the information provided by the printout. In the second part of this section, we use a linear programming problem, provide the printout, and answer a number of questions relevant to the decision maker.

In order to do so, we consider the following linear programming problem:

$$\begin{aligned}
\text{P}: \text{Max } z = 3x_1 &- x_2 \\
\text{s.t.} \quad x_1 &- x_2 = 2 \\
2x_1 &+ 3x_2 \leq 16 \\
5x_1 &+ x_2 \geq 15 \\
x_1, \quad x_2 &\geq 0.
\end{aligned}$$

The standard printout for the problem is shown in Table 2.27, while information concerning sensitivity analyses is provided in Table 2.28.

The *Summary of Results* shown in Table 2.27 subdivides in three parts. The first part shows the optimal value of the objective function, in our example $\bar{z} = 10.8$. The second part shows the optimal values of the decision variables as well as their *opportunity costs*. In our examples, the information provided indicates that at optimum, $\bar{x}_1 = 4.4$ and $\bar{x}_2 = 2.4$. The opportunity cost (sometimes also called

Table 2.27 *Summary of Results* for sample problem

```
                    SUMMARY OF RESULTS

VALUE OF THE OBJECTIVE FUNCTION 10.8000

    DECISION          VALUE AT        OPPORTUNITY
    VARIABLE          OPTIMUM         COST

    X1                4.4000          0.0000
    X2                2.4000          0.0000


    SLACK/EXCESS      CONSTRAINT      OPTIMAL      SHADOW PRICE
    VARIABLE          TYPE            VALUE

    CONSTRAINT 1:     EQ              0.0000       2.2000
    CONSTRAINT 2:     LE              0.0000       0.4000
    CONSTRAINT 3:     GE              9.4000       0.0000
```

reduced cost) of a variable indicates how far the price or cost of a variable is away from being included in the optimal solution. In our example, both opportunity costs are zero, as both variables are included in the solution with a positive value. Suppose now that there were a variable in a profit-maximizing problem, whose price is $5 and whose opportunity costs are 3. Furthermore, suppose that the variable equals zero at optimum, i.e., it is not included in the optimal solution with a positive value. The main reason for the variable not being part of the optimal solution is that it is not profitable enough. More specifically, its price is not high enough. The opportunity cost now indicates that the lowest price at which the variable would be included in the solution is its present price plus the opportunity cost, in our example $5 + $3 = $8. In other words, $8 is the lowest price at which we consider including the variable in an optimal solution. The interpretation of opportunity cost in cost minimization problems is similar. In those problems, it indicates by how much the cost of a variable has to decrease, before it will become part of an optimal solution.

Consider now the third and last part of the *Summary of Results* that is headed by the words "slack or excess variable." As discussed earlier in this book, the solver has automatically added a slack variable to the left-hand side of each $\leq$ inequality and subtracted an excess or surplus variable from the left-hand side of each $\geq$ inequality.

The printout first provides the number of the constraint these variables are added to, followed by a column that specifies the type of constraint (and with it the type of additional variable) that was automatically added. Here, we use the abbreviations *EQ* for an equation, *LE* for a "$\leq$" constraint, and *GE* for a "$\geq$" constraint. The optimal values of the slack, excess, and artificial variables are provided in the next column. In our example, they are 0, 0, and 9.4. These numbers indicate that the variable

2.4 Postoptimality Analyses

Table 2.28 *Sensitivity Analyses* for sample problem

```
                    SENSITIVITY ANALYSES

            COEFFICIENTS OF THE OBJECTIVE FUNCTION

VARIABLE       LOWEST            ORIGINAL           HIGHEST
               ALLOWABLE         VALUE              ALLOWABLE
               VALUE                                VALUE

   X1          1.0000            3.0000             INFINITY
   X2         -3.0000           -1.0000             INFINITY


                    RIGHT-HAND SIDE VALUES

CONSTRAINT     LOWEST            ORIGINAL           HIGHEST
NUMBER         ALLOWABLE         VALUE              ALLOWABLE
               VALUE                                VALUE

CONSTRAINT 1   -1.6154           2.0000             8.0000
CONSTRAINT 2    8.1667          16.0000             INFINITY
CONSTRAINT 3   -INFINITY        15.0000            24.4000
```

associated with the first constraint has an optimal value of 0. This is obvious, as the first constraint is an equation, which does not allow left- and right-hand sides to differ. The second constraint is of type "$\leq$," meaning that the system automatically added a slack variable to the left-hand side of the relation. The optimal value of this slack variable is 0, indicating that the constraint is actually satisfied as an equation at optimum, meaning that the constraint is binding at optimum. A constraint that is binding at optimum constitutes a *bottleneck* of the problem. Bottlenecks require special attention as any changes in the right-hand side value of such equations will immediately result in changes of the optimal solution. Finally, the third constraint is of type "$\geq$" and thus had an excess variable subtracted from its left-hand side. The optimal value of this excess variable equals 9.4, indicating that at optimum, the left-hand side exceeded the right-hand side value by 9.4. This constraint does not constitute a bottleneck, the constraint is not binding at optimum, so that smaller changes of the third right-hand side value will leave the solution unchanged.

The last column in Table 2.27 shows the *shadow prices* associated with the constraints. The shadow price associated with a constraint indicates the change of the value of the objective function, given that the right-hand side value of that constraint increases by one unit. It does not provide any information about how the solution itself, i.e., the values of the variables, would change, though.

In our example, the shadow price of the first constraint (the equation) equals 2.2. This means that as the first right-hand side value b_1 increases from its original value of 2 by one unit to 3, the value of the objective function increases from 10.8 by 2.2 to 13. Similarly, the increase of the second right-hand side value by one unit from its original value of 16 to 17 will result in a new optimal objective value that is 0.4 higher than the original objective value of 10.8, i.e., $\bar{z} = 11.2$. Finally, the third constraint has a shadow price of 0. This indicates that a change of the third right-hand side value by one unit will neither change the optimal solution, nor the value of the objective function at optimum. This is a result of the fact that in the present solution, the third left-hand side already exceeds the right-hand side value by 9.4, thus the constraint is not binding at optimum, so that smaller changes of the right-hand side value will have no effect on the solution.

It is also possible to identify some of the "special cases" on the printout. In case no feasible solution exists, some solvers will provide the analyst with the numbers of the constraints that are incompatible and thus cause the infeasibility. Such a message greatly simplifies the work of the analyst, who can then immediately attempt to loosen up some of the constraints that are involved in the problem. In the case of unbounded "optimal" solutions, a simple message to that effect will appear. Dual degeneracy can be detected only if it occurs at optimum, as intermediate solutions are not displayed. In this case, alternative optimal solutions exist, and they can be detected on the printout, if a variable that is not included in the solution, i.e., it has a zero value, has an opportunity cost of zero as well. The case of primal degeneracy can also be detected if it occurs at the optimal solution. If it does, there will be strictly less than m (the number of constraints, not counting the nonnegativity constraints) variables (decision, slack, and excess variables) with a positive value.

Additional information is available when choosing the "Sensitivity Analyses" option. The simulated printout for our example is shown in Table 2.28.

The *Sensitivity Analyses* option consists of two parts. The upper part headed by *Coefficients of the Objective Function* analyzes changes of the objective function coefficients c_j, while the part headed by *Right-Hand Side Values* provides information concerning changes of the right-hand side values b_i.

In Table 2.28, the line headed by x_1 specifies the original value of $c_1 = 3$, and also the interval $= [1, \infty[$. This interval indicates that as long as the value of c_1 is equal or larger than 1, the optimal solution shown in Table 2.27 will remain optimal. Similarly, we can interpret the numbers in the row headed by x_2. The original value of the objective function coefficient of this variable is $c_2 = -1$. The range specified here is $= [-3, \infty[$, and it indicates that as long as c_2 is -3 or larger, the solution shown in Table 2.27 will remain optimal.

Consider now the bottom part of Table 2.28. The row headed by "Constraint 1" specifies the original right-hand side value of the first constraint (which was $b_1 = 2$), as well as the interval $[-1.6154, 8.0000]$. This interval indicates that as long as the first right-hand side value remains between these bounds, the optimal basis will remain unchanged. In other words, within this range, the same constraints remain

2.4 Postoptimality Analyses

Table 2.29 Data for shoe production problem

Input material	Product		
	Walker	Hiker	Backpacker
NOwater (sq. ft per pair)	1/2	1	$2^{1}/_{2}$
Fabrinsula (oz per pair)	1/2	2/3	4/3
Lugster (pair of soles)	1	1	1

binding at optimum. The solution, however, may very well change, even within this range. However, it is not possible to determine from the printout what these changes will be. The remaining rows for the other two constraints are interpreted in a similar fashion.

The remainder of this section will present a linear programming problem, formulate it, and then interpret the results shown in a printout. The production planning model we will use throughout the remainder of this section is as follows.

Example: A footwear manufacturer is planning next year's product line. Part of that line are three types of hiking boots, called "Walker," "Hiker," and "Backpacker." Among the many raw materials used in the production are three particularly important and costly materials: NOwater (a lining that waterproofs the boots), Fabrinsula (fabric insulation for warmth), and Lugster (lug) soles. Currently, 10,000 sq. ft of NOwater are available at $20 per sq. ft. Similarly, the manufacturer has access to up to 5000 oz of Fabrinsula at $15 per ounce and up to 7000 pairs of Lugster soles at $10 per pair. The requirements for manufacturing the different types of boots are given in Table 2.29.

Contracts have been signed for the delivery of at least 3000 pairs of "Walkers," at least 2000 pairs of "Hikers," and at least 1000 pairs of "Backpackers." Customers are prepared to purchase additional quantities should they become available. The agreed-upon prices are $40, $65, and $110 per pair of the respective types. (Note: The unit profits in the objective function in the formulation below have been computed as the price minus the costs for the required amounts of NOwater, Fabrinsula, and Lugster, as will be explained below.)

In order to formulate the problem, we first define variables x_j as the quantity of the three types of boots that we made and sell. The formulation of the problem is then as follows.

P : Max $z = 12.5x_1 + 25x_2 + 30x_3$

s.t. $\frac{1}{2}x_1 + x_2 + 2\frac{1}{2}x_3 \leq 10,000$ (NOwater availability)

$\frac{1}{2}x_1 + \frac{2}{3}x_2 + 1\frac{1}{3}x_3 \leq 5,000$ (Fabrinsula availability)

$x_1 + x_2 + x_3 \leq 7,000$ (Lugster soles availability)

$x_1 \geq 3,000$ (Walker requirement)

$x_2 \geq 2,000$ (Hiker requirement)

$x_3 \geq 1,000$ (Backpacker requirement)

$x_1, x_2, x_3 \geq 0$.

Table 2.30 *Summary of Results* for shoe production problem

```
                    SUMMARY OF RESULTS

VALUE OF THE OBJECTIVE FUNCTION 143,750.0000

     DECISION         VALUE AT          OPPORTUNITY
     VARIABLE         OPTIMUM           COST

     WALKER           3,000.0000        0.0000
     HIKER            2,750.0750        0.0000
     BACKPACKER       1,250.0000        0.0000


     SLACK/EXCESS     CONSTRAINT        OPTIMAL     SHADOW PRICE
     VARIABLE         TYPE              VALUE

     NOWATER:         LE                2,625.0000          0.0000
     FABRINSULA:      LE                0.0000              7.5000
     LUGSTER:         LE                0.0000             20.0000
     WALKER:          GE                0.0000            -11.2500
     HIKER:           GE                750.0000            0.0000
     BACKPACKER:      GE                250.0000            0.0000
```

Before we provide the printout and continue with our interpretations, some comments regarding the formulation are in order. While the constraints are straightforward, the objective function coefficients have been obtained as follows. Consider the "Walker" boots. They sell for $40 a pair, from which we have to deduct the costs of ½ sq. ft of NOwater ($10), ½ oz of Fabrinsula ($7.50) and one pair of soles ($10), leaving us with a per-unit profit of $12.50. This is the coefficient found in the above formulation. The unit profits of the other two boots are computed in a similar fashion.

Tables 2.30 and 2.31 provide the usual printouts with the sensitivity option.

The following is a simulated dialog between an analyst, who modeled the problem and provides the information from the printouts and the decision maker, who asks the questions that are of managerial interest.

Q1: How many pairs of the boots should we manufacture and what will be the associated profit?

A1: The solution suggests that we make 3000 pairs of Walkers, 2750 pairs of Hikers, and 1250 pairs of Backpackers. Given this production plan, we can expect a profit of $143,750.

2.4 Postoptimality Analyses

Table 2.31 *Sensitivity Analyses* for shoe production problem

```
                    SENSITIVITY ANALYSES

              COEFFICIENTS OF THE OBJECTIVE FUNCTION

VARIABLE        LOWEST          ORIGINAL         HIGHEST
                ALLOWABLE       VALUE            ALLOWABLE
                VALUE                            VALUE

WALKER          -INFINITY       12.50            23.75
HIKER           16.00           25.00            30.00
BACKPACKER      25.00           30.00            50.00


                    RIGHT-HAND SIDE VALUES

CONSTRAINT      LOWEST          ORIGINAL         HIGHEST
                ALLOWABLE       VALUE            ALLOWABLE
                VALUE                            VALUE

NOWATER:        7,375.0000      10,000.00        INFINITY
FABRINSULA:     4,833.3333      5,000.00         5,500.00
LUGSTER:        6,625.00        7,000.00         7,250.00
WALKER:         2,000.00        3,000.00         3,600.00
HIKER:          -INFINITY       2,000.00         2,750.00
BACKPACKER:     -INFINITY       1,000.00         1,250.00
```

Q2: You are undoubtedly aware of the fact that NOwater, Fabrinsula, and Lugster are critical resources. How many of these do we use in the suggested plan and how much is left over?

A2: The printout tells me that we will have 2625 sq. ft of NOwater left over, while Fabrinsula and Lugster soles are both completely used. The latter two are obvious bottlenecks in the process. In other words, we are using 7375 sq. ft of NOwater, the complete supply of 5000 oz of Fabrinsula, and all of the 7000 pairs of Lugster soles.

Q3: I just was informed by our marketing research group that the Hiker boots are very popular and just about all of our customers would be prepared to pay an additional $10 to get a pair of these. Would such a price hike change the optimal solution?

A3 (aside to himself): Ahh, a sensitivity analysis on the objective function coefficient c. The range within which the present optimal solution remains optimal is [16, 30]. A price

increase by $10 leads to a price of $35 rather than the original $25, which is not in the interval, thus the solution will change.

(to the Decision Maker) If we hike the price by more than $5, our solution will change. If you need more details, I have to resolve the problem with the new price.

Q4: No no, that's OK. But how about this. We are presently making 3000 Walkers, just enough to satisfy one of our requirements. I wonder if it would be worthwhile to lower the price. Would that lead to increased sales?

A4 (to himself): Another sensitivity analysis on an objective function coefficient. Given the present unit profit of $12.50 for a pair of Walkers, the Sensitivity Analyses part of the printout indicates that as long as the unit profit remains in the interval $]-\infty, 23.75]$, the optimal solution will not change.

(to the Decision maker): It does not matter by how much you decrease the price of Walkers, we will not sell any more of them.

Q5: All right then. An interesting thing happened the other day. I met a salesman who offered me an additional 100 oz of Fabrinsula for $8.50 per ounce. Should we purchase that? You told me that we used up all the Fabrinsula that we have. What if I squeeze the man a bit and get it for $5? Will we take it for that price? And what will happen to our profit?

A5 (to himself): A sensitivity analysis on the second right-hand side value b_2. The Summary of Results tells me that the shadow price of Fabrinsula is $7.50, so we should not purchase it for more than that. Also, the basis will not change if we buy up to 500 ounces of Fabrinsula.

(to the Decision Maker): No, don't buy it for $8.50. As a matter of fact, you should not pay more than $7.50 for an extra ounce of Fabrinsula. If you can get it for $5, you can buy up to 500 oz of it. For each ounce that you get on top of what we have, our profit will increase by $2.50. What happens beyond 500 extra ounces, though, requires additional calculations.

Q6: Wonderful. I have also considered an alternative, though. In one of our trade magazines, I just read an offer for Lugster Soles that we could get for $19 a pair. Would you consider that?

A6 (to himself): The shadow price for Lugster Soles is $20, and the Sensitivity Analyses tell me that this holds for an increase of up to 250 pairs.

(to the Decision Maker): If you can get a pair for $19, get them. They are worth $20 to us, so for each extra pair, we make $1 in addition to our usual profit. You can get up to 250 pairs.

Q7: Thank you so much, Mr. Analyst. Your advice was very helpful. (Putters around with his cell phone). Wait a minute—stop the presses! Our production manager just texted me that 200 pairs of Lugster soles have been damaged in our warehouse and can no longer be used. What are we going to do? How much is that going to cost me?

A7 (to himself): The optimal basis remains unchanged as long as we have between 6625 and 7250 pairs of soles. So 200 pairs less will not change the basis, but it will change the

2.4 Postoptimality Analyses

solution. And since the shadow price for Lugster Soles is $20, our profit will decrease by 200 (20) = $4000.

(to the Decision Maker): I have good news and bad news. The bad news is that our profit will decrease by $4000, which is 2.78%. After solving the problem again with the new data, I found that not all is bad. The good news is that we do not have to rearrange the production plan altogether. We should still make 3000 pairs of Walkers, but we should make 400 pairs less of the Hikers, and 200 pairs more of the Backpackers.

The Decision Maker: I knew it. Why are they doing this to me? (Disappears into the bowels of the Administration Building).

Exercises

Problem 1 (a diet problem): A planner considers designing a diet that consists of Coke, garlic fingers, spring rolls, pita pockets, and apples. As far as nutrients go, the planner includes calories, riboflavin, and vitamin A, of which at most 1500, at least 98, and at least 27 must be included in the diet. The cost minimization problem was subsequently formulated as follows:

$$\text{Min } z = .9x_1 + 2.8x_2 + 3.2x_3 + 5.6x_4 + 3.6x_5$$

$$\begin{aligned}
\text{s.t.} \quad 80x_1 + 310x_2 + 340x_3 + 460x_4 + 20x_5 &\leq 1500 \\
2x_1 + 9x_2 + 25x_3 + 16x_4 + 5x_5 &\geq 98 \\
5x_2 + 4x_3 + 3x_4 + 10x_5 &\geq 27 \\
x_1, \quad x_2, \quad x_3, \quad x_4, \quad x_5 &\geq 0
\end{aligned}$$

The printout of the problem is shown in Table 2.32.

(a) What does the diet of the planner consist of? (Indicate type of food and quantity).
(b) How much does the diet cost?
(c) How many calories does the diet include? How much riboflavin? How much vitamin A?
(d) What is the highest price that will put Coke into the solution?
(e) What is the effect of increasing the riboflavin requirement from 98 to 99?

Solution:

(a) The diet consists of no Coke, no garlic fingers, 3.67 spring rolls, no pita pockets, and 1.23 apples.
(b) The cost of the diet is $16.19.
(c) The diet includes $1500 - 226.26 = 1273.74$ calories, $98 + 0 = 98$ units of riboflavin, and $27 + 0 = 27$ units of vitamin A.

Table 2.32 *Summary of Results* for diet problem

```
               SUMMARY OF RESULTS

VALUE OF THE OBJECTIVE FUNCTION 16.18609

    DECISION            VALUE AT            OPPORTUNITY
    VARIABLE            OPTIMUM             COST

    COKE                0.0000              0.7470
    GARLIC FINGERS      0.0000              0.5026
    SPRING ROLLS        3.6739              0.0000
    PITA POCKETS        0.0000              3.4104
    APPLES              1.2304              0.0000


    SLACK/EXCESS        CONSTRAINT          OPTIMAL       SHADOW PRICE
    VARIABLE            TYPE                VALUE

    CALORIES:           LE                  226.2609      0.0000
    RIBOFLAVIN:         GE                  0.0000        -0.0765
    VITAMIN A:          GE                  0.0000        -0.3217
```

Table 2.33 Risk and rate of return for the sample problem

Assets	Units of risk per Dollar invested	Expected rate of return
Northern Mines Shares	4	.15
Bucklin Automobiles	3.5	.11
Royal Bank Shares	2	.07
EE Savings Bonds	1	.05

(d) Presently, Coke costs 90¢ which is apparently too much for it to be included in the solution. Its opportunity cost is 74.7¢, so that its price will have be decrease by at least that amount. In other words, Coke will be included in the diet if its price is no higher than $90 - 74.7 = 15.3$¢.

(e) The cost will increase by 7.65¢.

Problem 2 (an investment problem): An investment agency has been asked to advise one of its clients how to invest all of his $100,000 among the four assets shown in Table 2.33.

The client would like as high an annual return as is possible to receive while incurring of an average of no more than 2.5 risk units per dollar invested. The

2.4 Postoptimality Analyses

Table 2.34 *Summary of Results* for Problem 2

```
                    SUMMARY OF RESULTS

VALUE OF THE OBJECTIVE FUNCTION 9,733.333

   DECISION            VALUE AT           OPPORTUNITY
   VARIABLE            OPTIMUM            COST

   MINES               43,333.33          0.0000
   BUCKLIN             0.0000             0.01
   ROYAL BANK          20,000.00          0.0000
   SAVINGS             36,666.67          0.0000


   SLACK/EXCESS    CONSTRAINT    OPTIMAL      SHADOW PRICE
   VARIABLE        TYPE          VALUE

   RISK            LE            0.0000        0.0333
   INVESTMENT      LE            0.0000        0.1000
   BANK LIMIT      LE            20,000.00     0.0000
   BANK & AUTO     GE            0.0000       -0.0133
```

amount invested in Royal Bank cannot exceed $40,000. Furthermore, the investment in Automobiles and Banks combined must be at least $20,000.

Defining x_1, x_2, x_3, and x_4 for the amount invested in the four alternatives, the problem can be formulated as follows.

$$P: \text{Max } z = .15x_1 + .11x_2 + .07x_3 + .05x_4$$
$$\text{s.t.} \quad 1.5x_1 + x_2 - .5x_3 - 1.5x_4 \leq 0$$
$$x_1 + x_2 + x_3 + x_4 \leq 100,000$$
$$x_3 \leq 40,000$$
$$x_2 + x_3 \geq 20,000$$
$$x_1, x_2, x_3, x_4 \geq 0.$$

The printout is shown in Tables 2.34 and 2.35.

(a) How much is invested in each of the alternatives and what is the average rate of return?
(b) Identify the bottlenecks in the investment plan.
(c) If we could borrow some additional funds at 11%, would it be worth our while? Explain in one short sentence.

Table 2.35 *Sensitivity Analyses* for Problem 2

```
                    SENSITIVITY ANALYSES

         COEFFICIENTS OF THE OBJECTIVE FUNCTION
```

VARIABLE	LOWEST ALLOWABLE VALUE	ORIGINAL VALUE	HIGHEST ALLOWABLE VALUE
MINES	0.1300	0.1500	INFINITY
BUCKLIN	-INFINITY	0.1100	0.1200
ROYAL BANK	0.0600	0.0700	0.0833
SAVINGS	0.0300	0.0500	0.0700

```
                    RIGHT-HAND SIDE VALUES
```

CONSTRAINT	LOWEST ALLOWABLE VALUE	ORIGINAL VALUE	HIGHEST ALLOWABLE VALUE
RISK	-130,000.00	0.0000	110,000.00
INVESTMENT	26,666.67	100,000.00	INFINITY
BANK LIMIT	20,000.00	40,000.00	INFINITY
BANK & AUTO	0.00	20,000.00	40,000.00

(d) There is a rumor that the return of the Royal Bank will increase to 8%. Will this change the investor's plans? What if it decreases to 5.5%? Explain in one short sentence.

Solution:

(a) The solution prescribes investments as follows: Mines: $43,333.33, Automobile: $0, Bank: $20,000, and EE Savings: $36,666.66. The average rate of return is 9.7333%.

(b) The risk constraint (constraint 1), the total investment level (constraint 2), and the requirement to invest at least $20,000 in "Auto and Bank" (constraint 4) are all tight.

(c) The shadow price of total investment (constraint 2) is 0.1, i.e., the benefit of an extra dollar is 10¢. Given the interest rate of 11% on loans, it is not worthwhile to borrow.

2.4 Postoptimality Analyses

(d) Sensitivity on objective function coefficients. The range for the bank shares extends from 6% to 8.33%, so an increase from 7% to 8% will not change the investment plan. A decrease to 5.5% will.

Problem 3 (a transportation problem): Consider a transportation problem with two origins (warehouses) and three destinations (customers). The supplies at the origins are 60 and 80 units, respectively, while the demand at the destinations is exactly 30, 50, and 40, respectively. The problem has been formulated as follows:

$$P : \text{Min } z = 3x_{11} + 7x_{12} + 4x_{13} + 9x_{21} + 2x_{22} + 5x_{23}$$

$$\begin{array}{llllllr}
\text{s.t.} & x_{11} + & x_{12} + & x_{13} & & & \leq 60 \\
& & & & x_{21} + x_{22} + x_{23} & \leq 80 \\
& x_{11} + & & & x_{21} & = 30 \\
& & x_{12} + & & x_{22} & = 50 \\
& & & x_{13} + & & x_{23} = 40 \\
& x_{11}, & x_{12}, & x_{13}, & x_{21}, & x_{22}, & x_{23} \geq 0.
\end{array}$$

The printout is shown in Tables 2.36 and 2.37.
(a) What is the shipment plan and what are the associated costs?
(b) Which of the origins are fully used and which have still some units of the product in them (and how many)?
(c) What if the per-unit-cost of a shipment from origin 2 to destination 3 were to increase by $2, would that change the optimal solution? What if the cost were to decrease by $2?
(d) What would happen if the number of units available at the first origin were to be reduced by one unit? (Include cost considerations).
(e) What if we were offered extra units delivered to origin 2 at a rate of $2 per unit?

Solution:

(a) The shipments from the Origin 1 to the three destinations are 30, 0. and 30, while the shipments from Origin 2 to the three destinations are 0, 50, and 10. The total transportation costs are $360.
(b) All units in Origin 1 are shipped out, but 20 units are left over in Origin 2.
(c) The range for the unit transportation costs c_{23} is [4, 10]. An increase by $2 puts the unit transportation cost at $5 + 2 = \$7$, which is in the interval. As a result, there will be no change in the optimal transportation plan, but the costs will increase by $2x_{23} = 20$. A decrease of $2 would put the unit transportation cost at $5 - 2 = 3$, which is outside of the interval, so that the optimal transportation plan (and its costs) will change.
(d) As the constraint that belongs to Origin 1 has zero slack, the solution will change. As the shadow price is 1, the total cost will increase by $1.

Table 2.36 *Summary of Results* for Problem 3

```
                SUMMARY OF RESULTS

VALUE OF THE OBJECTIVE FUNCTION 360.0000

    DECISION            VALUE AT           OPPORTUNITY
    VARIABLE            OPTIMUM            COST

    X11                 30.0000            0.0000
    X12                 0.0000             6.0000
    X13                 30.0000            0.0000
    X21                 0.0000             5.0000
    X22                 50.0000            0.0000
    X23                 10.0000            0.0000


    SLACK/EXCESS        CONSTRAINT         OPTIMAL       SHADOW PRICE
    VARIABLE            TYPE               VALUE

    ORIGIN 1            LE                 0.0000        1.0000
    ORIGIN 2            LE                 20.0000       0.0000
    DESTINATION 1       EQ                 0.0000        -4.0000
    DESTINATION 2       EQ                 0.0000        -2.0000
    DESTINATION 3       EQ                 0.0000        -5.0000
```

(e) There are still 20 units left at Origin 2, so that no additional units are needed, and we decline the offer.

2.5 Duality

This section explores some aspects of duality theory, the theory behind linear programming that explores how and why solution methods work. Given the scope of this book, we will restrict ourselves to some of the relations and interpretations, without getting into technical details. For more details, interested readers are referred to the standard advanced texts such as Dantzig (1963), Dantzig and Thapa (1997), or Eiselt and Sandblom (2007).

To every linear programming problem, which we will call a *primal problem*, we can now assign another linear programming problem, called the *dual problem*. The variables in the primal problem are also referred to as the *primal variables*, while the variables in the dual problem are the dual variables.

In order to simplify our discussion, we will base our arguments on a production problem similar to that presented in Sect. 2.2.1 when we introduced linear

2.5 Duality

Table 2.37 *Sensitivity Analyses* for Problem 3

```
                    SENSITIVITY ANALYSES

          COEFFICIENTS OF THE OBJECTIVE FUNCTION

VARIABLE       LOWEST           ORIGINAL          HIGHEST
               ALLOWABLE        VALUE             ALLOWABLE
               VALUE                              VALUE

X11            -INFINITY        3.00              8.00
X12            1.00             7.00              INFINITY
X13            -1.00            4.00              5.00
X21            4.00             9.00              INFINITY
X22            -INFINITY        2.00              8.00
X23            4.00             5.00              10.00


                    RIGHT-HAND SIDE VALUES

CONSTRAINT     LOWEST           ORIGINAL          HIGHEST
               ALLOWABLE        VALUE             ALLOWABLE
               VALUE                              VALUE

ORIGIN 1       40.00            60.00             70
ORIGIN 2       60.00            80.00             INFINITY
DESTINATION 1  20.00            30.00             50.00
DESTINATION 2  0.00             50.00             70.00
DESTINATION 3  30.00            40.00             60.00
```

programming. In this case, there are two products, floorboards, and spindles, which are processed on three machines: a saw, a router, and a sander. The unit profits of the two products are $1 per floorboard and $4 per spindle. The machines have capacities (in seconds) of 50,000, 450,000, and 600,000. It takes 5 sec to saw the floorboard or the spindle. Thirty seconds are needed to process a floorboard on the router, while a spindle takes 90 sec on this machine. One floorboard requires 20 sec on the sander, while a spindle needs 100 sec for processing. The objective of the firm is to maximize its profits from the floorboard and spindle production.

Defining variables x_1 and x_2 as the number of floorboards and spindles that we make and sell, respectively, we can formulate the primal problem as follows.

$$P : \text{Max } z = 1x_1 + 4x_2 \quad (\text{max profit})$$
$$\text{s.t.} \quad 5x_1 + 5x_2 \leq 50{,}000 \quad (\text{saw})$$
$$30x_1 + 90x_2 \leq 450{,}000 \quad (\text{router})$$
$$20x_1 + 100x_2 \leq 600{,}000 \quad (\text{sander})$$
$$x_1, \quad x_2, \quad \geq 0. \quad (\text{nonnegativity})$$

In order to explain the dual problem, we will refer to the firm that has the primal problem as its planning model as the *Manufacturer*. Suppose now that there is another company that, for reasons to become clear as we proceed, we will refer to as the *Lessor*. The *Lessor* actually owns the saws, routers, and sanders, which he can use to either manufacture floor boards and spindles himself or lease out the machine time to the *Manufacturer* or Lessee. The task for the *Manufacturer* is now to set up a pricing system that will minimize its own overall production costs, while making it interesting to the *Lessor* to rent machine time to the *Manufacturer* rather than make the products himself. Note that the *Lessor* will not rent out an hour here or there, he either rents the entire time—50,000 sec on the saw, 450,000 sec on the router, and 600,000 sec on the sander—or not at all.

In order to determine such a pricing system, the *Manufacturer* will set up a pricing system that defines u_1 as the price per second on the saw, u_2 as the price per second on the router, and u_3 as the price per second on the sander. Clearly, the *Manufacturer* wishes to find prices that will minimize the total cost of leasing the equipment. Leasing the machine time on the saw will cost u_1 dollars per second, and the time to lease is 50,000 sec, and similar for the other two machines. Hence the objective is to minimize $50{,}000u_1 + 450{,}000u_2 + 600{,}000u_3$. The next task is to fix the value of the price u_1 so as to make it interesting to the *Lessor* to rent out the machine time rather than running the manufacturing procedure himself. This can be achieved by ensuring that the price charged for the rental time required to make one unit of a product is at least as much as it would cost by making and selling the product directly. In order to illustrate, consider the floorboards. It takes 5 sec on the saw to make it, which, given the pricing system, is evaluated at a cost of $5u_1$. In addition, we also use time on the router and the sander, and these times are evaluated at costs of $30u_2$ and $20u_3$, respectively. The sum of all of these costs—the rental cost equivalent to making one floorboard—should be at least as much as the profit of making a floorboard. If it were not, the *Lessor* might as well go into the manufacturing business himself. This means that we have to require that $5u_1 + 30u_2 + 20u_3 \geq 1$, where the right-hand side corresponds to the $1 profit for making one floor board. A similar constraint needs to be written for the spindles, again in order to make it as least as attractive for the *Lessor* to rent out machine time rather than manufacture himself. That constraint is $5u_1 + 90u_2 + 100u_3 \geq 4$. Adding the fact that the prices must all be nonnegative, we now have formulated a complete dual problem, which we will call P_D. Below, we show the primal problem and the dual problem next to each other.

2.5 Duality

P : Max $z = 1x_1 + 4x_2$
s.t. $5x_1 + 5x_2 \leq 50,000$
$30x_1 + 90x_2 \leq 450,000$
$20x_1 + 100x_2 \leq 600,000$
$x_1, x_2 \geq 0$.

P_D : Min $z_D = 50,000u_1 + 450,000u_2 + 600,000u_3$
s.t. $5u_1 + 30u_2 + 20u_3 \geq 1$
$5u_1 + 90u_2 + 100u_3 \geq 4$
$u_1, u_2, u_3 \geq 0$.

Before we continue discussing the relations between a primal and its associated dual problem, we would like to point out that the roles of variables and constraints in the two problems are exchanged. The objective function coefficients of the primal are found on the right-hand sides of the dual, while the right-hand side values of the primal are coefficients in the objective function of the dual. Similarly, the technological coefficients on the left-hand sides are the same, but with rows and columns exchanged.

Table 2.38a, b show the simulated printout of the optimal solutions of the primal and dual problem, respectively. While it is optimal for the *Manufacturer* to make no floorboards and 5000 spindles (for a total profit of $20,000), the Lessor should charge nothing for the saw and the sander, but 4.44¢ for each second of the router. Given that he leases the machines to the *Manufacturer* rather than making floorboards and spindles himself, his profit from the leasing will then be $20,000.

Comparing the primal and dual solutions, we also note that the optimal values of the primal variables are found in the dual solution as shadow prices of resources, while the optimal values of the primal slack and excess variables are the opportunity costs of the dual variables. Similarly, the shadow prices of the primal resources equal the optimal values of the dual decision variables, and the opportunity costs of the primal variables equal the optimal values of the slack and excess variables in the dual. This means, of course, that we can solve either the primal or the dual solution and have both optimal solutions available in one *Summary of Results*.

In order to explain some further relations between the primal and its associated dual problem, let us multiply each primal constraint with its associated dual variable

Table 2.38 Optimal solution of primal and dual problem

SUMMARY OF RESULTS - PRIMAL				SUMMARY OF RESULTS - DUAL		
VALUE OF THE OBJECTIVE FUNCTION 20,000.00				VALUE OF THE OBJECTIVE FUNCTION 20,000.00		
DECISION VARIABLE	VALUE AT OPTIMUM	OPPORTUNITY COST		DECISION VARIABLE	VALUE AT OPTIMUM	OPPORTUNITY COST
FLOOR BOARDS	0.0000	0.3333		PRICE SAW	0.0000	25,000.00
SPINDLES	5,000.00	0.0000		PRICE ROUTER	0.0444	0.0000
				PRICE SANDER	0.0000	100,000.00
SLACK/EXCESS VARIABLE	OPTIMAL VALUE	SHADOW PRICE		SLACK/EXCESS VARIABLE	OPTIMAL VALUE	SHADOW PRICE
SAW	25,000.00	0.0000		FLOOR BOARDS	0.3333	0.0000
ROUTER	0.0000	0.0444		SPINDLES	0.0000	5,000.00
SANDER	100,000.00	0.0000				
(a)				(b)		

and each dual constraint with the primal variable it is associated with. (Note that due to the nonnegativity constraints on the variables, the inequalities do not change.) We then obtain the following system **S**:

$$\mathbf{S}: z = 1x_1 + 4x_2 \text{(the primal objective function)}$$

$5x_1u_1 + 5x_2u_1 \leq 50,000u_1$ (the first primal constraint multiplied by u_1)
$30x_1u_2 + 90x_2u_2 \leq 450,000u_2$ (the second primal constraint multiplied by u_2)
$20x_1u_3 + 100x_2u_3 \leq 600,000u_3$ (the third primal constraint multiplied by u_3)

$5u_1x_1 + 30u_2x_1 + 20u_3x_1 \geq 1x_1$ (the first dual constraint multiplied by x_1)
$5u_1x_2 + 90u_2x_2 + 100u_3x_2 \geq 4x_2$ (the second dual constraint multiplied by x_1)

$$z_D = 50,000u_1 + 450,000u_2 + 600,000u_3 \text{(the dual objective function)}$$

$$x_1, x_2 \geq 0, u_1, u_2, u_3 \geq 0. \quad \text{(nonnegativity constraints)}$$

Adding all primal constraints in system **S**, we obtain $5x_1u_1 + 5x_2u_1 + 30x_1u_2 + 90x_2u_2 + 20x_1u_3 + 100x_2u_3 \leq 50{,}000u_1 + 450{,}000u_2 + 600{,}000u_3$, where we note that the right-hand side value equals z_D. Similarly, adding all dual constraints in system **S**, we obtain $5u_1x_1 + 30u_2x_1 + 20u_3x_1 + 5u_1x_2 + 90u_2x_2 + 100u_3x_2 \geq 1x_1 + 4x_2$, where we note that the right-hand side value of the aggregate constraint equals the primal value of the objective function z. Putting the two expressions together, we obtain $z_D = 50{,}000u_1 + 450{,}000u_2 + 600{,}000u_3 \geq 1x_1 + 4x_2 = z$, or simply $z_D \geq z$. This property is commonly referred to as *weak duality*. Given that the primal problem maximizes z and the dual problem minimizes z_D, the two values will move towards each other and be equal when they reach an optimal solution. This situation is shown in Fig. 2.23.

In order for the two objective values to be equal at optimum, it is necessary that *all* of the primal inequalities multiplied by their respective dual variables and all dual constraints multiplied by their respective primal variables are satisfied as equations. In our example, this means that if $\bar{x}_1, \bar{x}_2, \bar{x}_3$ is an optimal solution of the primal problem and $\bar{u}_1, \bar{u}_2$ is an optimal solution of the dual problem, then

$5\bar{x}_1\bar{u}_1 + 30\bar{x}_2\bar{u}_1 = 50,000\bar{u}_1,$
$30\bar{x}_1\bar{u}_2 + 90\bar{x}_2\bar{u}_2 = 450,000\bar{u}_2, \quad$ and
$20\bar{x}_1\bar{u}_3 + 100\bar{x}_2\bar{u}_3 = 600,000\bar{u}_3 \quad$ for the primal constraints, and

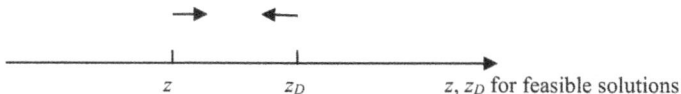

Fig. 2.23 Weak duality

$$50\overline{u}_1\overline{x}_1 + 30\overline{u}_2\overline{x}_1 + 20\overline{u}_3\overline{x}_1 = 1\overline{x}_1, \quad \text{and}$$
$$5\overline{u}_1\overline{x}_2 + 90\overline{u}_2\overline{x}_2 + 100\overline{u}_3\overline{x}_2 = 4\overline{x}_2 \quad \text{for the dual constraints.}$$

If we were to define slack variables S_1, S_2, and S_3 for the three primal constraints and excess variables E_1^D and E_2^D for the two dual constraints, and their optimal values are also indicated by a bar over the variables, we can rewrite the above equations as $\overline{S}_1\overline{u}_1 = 0$, $\overline{S}_2\overline{u}_2 = 0$, and $\overline{S}_3\overline{u}_3 = 0$ for the primal problem, and $\overline{E}_1^D\overline{x}_1 = 0$ and $\overline{E}_2^D\overline{x}_2 = 0$ for the dual problem. These conditions are usually called (*weak*) *complementary slackness conditions*.

These conditions mean that if an inequality constraint is not satisfied as an equation at optimum, then its dual variable must be equal to zero. If a constraint is satisfied as an equation, then its dual variable may be zero or positive. In terms of our example, this means that if we do not fully use a resource (here: machine capacities), then the dual variable (the shadow price of the resource) must equal zero. If, on the other hand, a resource is fully used, then the dual variable may be positive. Similarly, if a variable is positive, then its opportunity cost must be zero; if a variable is zero, then its opportunity cost may be positive. This corresponds with the interpretation we provided in Sect. 2.5 on postoptimality analyses.

It is worth noting that the dual of the dual problem is again the primal problem.

The remainder of this section will demonstrate some further relations between a pair of primal and dual problems. In general, we have three possible cases:

1. The primal problem and its dual both have finite optimal solutions, in which case $\overline{z} = \overline{z}_D$.
2. One of the two problems has no feasible solution, while the other has unbounded "optimal" solutions.
3. Both problems have no feasible solutions.

We already have an example of the first case: the numerical illustration used throughout this section belongs into that category.

In order to demonstrate the second case, consider the following pair of primal (P) and dual (P_D) problems:

$$\begin{array}{ll}
\text{P : Max } z = 3x_1 + 2x_2 & \text{P}_D : \text{Min } z_D = -u_1 + 2u_2 \\
\text{s.t.} \quad x_1 - x_2 \leq 1 & \text{s.t.} \quad u_1 - 2u_2 \geq 3 \\
\quad\quad -2x_1 + 1x_2 \leq 2 & \quad\quad -u_1 + u_2 \geq 2 \\
\quad\quad x_1, \quad x_2 \geq 0 & \quad\quad u_1, \quad u_2 \geq 0.
\end{array}$$

The graphical representations are shown in Fig. 2.24a (for P) and b (for P_D). It is apparent that while the primal problem has unbounded "optimal" solutions, its dual problem has no feasible solution.

Finally, consider the third case, in which neither problem has a feasible solution. As a numerical example, consider the following pair of primal (P) and dual (P_D) problems:

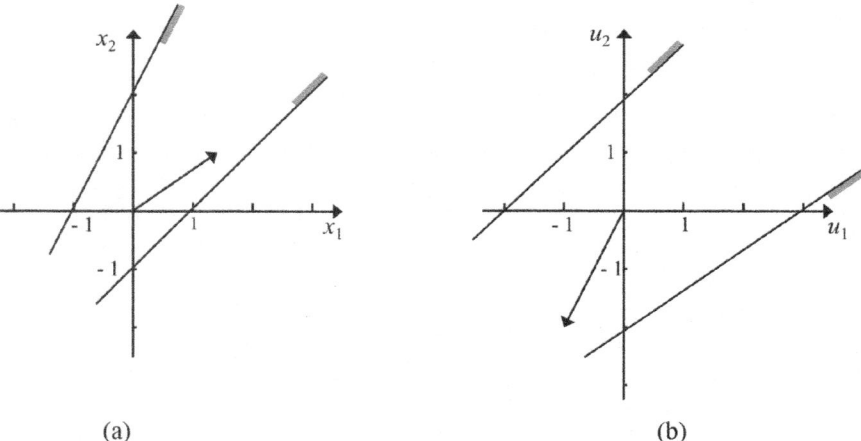

Fig. 2.24 Pairs of dual programs

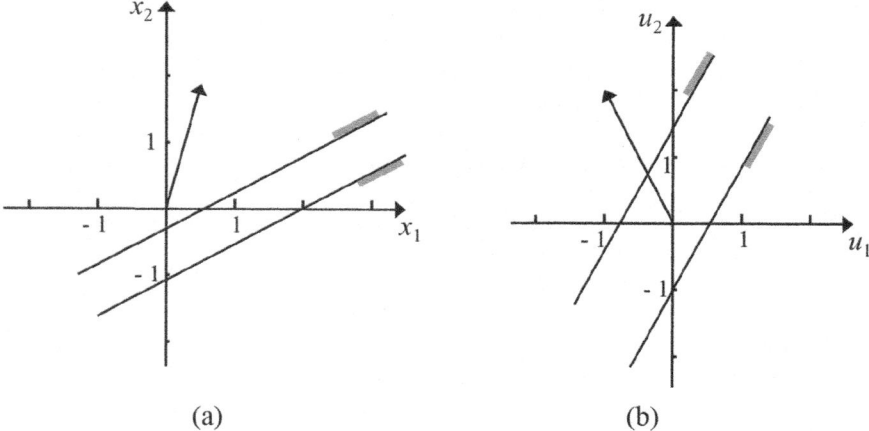

Fig. 2.25 Primal and dual both lack feasible solutions

$$P : \text{Max } z = x_1 + 3x_2 \qquad P_D : \text{Min } z_D = u_1 - 2u_2$$
$$\text{s.t.} \quad 2x_1 - 4x_2 \leq 1 \qquad \text{s.t.} \quad 2u_1 - u_2 \geq 1$$
$$-x_1 + 2x_2 \leq -2 \qquad\qquad -4u_1 + 2u_2 \geq 3$$
$$x_1, \quad x_2 \geq 0 \qquad\qquad u_1, \quad u_2 \geq 0.$$

The graphical representation of the two problems is shown in Fig. 2.25a (for P) and b (for P_D). It is apparent that the constraints in the primal and the dual problems are parallel to each other, so that the feasible set is empty.

2.5 Duality

Exercises

Problem 1 (setting up the dual problem): Consider the following primal linear programming problem P:

$$P : \text{Min } z = -3x_1 + 4x_2$$
$$\text{s.t.} \quad x_1 + 2x_2 = 5$$
$$5x_1 - x_2 \geq 2$$
$$x_1, \quad x_2 \geq 0.$$

Set up the dual problem P_D.

Solution: Since we have not provided any rules for the transformation other than for maximization problems with "$\leq$" constraints, so we simply bring the problem P into that form. Changing a minimization problem to a maximization problem and changing the direction of an inequality is standard. As far as the equation $x_1 + 2x_2 = 5$ is concerned, we replace it by two opposing inequalities $x_1 + 2x_2 \leq 5$ and $x_1 + 2x_2 \geq 5$. We can then define an equivalent version of the problem description P as

$$P' : \text{Max} - z = 3x_1 - 4x_2$$
$$\text{s.t.} \quad x_1 + 2x_2 \leq 5$$
$$-x_1 - 2x_2 \leq -5$$
$$-5x_1 + x_2 \leq -2$$
$$x_1, \quad x_2 \geq 0.$$

The dual of this problem is then

$$P''_D : \text{Min} - z_D = 5u_1 - 5u_2 - 2u_3$$
$$\text{s.t.} \quad 2u_1 - u_2 - 5u_3 \geq 3$$
$$2u_1 - 2u_2 + u_3 \geq -4$$
$$u_1, \quad u_2, \quad u_3 \geq 0.$$

If desired, we can clean up this problem a bit and write it in its equivalent form

$$P'_D : \text{Max } z_D = -5u_1 + 5u_2 + 2u_3$$
$$\text{s.t.} \quad u_1 - u_2 - 5u_3 \geq 3$$
$$-2u_1 + 2u_2 - u_3 \leq 4$$
$$u_1, \quad u_2, \quad u_3 \geq 0.$$

We notice that in the dual problem, the variables u_1 and u_2 always appear with the same coefficients but with opposite signs. We may thus define a new variable $u_{12} = u_1 - u_2$, which can then replace the original variables u_1 and u_2. Note that the new variable u_{12} is now unrestricted in sign. We then obtain the dual

$$P_D : \text{Max } z_D = -5u_{12} + 2u_3$$
$$\text{s.t.} \quad u_{12} - 5u_3 \geq 3$$
$$-2u_{12} - u_3 \leq 4$$
$$u_3 \geq 0.$$

Problem 2 (duality for a single-constraint primal problem): Consider the following single-constraint primal linear programming problem:

$$P : \text{Max } z = 2x_1 + 3x_2 + 8x_3 + 7x_4 + 6x_5$$
$$\text{s.t.} \quad x_1 + 6x_2 + 4x_3 + 3x_4 + 3x_5 \leq 11$$
$$x_1, \quad x_2, \quad x_3, \quad x_4, \quad x_5 \geq 0.$$

(a) Formulate the dual problem P_D. How many variables does it have?
(b) Show that P_D can be solved by simple inspection. State the optimal solution and objective function value of P_D. Explain why $\bar{z} = \bar{z}_D$.
(c) Using duality relationships, find the unique optimal solution to P.

Solution:

(a) The problem under consideration is a continuous knapsack problem similar to those discussed in Chap. 5 of this book. Since there is only a single constraint in the primal problem, the dual problem features only a single variable. The dual is

$$P_D : \text{Min } z_D = 11u$$
$$\text{s.t.} \quad u \geq 2$$
$$6u \geq 3$$
$$4u \geq 8$$
$$3u \geq 7$$
$$3u \geq 6$$
$$u \geq 0.$$

(b) With $z_D = 11u$, the optimal value of the variable u must be as small as possible. Considering the lower bounds specified in the constraints, we determine that $\bar{u} = \max\{2, \frac{3}{6}, \frac{8}{4}, \frac{7}{3}, \frac{6}{3}, 0\} = 2\frac{1}{3}$, so that $\bar{z}_D = 11\bar{u} = 25\frac{2}{3}$. Since P has feasible solutions (for instance, $x_1 = x_2 = \ldots = x_5 = 0$), and $\bar{z}_D$ exists, $\bar{z}$ must exist as well and $\bar{z} = \bar{z}_D = 25\frac{2}{3}$.

(c) Since $3u \geq 7$ is the only constraint in the dual problem P_D that is tight (binding) at optimum, the excess variables of all other dual constraints are strictly positive at optimum. The complementary slackness conditions then require that $\bar{x}_1 = \bar{x}_2 = \bar{x}_3 = \bar{x}_5 = 0$. Since $\bar{u} = 2\frac{1}{3} > 0$, the slack variable of the corresponding primal constraint must be zero at optimum, again due to complementary slackness. The primal constraint is then binding at optimum, so that $3\bar{x}_4 = 11$ or $\bar{x}_4 = 3\frac{2}{3}$.

References

Dantzig GB (1963) Linear programming and extensions. Princeton University Press, Princeton, NJ
Dantzig GB, Thapa MN (1997) Linear programming: introduction. Springer, New York
Eiselt HA, Sandblom C-L (2007) Linear programming and its applications. Springer, Berlin
Garner Garille S, Gass SI (2001) Stigler's diet problem revisited. Oper Res 49:1–13

3 Multiobjective Programming

As diverse as the problems in the previous chapters have been, they share one common feature: they all have one single objective function and the result is an optimal solution (or multiple optima, in case of dual degeneracy). However, the concept of optimality applies only in case of a single objective. If we state that something is "the best" or optimal, we always have an objective in mind: the fastest car, the most comfortable vehicle, the automobile that is cheapest to operate, and so forth. Whenever a second or even more objectives are included in a problem, the concept of optimality no longer applies. For instance, if the top speed of a vehicle and its gas mileage are relevant concerns, then the comparison between a car, whose speed may top 110 miles per hour and which gives 20 miles to the gallon (highway rating) and a vehicle that can go up to 90 miles per hour and which gives 25 miles to the gallon is no longer a simple one: the former car is faster at the expense of fuel efficiency. It will now depend on the decision maker which of the two criteria is considered more important. In other words, the decision maker will—sooner or later—have to specify a *tradeoff* between the criteria. This is the type of problems considered in this chapter.

There are two main reasons that recommend multiobjective optimization, despite its added conceptual and technical difficulties. First, consider a simple linear programming problem. For mathematical reasons, the number of variables that have positive values in an optimal solution cannot exceed the number of constraints. For instance, if we had a diet problem with, say, 10,000 foods but only two constraints, then an optimal diet will have no more than two foodstuffs. This is clearly extreme (not to be confused with the concept of extreme points, which is different, albeit related). We would like to point out that this "extreme" property is particular to linear programming; it does not occur in integer linear or nonlinear programming. Secondly, solutions found with single-objective optimization methods tend to be extreme, as, by definition, they have only a single concern. For instance, if the task is to schedule employees efficiently with a cost minimization objective, then the solutions will completely disregard the convenience of the schedule to the employees and other issues, provided, of course, that there are no constraints to that effect. This is different when multiple objectives are included; their very essence is the concept of compromise.

While the terminology in this field is not quite standardized, we typically refer to problems with multiple concerns (objective functions or criteria) as *multicriteria decision-making (MCDM)* problems. This big umbrella has two main subcategories: problems with multiple objectives and an infinite number of possible decisions (and linear functions) which are referred to as *multiobjective (linear) programming* problems *(MOLP)*, while problems with multiple evaluation criteria and a finite number of possible decisions are usually called *multiattribute decision-making (MADM)* problems. In this chapter, we will exclusively deal with *MOLP*, while Chap. 11 discusses *MADM* problems.

Interestingly enough, *MOLP* have been around almost as long as linear programming problems. Similarly, the replacement of the concept of optimality by the (much weaker) concept of pareto-optimality has been known to economists for more than 100 years. (It is named after the Italian economist Vilfredo Pareto, 1848–1923.) In the sections below, we discuss two of the main approaches to multiobjective linear programming: the *vector optimization* problem and *goal programming* problems. Their main distinction is rooted in the decision maker's input. In vector optimization problems, the decision maker does not provide any input regarding the tradeoff between objectives. This means, of course, that many pareto-optimal solutions will be generated, which then must be compared manually by the decision maker—a process, in which he will have to use some (probably implicit) tradeoffs. Another possibility is to openly define tradeoffs, which then allows the analyst to reduce the problem to a standard linear programming problem. The problem with this approach is that it will likely be impossible for any decision maker to specify with any degree of certainty that one objective is, say, 2.7 times as important as another. Such an approach will have to rely very heavily on sensitivity analyses. Finally, we present *goal programming*, an approach whose main feature is that it blurs the distinction between objectives and constraints.

3.1 Vector Optimization

The most logical way to introduce vector optimization problems starts with linear programming and then extends the analysis to multiple objectives. As usual, we will give preference to intuitive reasoning based on graphical arguments.

As far as the formal statement of a vector optimization problem is concerned, consider the following example that we will also use in this in the next section.

$$\text{Max } z_1 = 3x_1 + x_2$$

$$\text{Max } z_2 = -2x_1 + x_2$$

Fig. 3.1 Single objective function

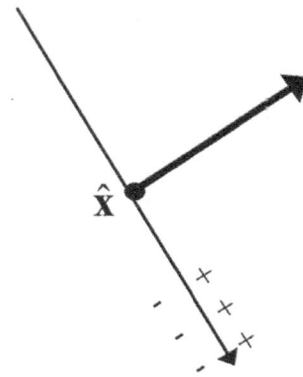

$$\text{s.t.} \ -x_1 + x_2 \leq 3 \quad \text{(I)}$$

$$x_2 \leq 4 \quad \text{(II)}$$

$$x_1 + x_2 \leq 6 \quad \text{(III)}$$

$$x_1 \leq 5 \quad \text{(IV)}$$

$$x_1, x_2 \geq 0.$$

First of all, the name "vector optimization" stems from the fact that rather than a single objective, we have a "vector" of objectives.

For now, consider a single objective, whose gradient of the objective function and iso-profit line through some point "x" are shown in Fig. 3.1.

To recap from our discussion in linear programming, the "z" in Fig. 3.1 indicates the gradient of the objective function, while the line with flags "+ + +" and "– – –" is the iso-profit line through an arbitrary point $\hat{x}$. This iso-profit line subdivides the space into two halfspaces: all points in the halfspace flagged with "+ + +" have objective function values better (i.e., higher for maximization problems and lower for minimization problems) than the point $\hat{x}$, while all points in the halfspace flagged with "– – –" have objective function values worse (i.e., lower for maximization and higher for minimization) than $\hat{x}$.

Consider now a problem with two objective functions. We can then take an arbitrary point $\hat{x}$, anchor both objective function gradients at that point, and plot iso-profit lines through it. This is shown in Fig. 3.2.

Figure 3.2 is based on a problem with two objectives, whose gradients are shown by the arrows marked with z_1 and z_2, respectively, and their two iso-profit lines. The two iso-profit lines subdivide the plane into four parts, labeled with C^{++}, C^{+-}, C^{-+}, and C^{--}. In the following analysis, we will compare points in these four parts of the plane with $\hat{x}$.

First consider any point in the set labeled C^{--}. As compared to $\hat{x}$, this point will be worse than $\hat{x}$ with respective to the first objective and also worse with respect to

Fig. 3.2 Two objective functions

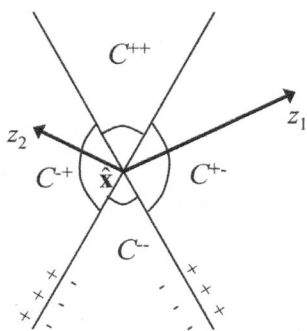

the second objective, as it is located in the intersection of the two halfplanes flagged with "– – –." This means that if we had realized point $\hat{x}$, there would be no reason to move to any point in the set C^{--}, as the new point will be worse than $\hat{x}$ with respect to both objectives. Next, consider a point in the set C^{-+}. Here, things are a bit more difficult, as any point in that set will be worse than $\hat{x}$ with respect to the first objective but better than $\hat{x}$ with respect to the second objective. This means that points in C^{-+} are not comparable to $\hat{x}$.

A similar argument applies to all points in the set C^{+-}. All points in this set are better than $\hat{x}$ with respect to the first objective but worse than $\hat{x}$ with respect to the second objective. So again, no comparison is possible. Finally, consider the set C^{++}. Any point in this set is better than $\hat{x}$ with respect to both objectives, so that we would move out of the present solution $\hat{x}$ into C^{++} whenever possible, i.e., if not restricted to do so by the constraints. As all points in C^{++} are better than $\hat{x}$, we will call C^{++} the *improvement cone* (rooted at $\hat{x}$).

Before using this concept to solve vector optimization problems, it is useful to discuss the relation between objectives. Suppose that two objectives are very similar, for example, Max $z_1 = 2x_1 + 5x_2$ and Max $z_2 = 2x_1 + 6x_2$. It is apparent that there is only little conflict between the two. Such a case is shown in Fig. 3.3a, where the angle between the two gradients is very small. This results in an improvement cone with a very large angle. Actually, in the extreme case of two identical objectives (i.e., no conflict), the improvement cone is then the same as that shown in Fig. 3.1 with the halfplane labeled "+ + +."

On the other hand, consider two objectives that show extensive conflict. This situation is depicted in Fig. 3.3b. Here, the angle between the two gradients is large, and the angle of the improvement cone is very small, indicating that there is only limited potential for improvement. In the limiting case, one gradient would be diametrically opposed to the other, a case of total conflict. In such a case, the improvement cone is empty and there is no potential for any improvement.

However, it may still be possible in such a case to find a compromise after all. One possibility to do so is to introduce additional criteria. As an example, consider a couple who is planning this year's vacation. Suppose that the husband would like to maximize the amount of time the couple spends on the beach to relax (and watch other people), the wife would like to minimize the time on the beach and go boating

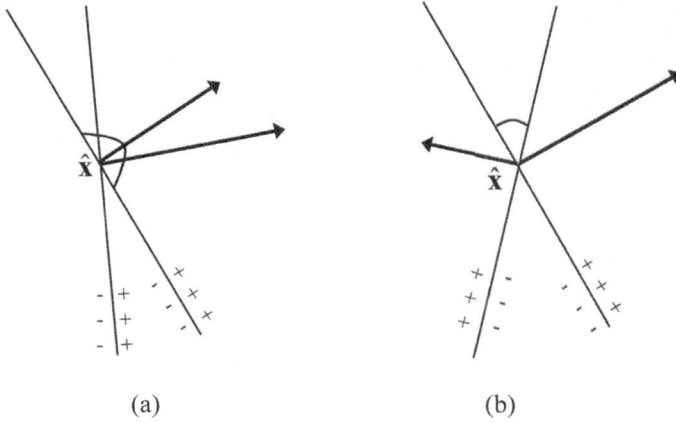

Fig. 3.3 Little (**a**) and extensive (**b**) conflict between objectives

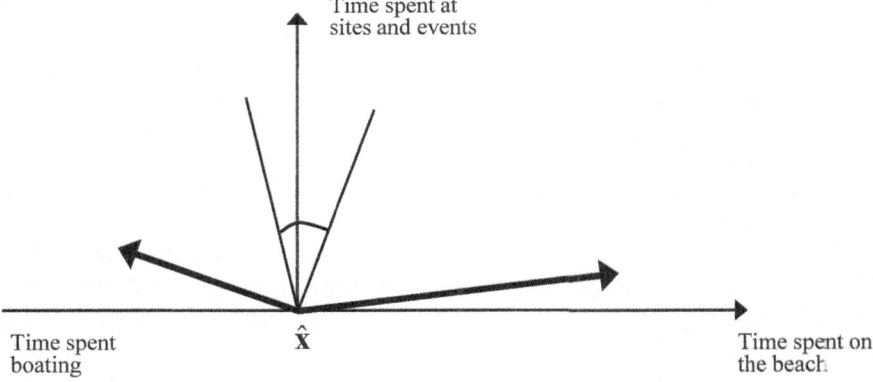

Fig. 3.4 Possibility for compromise

instead, something the husband has absolutely no interest in. (A somewhat similar situation has been dealt with in game theory under the name "battle of the sexes"). This has escalated to a major fight, and as it is, there is no room for compromise. Suppose now that we introduce another activity, e.g., visiting events, festivals, zoos, or museums, which we will call sites and events, something both are interested in, at least to some extent. The objective functions given two dimensions are shown in Fig. 3.4, indicating that there is now an actual possibility for compromise. Similar situations occur in labor negotiations, in which management wants to minimize the amount spent on wages and salaries, while labor wants to maximize it. Issues added to the list of topics to be negotiated could include concerns such as work conditions, something both parties are interested in.

Fig. 3.5 Improvement cones and nondominated frontier

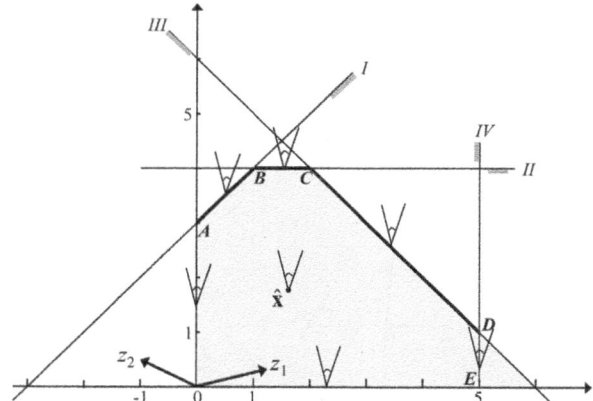

Back to the improvement cone. We stated above that whenever possible, we would try to move out of an existing point $\hat{x}$ into the improvement cone. The only thing that could prevent us from doing so are the constraints. In case it is possible to move out of a feasible point $\hat{x}$ to another feasible point in the improvement cone rooted at $\hat{x}$, we should do so, indicating that the point $\hat{x}$ is not a point to be considered further, as there are other points that are better than $\hat{x}$ with respect to all objectives. In other words, if the intersection of the improvement cone and the feasible set includes at least one more point other than $\hat{x}$, then $\hat{x}$ does no longer have to be considered.

On the other hand, if moving out of $\hat{x}$ into its improvement cone will always result in the loss of feasibility, then the point $\hat{x}$ is called *nondominated* (or, alternatively, *noninferior*, *efficient*, or *pareto-optimal*). We will use these terms interchangeably. The collection of all noninferior points is called the *nondominated frontier*. It is apparent that points in the interior of the feasible set cannot be nondominated, as it is possible to move out of them into any direction and stay feasible, at least for some sufficiently small distance. This is shown in Fig. 3.5, in which the feasible set of our example problem is shaded, and the extreme points are *0, A, B, C, D*, and *E*. The gradients of the two objectives are labeled z_1 and z_2, and improvement cones are shown anchored at all extreme points as well as at some interior point $\hat{x}$. From any of these points, we will try to move into its improvement cone and determine whether or not it is possible to do so and stay within the feasible region. Clearly, at the interior point $\hat{x}$, this is feasible.

Consider now the line segments that border the feasible set. All points on the line segment (*O, A*) are dominated, as it is possible from any of them to move into the improvement cone and stay feasible. The same is true for all points on the line segment (*O, E*) and the segment (*D, E*). On the other hand, all points on each of the line segments (*A, B*), (*B, C*), and (*C, D*) is nondominated, as, the moment that we move into the improvement cone constructed with the point at its vertex, we lose feasibility. The result of our analysis is then the nondominated frontier, shown in Fig. 3.5 by bold lines.

It is now possible to prove that the nondominated set is connected. The union of all nondominated line segments is the *nondominated frontier*. All points on this

3.2 Solution Approaches to Vector Optimization Problems

frontier are of interest to the decision maker, whose task is now to determine which of these solutions to choose. The choice will depend on criteria other than those already included in the model.

3.2 Solution Approaches to Vector Optimization Problems

This section will examine techniques that can be used to approximate the efficient frontier. While it is possible to use a modified simplex method to determine all extreme points on the efficient frontier, this is not only a lengthy process, but also something that leaves the decision maker with tons of solutions to manually compare. This is clearly out of the question. As a result, analysts typically determine a few solutions, and based on the decision maker's response to those, will generate more solutions that attempt to reflect the decision maker's comments.

Two methods for this purpose stand out. One is the *weighting method*, and the other is called the *constraint method*. The basic idea of the weighting method is to first assign positive weights w_1, w_2, ..., w_p to the p given objectives, and then aggregate them into a single new *composite objective*. The result is then a linear programming problem that can easily be solved. It is then possible to prove that an optimal solution to this linear programming problem is always a point on the nondominated frontier.

As a numerical illustration, consider the example introduced in the previous section. Recall that the two objectives were Max $z_1 = 3x_1 + x_2$ and Max $z_2 = -2x_1 + x_2$. Suppose that we choose $w_1 = 5$ and $w_2 = 1$. This means that one unit of whatever the first objective measures is considered five times as important than one unit of what the second objective measures. Using these weights, the composite objective is then Max $z = w_1z_1 + w_2z_2 = 5(3x_1 + x_2) + 1(-2x_1 + x_2) = 13x_1 + 6x_2$. Using this objective in conjunction with the constraints of the problems results in the optimal solution $\bar{x}_1 = 5$, $\bar{x}_2 = 1$ (which is point D in Fig. 3.5), with an objective value of $\bar{z} = 71$. For our purposes, the value of the aggregated objective function is of no interest, it is more useful to take the solution obtained in the optimization and insert the optimal values into the two individual objective functions, resulting in $\bar{z}_1 = 16$ and $\bar{z}_2 = -9$. Table 3.1 provides a listing of different selected weight combinations and the nondominated points they generate. Note that all nondominated solutions that this technique generates are at extreme points.

Table 3.1 Nondominated solutions generated by the weighting method

(w_1, w_2)	$z = w_1z_1 + w_2z_2$	$(\bar{x}_1, \bar{x}_2)$	$(\bar{z}_1, \bar{z}_2)$
(5, 1)	$13x_1 + 6x_2$	$D = (5, 1)$	$(16, -9)$
(3, 1)	$7x_1 + 4x_2$	$D = (5, 1)$	$(16, -9)$
(1, 1)	$x_1 + 2x_2$	$C = (2, 4)$	$(10, 0)$
(1, 3)	$-3x_1 + 4x_2$	$B = (1, 3)$	$(7, 2)$
(1, 5)	$-7x_1 + 6x_2$	$A = (0, 3)$	$(3, 3)$

Fig. 3.6 The weighting method

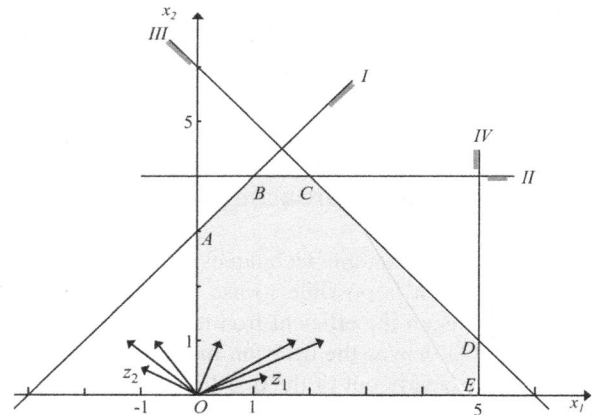

A graphical representation for our numerical problem is shown in Fig. 3.6. Notice the objectives z_1 and z_2 as limits, the gradients of all of their selected combinations are between them. In particular, the gradients shown here are z_1, those generated by the weight combinations (5, 1), (3, 1), (1, 1), (1, 3), (1, 5), and z_2, in counterclockwise direction.

The second approach to generate nondominated solutions is the *constraint method*. It can be described as a technique that keeps one of the objective functions, while using the others as constraints with variable right-hand side values. This is done by first designating one of the p objectives as objective, while all others are reformulated as constraints. Each objective of the type "Max z_k" will be rewritten as a constraint $z_k \geq b_k$ with a yet-to-be-determined value of b_k, while each original objective of the type "Min z_k" will be rewritten as a constraint of the type $z_k \leq b_k$ with variable values of b_k. The resulting linear programming can then easily be solved. The solution process is repeated for a number of combinations of selected values b_k, $k = 1, \ldots, p$. All solutions that are generated in this fashion are nondominated. However, these solutions are not necessarily extreme points of the feasible set.

As an illustration, consider again the example introduced in the previous section. We will (arbitrarily) retain the first objective as an objective and reformulate the second objective as a constraint. The problem can then be written as

$$\text{Max } z_1 = 3x_1 + x_2$$
$$\text{s.t.} \quad -x_1 + x_2 \leq 3$$
$$x_2 \leq 4$$
$$x_1 + x_2 \leq 6$$
$$x_1 \leq 5$$
$$-2x_1 + x_2 \geq b_2$$
$$x_1, \ x_2 \geq 0.$$

Table 3.2 Nondominated solutions determined with the constraint method

b_2	$(\bar{x}_1, \bar{x}_2)$	$\bar{z}_1$
5	No feasible solution	
0	(2, 4)	10
−1	(2⅓, 3⅔)	10⅔
−2	(2⅔, 3⅓)	11⅓
−3	(3, 3)	12
−4	(3⅓, 2⅔)	12⅔
−5	(3⅔, 2⅓)	13⅓
−6	(4, 2)	14
−7	(4⅓, 1⅔)	14⅔
−8	(4⅔, 1⅓)	15⅓
−9	(5, 1)	16
−10	(5, 1)	16

At this point, we can solve the problem for a variety of values of b_2. A summary of solutions for some chosen values is displayed in Table 3.2.

For $b_2 = 0$, we obtain point *C* in Fig. 3.5, and for $b_2 = -1, -2, \ldots, -8$, points between *C* and *D* are found, while $b_2 = -9$ and $b_2 = -10$ result in point *D*. Note that this selection of b_2 values misses the nondominated solutions *A* and *B*. This is why both, the weighting method and the constraint methods, are referred to as approximation methods. As a matter of fact, using $b_2 = 1$ results in the solution $(\bar{x}_1, \bar{x}_2) = (1\frac{1}{2}, 4)$, which is the point halfway between *B* and *C*, a value of $b_2 = 2$ generates the solution $(\bar{x}_1, \bar{x}_2) = (1, 4)$, which is point *B*, a value of $b_2 = 3$ results in a solution $(\bar{x}_1, \bar{x}_2) = (0, 3)$, which is point *A*, and for values in excess of $b_4 = 4$, the problem has no feasible solution. Figure 3.7 shows the feasible set of the problem and the constraints based on the second objective for various values of b_2. The bold points are the resulting optimal solutions.

3.3 Goal Programming

When introducing goal programming, it is useful to return to the basic discussion in Chap. 1 about constraints and objective functions. Recall that constraints express requirements that *must* be satisfied, while objective functions are for requirements that *should* be satisfied, if possible. While this distinction appears clear, the difference between "required" and "desired" is blurred in reality. Consider a simple budget constraint that expresses the condition that we cannot spend more than we have. While it may not be a wise choice, we could consider the possibility to borrow money. Consider a variety of similar examples: All offices must fit into the space that we own—but we can rent some more space. Do not use more employees than are available—but that is what temp agencies are there for. Payments are due on a specific date—but we may be able to defer them. All of this is introduced into the discussion to demonstrate that many requirements are much softer than they appear. And this is why modelers should take precautions before formulating constraints, as

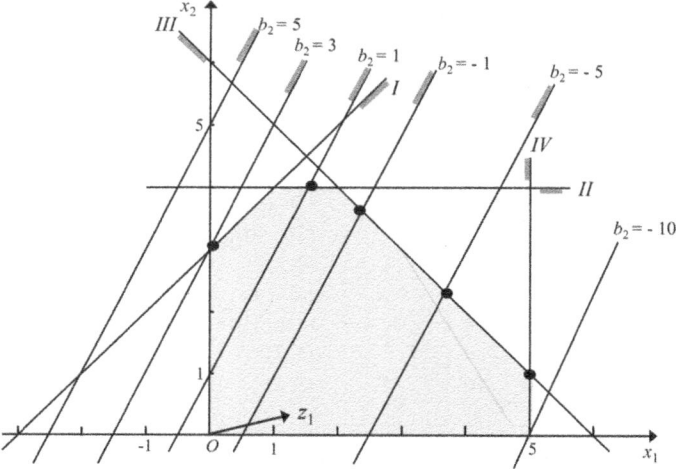

Fig. 3.7 The constraint method

Table 3.3 Formulation of soft constraints as goal constraints

Desired situation	Formulation of goal constraint	Contribution to the objective function
LHS $\leq$ RHS	LHS + d_ℓ^- − d_ℓ^+ = RHS	Min d_ℓ^+
LHS = RHS	$d_\ell^-, d_\ell^+ \geq 0$	Min $d_\ell^- + d_\ell^+$
LHS $\geq$ RHS		Min d_ℓ^-

they are absolute: if they cannot be satisfied by the given data, the solver will return a message indicating that there is no feasible solution.

Goal programming is one tool that attempts to deal with "soft constraints." The general idea was developed by the later Nobel laureate Herbert A. Simon in 1957, who introduced the concept of *satisficing*, a composite word that joins the concepts of "satisfy" and "suffice." The concept as applied to goal programming works as follows. All regular (absolute) constraints are written as usual. Continuing, a soft constraint is first formulated as a regular constraint. It is then reformulated as a goal constraint with the help of a *target value* (or *aspiration level*) and *deviational variables*. The target value is a number that expresses how many resources we have, what output we would like to achieve, or similar measures. The nonnegative deviational variables d_ℓ^+ and d_ℓ^- measure over- and underachievements, respectively. The way to formulate goal constraints is to first write the requirement as a regular constraint, and then reformulate it as shown in Table 3.3.

The deviational variable d_ℓ^- is similar to a slack variable, and the deviational variable d_ℓ^+ resembles an excess variable. While it appears counterintuitive that slack and excess variables should appear in the same constraint, it really is not. As an example, consider a budget constraint that states that the actual expenditures should not exceed the available amount. First of all, the appearance of slack and surplus in this constraint indicates that the actual expenditures may be smaller than the available budget (underspending) or may exceed the actual budget (overspending). Secondly, technical reasons will ensure that at most one of the deviational variables can be positive, so that we cannot have over- and underspending at the same time.

As a numerical example, suppose that we have $100 that we can spend on two items, food and entertainment (*panem et circensis*, as the Romans would have it). The amounts spent on the two items are x_1 and x_2, respectively, and the budget constraint would be $x_1 + x_2 \leq 100$. Reformulating it as a goal constraint leads to $x_1 + x_2 + d_\ell^- - d_\ell^+ = 100$. Suppose now that we have decided to spend $60 of food and $30 on entertainment. This means that the goal constraint is then $90 + d_\ell^- - d_\ell^+ = 100$ or $d_\ell^- - d_\ell^+ = 10$. Given that the deviational variables must satisfy the nonnegativity constraints, and that at most one of them can be positive, this implies that $d_\ell^- = 10$. The meaning is that the present budget is "underachieved" by $10, or, in more standard terms, there are $10 left over.

If, on the other hand, we spend a total of, say, $90 on food and $30 on entertainment, the goal constraint reads $120 + d_\ell^- - d_\ell^+ = 100$ or $d_\ell^- - d_\ell^+ = -20$. Again, given the nonnegativity of the deviational variables, the result is that $d_\ell^+ = 20$, indicating an "overachievement" (or, similarly, overspending) of the budget by $20.

The last column in Table 3.3 then indicates the contribution to the overall objective made by the deviational variables introduced in a goal constraint. In the aforementioned budget constraint, the relation "actual amount spent $\leq$ amount available" was desired, so that goal programming, after rewriting the constraint as a goal constraint, will minimize the overachievement, i.e. overspending. Note that this approach does justice to the "softness" of the budget constraint by allowing overspending but trying to minimize it. In practice, absolute or rigid constraints and soft constraints can often be distinguished by the way the requirements are worded. A telltale sign is the expression "if possible." Whenever it is appended to a requirement, it clearly indicates that formulation as a goal constraint is in order.

The next issue is then how to aggregate the deviational constraints into a single objective function. The original version of goal programming has multiple levels, each of them assumed to be infinitely more important than the next. This structure has been criticized profusely in the literature, even though the principle is common, even in linear programming: the absolute constraints are infinitely more important than the objective. This can be seen that if a constraint cannot be satisfied, we will obtain the signal "there exists no feasible solution" from the solver, regardless how good the objective function value might or could be.

In this book, we will restrict ourselves to a single level, on which we aggregate the deviational variables similar to the way we aggregated objective functions in the weighting method in the previous section of this chapter. The problem with such a procedure is commensurability. For instance, if one deviational variable expresses the overexpenditure of the budget (measured in dollars), while another expresses the underuse of manpower (measured in the number of employees), we cannot simply add these two together. By using weights, we express a tradeoff between the units, so that a weight, which is multiplied by the overexpenditure of the budget, will have to express the importance of one dollar of overexpenditure in relation to the underuse of one employee.

In order to illustrate the modeling process, consider the following numerical

Example The owner of a chain of jewelry stores has to decide how to distribute parts of a new shipment of diamonds to five stores in a region. The first three stores of the chain are located in shopping malls. The following conditions have to be observed.

Absolutely necessary:

(a) Allocate between 1000 and 1200 carats in total to the five stores.
(b) Store 5 must receive at least 300 carats of diamonds.

Desired properties of the allocation:

(c) The stores in the malls should receive at least 80% of all the diamonds, if possible.
(d) The allocations to the stores in the malls should be equal to each other, if possible.
(e) The probabilities of theft in the stores have been estimated to be 0.1%, 0.1%, 9%, 2%, and 3%, respectively. The owner would like to minimize the expected loss.

Requirement (e) takes priority in the list and is considered to be 25 times as important as requirement (d), which, in turn, is considered twice as important as (c).

In order to formulate the problem, we first define decision variables $x_1, x_2, \ldots, x_5$ as the quantity of diamonds allocated to stores 1, 2, 3, 4, and 5, respectively. The absolute constraints (a) and (b) can then be written as

$$x_1 + x_2 + x_3 + x_4 + x_5 \geq 1000 \quad (3.1)$$

$$x_1 + x_2 + x_3 + x_4 + x_5 \leq 1200 \quad (3.2)$$

$$x_5 \geq 300. \quad (3.3)$$

Consider now requirement (c). Written as a constraint, the requirement can be formulated as $x_1 + x_2 + x_3 \geq 0.8(x_1 + x_2 + x_3 + x_4 + x_5)$ or, equivalently, as

$$0.2x_1 + 0.2x_2 + 0.2x_3 - 0.8x_4 - 0.8x_5 \geq 0.$$

Rewriting the requirement as a goal constraint with deviational variables, we obtain

$$0.2x_1 + 0.2x_2 + 0.2x_3 - 0.8x_4 - 0.8x_5 + d_1^- - d_1^+ = 0 \quad (3.4)$$

with Min d_1^- as the contribution to the objective function.

3.3 Goal Programming

Next consider requirement (d). The average allocation to a store in the mall is $\frac{1}{3}(x_1 + x_2 + x_3)$, so that we would like to see $x_1 = \frac{1}{3}(x_1 + x_2 + x_3)$, $x_2 = \frac{1}{3} \times (x_1 + x_2 + x_3)$, and $x_3 = \frac{1}{3}(x_1 + x_2 + x_3)$. Rewriting the first of these constraints results in $\frac{2}{3}x_1 - \frac{1}{3}x_2 - \frac{1}{3}x_3 = 0$. As a goal constraint, we write

$$\frac{2}{3}x_1 - \frac{1}{3}x_2 - \frac{1}{3}x_3 + d_2^- - d_2^+ = 0 \tag{3.5}$$

with the objective function contribution Min $d_2^- + d_2^+$. Similarly, we obtain the other two goal constraints

$$-\frac{1}{3}x_1 + \frac{2}{3}x_2 - \frac{1}{3}x_3 + d_3^- - d_3^+ = 0 \tag{3.6}$$

and

$$-\frac{1}{3}x_1 - \frac{1}{3}x_2 + \frac{2}{3}x_3 + d_4^- - d_4^+ = 0 \tag{3.7}$$

with the objective function contributions Min $d_3^- + d_3^+$ and Min $d_4^- + d_4^+$, respectively.

Finally, consider requirement (e) The original objective is written as

$$\text{Min } z = 0.001x_1 + 0.001x_2 + 0.09x_3 + 0.02x_4 + 0.03x_5.$$

Setting the expected loss at some unattainably low level, e.g., $z = 0$, we can then require that the expected loss is no larger than that level, if possible (which it cannot, so that we minimize the overachievement). This is then written as

$$0.001x_1 + 0.001x_2 + 0.09x_3 + 0.02x_4 + 0.03x_5 + d_5^- - d_5^+ = 0 \tag{3.8}$$

with the objective function contribution Min d_5^+.

The problem can then be written as

$$\text{Min } z = 50d_5^+ + 2\left(d_2^- + d_2^+\right) + 2\left(d_3^- + d_3^+\right) + 2\left(d_4^- + d_4^+\right) + d_1^-$$

s.t. constraints (3.1)–(3.8) and the nonnegativity constraints for all variables.

The optimal solution allocates 466.69 carats of diamonds to store 1, another 233.31 carats of diamonds to store 2, no diamonds to stores 3 and 4, and the minimally required 300 carats to store 5. The total allocation is 1000 carats, the lower bound of the (absolute) constraints. The goal constraint with the highest weight (the expected loss) is measured by the overachievement d_5^+, whose optimal value is 9.7. The second-ranking goal constraint that attempts to make the mall allocations equal is violated by 466.67 carats, and the third-ranking goal constraint that wants at least 80% of the allocations in malls is violated by 10 percentage points.

Exercises

Problem 1 (improvement cone): Consider the following two vector optimization problems (to simplify matters, the constraints have been ignored):

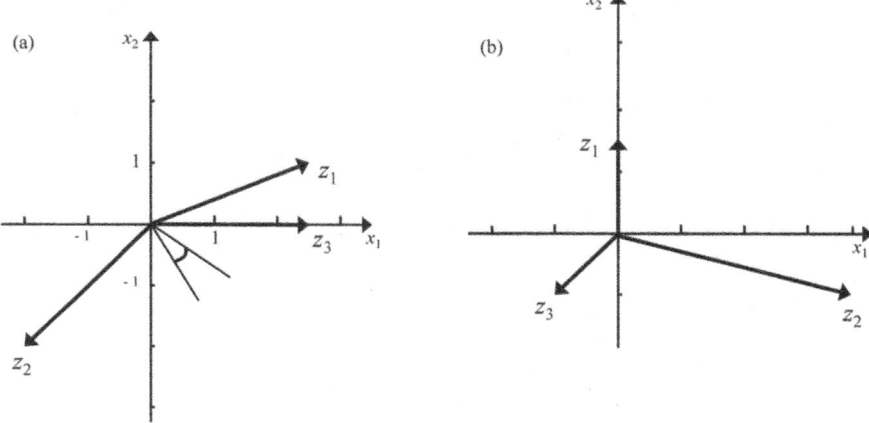

Fig. 3.8 Improvement cones

(a) P_1 : Max $z_1 = 5x_1 + 2x_2$, Max $z_2 = -2x_1 - 2x_2$, Max $z_3 = 3x_1$, and
(b) P_2 : Max $z_1 = x_2$, Max $z_2 = 4x_1 - x_2$, Min $z_3 = x_1 + x_2$.

Plot each of these two problems individually and determine the improvement cone.

Solution: The improvement cone for (a) is shown in Fig. 3.8a, the improvement cone for (b) is empty as shown in Fig. 3.8b.

Problem 2 (nondominated frontier and composite objective, graphical): Consider the following linear programming problem:

$$P : \text{Max } z_1 = x_1 + 2x_2$$
$$\text{s.t.} -2x_1 + x_2 \leq 2$$
$$x_1 + x_2 \leq 5$$
$$x_1 \leq 3$$
$$x_1, x_2 \geq 0.$$

(a) Plot the constraints and determine the feasible set.
(b) Graph the gradient of the objective function and use the graphical solution technique to determine the optimal point. Compute the exact coordinates of the optimal point and its value of the objective function.
(c) Consider a second objective function Max $z_2 = 2x_1 - x_2$. Ignoring the first objective, what is the optimal point? Compute its exact coordinates and its value of the objective function.
(d) Determine the nondominated frontier given the two objectives.
(e) Use the two objectives above to construct the composite objective function with weights $w_1 = ¾$ and $w_2 = ¼$. What is the optimal solution with this objective?

Fig. 3.9 Composite objective function and efficient frontier

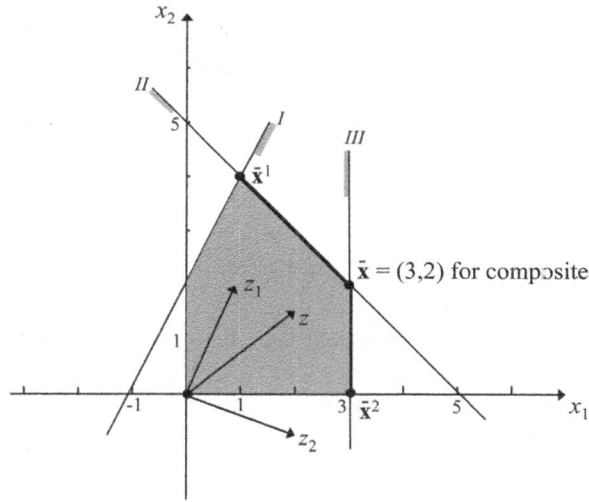

Solution:

(a) The solutions are based on Fig. 3.9.
(b) The exact coordinates of the optimal solution are $\bar{x}^1 = (1,4)$ with value of the objective function $\bar{z}_1 = 9$.
(c) The optimal solution with the second objective is $\bar{x}^2 = (3,0)$ with $\bar{z}_2 = 6$.
(d) The efficient frontier is shown by the bold line in Fig. 3.9.
(e) The composite objective function is Max $z = \frac{5}{4}x_1 + \frac{5}{4}x_2$, it is shown as the gradient labeled "z" in Fig. 3.9. The optimal solution with the composite objective is $\bar{x} = (3,2)$, and the values of the objective functions at that point are $\bar{z}_1 = 7$ and $\bar{z}_2 = 4$.

Problem 3 (vector optimization, nondominated frontier): Consider the following vector optimization problem:

$$P : \text{Min } z_1 = x_1 + x_2$$
$$\text{Max } z_2 = 2x_1 + x_2$$

$$\text{s.t. } x_1 \leq 3$$
$$x_2 \leq 2$$
$$-x_1 + x_2 \leq 1$$
$$x_1, x_2 \geq 0.$$

Fig. 3.10 Solution for Problems 3(a) and 3(c)

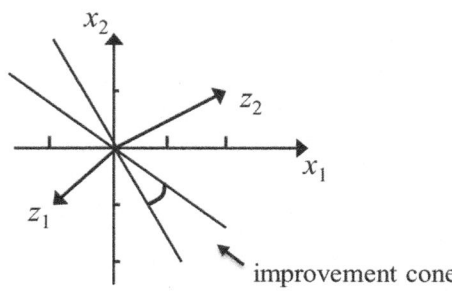

(a) Graph the directions of both objective functions and determine the improvement cone.
(b) Graph the constraints and clearly indicate the feasible set. Plot the improvement cone at each outside boundary of the feasible set and determine the efficient frontier. Clearly describe the efficient frontier.

Solution:

(a) Figure 3.10 shows the gradients of the two objective functions and the improvement cone.
(b) Figure 3.11 shows the feasible set (the edges of the set are shaded in gray), the improvement cones at each boundary segment of the feasible set. The nondominated set is shown as the bold line segments (0, 0) to (3, 0) and (3, 0) to (0, 2).

Problem 4 (weighting method): Consider again the vector optimization problem in Problem 3.

(a) Use the weighting method with weight combinations $\mathbf{w} = (5, 1), (3, 1), (1, 1), (1, 3)$, and $(1, 5)$ to determine nondominated solutions.
(b) Use the constraint method by keeping the first objective and using the second objective as constraint with a variable right-hand side value b_2.
(c) Repeat (b) by keeping the second objective and using the first objective as a constraint with variable right-hand side b_1.

Solution:

(a) The different weight combinations result in the solutions shown below.

Weights **w**	Composite objective	Solution $\bar{\mathbf{x}}$
5, 1	Max $z = -3x_1 - 4x_2$	0, 0
3, 1	Max $z = -1x_1 - 2x_2$	0, 0
1, 1	Max $z = 1x_1$	3, 0 and 3, 2
1, 3	Max $z = 5x_1 + 2x_2$	3, 2
1, 5	Max $z = 9x_1 + 4x_2$	3, 2

3.3 Goal Programming

Fig. 3.11 Improvement cones and nondominated frontier

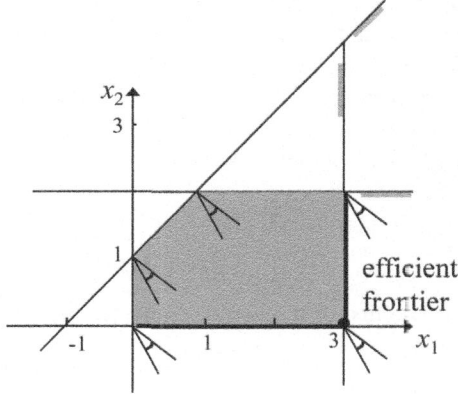

(b) The method generates the noninferior solutions shown below for various values of b_2 in the constraint $2x_1 + x_2 \geq b_2$.

b_2	$\bar{\mathbf{x}}$	$\bar{z}_1$
-10	0, 0	0
0	0, 0	0
5	2.5, 0	2.5
6	3, 0	3
7	3, 1	4
8	3, 2	5
9	No feasible solution	

(c) The method generates the noninferior solutions shown below for various values of b_1 in the constraint $x_1 + x_2 \leq b_1$.

b_1	$\bar{\mathbf{x}}$	$\bar{z}_2$
10	3, 2	8
5	3, 2	8
4	3, 1	7
3	3, 0	6
2	2, 0	4
1	1, 0	2
0	0, 0	0
-1	No feasible solution	

Problem 5 (goal programming formulation): A good B is to be blended from three ingredients $I_1, I_2,$ and I_3. Firm requirements dictate that at least 100 lbs of B are to be blended, and that the average cost of the blend per pound do not exceed \$2.80, given that one pound of the three ingredients costs \$5, \$3, and \$2 for $I_1, I_2,$ and I_3, respectively. In addition, it would be desirable if 20% of B were to be I_1. Similarly, the decision maker would like to have B consist of no more than 50% of the cheap ingredient I_3, if possible. This desirable feature is about half as important as the former desirable feature. Formulate as a goal programming problem.

Solution: As there is only a single product, we only need variables with a single subscript. Define variables x_1, x_2, and x_3 as the quantities of the three respective ingredients in the product. The constraints can then be written as

$$x_1 + x_2 + x_3 \geq 100 \text{ and}$$

$$5x_1 + 3x_2 + 2x_3 \leq 2.8(x_1 + x_2 + x_3).$$

The first of the two desirable properties, written as a constraint is

$$x_1 \geq 0.2(x_1 + x_2 + x_3).$$

Rewriting as a goal constraint, we obtain

$$x_1 + d_1^- - d_1^+ = .2(x_1 + x_2 + x_3)$$

with objective function contribution Min d_1^-. The latter desirable property, written as a constraint, is

$$x_3 \leq .5(x_1 + x_2 + x_3).$$

Rewritten as a goal constraint, we obtain

$$x_3 + d_2^- - d_2^+ = .5(x_1 + x_2 + x_3)$$

with objective function contribution Min d_2^+. The objective function is

$$\text{Min } z = 2d_1^- + d_2^+.$$

Solving the problem results in 15, 35, and 50 lbs of the three ingredients being used, so that exactly 100 lbs are blended, whose price is exactly equal to the required value of 2.8. It is apparent that the product includes 15% I_1, 5% short of the desired target. On the other hand, the upper limit of 50% I_3 is satisfied as equation.

Nonlinear Programming 4

In the Introduction to Linear Programming in Sect. 2.1 in this volume, we outlined that the objective function(s) and the constraints in linear programming are assumed to be linear functions in the variables. In this chapter, we drop this assumption and only assume divisibility and the deterministic property. Given that, we can view nonlinear programming as a generalization of linear programming. Another important distinction between linear and nonlinear programming is that in nonlinear programming, constraints are not necessarily needed to ensure finite optima as is the case in linear programming. For instance, the nonlinear objective

$$\text{Min } z = x^2 - 6x + 4y^2 - 16y + 21$$

has the optimal solution $(\bar{x}, \bar{y}) = (3, 2)$, which is at the center of the ellipse with major axis of 4 and minor axis of 2. Given this, it stands to reason to first investigate unconstrained nonlinear optimization, before constraints are then introduced later.

We will start our discussion with an introductory section, followed by a number of applications of nonlinear programming. The third section focuses on formal aspects, and Sect. 4.4 describes techniques for the solution of unconstrained and constrained nonlinear programs. An in-depth discussion of nonlinear programming can be found in Eiselt and Sandblom (2019).

4.1 Introduction

The historical roots of nonlinear programming are difficult to isolate since the study of many nonlinear mathematical problems can be put into the context of nonlinear optimization. Noteworthy are the Newton (1643–1727) method for finding roots of polynomials (used in unconstrained optimization, to be discussed in Sect. 4.4 of this chapter), as well as the introduction of Lagrange multipliers (Lagrange, 1736–1813) for constrained problems (closely related to the dual variables in linear programming). As a separate field of study, nonlinear optimization developed quickly since the 1950s, roughly in lockstep with linear programming. Particularly noteworthy are the contributions by Kuhn and Tucker (1951) with some of their work based on a contribution by Karush (1939); see also Davidon (1959), Fletcher and Powell (1963), and many others.

To formalize our discussion, we will write a general nonlinear optimization problem as

$$\text{Min } z = f(x_1, x_2, \ldots, x_n)$$

$$\text{s.t. } g_i(x_1, x_2, \ldots, x_n) \; R_i \; b_i \quad \text{for} \quad i = 1, \ldots, m,$$

where f and g are some functions, R_i are relations of the type $\leq$, $=$, or $\geq$, and b_i are right-hand side constants. Clearly, in case the functions f and g_i, $i = 1, \ldots, m$ are all linear, we are dealing with a linear programming problem. To simplify the notation, we will write nonlinear programming problems as Min $z = f(x)$, s.t. $g(x) \; R \; b$. Note that we typically write nonlinear programs as minimization problems, whereas we usually maximize in linear programming. Also, we usually assume that the relations in nonlinear programming are of the type $\leq$. This is not restrictive, of course, as each maximization problem can be converted into an equivalent minimization problem, just as $\geq$ constraints can be converted to equivalent $\leq$ constraints. Some specific transformations are as follows:

- If a variable x_j is required to be nonnegative, we simply add the constraint $-x_j \leq 0$,
- a constraint $g_i(x) \leq b_i$ with a nonzero right-hand side b_i can be rewritten as $g_i(x) - b_i \leq 0$,
- a constraint $g(x) = 0$ can be replaced by the two opposing constraints $g(x) \leq 0$ and $-g(x) \leq 0$.

As a very simple example of a nonlinear optimization model, consider a single product, whose price depends on the quantity that is produced and sold. More specifically, let p denote the unit price of the product, let a and b be two positive parameters, and denote by x the quantity that is produced and sold. The price–quantity relation could then be $p = a - bx$, indicating that the more of the good we produce, the lower its unit price will be. The revenue function is then $R(x) = (a - bx)x$, which, if coupled with a simple cost function $C(x) = cx$ with per-unit costs of c, results in the profit function $P = (a - bx)x - cx$, where we have to ensure that we only consider quantities, which are (1) nonnegative, and (2) for which the price remains nonnegative. This results in the nonlinear optimization problem

$$\text{P} : \text{Max } z = (a - c)x - bx^2$$

$$\text{s.t. } 0 \leq x \leq a/b.$$

The problem can be visualized in Fig. 4.1, where $R(x)$ is the revenue function, $C(x)$ the cost function, and $z = P(x)$ denotes the profit function.

4.1 Introduction

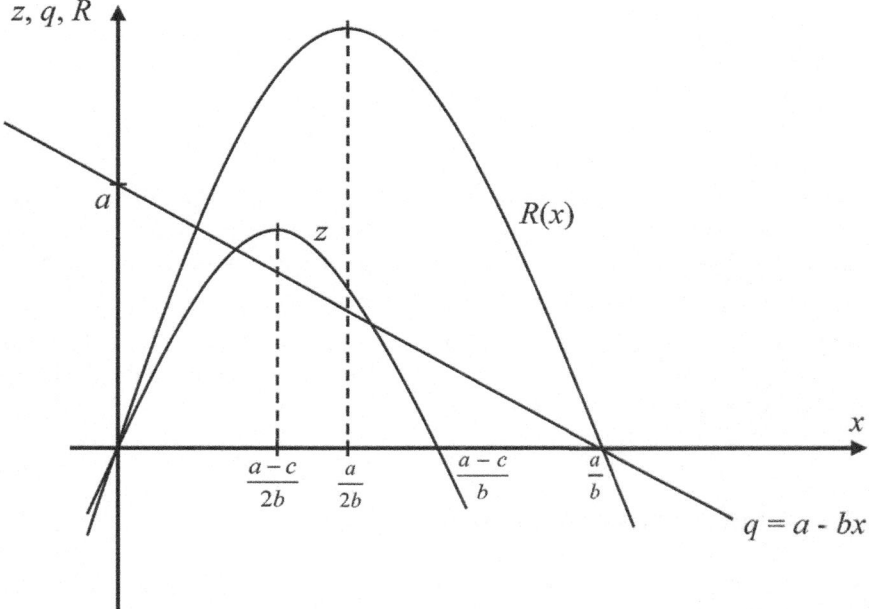

Fig. 4.1 Revenue and profit functions

Fig. 4.2 Open cylindrical tank

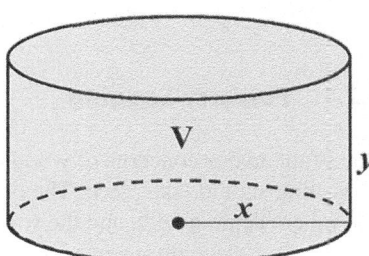

Simple differential calculus reveals that the point of maximal revenue is $\widehat{x} = \frac{a}{2b}$, whereas the point of maximal profit is $\bar{x} = \frac{a-c}{2b}$, which is always lower than the point of maximal revenue. Furthermore, in order to ensure nonnegative profits, we need to have $0 \leq x \leq \frac{a-c}{b}$.

Another simple example concerns the problem of designing the shape of an open cylindrical tank, given a prescribed volume V. Assuming that the material for the bottom of the tank and its sides is the same, we wish to determine the shape of the tank that minimizes the material used. In order to formulate the problem, denote by x the radius and by y the height of the cylinder. This can be visualized in Fig. 4.2.

The volume of the cylinder can then be expressed as $V = \pi x^2 y$. The area of the bottom is πx^2, and the area of the side is $2\pi xy$, so that the problem can be written as

$$P : \text{Min } z = \pi x^2 + 2\pi xy$$

$$\text{s.t. } \pi x^2 y = V,$$

$$x, y \geq 0,$$

which is a nonlinear optimization problem with the two variables x and y. Solving the equality constraint for y and replacing the variable in the objective function leads to the objective Min $z = \pi x^2 + 2Vx^{-1}$. The function has a unique optimum at $\bar{x} = \sqrt[3]{\frac{V}{\pi}}$. Substituting this value in the volume equation, we obtain $\bar{y} = \bar{x} = \sqrt[3]{\frac{V}{\pi}}$. In other words, for an optimal open cylinder, its height should equal its radius. The next section will examine a number of applications that demonstrate the flexibility and usefulness of nonlinear formulations.

4.2 Applications of Nonlinear Programming

This section presents some selected applications of nonlinear programming. They are somewhat closer to reality as compared to the examples presented in the introductory section. For a more detailed discussion, readers are referred to Eiselt and Sandblom (2019).

4.2.1 Forestry Logging

One of the major concerns of woodlot management deals with the question of when harvesting should take place. The optimal time to harvest will depend on the type of tree we are dealing with, and the soil and climate conditions. To simplify matters, we assume that clearcutting is used, and monoculture has been chosen, i.e. we will replant with a single species. We also assume that our objective is to maximize the average annual biomass harvested from the trees, taken over an infinite planning horizon. As soon as the parcel is harvested, we will replant, and growth starts again. Although simplified, the model should shed some light on the planning process in forest management. Typically, the useful biomass of a tree develops over time according to an "S"-shaped sigmoidal function such as the one shown in Fig. 4.3. Starting from zero, the biomass grows first slowly, then rapidly, until growth slows down, then reaches a peak, and then decay sets in until nothing is left.

Denoting by $f(t)$ a tree's biomass at a time t, the objective here is to determine the optimal age $\bar{t}$ of a tree, when logging should take place. Since we are interested in the annual production of biomass, averaged over time, our optimization problem is

4.2 Applications of Nonlinear Programming

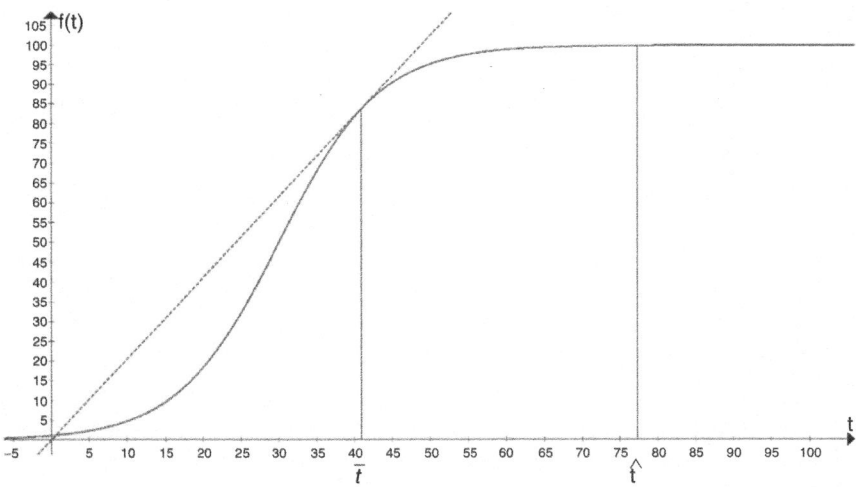

Fig. 4.3 Biomass of a tree over time

$$P : \underset{t>0}{\text{Max}} \ z = f(t)/t.$$

One can show that an optimal solution $\bar{t}$ to problem P must satisfy the equation $\bar{t} f'(\bar{t}) = f(\bar{t})$ and that $\bar{t}$ has an easy geometric interpretation: its value is determined by constructing a straight line through the origin of Fig. 4.3 that touches the growth function and has the smallest possible slope. This point maximizes the biomass (the value on the ordinate) divided by time (the value on the abscissa), which is precisely the tangent of the straight line we constructed. No straight line with a larger slope touches the growth function and can thus not be realized. Denoting the time, at which a tree obtains the maximal biomass by $\hat{t}$ (the maximum of the growth function), then it becomes clear that $\bar{t} < \hat{t}$, i.e. the best time to harvest is before a tree has reached its maximal growth. The reason is the slower growth near the point of maximum biomass.

This type of optimization model occurs in other contexts as well, e.g., the purchase and maintenance of equipment that deteriorates over time (e.g., school buses), for which we want to determine the optimal time of replacement.

4.2.2 Maximizing Tax Revenues

Consider a state, in which taxpayers are levied a flat income tax rate x, with x ranging from 0 (no income tax at all) to 1 (i.e., 100%, *viz.*, confiscation of income). The state now wishes to determine the tax rate x that maximizes the state's income $R(x)$. Clearly, the shape of the function $R(x)$ is crucial. All we know at this point is that

Fig. 4.4 Laffer curve

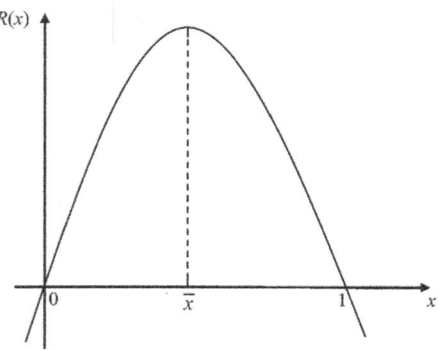

$R(0) = 0$, i.e., with a tax rate of zero, the state's income is zero as well, and that $R(1) = 0$, i.e., with a tax rate of 100%, the state's income will also be zero, as there is no longer any reason for an individual to work. Small tax rates are not much of a disincentive to work, whereas for larger tax rates, individuals will choose to spend their time doing other things, i.e. spend time with their families or including moonlighting. The shape of the state's revenue curve will roughly resemble that of an inverted parabola with a maximal point somewhere between $x = 0$ and $x = 1$. The function $R(x)$ is called the *Laffer curve*, named after the economist Arthur Laffer, who famously drew it in 1974 on a napkin (which is on display at the National Museum of American History in Washington, DC), when discussing American taxation policy. A Laffer curve is shown in Fig. 4.4.

Without presenting any technical details, the Laffer curve can be constructed by considering the amount of time an individual is prepared to work as opposed to spend in leisure pursuits for any given tax rate. Once the curve has been constructed (this is the difficult part), the state's optimization problem is then

$$P : \text{Max } z = R(x).$$

Assume now that $\bar{x}$ is the unique optimal taxation rate. If the current taxation rate is less than $\bar{x}$, an increase of the taxation rate (to a maximum at $\bar{x}$) will increase the state's income, while if the present rate of taxation is higher than $\bar{x}$, any further increase will actually lower the state's income. We should note that there is no universal optimal taxation rate, different people and nations have different acceptances of pain.

4.2.3 Optimizing Inventory and Queuing Models

Inventory models, to be discussed in Chap. 12, deal with the problem of determining the optimal size x of an order, so as to minimize some total cost function $TC(x)$, which includes the costs of placing an order and receiving the shipment, on the one hand, and the cost of keeping units in stock. Deferring the discussion of details to the chapter on the subject, we attempt to solve the optimization problem

4.3 Properties of Nonlinear Programming Problems

$$P : \text{Min } TC(x) = \frac{a_1}{x} + b_1 x,$$

where x denotes the order quantity, the parameter a_1 is proportional to the unit ordering costs, and the parameter b_1 is proportional to the unit holding cost. The unconstrained problem can be solved by differential calculus, resulting in $\bar{x} = \sqrt{a_1/b_1}$. This is the economic order quantity (EOQ), which will be properly introduced in the chapter on inventory management.

One extension to the basic model allows backorders, which leads to the two-variable unconstrained problem

$$P : \text{Min } z = \frac{a_2}{x} + b_2 \frac{(x-y)^2}{x} + c_2 \frac{y^2}{x},$$

with y denoting the maximal allowable shortage and a_2, b_2, and c_2 being known constants. Partial differentiation (differentiation with respect to one variable at a time) and setting the expression to zero then determines the optimal solution $\bar{x} = \sqrt{\frac{a_2}{c_2} \frac{b_2+c_2}{b_2}}$ and $\bar{y} = \sqrt{\frac{a_2}{c_2} \frac{b_2}{b_2+c_2}}$. For a full discussion, see Chap. 12 of this volume.

A final application is found in the area of queuing. A more detailed discussion is found in Chap. 15 of this book. Suppose that on average, λ customers arrive in 1 hr at a single service station. Based on training and other factors, the service station is manned by an agent, who can deal with x customers per hour (the service rate). In order to be feasible, $x > \lambda$ is required. Knowing that the expected number of customers in the system is $\frac{\lambda}{x-\lambda}$, a cost of ax for the service station (capturing the fact that a faster server is more expensive) and unit costs c for keeping a customer in the system for an hour, minimizing the total cost of the system results in the problem

$$P : \text{Min } TC(x) = ax + \frac{c\lambda}{x-\lambda}, \text{ s.t. } x > \lambda.$$

Using again differential calculus, we find that the unique optimal solution is given by $\bar{x} = \lambda + \sqrt{\frac{c\lambda}{a}}$, which also satisfies the single constraint $\bar{x} > \lambda$.

4.3 Properties of Nonlinear Programming Problems

In dealing with a nonlinear optimization problem, there are two important questions to be answered.

- Given a feasible solution, how can we test it for optimality? and
- How do we find an optimal solution?

We will relegate a discussion of the second question to the next section and discuss here how to perform an optimality test. First, consider the unconstrained

Fig. 4.5 The function $f(x) = x^3 - 3x$

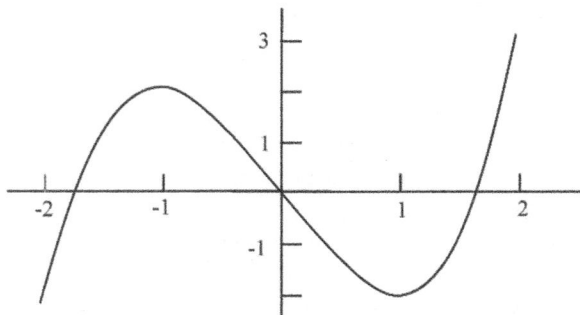

case, i.e., the case, in which a problem has an objective function but no constraints. For simplicity, assume that we are minimizing a differentiable function $f(x)$ of a single variable x. A fundamental result from differential calculus is then that if $\bar{x}$ satisfies $f'(\bar{x}) = 0$ and $f''(\bar{x}) > 0$, then the function f has a local minimum (for a definition, see, e.g., Chap. 17) at $\bar{x}$. As an example, consider the function $f(x) = x^3 - 3x$, pictured in Fig. 4.5.

This function has a local minimum at $\bar{x} = 1$, for which $f(\bar{x}) = f(1) = -2$, but it is apparent that there are values of x, for which its functional value is less than -2, for instance, $x = -3$, for which $f(-3) = -18$. In fact, the function $f(x)$ does not even have a global minimum (a smallest value that the function f can take), since with ever smaller negative values of x, there is no lower bound on the value of $f(x)$. However, for a convex function (i.e., a function, with the property that all points on a linear line segment—the chord—between any of its two points is located above the function itself, see, e.g., Fig. 4.6), any local minimum will necessarily be global. In our case, the function is convex in $[0, \infty[$, where $\bar{x} = 1$ is the minimal point.

In many situations that involve nonconvex functions we will consider that part of the function in the region within which it is convex. For functions of several variables, there are results similar to and extensions of the one-dimensional case.

Consider now the case, in which constraints are present. For pedagogical reasons, we will assume that the objective function is quadratic, and the constraints are linear. Such a case is referred to as a *quadratic programming problem*. Furthermore, let us consider a problem with two variables and a convex objective. Momentarily disregarding the constraints, the overall minimal point is $\hat{x}$, and we define the optimal solution that respects the constraints by $\bar{x}$. Figure 4.7 then shows three possible cases: Either $\hat{x}$ is located in the feasible region (so that $\hat{x} = \bar{x}$), in which case the constraints are irrelevant as they are satisfied anyway, or $\bar{x}$ is located on the boundary of the feasible region, or $\bar{x}$ is at a corner point of the feasible set. The latter case occurs in linear programming.

A test that determines the optimality of a solution is provided by the so-called *Karush–Kuhn–Tucker (KKT) conditions*. They are necessary and sufficient conditions for the optimality of a convex differentiable nonlinear programming problem.

4.3 Properties of Nonlinear Programming Problems

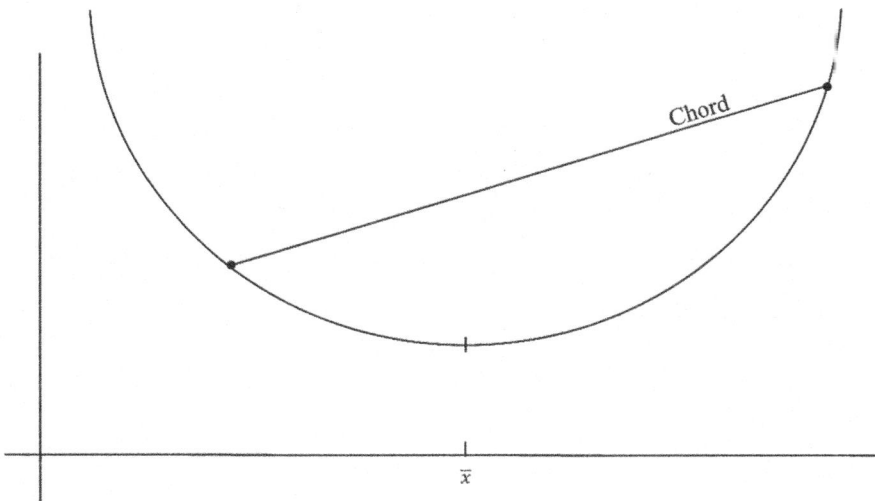

Fig. 4.6 Convex function and global minimum at $\bar{x}$

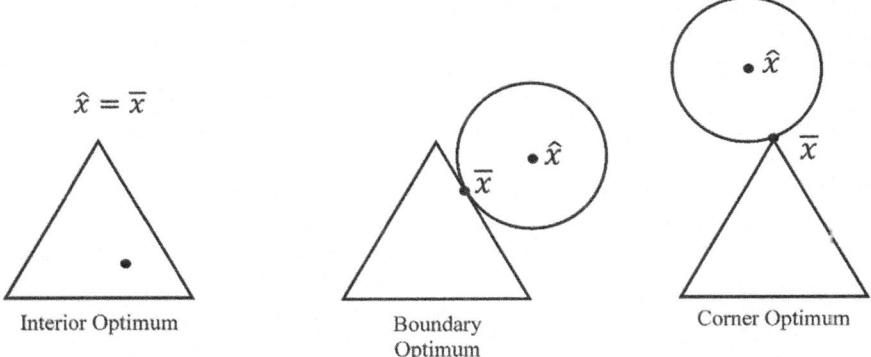

Fig. 4.7 Different cases for quadratic programming

In the simplest case, we start with a problem of the type Min $z = f(x_1, x_2, \ldots, x_n)$, s.t. $g_i(x_1, x_2, \ldots, x_n) = 0$ for all i, where f is a convex function and g_i denotes the difference between the left-hand side and the right-hand side of the i-th constraint, where all m constraints are assumed to be equations. We then set up the *Lagrangean function* $L(x_1, x_2, \ldots, x_n; u_1, u_2, \ldots, u_m)$, which is the sum of the minimization objective minus the sum of the constraints, each of which is weighted by a *Lagrangean multiplier* u_i. The *KKT* necessary optimality conditions in this case simply state that the derivative of the Lagrangean function with respect to each of the variables including the Lagrangean multipliers must equal zero. Solving this system of simultaneous equations then finds an optimal solution.

As an example, consider the problem

$$P : \text{Min } z = 4x_1^2 - 6x_1x_2 + 3x_2^2 - 12x_1 - 6x_2$$

$$\text{s.t. } x_1 + x_2 - 12 = 0.$$

Note that in this formulation, the variables are unconstrained, i.e., they may be positive or negative. The Lagrangean function is then

$$L(x_1, x_2, u) = 4x_1^2 - 6x_1x_2 + 3x_2^2 - 12x_1 - 6x_2 - (x_1 + x_2 - 12)u.$$

The *KKT* conditions are then

$$8x_1 - 6x_2 - 12 - u = 0,$$

$$-6x_1 + 6x_2 - 6 - u = 0, \quad \text{and}$$

$$x_1 + x_2 - 12 = 0,$$

where the last equation is nothing but the original constraint. Solving this system of simultaneous linear equations results in the solution $\bar{x}_1 = 5.7692$, $\bar{x}_2 = 6.2308$ with an objective value of $\bar{z} = -72.6923$.

The Karush–Kuhn–Tucker conditions can also be formulated for the considerably more general case, in which the constraints are inequalities rather than equations. A rich and fruitful theory can then be built around these conditions involving sensitivity, postoptimality, and duality aspects. Such considerations are beyond the scope of this book and readers are referred to specialized texts on the subject, see, e.g., Eiselt and Sandblom (2019).

As mentioned earlier, many other solution techniques exist. Of special interest may be *SUMT* (*S*equential *U*nconstrained *M*inimization *T*echnique) methods, whose basic idea is to eliminate the constraints in a model by allowing them to be violated, albeit with a penalty term in the objective function. This idea is reminiscent of goal programming, where we minimize the over- and/or underachievement of a constraint in the objective function as well; see, e.g., Sect. 3.3 in this volume.

4.4 Solution Methods for Nonlinear Optimization

Unlike linear programming, where there are only a few practically relevant solution techniques, there are myriads of different solution methods for nonlinear optimization problems. Many algorithms exist for constrained and unconstrained problems, most tailored to a specific type of nonlinear function. This volume describes two techniques, Newton's method for unconstrained minimization and the Frank and Wolfe method for linearly constrained convex models.

4.4.1 Newton's Method for Unconstrained Optimization

Consider the unconstrained problem

$$P : \text{Min } z = f(x),$$

where f is a single-variable, twice differentiable convex function. Our description of Newton's method will rely heavily on geometrical arguments; details are shown in Fig. 4.8. The general idea is to generate a sequence of points $x_1, x_2, x_3, \ldots$, which are estimates of the true optimal solution $\bar{x}$. Since $\bar{x}$ is characterized by $f'(\bar{x}) = 0$, we attempt to solve this equation.

Given some starting estimate x_1, we construct the line ℓ_1 as a tangent to the curve $f'(x)$ at the point $(x_1, f'(x_1))$. The next estimate of $\bar{x}$ is at x_2, where the line ℓ_1 intersects the abscissa. Pursuing this argument further, we find the next approximation x_3 at the intersection of the tangent line ℓ_2 with the abscissa. This process terminates whenever the steps between successive approximations are sufficiently small, or some other termination criterion is satisfied. The general iteration formula is

$$x_{k+1} = x_k - \frac{f'(x_k)}{f''(x_k)}.$$

As a numerical example, consider again the function $f(x) = x^3 - 3x$ in Fig. 4.5. As $f'(x) = 3x^2 - 3$ and $f''(x) = 6x$, Newton's iteration formula is

$$x_{k+1} = x_k - \frac{3x_k^2 - 3}{6x_k} = \tfrac{1}{2}x_k + \frac{1}{2x_k}.$$

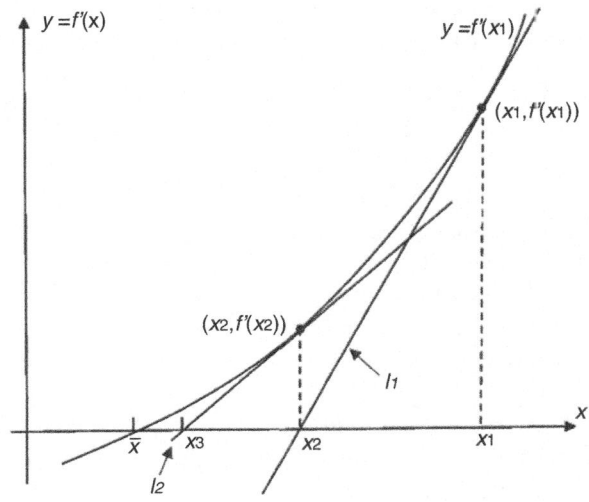

Fig. 4.8 Newton's minimization method

Arbitrarily starting with the initial estimate $x_1 = \frac{1}{2}$, we obtain $x_2 = 1.25$, $x_3 = 1.025$, $x_4 = 1.0003049$, and we see that the sequence converges to the true value of 1 very rapidly.

We should note that Newton's method for minimization problems described above is equivalent to the Newton–Raphson method for solving equations—here, we actually solve the equation $f'(x) = 0$ for x.

4.4.2 Constrained Optimization

This section describes the Frank and Wolfe (1956) method for the solution of linearly constrained optimization problems. To formalize, consider the problem

$$P : \text{Min } z = f(x_1, x_2, \ldots, x_n)$$
$$\text{s.t. } a_{11}x_1 + a_{12}x_2 + \cdots + a_{1n}x_n \leq b_1,$$
$$a_{21}x_1 + a_{22}x_2 + \cdots + a_{2n}x_n \leq b_2,$$
$$\vdots$$
$$a_{m1}x_1 + a_{m2}x_2 + \cdots + a_{mn}x_n \leq b_m.$$

For simplicity, we will consider the case of two variables, the extension to n variables is straightforward. We will assume that the function f is convex and differentiable, and that the feasible set is bounded. Each iteration of the method proceeds in two steps. The first step linearizes the objective function in the vicinity of the present interior point, and then solves the resulting linear programming problem, while in the second step we solve a single-dimensional interpolation problem that finds the next improved trial point.

Starting with an arbitrary initial interior feasible point (x_1^1, x_2^1), we first solve the auxiliary linear programming problem P^ℓ, which is obtained by linearizing the objective function in the vicinity of the initial interior point (x_1^1, x_2^1), i.e., we solve

$$P_1^\ell : \text{Min } z_1^\ell = \left(\frac{\partial f(x_1^1, x_2^1)}{\partial x_1}\right) x_1 + \left(\frac{\partial f(x_1^1, x_2^1)}{\partial x_2}\right) x_2$$
s.t. the given constraints,

where, as usual, $\partial f/\partial x$ denotes the partial derivative of the function f with respect to the variable x. This is a linear programming problem whose solution we denote by $(\tilde{x}_1^1, \tilde{x}_2^1)$, which by virtue of Dantzig's theorem, must be on the boundary of the feasible set. Let us assume the problem is nondegenerate, so that we have an optimal solution at an extreme point.

The next step performs an interpolation search between the points (x_1^1, x_2^1) and $(\tilde{x}_1^1, \tilde{x}_2^1)$, i.e. we determine the point on the line segment between these two points that has the smallest value of the original objective function f. This is accomplished by solving the one-dimensional interpolation minimization problem

4.4 Solution Methods for Nonlinear Optimization

$$P_1^\lambda : \underset{0\leq\lambda\leq 1}{\text{Min}}\ z_1^\lambda = f\left(\lambda x_1^1 + (1-\lambda)\tilde{x}_1^1, \lambda x_2^1 + (1-\lambda)\tilde{x}_2^1\right).$$

Denoting the solution to this problem by λ_1, we then define the next estimate (x_1^2, x_2^2) of an optimal solution to P by

$$x_1^2 := \lambda_1 x_1^1 + (1-\lambda_1)\tilde{x}_1^1 \quad \text{and}$$

$$x_2^2 := \lambda_1 x_2^1 + (1-\lambda_1)\tilde{x}_2^1$$

and the process is repeated, until some stop criterion is satisfied.

As an illustration of the procedure, we will use the convex optimization problem

$$P : \text{Min}\ z = x_1^2 + x_2^2 + \frac{16}{x_1} - 4x_2$$

$$\text{s.t.}\ -2x_1 + x_2 \leq 0$$

$$x_1 + x_2 \leq 3$$

$$x_1, x_2 \geq 0.$$

Disregarding the constraints, the unconstrained optimal solution is $(\hat{x}_1, \hat{x}_2) = (2, 2)$, which violates the second constraint, i.e. it is not a feasible point. Hence, the true optimal solution must be located on the boundary of the feasible set. We now arbitrarily select $(x_1^1, x_2^1) = (1, 1)$ as our interior starting point. We find $\frac{\partial f}{\partial x_1} = 2x_1 - \frac{16}{x_1^2}$ and $\frac{\partial f}{\partial x_2} = 2x_2 - 4$, so that the first auxiliary linear programming problem is

$$P_1^\ell : \text{Min}\ z_1^\ell = -14x_1 - 2x_2$$

$$\text{s.t.}\ -2x_1 + x_2 \leq 0$$

$$x_1 + x_2 \leq 3$$

$$x_1, x_2 \geq 0,$$

which has the unique optimal solution $\tilde{x}_1^1 = 3$ and $\tilde{x}_2^1 = 0$. Therefore, the first interpolation problem is

$$P_1^\lambda : \underset{0\leq\lambda\leq 1}{\text{Min}}\ z_1^\lambda = f(\lambda + (1-\lambda)3, \lambda).$$

We find that $z_1^\lambda = f(3 - 2\lambda, \lambda) = (3 - 2\lambda)^2 + \lambda^2 + \frac{16}{3-2\lambda} - 4\lambda$, which attains its minimal value at $\lambda_1 = 0.604$. Therefore, the next estimate of the optimal solution will be $x_1^2 = 1\lambda_1 + 3(1 - \lambda_1) = 1.792$, and $x_2^2 = 1\lambda_1 + 0(1 - \lambda_1) = 0.604$.

The next iteration computes $\frac{\partial f}{\partial x_1} = (2)(1.792) - \frac{16}{1.792} \approx -1.3985$ and $\frac{\partial f}{\partial x_2} = (2)(0.604) - 4 = -2.792$; the second auxiliary linear programming problem is then

Fig. 4.9 Example for the Frank and Wolfe method

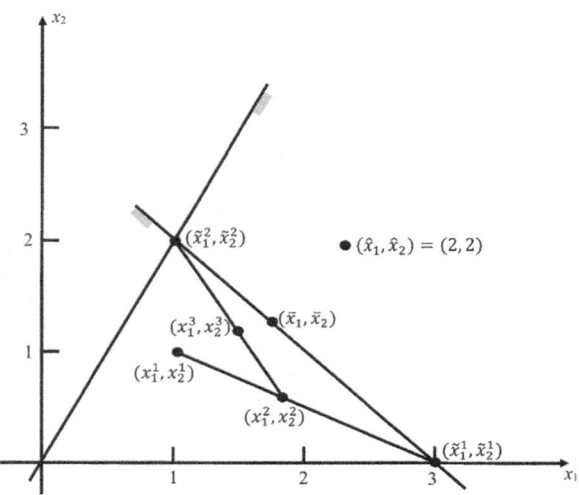

$$P_2^\ell : \text{Min } z_2^\ell = -1.3985x_1 - 2.792x_2$$

$$\text{s.t. } -2x_1 + x_2 \leq 0$$

$$x_1 + x_2 \leq 3$$

$$x_1, x_2 \geq 0,$$

which has the unique optimal solution $\tilde{x}_1^2 = 1$ and $\tilde{x}_2^2 = 2$. The second interpolation problem is then

$$P_2^\lambda : \underset{0 \leq \lambda \leq 1}{\text{Min}} \; z_2^\lambda = f(1.792\lambda + (1-\lambda)1, 0.604\lambda + (1-\lambda)2),$$

so that $z_2^\lambda = f(1 + 0.792\lambda, 2 - 1.596\lambda)$, and the process continues as shown in Fig. 4.9. It should be noted that the convergence to the true optimal solution at $\bar{x}_1 = 1.7728$ and $\bar{x}_2 = 1.2272$ with $\bar{z} = 8.7653$ on the boundary of the feasible region is rather slow.

Exercises
Problem 1 (Lagrangean function and KKT conditions): Consider the problem

$$P : \text{Min } z = (x_1 - 1)^2 + (x_2 - 2)^2 + (x_3 - 3)^2$$

$$\text{s.t. } x_1^2 + x_2^2 + x_3^2 = 1$$

and solve it, using the Karush–Kuhn–Tucker conditions.
Solution: The Lagrangean function is

4.4 Solution Methods for Nonlinear Optimization

$$L(x_1, x_2, x_3, u) = (x_1 - 1)^2 + (x_2 - 2)^2 + (x_3 - 3)^2 - (x_1^2 + x_2^2 + x_3^2 - 1)u,$$

so that the Karush–Kuhn–Tucker conditions can be written as

$$2(x_1 - 1) - 2x_1 u = 0,$$
$$2(x_2 - 2) - 2x_2 u = 0,$$
$$2(x_3 - 3) - 2x_3 u = 0,$$
$$x_1^2 + x_2^2 + x_3^2 - 1 = 0,$$

which can be simplified to $x_1 = \frac{1}{1-u}$, $x_2 = \frac{2}{1-u}$, $x_3 = \frac{3}{1-u}$, and $x_1^2 + x_2^2 + x_3^2 = 1$, which has the solution $x_1 = \frac{1}{\sqrt{14}}$, $x_2 = \frac{2}{\sqrt{14}}$, $x_3 = \frac{3}{\sqrt{14}}$ (as well as the negative values thereof). Inserting each of the two solutions into the objective function, we determine that the problem has a unique optimal solution with all values of the variables being positive, i.e., $\bar{x}_1 = 1/\sqrt{14} \approx 0.26726$, $\bar{x}_2 = 2/\sqrt{14} \approx 0.53452$, and $\bar{x}_3 = 3/\sqrt{14} \approx 0.80178$ with the objective function value $\bar{z} = 7.5167$.

Problem 2 (Solve a one-dimensional unconstrained optimization problem): Using the Newton method, find the solution to the unconstrained problem

$$P : \text{Min } z = f(x) = \frac{x^3}{6} - 2x^2 - \tfrac{1}{2}x + \pi,$$

using the starting point $x_1 = 5$ and follow the process until the method converges, using four decimal places for accuracy.

Solution: Since $f'(x) = \tfrac{1}{2}x^2 - 4x - \tfrac{1}{2}$ and $f''(x) = x - 4$, Newton's iteration formula will take the form $x_{k+1} = x_k - \frac{f'(x_k)}{f''(x_k)} = x_k - \frac{\tfrac{1}{2}x_k^2 - 4x_k - \tfrac{1}{2}}{x_k - 4} = \frac{x_k^2 + 1}{2x_k - 8}$. Setting $x_1 = 5$, we obtain $x_2 = 13$, $x_3 = 9.4444$, $x_4 = 8.2834$, $x_5 = 8.1261$, $x_6 = 8.1231$, and $x_7 = 8.1231$, and the method has converged.

Problem 3 (Solve a convex constrained minimization problem): Consider the following constrained nonlinear optimization problem

$$P : \text{Min } z = 2x_1 - x_2 + \frac{3}{2x_1} + \frac{x_2^2}{2x_1}$$

$$\text{s.t.} -4x_1 + x_2 \leq 0$$

$$6 \leq x_1 \leq 10$$

$$x_2 \geq 0.$$

Solve the problem P via the Frank and Wolfe method, starting with the interior feasible point $(x_1^1, x_2^1) = (8, 15)$ and perform two complete iterations, i.e. compute the points (x_1^2, x_2^2) and (x_1^3, x_2^3).

Fig. 4.10 Illustration of the Frank and Wolfe problem

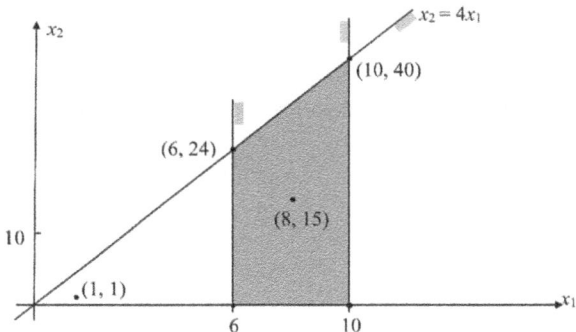

Solution: We find $\frac{\partial f}{\partial x_1} = 2 - \frac{3}{2x_1^2} - \frac{x_2^2}{2x_1^2}$ and $\frac{\partial f}{\partial x_2} = -1 + \frac{x_2}{x_1}$, and therefore the unconstrained minimal point for the problem P is $(\hat{x}_1, \hat{x}_2) = (1, 1)$, which is not feasible. On the other hand, we can easily check that $(x_1^1, x_2^1) = (8, 15)$ is an interior feasible point, see Fig. 4.10.

At the point $(x_1^1, x_2^1) = (8, 15)$, we obtain $\frac{\partial f}{\partial x_1} = 7/32$ $\frac{\partial f}{\partial x_2} = 7/8$, so that after scaling the objective function, the first auxiliary linear programming problem is

$$P_1^\ell : \text{Min } z_1^\ell = x_1 + 4x_2$$

$$\text{s.t.} - 4x_1 + x_2 \leq 0$$

$$6 \leq x_1 \leq 10$$

$$x_2 \geq 0,$$

which has the unique optimal solution $(\tilde{x}_1^1, \tilde{x}_2^1) = (6, 0)$, which is the point located in the southwest corner of the feasible region. The interpolation search is then

$$P_1^\lambda : \underset{0 \leq \lambda \leq 1}{\text{Min }} z_1^\lambda = f(6 + 2\lambda, 15\lambda),$$

which has the solution $\lambda_1 = 0.3473$, so that $(x_1^2, x_2^2) = (6.6946, 5.2095)$. At this point, we find that $\frac{\partial f}{\partial x_1} = 1.664$ and $\frac{\partial f}{\partial x_2} = -0.222$. The resulting second auxiliary problem is then

$$P_2^\ell : \text{Min } z_2^\ell = 1.664x_1 - 0.222x_2$$

$$\text{s.t.} - 4x_1 + x_2 \leq 0$$

$$6 \leq x_1 \leq 10$$

$$x_2 \geq 0,$$

which has a unique optimal solution $(\tilde{x}_1^2, \tilde{x}_2^2) = (6, 24)$ in the northwest corner of the feasible set. The interpolation search is then

$$P_1^\lambda : \underset{0 \leq \lambda \leq 1}{\text{Min}} \ z_2^\lambda = f(6 + 0.6946\lambda, 24 - 18.7905\lambda),$$

which has a solution $\lambda_2 = 0.90602$, from which we calculate $(x_1^3, x_2^3) = (6.6293, 6.9754)$, at which time we stop, having completed two full iterations of the technique. It should be noted that the true optimal solution is $(\bar{x}_1, \bar{x}_2) = (6, 6)$.

References

Davidon WC (1959) Variable metric method for minimization. Research and Development Report ANL-5990 (Rev.) Argonne National Laboratory, U. S. Atomic Energy Commission

Eiselt HA, Sandblom C-L (2019) Nonlinear programming. Springer, Berlin

Fletcher R, Powell MJD (1963) A rapidly convergent descent method for minimization. Comput J 6:163–168

Frank M, Wolfe P (1956) An algorithm for quadratic programming. Nav Res Logist Q 3:95–110

Karush W (1939) Minima of functions of several variables with inequalities as side conditions. MS Thesis, Department of Mathematics, University of Chicago, IL

Kuhn HW, Tucker AW (1951) Nonlinear programming. In: Proceedings of 2nd Berkeley symposium. University of California Press, Berkeley, CA, pp 481–492

Integer Linear Programming

Not too long after more and more applications of linear programming were developed it became apparent that in some of these applications, the variables would not be able to attain just any (nonnegative) value but should be integers. As a simple example, if a variable has been defined to denote the number of cans of beans manufactured in the planning period, then surely it would make no sense to make, say, 1,305,557.3 cans: the last 0.3 cans would have to be rounded up or down. While this may be an acceptable practice when dealing with this application (after all, it makes very little difference whether or not we make 0.3 cans more or less), in other applications this may make a huge difference. For instance, assigning airplanes to routes or trucks to deliveries may very well make the difference between gain and loss. Furthermore, simply rounding up or down a noninteger (usually referred to as a continuous solution) will not necessarily result in an optimal integer solution. We will demonstrate this fact below.

Even though the difference may be blurry, it may be useful to distinguish between variables that are naturally required to be integer (such as the number of trucks to be used for deliveries, the number of work crews dispatched to a construction site, or the number of drums of hazardous material shipped from one site to another), and the so-called logical variables that also must be integer and are introduced for logical reasons. Below, we will provide a number of examples of the latter.

Integer programming problems were first discussed by Ralph E. Gomory in the 1950s, who also devised a solution technique for them. Gomory's class of techniques is called *cutting plane techniques*, and we will describe their basic idea below. A subsequent breakthrough is the 1961 contribution by Land and Doig, whose *branch-and-bound* idea remains the basis of most solution methods for integer programming problems to this day.

5.1 Definitions and Basic Concepts

In order to define integer programming problems, we will start with a standard linear programming problem. As an example, consider the problem

$$P_1 : \text{Max } z = 10x_1 + 11x_2$$
$$\text{s.t.} \quad 3x_1 + 5x_2 \leq 15$$
$$5x_1 + 2x_2 \leq 10$$
$$x_1, x_2 \geq 0.$$

The only difference between the standard linear programming problem above and an integer programming problem is that some or all of the variables, in addition to be required to be nonnegative, are *also* required to be integer. It is very helpful to think of the integrality condition as an *additional* requirement. All-integer programming problems that require some, but not all, variables, to be integer, are called *mixed-integer linear programming* (or *MILPs*) problems, while problems, in which all variables are required to assume integer values, are called *all-integer linear programming* (or *AILP*) problems.

If we replace the two nonnegativity constraints in the above example by $x_1 \geq 0$, and $x_2 \geq 0$ *and* integer, then we have a mixed-integer programming problem, which we may call P_2. Similarly, if we replace the nonnegativity constraints in P_1 by the conditions $x_1 \geq 0$ *and* integer and $x_2 \geq 0$, we have another mixed-integer linear programming problem P_3. Finally, if we replace in P_1 both nonnegativity constraints by $x_1 \geq 0$ *and* integer as well as $x_2 \geq 0$ *and* integer, we then have the all-integer linear programming problem P_4.

The graphical representations of the feasible sets of the four problems P_1, P_2, P_3, and P_4 are shown in Figs. 5.1a–d. Figure 5.1a shows the linear programming problem P_1 with the feasible set shaded. The *mixed-integer linear programming problem* (*MILP*) P_2 is shown in Fig. 5.1b; here, only the bold horizontal bars are feasible, as only those guarantee that the variable x_2 is integer, while in addition, respect the given constraints. Similarly, only the bold vertical bars in Fig. 5.1c are feasible for the mixed-integer linear programming problem P_3. Finally, the feasible set of the *all-integer linear programming* (*AILP*) problem P_4 consists exclusively of the grid points shown in Fig. 5.1d. Clearly, the feasible set of this problem is smallest, as it is the most constrained. Incidentally, the optimal solutions of the four problems are as follows:

$$P_1 : \bar{x} = (1.0526, 2.3684) \quad \text{with } \bar{z} = 36.5790$$
$$P_2 : \bar{x} = (1.2, 2) \quad \text{with } \bar{z} = 34,$$
$$P_3 : \bar{x} = (1, 2.4) \quad \text{with } \bar{z} = 36.4, \text{ and}$$
$$P_4 : \bar{x} = (0, 3) \quad \text{with } \bar{z} = 33.$$

Again, it is apparent that moving from the least restricted problem P_1 on to the more restricted problems P_2 and P_3 to the most restricted problem P_4, the values of

5.1 Definitions and Basic Concepts

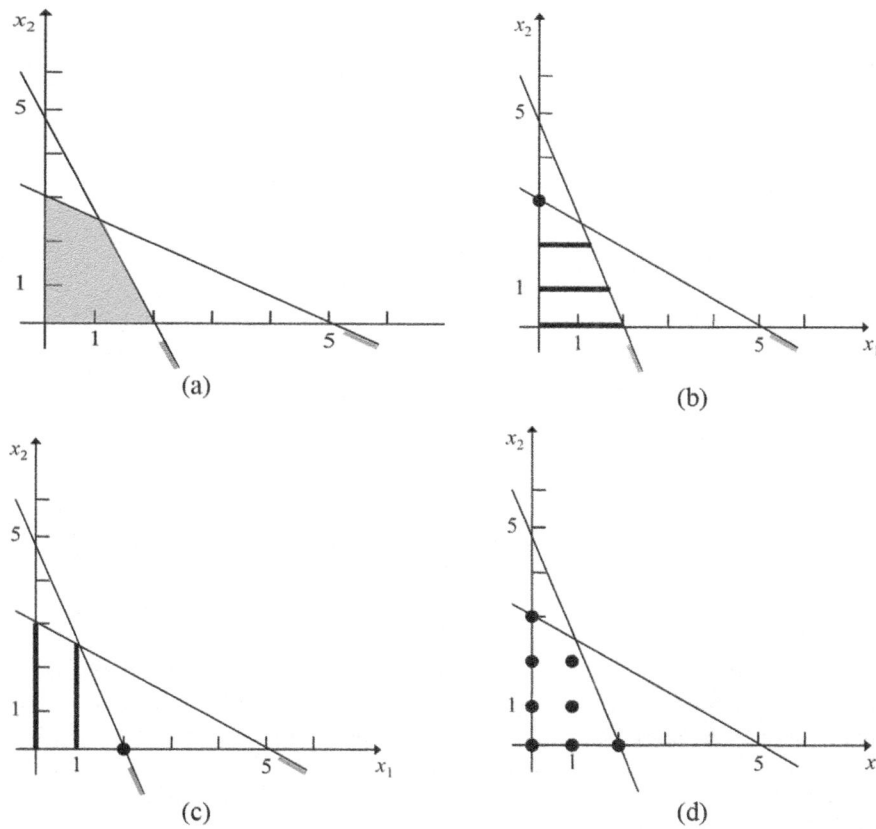

Fig. 5.1 (a)–(d) Feasible sets for P_1, P_2, P_3, and P_4

the objective function are getting worse, i.e., they decrease in maximization problems, while they would increase in minimization problems. Also, even in this small example, we notice that simple rounding of the continuous solution of P_1 does not result in the all-integer solution of P_4.

Furthermore, it is important to realize that none of the above integer programming problems has its optimal solution at an extreme point, which is always the case in linear programming (recall Dantzig's corner point theorem) from Sect. 2.3.1.

In order to highlight the differences between linear and integer linear optimization, consider the following all-integer linear programming problem P_5:

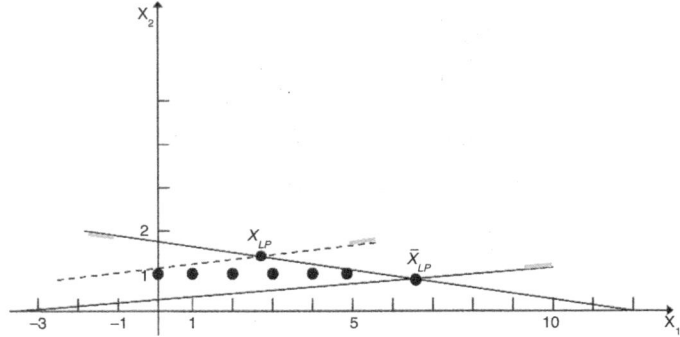

Fig. 5.2 Graph for Problem P$_5$

$$P_5 : \text{Max } z = 2x_1 + x_2$$
$$\text{s.t.} \quad 7x_1 + 48x_2 \leq 84$$
$$-x_1 + 12x_2 \geq 3$$
$$x_1, \quad x_2 \geq 0 \text{ and integer.}$$

The two solid lines in Fig. 5.2 show the two constraints (ignore the broken line for now), and the bold dots indicate the all-integer solutions. The optimal solution of the linear programming problem without the integrality requirements is $\bar{\mathbf{x}} = (6.5455, 0.7955)$ with $\bar{z} = 13.8864$, while the optimal solution of the all-integer programming problem is $\bar{\mathbf{x}} = (5, 1)$ with $\bar{z} = 11$. This is a solution that cannot be obtained by simple rounding.

Worse yet, if we were to change the right-hand side value of the second constraint from "3" to "13," one of the constraints in Fig. 5.2 moves in parallel fashion to the broken line. The feasible set for the linear programming problem without integrality conditions is the triangle with the vertices $(0, 1\frac{1}{12})$, $(0, 1\frac{3}{4})$, and the point marked with $\bar{x}'_{LP}$, which has coordinates $(2.9091, 1.3258)$ and a value of the objective function of 7.1439. The all-integer linear programming problem does, however, have no feasible solution (notice that there are no bold dots in the triangle described above).

Finally, consider the feasible set generated by the following constraints:

$$x_1 + x_2 \leq 10$$
$$5x_1 + 3x_2 \geq 15$$
$$x_1 \leq 6$$
$$x_2 \leq 7$$
$$x_1, \quad x_2 \geq 0.$$

The feasible set is shown in Fig. 5.3.

Fig. 5.3 Feasible set with integer-valued extreme points

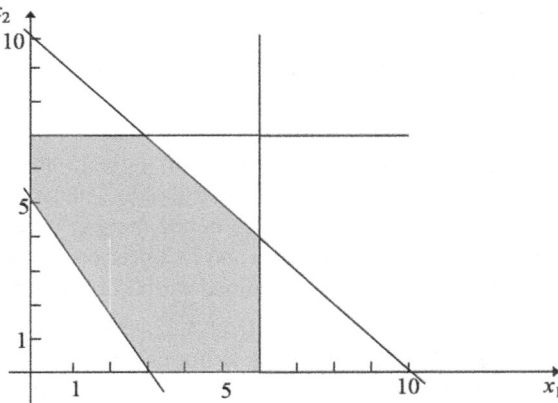

Inspection reveals that the feasible set has extreme points (3, 0), (6, 0), (6, 4), (3, 7), (0, 7), and (0, 5). In other words, all corner points of the feasible set are integer-valued. If we were to solve a linear programming problem with this feasible set and any arbitrary objective function, at least one optimal solution would be—as proved by Dantzig's "corner point theorem"—one of these points, and thus integer without us actually requiring it. There are some large classes of linear programming formulations that fall into this category, i.e. having integer optimal solutions without explicitly including integer requirements. This is a very appealing feature, since linear programming problems are generally much easier to solve than their integer programming counterparts. We will encounter this type of problem in the chapter on network models.

The remainder of this section will discuss the relations between the objective function values of linear programming problems and those of their integer linear programming problems. In order to facilitate the discussion, we need to introduce some terminology. Consider any mixed- or all-integer linear programming problem P. Its *linear programming relaxation* P_{rel} is the very same problem P except with all of the integrality conditions deleted. Suppose that P is a maximization problem, and its optimal value of the objective function is $\bar{z}_{IP}$. Let now $\bar{z}_{LP}$ denote the optimal value of the objective function of its linear programming relaxation, then we find that

$$\bar{z}_{IP} \leq \bar{z}_{LP}.$$

The reason for this relation is easy to see. Starting with the relaxation, we obtain the integer programming problem by adding constraints, namely the integrality constraints. And whenever constraints are added to a problem, the optimal value of the objective function gets worse, i.e. lower in case of maximization problems and higher in case of minimization problems. And that is exactly what is captured by the above relation. Note that the inequality is reversed in case of minimization problems. In other words, the linear programming relaxation provides an upper bound on the optimal value of the objective function of the integer programming problem in case of maximization problems, while it is a lower bound in case of minimization problems.

As a numerical example, consider the all-integer linear programming problem P_5 introduced earlier in this section. The value of the objective function of the all-integer solution was found to be $\bar{z}_{IP} = 11$, while the linear programming relaxation had an optimal objective value of $\bar{z}_{LP} = 13.8864$.

Not only does the numerical example satisfy the inequality shown above as expected, but it also leads us to define the gap between the objective values of an integer programming problem and its linear programming relaxation. Formally, the *absolute integrality gap* is defined as the difference $|\bar{z}_{LP} - \bar{z}_{IP}|$, while the *relative integrality gap* is defined as $\frac{|\bar{z}_{LP} - \bar{z}_{IP}|}{\max\{|\bar{z}_{LP}|, |\bar{z}_{IP}|\}}$ with the relative gap defined as 0, in case $\bar{z}_{LP} = \bar{z}_{IP} = 0$. In case we want to normalize the measure, we can use the expression $\frac{|\bar{z}_{LP} - \bar{z}_{IP}|}{\max\{|\bar{z}_{LP}|, |\bar{z}_{IP}|, |\bar{z}_{LP} - \bar{z}_{IP}|\}}$, which is always defined in the [0, 1] interval.

In the numerical example, the absolute integrality gap is $13.8864 - 11 = 2.8864$, while the relative integrality gap is $2.8864/13.8864 = .2079$. The relative integrality gap is usually a good indicator of the degree of difficulty of the problem. Any problem with a relative integrality gap in excess of 0.1 or 10% is fairly difficult, while problems with integrality gaps in excess of 0.5 or 50% are typically really difficult. The integrality gap can, of course, only be computed after the problem and its relaxations have been solved, thus diminishing its usefulness. However, often good approximations for $\bar{z}_{IP}$ are known (or $\bar{z}_{IP}$ is replaced by the best-known integer solution) and $\bar{z}_{LP}$ is usually not too difficult to compute.

A related expression is the *optimality gap*, which, in case a problem was not or could not be solved to optimality, measures how far the best-known solution deviates from the best-known bound (upper for maximization and lower for minimization). The actual expression replaces $\bar{z}_{LP}$ and $\bar{z}_{IP}$ by the best-known bound and best-known integer solution in the above expression.

While the linear programming relaxation is usually the first step to solve an integer programming problem, other types of relaxations exist. Most prominent among them is the so-called *Lagrangean relaxation* P_{Lagr}. In order to set up this relaxed problem (which we assume has a maximization objective), we first select some or all of the given structural constraints. These constraints, usually called *dualized constraints*, are rearranged to be in the form $LHS \leq 0$ and then multiplied by some preselected nonnegative constants (referred to as the *Lagrangean multipliers* or *dual variables*) and subtracted from the objective function. Furthermore, the dualized constraints are removed from the set of constraints.

In order to illustrate the process, we will again refer to problem P_4, repeated here for convenience:

5.1 Definitions and Basic Concepts

$$P_4 : \text{Max } z = 10x_1 + 11x_2$$
$$\text{s.t. } 3x_1 + 5x_2 \leq 15$$
$$5x_1 + 2x_2 \leq 10$$
$$x_1, x_2 \geq 0 \text{ and integer.}$$

First we set up its Lagrangean relaxation by dualizing the first constraint, using the Lagrangean multiplier $\widehat{u}_1 = 2$. The relaxed problem is then

$$P^1_{Lagr} : \text{Max } z^1_{Lagr} = 10x_1 + 11x_2 - 2(3x_1 + 5x_2 - 15)$$
$$\text{s.t. } 5x_1 + 2x_2 \leq 10$$
$$x_1, x_2 \geq 0 \text{ and integer.}$$

This problem has an optimal solution $\widehat{x}_1 = 2$, $\widehat{x}_2 = 0$ with a value of the objective function of $\widehat{z}^1_{Lagr} = 38$.

Alternatively, if we were to dualize the second constraint instead with, say, a Lagrangean multiplier of $\widehat{u}_2 = 3$, we would obtain a different relaxed problem, viz.,

$$P^2_{Lagr} : \text{Max } z^2_{Lagr} = 10x_1 + 11x_2 - 3(5x_1 + 2x_2 - 10)$$
$$\text{s.t. } 3x_1 + 5x_2 \leq 15$$
$$x_1, x_2 \geq 0 \text{ and integer.}$$

The optimal solution for this problem is $\widehat{x}_1 = 0$, $\widehat{x}_2 = 3$ with a value of the objective function of $\widehat{z}^2_{Lagr} = 45$.

Yet another possibility would be to dualize both constraints, which would leave us with a problem that maximizes the objective function $10x_1 + 11x_2 - \widehat{u}_1(3x_1 + 5x_2 - 15) - \widehat{u}_2(5x_1 + 2x_2 - 10)$ subject to only the nonnegativity constraints and the integrality requirements for both variables. If we were to choose multipliers $\widehat{u}_1 = 2$ and $\widehat{u}_2 = 1$, the objective function reduces to Max $z^3_{Lagr} = -x_1 - x_2 + 40$, which has an optimal solution $\widehat{x}_1 = 0$ and $\widehat{x}_2 = 0$ with objective function value $\widehat{z}^3_{Lagr} = 40$. On the other hand, choosing multipliers $\widehat{u}_1 = 1$ and $\widehat{u}_2 = 2$ results in a problem with an objective function Max $z^4_{Lagr} = -3x_1 + 2x_2 + 35$, which has unbounded "optimal" solutions (as the variable x_2 can be chosen arbitrarily large).

It is easy to show that the optimal value of the objective function $\widehat{z}_{Lagr}$ of a Lagrangean relaxation is always greater than or equal to the value of the objective function $\bar{z}$ of the original (maximization) problem. The main idea is this. Dualizing any subset of constraints of the type $LHS \leq 0$ increases the feasible set, allowing solutions with higher value of the original objective function. In addition to the removal of some constraints from the problem, the objective function of the Lagrangean problem subtracts $\widehat{u}$ LHS from the original objective. By definition, $\widehat{u} \geq 0$ and $LHS \leq 0$, so that their product is a nonpositive number, which is subtracted, so that the value of the objective function can be larger than that of the original problem.

It has become apparent in the above discussion that the choice of the Lagrangean multipliers is crucial in determining how close the gap between $\widehat{z}_{Lagr}$ and $\bar{z}$, the optimal objective function value of the original problem P, will be. In the context of linear programming, we can show that by selecting the Lagrangean multipliers to be equal to the optimal values of the dual variables, the inequality $\widehat{z}_{Lagr} \geq \bar{z}$ is actually satisfied as an equation, i.e. $\widehat{z}_{Lagr} = \bar{z}$. We will exploit this idea further in a heuristic method in Sect. 5.3.3 below.

In some problems, integrality does not occur naturally but is a result of the way we must formulate constraints. Some of these formulations use the so-called *logical variables*, which are zero-one variables that are introduced to model logical implications that cannot be modeled in the usual variables.

In order to explain some of the intricacies, assume that we are facing a manufacturing problem, in which one of the tasks is to decide which k of the p possible machines we should lease or purchase. Defining the variables is easy: For each of the p machines, we define a binary variable y_j, $j = 1, \ldots, p$, which assumes a value of one, if we purchase/lease the machine and zero otherwise. This part of the formulation is straightforward. However, we need to ensure that the constraints related to each of the machines, e.g., capacity constraints and others, only apply if we have purchased/leased the machine, otherwise they are obviously irrelevant. This can be achieved as follows. If there exists a constraint relating to any one of the p machines, we add to/subtract from the right-hand side a specific amount as follows: If the original constraint is $LHS \leq RHS$, we reformulate it as $LHS \leq RHS + (y - 1)M$, and if the original constraint were $LHS \geq RHS$, we reformulate it as $LHS \leq RHS - (y - 1)M$, where M is a sufficiently large number.

The reason for this reformulation is as follows. Suppose that we purchase/lease a machine. The variable associated with this machine then assumes a value of $y = 1$, in which case the constraints for this machine appear after the reformulation exactly as before, since the added/subtracted term equals zero. In other words, the machine constraints must hold when we purchase/lease the machine. Suppose now that the solution does not include a specific machine. This means that the variable y assumes a value of 0, and the reformulation guarantees that the constraints, which belong to this machine are redundant. This is the idea behind the reformulation.

In general, we distinguish between different classes of logical constraints, i.e., constraints involving logical variables. In case of a dichotomy, i.e., either one constraint holds or another constraint, we talk about *either-or constraints*, or, in case that we need to choose k out of p machines, we refer to "k out of p constraints." Clearly, for $k = 1$ and $p = 2$, either-or constraints result as a special case. Another class is that of *conditional constraints*. They can be spotted by their unique wording "*if* this, *then* that." Examples of these conditions will follow.

One simple way to formulate problems involving logical constraints uses a small table that enumerates all possible solutions of the variables or activities involved in a problem and determines whether or not they are acceptable. This will allow us to decide which solutions to exclude. In order to demonstrate this, consider first the case of two zero-one variables y_1 and y_2 which take the value of one if the

5.1 Definitions and Basic Concepts

Table 5.1 All solutions for two binary variables/activities

#	y_1	y_2
1	0	0
2a	0	1
2b	1	0
3	1	1

Table 5.2 Exclude a single solution in case of two variables

Solution to be excluded	Formulation	Wording
1	$y_1 + y_2 \geq 1$	"At least one of the two activities needs to be chosen"
2a	$y_1 \geq y_2$	"If activity y_2 is chosen, then activity y_1 must also be chosen"
2b	$y_1 \leq y_2$	"If activity y_1 is chosen, then activity y_2 must also be chosen"
3	$y_1 + y_2 \leq 1$	"At most one of the activities can be chosen"

Table 5.3 All solutions for three binary variables/activities

#	y_1	y_2	y_3
1	0	0	0
2a	0	0	1
2b	0	1	0
2c	1	0	0
3a	0	1	1
3b	1	0	1
3c	1	1	0
4	1	1	1

corresponding activity is undertaken and zero if it is not. There are only four possible solutions and they are shown in Table 5.1.

Table 5.2 shows the *exclusion constraints* that force the exclusion of one specific solution without affecting any of the others.

Similarly, in the case of three variables (i.e., activities), we have the following eight solutions, shown in Table 5.3.

Table 5.4 shows how to formulate a logical constraint causing a single solution to be excluded.

Note that the wording offered in Table 5.4 is not necessarily unique. For instance, in case Solution 1 is to be excluded, this could also be based on the statement "If y_1 and y_2 are both not chosen, then y_3 must be chosen." Or, for instance, if Solution 3a is to be excluded, we could state that "If y_2 and y_3 are both chosen, then y_1 must be chosen as well."

In case more than a single solution is to be excluded, we simply combine the pertinent logical constraints. As an example, consider the case of two variables and exclude the solutions #1 and #2a. This task can be achieved by employing the two constraints $y_1 + y_2 \geq 1$ and $y_1 \geq y_2$. Simple inspection reveals that these two constraints can be collapsed into the single constraint $y_1 = 1$. In such a case, we

Table 5.4 Exclude a single solution in case of three variables

Solution to be excluded	Formulation	Wording
1	$y_1 + y_2 + y_3 \geq 1$	"At least one of the three activities must be chosen"
2a	$y_1 + y_2 \geq y_3$	"If neither y_1 nor y_2 are chosen, then y_3 cannot be chosen either"
2b	$y_1 + y_3 \geq y_2$	"If neither y_1 nor y_3 are chosen, then y_2 cannot be chosen either"
2c	$y_2 + y_3 \geq y_1$	"If neither y_2 nor y_3 are chosen, then y_1 cannot be chosen either"
3a	$-y_1 + y_2 + y_3 \leq 1$	"If y_1 is not chosen, then y_2 and y_3 cannot both be chosen"
3b	$y_1 - y_2 + y_3 \leq 1$	"If y_2 is not chosen, then y_1 and y_3 cannot both be chosen"
3c	$y_1 + y_2 - y_3 \leq 1$	"If y_3 is not chosen, then y_1 and y_2 cannot both be chosen"
4	$y_1 + y_2 + y_3 \leq 2$	"No more than two of the three activities can be chosen"

could simply replace y_1 by the value of 1 everywhere in the problem and thus reduce the size of the model.

Another logical constraint may require that an activity y_0 can only be chosen if at least k activities in a set $J = \{y_1, y_2, \ldots, y_p\}$ are also chosen. This can be formulated as

$$y_0 \leq \frac{1}{k}\left(y_1 + y_2 + \ldots + y_p\right).$$

5.2 Applications of Integer Programming

This section will introduce some classes and principles of applications of integer programming problems. The applications discussed here are not simply the ones that require integrality (those have been discussed in Sect. 2.2 of this volume), but those, in which integrality takes center stage. Those are the applications that require zero-one variables. The simplest integer programming problems in this class are the so-called *knapsack problems*. Formally, knapsack problems have an objective and a single constraint. The story behind the name can be told as follows. A backpacker wants to decide which items to take into the woods. Items to be considered include a tent, a sleeping bag, a stove, stove fuel, map, compass, and so forth. With each item, the backpacker associates two numbers: one that expresses the usefulness or value of the item, and the other its weight. The problem is then to choose items, so as to maximize their total value to the backpacker, while the total weight should not exceed a prespecified limit.

5.2 Applications of Integer Programming

Table 5.5 Profit contributions and resource consumption of projects

	High rise	Shopping mall	Amusement park	Warehouses	Airport
Profit contribution (in m$)	10	6	12	2	7
Resource consumption	4	2	5	1	3

There are a number of different versions of knapsack problems. One version only allows the backpacker to either pack an item or leave it at home, while another version permits the hiker to take any (integer) number of units of an item. The definition of variables is similar regardless of the application. In case we can only decide to either pack an item or not, we will define a logical variable for item j, such that $y_j = 1$, if we include the item in our pack, and 0 otherwise. In contrast, if we are allowed to take any number of units of item j, we will define variables y_j that are defined as the number of items of type j that we include in our pack, so that $y_j \geq 0$ and integer. (As a matter of fact, in the former case, we can also think of the variable in terms of the number of items taken, only that this number is restricted to no more than one.) Note that in order to distinguish integer variables from those that do not have to satisfy integrality, we will use y_j for integer variables and x_j for continuous variables.

Knapsack problems occur in a variety of guises. For continuous knapsack or *cargo loading* problems (i.e., those, in which the variables do not have to be integers or even zeroes and ones), an example is provided in Problem 2 in Sect. 2.5. Another popular example of a knapsack problem is *capital budgeting*. As a numerical illustration, suppose that a developer can engage in five different projects, viz., a high rise building, a shopping mall, an amusement park, a warehouse, and an airport. The expected profit contributions and resource consumption (e.g., the number of construction workers needed for the respective projects) are shown in Table 5.5.

Assume now that the developer has seven work crews at his disposal and suppose that it is not possible to hire additional work crews. We can then define variables $y_j = 1$, if the developer is to engage in project j, and 0 otherwise (clearly, engaging in partial projects is not feasible and none of the projects can be performed more than once), so that the problem can be formulated as follows:

$$P : \text{Max } z = 10y_1 + 6y_2 + 12y_3 + 2y_4 + 7y_5$$
$$\text{s.t.} \quad 4y_1 + 2y_2 + 5y_3 + 1y_4 + 3y_5 \leq 7$$
$$y_1, y_2, y_3, y_4, y_5 = 0 \text{ or } 1.$$

The linear programming relaxation that includes just the objective, the single constraint, and the nonnegativity constraints, has the optimal solution $\bar{y}_2 = 3.5$, $\bar{y}_1 = \bar{y}_3 = \bar{y}_4 = \bar{y}_5 = 0$ with the objective value $\bar{z} = 21$. Adding upper bounds $y_1 \leq 1$, $y_2 \leq 1$, $y_3 \leq 1$, $y_4 \leq 1$, and $y_5 \leq 1$ results in $\bar{y}_1 = \bar{y}_2 = 1$, $\bar{y}_3 = 0.8$, and $\bar{y}_4 = \bar{y}_5 = 0$ with an objective value of $\bar{z} = 18.4$. Finally, requiring integrality of all variables, we obtain the optimal solutions $\bar{y}_1 = \bar{y}_2 = \bar{y}_4 = 1$, $\bar{y}_3 = \bar{y}_5 = 0$ with the objective value $\bar{z} = 18$.

Many beginners might think that knapsack problems—and particularly zero-one knapsack problems—are so simple that modern computational equipment must surely be able to solve such problems by simply enumerating all solutions, without having to resort to complicated algorithms. Nothing could be further from the truth. In order to demonstrate this, let us compute the number of different solutions of a zero-one integer programming problem. A problem with a single variable has two solutions, as the variable can assume the values of zero and one. With two variables, we have four solutions: (0, 0), (0, 1), (1, 0), and (1, 1). With 3 variables, we already have 8 different solutions, with 4 variables, there are 16 different solutions, and so on. As a matter of fact, adding a single variable will double the number of solutions of the problem. The reason is this: the variable that we add can have a value of zero or one. If it were zero, together with all possible solutions of the other variables, we have just as many solutions as before. The same applies if the new variable were equal to one, so that adding a variable doubles the number of solutions. This means that in case of n variables, we will have 2^n different solutions that have to be examined. This means that for $n = 10$, we have 1024 solutions, for $n = 20$, there will be about a million solutions, for $n = 30$ there is a billion, for $n = 40$ a trillion, and so forth. Clearly, even if a computer could examine ten quadrillion solutions within a single second, (which is pretty much the limit by today's standards), a problem with 100 variables (a tiny problem by today's standards) would require more than 4 million years for the computer to examine all solutions. For most business problems, such a time frame appears excessive.

It is no wonder that many users resort to heuristic algorithms for the solution of integer programming problems. We will discuss exact and heuristic solution techniques in Sect. 5.3 of this chapter.

5.2.1 Diet Problems Revisited

First, recall the standard diet problem in linear programming. To simplify matters, consider only two foodstuffs and a single nutrient. The quantities of the two foods are defined as x_1 and x_2, respectively, and at least five units of the nutrient are required in the diet. We assume that the problem has been formulated as follows:

$$\begin{align} \text{P : Min } z &= 3x_1 + 4x_2 \\ \text{s.t.} \quad x_1 + 2x_2 &\geq 5 \\ x_1, \quad x_2 &\geq 0. \end{align}$$

Suppose now that the additional requirement is that if food 1 is included in the diet (in any quantity), then food 2 should not be. (One reason for this may be incompatibilities due to taste such as ice cream and mustard, or unfortunate side effects of incompatible foods, such as water and green apples, or, worse, yoghurt and yeast.) This is a conditional constraint of the type "if food 1 is included, then food

5.2 Applications of Integer Programming

Table 5.6 Decision table for the diet problem

y_1	y_2	OK?
0	0	Yes
0	1	Yes
1	0	Yes
1	1	No

2 should not be." We first must define logical zero-one variables, one for each foodstuff. These new variables y_1 (and y_2) are defined as being one, if food 1 (food 2) is included in the diet, and zero otherwise. We will need these variables *in addition to* the variables x_1 and x_2 that denote the quantities of the two foods that are included in the diet.

For instance, the first row in Table 5.6 includes neither of the two foods, and while it may leave us hungry, it does not violate the condition. Here, we find that only the solution that has y_1 and y_2 both equal to one (the case in which both foods are in the diet) is prohibited. We can now use the exclusion constraints introduced earlier.

Eliminating this solution from consideration is achieved by writing the constraint

$$y_1 + y_2 \leq 1.$$

This surely excludes only the case of both foods in the diet and adding this constraint to our formulation should do it. However, it does not. The reason is that by adding this constraint to the formulation, our new problem now has two sets of constraints; one that includes *only* the continuous variables x_1 and x_2, and another, completely separate part that includes *only* the variables y_1 and y_2. This would allow the two types of variables to change their values independent of each other, for instance, allowing solutions that have $y_1 = 0$ and $x_1 = 7$. This does not make sense, as $y_1 = 0$ states that "food 1 is not included in the diet," while $x_1 = 7$ says that "there are 7 units of food 1 in the diet."

The remedy is to include additional *linking constraints*, i.e. constraints that include the continuous variables x_1 and x_2 as well as the logical variables y_1 and y_2. In this problem, the linking constraints are

$$x_1 \leq My_1 \quad \text{and}$$

$$x_2 \leq My_2,$$

where M is a "sufficiently large" constant. In order to understand the workings of these linking constraints, consider the first of these constraints and use the two possible solutions $y_1 = 1$ and $y_1 = 0$. If $y_1 = 1$, then the constraint reads $x_1 \leq M$, which, with M being sufficiently large, is a redundant constraint that does not affect the solution. If, on the other hand, $y_1 = 0$, then the constraint reads $x_1 \leq 0$, which, in conjunction with the nonnegativity constraint $x_1 \geq 0$, forces x_1 to be equal to zero. This is exactly the desired effect, as if food 1 is not included in the diet (i.e., $y_1 = 0$), then its quantity in the diet (x_1) must be zero as well. And this is what these linking constraints guarantee.

Table 5.7 Decision table for the modified diet problem

y_1	y_2	OK?
0	0	Yes
0	1	Yes
1	0	No
1	1	Yes

Table 5.8 Decision table for the simple land use problem

y_1	y_2	OK?
0	0	Yes
0	1	Yes
1	0	Yes
1	1	No

Consider now a few extensions of the model. If, for instance, it would not be acceptable to have a diet without any food (i.e., we were to consider the solution $y_1 = y_2 = 0$ unacceptable), then the constraint $y_1 + y_2 = 1$ would guarantee that while both foods cannot be together in the diet, at least one of them has to be.

Consider now the diet problem with the different conditional constraint "if food 1 is included in the diet, then food 2 must be included in the diet as well." The decision table for this condition is shown in Table 5.7.

For this scenario, the constraint

$$y_1 \leq y_2$$

must be included (together with the linking constraints). The only solution that this constraints excludes is $\mathbf{y} = [0, 1]$, which is the desired effect.

5.2.2 Land Use

The next example deals with land use. Suppose that a land owner owns a parcel of land that he has to decide what to do with. He has narrowed down his decisions to two: sell stumpage, i.e., harvest the land, or build an animal sanctuary, but not both. Here, we will formulate only this aspect of the model and ignore all other considerations. We will need a decision variable for each possible decision, so that we define $y_1 = 1$, if we decide to harvest the parcel, and 0 otherwise, and $y_2 = 1$, if we decide to build an animal sanctuary, and 0 otherwise. The decision table for this problem is then shown in Table 5.8.

Once formalized as done here, we see that the situation is the same as in the diet problem by not allowing both options at the same time, which is modeled as $y_1 + y_2 \leq 1$. Since this problem does not appear to need quantitative variables x_1 and x_2 as was the case in the diet problem, we do not need linking variables, as there is nothing to link.

Things may get much more complicated when more options exist. Suppose now that for the parcel in question, three choices have been identified: Harvest (decision

5.2 Applications of Integer Programming

Table 5.9 Decision table for the extended land use problem

y_1	y_2	y_3	OK?
0	0	0	Yes
0	0	1	Yes
0	1	0	Yes
1	0	0	Yes
0	1	1	Yes
1	0	1	No
1	1	0	No
1	1	1	No

variable y_1), build a sanctuary (decision variable y_2), or allow the building of a municipal well (decision variable y_3). As in the previous land use example, it is not possible to harvest and have a sanctuary at the same time in the parcel in questions. Furthermore, the parcel cannot be harvested if there is a municipal well on the parcel, while we could very well have a well and a sanctuary on the same parcel. The decision table for this extended problem is shown in Table 5.9.

The modeling of this situation is considerably more complicated than that in the previous examples. As a matter of fact, we need two constraints to ensure that the bottom three solutions in Table 5.9 are excluded from consideration. Using the exclusion constraints in Table 5.4 at the end of Sect. 5.1, we can exclude the bottom two solutions from consideration by requiring $y_1 + y_2 - y_3 \leq 1$ and $y_1 + y_2 + y_3 \leq 2$. We could add the two constraints, resulting in $2y_1 + 2y_2 \leq 3$, or alternatively, $y_1 + y_2 \leq 1.5$, which, because both remaining variables are zero or one, can be tightened to $y_1 + y_2 \leq 1$. The third solution from the bottom can be eliminated by using the constraint $y_1 - y_2 + y_3 \leq 1$ as shown in Table 5.4. Instead, we may write $y_1 + y_3 \leq 1$, as this constraint additionally eliminates the bottom solution, which is not desired either (even though it was already excluded). In summary, we find that adding the constraints $y_1 + y_2 \leq 1$ and $y_1 + y_3 \leq 1$ leaves the first five solutions as options and disallows the last three, which is precisely what the decision maker had in mind.

5.2.3 Modeling Fixed Charges

The purpose of this section is to introduce decision models, which provide options regarding the choice of machines. One of the pertinent tasks is to ensure that machine-related constraints apply only to those machines that are purchased or leased. As an example, consider a publishing company that intends to produce its annual lineup of operations research texts. This year, they have the books by Gabby and Blabby (*GB*), Huff, Fluff, and Stuff (*HFS*), and the "Real OR" (*ROR*) texts. As usual nowadays, authors are required to do everything except for the printing, binding, and the subsequent marketing. A number of different machines are available for printing and binding, and from each type of machine, exactly one must be chosen. The three printing machines under consideration are P_1, P_2, and P_3, while

Table 5.10 Processing times for printing and binding machines

	Printing			Binding	
	P_1	P_2	P_3	B_4	B_5
GB	3	6	4	10	10
HFS	2	3	3	12	11
ROR	4	5	5	15	14

the two binding machines are B_4 and B_5. The processing times for the different books on the respective machines in minutes per book are shown in Table 5.10.

The capacities of the three printing machines are 120, 100, and 110 hrs (7200, 6000, and 6600 min), respectively, in the planning period. Similarly, the capacities of the binding machines are $333\frac{1}{3}$ and 300 hrs, respectively (or 20,000 and 18,000 min). The costs to lease the machines are independent of the number of books made with them. They are \$10,000, \$8000, \$9000, \$20,000, and \$23,000, respectively. The profit contributions of the three books (other than the leasing costs) have been identified as \$40, \$60, and \$70, respectively. It has also been determined that the publishing house should produce at least 500 copies of the landmark *ROR* book in order to maintain a good academic image.

We can formulate the problem by first defining variables x_1, x_2, and x_3 to indicate the number of books of the three types that are manufactured and sold. (As usual in single-period models, we assume that all units, which are manufactured can also be sold.) In addition, we also need zero-one variables that show whether or not a machine is leased. In particular, we define binary variables $y_1, y_2, \ldots, y_5$ that assume a value of one, if a machine is leased, and 0 otherwise. The objective function then consists of two parts: the sum of the profit contributions of the individual books (which are $40x_1$, $60x_2$, and $70x_3$), and the sum of the leasing costs, which are $10{,}000y_1$, $8000y_2$, $9000y_3$, $20{,}000y_4$, and $23{,}000y_5$, respectively.

First the easy constraints: making at least 500 copies of *ROR* is modeled as $x_3 \geq 500$, and the fact that we need exactly one printing and one binding machine can be written as $y_1 + y_2 + y_3 = 1$ and $y_4 + y_5 = 1$, respectively.

Consider now the capacity constraints of the machines. As usual, they are written as (machine usage) $\leq$ (machine capacity). For example, for the first printing machine, the capacity constraint is $3x_1 + 2x_2 + 4x_3 \leq 7200$. The problem here is that this constraint must hold only if the machine is actually leased. If it is not leased, the constraint can be ignored. The way to model this is the same technique that was applied in the diet problem in Sect. 5.2.1 to write the linking constraints. If the first machine is not leased, $y_1 = 0$, and the capacity constraint is made redundant, by having a sufficiently large right-hand side value. (Again, we will use the very large value M as introduced in linear programming.) However, if the first machine is leased, we have $y_1 = 1$, and the right-hand side of the capacity constraint should be 7200, the actual capacity of the first machine. We can formulate this by writing the right-hand side as $7200 + M(1 - y_1)$.

To demonstrate the validity of this formulation, let $y_1 = 0$. In this case the right-hand side value is $7200 + M(1 - y_1) = 7200 + M$, so that the constraint has a very large right-hand side, making it redundant. On the other hand, if we do lease the first

5.2 Applications of Integer Programming

machine and $y_1 = 1$, the right-hand side value equals $7200 + M(1 - y_1) = 7200$, which is the actual capacity of the first machine. The capacity constraints for the other machines are constructed similarly. The formulation can then be written as follows:

$$P : \text{Max } z = 40x_1 + 60x_2 + 70x_3 \\ - 10{,}000y_1 - 8000y_2 - 9000y_3 - 20{,}000y_4 - 23{,}000y_5$$

$$\text{s.t. } 3x_1 + 2x_2 + 4x_3 \leq 7200 + M(1 - y_1)$$

$$6x_1 + 3x_2 + 5x_3 \leq 6000 + M(1 - y_2)$$

$$4x_1 + 3x_2 + 5x_3 \leq 6600 + M(1 - y_3)$$

$$10x_1 + 12x_2 + 15x_3 \leq 20{,}000 + M(1 - y_4)$$

$$10x_1 + 11x_2 + 14x_3 \leq 18{,}000 + M(1 - y_5)$$

$$x_3 \geq 500$$

$$y_1 + y_2 + y_3 = 1$$

$$y_4 + y_5 = 1$$

$$x_1, x_2, x_3 \geq 0 \text{ and integer}$$

$$y_1, y_2, y_3, y_4, y_5 = 0 \text{ or } 1.$$

Using a large value for the constant M (here, we use $M = 1{,}000{,}000$), multiplying the brackets and sorting the variables, we obtain constraints such as $3x_1 + 2x_2 + 4x_3 + 1{,}000{,}000y_1 \leq 1{,}007{,}200$ for the first constraint, and similar for the other four capacity constraints. The solution of this all-integer programming problem is $\bar{y}_2 = \bar{y}_4 = 1$ and $\bar{y}_1 = \bar{y}_3 = \bar{y}_5 = 0$ (i.e., we lease the second printing and the first binding machine), and make $\bar{x}_1 = 0$ GB books, $\bar{x}_2 = 1039$ HFS books, and $\bar{x}_3 = 502$ ROR books. The profit associated with this plan is \$69,480. Note that the slack capacities on the printout will indicate huge (and meaningless) values for the machines that are not leased. This is due to the fact that their right-hand side values have the artificially high value of $M = 1{,}000{,}000$, from which some nonexistent usage is subtracted.

Suppose now that we allow more than one printing machine and/or more than one binding machine to be used. This will make the model considerably more complex. In particular, we will have to define additional variables x_{ij}, which denote the number of books of type i that are processed on machine j. The model can then be formulated as follows:

$$P : \text{Max } z = 40x_1 + 60x_2 + 70x_3 \\ - 10{,}000y_1 - 8000y_2 - 9000y_3 - 20{,}000y_4 - 23{,}000y_5$$

$$\text{s.t. } x_1 = x_{11} + x_{12} + x_{13}$$

$$x_2 = x_{21} + x_{22} + x_{23}$$

$$x_3 = x_{31} + x_{32} + x_{33}$$

$$3x_{11} + 2x_{21} + 4x_{31} \leq 7200y_1$$

$$6x_{12} + 3x_{22} + 5x_{32} \leq 6000y_2$$

$$4x_{13} + 3x_{23} + 5x_{33} \leq 6600y_3$$

$$10x_{14} + 12x_{24} + 15x_{34} \leq 20,000y_4$$

$$10x_{15} + 11x_{25} + 14x_{35} \leq 10,000y_5$$

$$x_3 \geq 500$$

$$x_{11} + x_{12} + x_{13} = x_{14} + x_{15}$$

$$x_{21} + x_{22} + x_{23} = x_{24} + x_{25}$$

$$x_{31} + x_{32} + x_{33} = x_{34} + x_{35}$$

$$y_1, \ldots, y_5 = 0 \text{ or } 1$$

$$x_1, x_2, x_3; x_{11}, x_{12}, \ldots, x_{35} \geq 0 \text{ and integer.}$$

We wish to point out that the former model can be written as a special case of this formulation. All we need to do is to include the two constraints $y_1 + y_2 + y_3 = 1$ and $y_4 + y_5 = 1$.

The objective function has not changed in comparison to the previous model. The first three constraints link the variables x_{ij} and the variables x_i with each other. In other words, they define the number of units of the products that are actually made. The next five constraints are the capacity constraints that restrict the number of units that we can make by the machine capacities (which are zero if we do not lease the machine). The next constraint ensures that we make at least 500 units of the *ROR* book. Finally, the last three structural constraints require that the number of books of each type that we print must also be bound. Note that we could, of course, replace all variables x_j by x_{ij} if so desired.

The optimal solution has the decision maker lease the first printing machine and both binding machines. As before, we will make no *GB* books, but now we make 2600 *HFS* books, and 500 *ROR* books. In other words, the only significant change is the increase of *HFS* books from 1039 to 2600. This results in a doubling of the profit from the original $69,480 to $138,000.

5.2 Applications of Integer Programming

5.2.4 Workload Balancing

The problem presented in this subsection deals with the allocation of tasks to employees, so as to ensure that none of the employees is overworked, while others are partially idle. We assume that tasks cannot be split, meaning that once an employee starts a job, he will have to finish it. (This scenario is reminiscent of the bin packing problem, see, e.g., Chap. 17.) Due to their different backgrounds and training, a job will take different amounts of time if different employees perform it. There are three workers W_1, W_2, and W_3, who will have to perform tasks $T_1, \ldots, T_5$. Table 5.11 shows the processing times (in hours) for all worker–task combinations.

In order to formulate the problem, we need to introduce zero-one variables, which are defined as $y_{ij} = 1$, if employee W_i is assigned to task T_j, and zero otherwise. The only constraints of the model ensure that each task is assigned to exactly one employee. Formally, we can write

$$y_{1j} + y_{2j} + y_{3j} = 1 \text{ for all } j = 1, \ldots, 5.$$

The more contentious issue concerns the objective function. First, we note that the actual working time of the employees can be written as

$$w_1 = 5y_{11} + 1y_{12} + 9y_{13} + 4y_{14} + 9y_{15},$$
$$w_2 = 4y_{21} + 3y_{22} + 8y_{23} + 3y_{24} + 8y_{25}, \text{ and}$$
$$w_3 = 7y_{31} + 5y_{32} + 6y_{33} + 4y_{34} + 7y_{35},$$

where the new variables w_1, w_2, and w_3 denote that time that employees W_1, W_2, and W_3 are busy working on the tasks. One possibility to ensure fairness in the solution is to attempt to make the longest working time as short as possible. In other words, the employee who works longest should have the shortest working hours possible. Formally for our problem, we can write this objective as

$$\text{Min } z = \max \{w_1, w_2, w_3\}.$$

Clearly, this is not part of a linear programming problem the way we have defined it. However, *minimax objective functions* of this type can easily be reformulated, resulting in standard linear (or integer) programming problems. This is done by introducing a single new variable, say z, which measures the highest workload of any employee, i.e., the maximum of the right-hand side of the above objective. We will then minimize this highest workload, but we must ensure that none of the actual workloads is higher, which is done by introducing the three constraints $z \geq w_1$,

Table 5.11 Processing times for worker–task combinations

	T_1	T_2	T_3	T_4	T_5
W_1	5	1	9	4	9
W_2	4	3	8	3	8
W_3	7	5	6	4	7

$z \geq w_2$, and $z \geq w_3$. Replacing w_1, w_2, and w_3 by the functions of y_{ij}, we can then write the model as

$$P : \text{Min } z$$

$$\text{s.t. } z \geq 5y_{11} + 1y_{12} + 9y_{13} + 4y_{14} + 9y_{15}$$

$$z \geq 4y_{21} + 3y_{22} + 8y_{23} + 3y_{24} + 8y_{25}$$

$$z \geq 7y_{31} + 5y_{32} + 6y_{33} + 4y_{34} + 7y_{35}$$

$$y_{11} + y_{21} + y_{31} = 1$$

$$y_{12} + y_{22} + y_{32} = 1$$

$$y_{13} + y_{23} + y_{33} = 1$$

$$y_{14} + y_{24} + y_{34} = 1$$

$$y_{15} + y_{25} + y_{35} = 1$$

$$y_{ij} = 0 \text{ or } 1 \text{ for } i = 1, 2, 3; j = 1, \ldots, 5.$$

One solution will have worker W_1 work on tasks T_2 and T_5, worker W_2 works on tasks T_1 and T_4, while worker W_3 is assigned task T_3. The resulting workloads for the three employees are then 10, 7, and 6 hrs, respectively (with the variable $z = 10$ denoting the highest of the workloads). Note that other solutions exist, e.g., worker W_1 works on tasks T_1, T_2, and 4, worker W_2 works on task T_5, and worker W_3 works on task T_3, so that the workloads are 10, 8, and 6, respectively. Again, the highest workload is $z = 10$ hrs. The total time the employees work is 24, 1 hr more than in the previous solution.

Other objective functions to ensure fairness have been used in the literature. For instance, we could define a variable $w = w_1 + w_2 + w_3$ as the total workload by all of the employees, so that ideally, each employee would work $\frac{1}{3}w$. We could not minimize the sum of deviations from this workload. Since positive and negative deviations will cancel out, we have to either minimize the sum of absolute values of the deviations or square the deviations (as done to derive the variance in statistics, which is the sum of squared deviations from the mean). This objective would then be

$$\text{Min } z = \left(\tfrac{1}{3}w - w_1\right)^2 + \left(\tfrac{1}{3}w - w_2\right)^2 + \left(\tfrac{1}{3}w - w_3\right)^2.$$

Unfortunately, this objective is nonlinear and the resulting problem will then be nonlinear and integer, a very difficult combination. Solving this problem results in workloads of 9, 11, and 11 hrs with a total work time of 31 hrs (much higher than in the previous solutions), but with a much smaller variance.

In general, fairness (or "equity") objectives should normally be combined with efficiency objectives. If not, a solution such as workloads of 12, 12, and 12 hrs for three employees would be preferred to workloads of 4, 9, and 8 hrs, because the former solution is "more equal," even though each employee would gain by moving from the former to the latter solution.

5.2.5 Political Districting

An application of considerable importance and controversy deals with the redesign of electoral districts. The controversy arises from the fact that each party will attempt to redesign the districts, so that it will win (possibly by a small margin) in as many districts as possible, while "in return," it accepts to lose big in a few districts. Here, we provide a simple prototype model for the problem. Consider an area with eight regions, which are to be put together into three electoral districts, so that a voter in each of these districts wielded approximately the same power as any other voter, thus satisfying the principle of "one person, one vote." The regions are shown in Fig. 5.4, and the number of eligible voters in the regions are 65, 29, 95, 30, 52, 57, 21, and 41, respectively, for a total of 390 registered voters. Given that the task is to aggregate regions to three electoral districts, ideally, each district comprises 130 voters. This is the basis of the formulation.

Defining binary variables x_{ij} which assume a value of 1, if region i is in district j, and 0 otherwise, as well as deviational variables (as in goal programming, see Sect. 3.3 of this book) d_j^+ and d_j^- as district j's deviation below (above) the ideal number of voters. We can then formulate the problem as

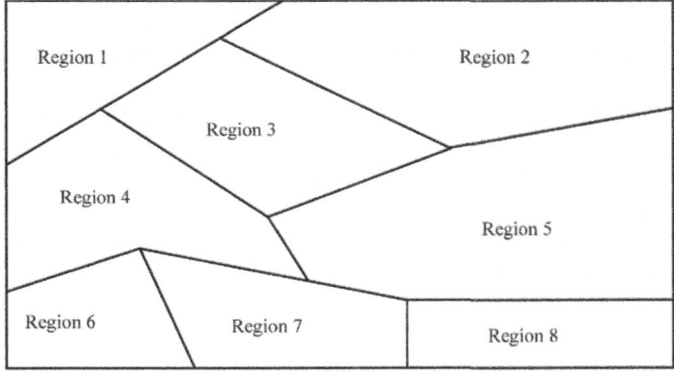

Fig. 5.4 Regions in the reapportionment problem

P : Min $z = d_1^+ + d_1^- + d_2^+ + d_2^- + d_3^+ + d_3^-$
s.t. $x_{11} + x_{12} + x_{13} = 1$
$x_{21} + x_{22} + x_{23} = 1$
$\vdots$
$x_{81} + x_{82} + x_{83} = 1$
$65x_{11} + 29x_{21} + 95x_{31} + 30x_{41} + 52x_{51} + 57x_{61} + 21x_{71} + 41x_{81} = 130 - d_1^+ + d_1^-$
$65x_{12} + 29x_{22} + 95x_{32} + 30x_{42} + 52x_{52} + 57x_{62} + 21x_{72} + 41x_{82} = d_2^+ + d_2^-$
$65x_{13} + 29x_{23} + 95x_{33} + 30x_{43} + 52x_{53} + 57x_{63} + 21x_{73} + 41x_{83} = d_3^+ + d_3^-$
$x_{ij} = 0$ or 1 for all i, j.

The optimal solution has $x_{11} = x_{21} = x_{81} = 1$, $x_{32} = x_{42} = 1$, and $x_{53} = x_{63} = x_{73} = 1$, i.e., the first district comprises Regions 1, 2, and 8, the second includes Regions 3 and 4, and the third district consist of Regions 5, 6, and 7, so that the number of voters in the respective districts is 135, 125, and 130, respectively with a total deviation of 10 from the idea size of 130. This is a nice result, except for the fact that the first district is not contiguous, Region 8 is not directly connected to Regions 1 and 2. We can then formulate the conditional constraint "If regions 2 and 8 are in a district, then region 5 must be in the district as well." Rather than formulating this only for the first district (which currently violates this condition), we formulate it for all three districts, as otherwise, the solver could simply switch the first and the second district and the same problem would appear, this time for the second district. Hence, we formulate

$$x_{21} + x_{81} - x_{51} \leq 1,$$
$$x_{22} + x_{82} - x_{52} \leq 1, \text{ and}$$
$$x_{23} + x_{83} - x_{53} \leq 1.$$

Adding these constraints to the formulation results in a solution, in which District 1 now consists of Regions 3 and 8 (not contiguous), District 2 comprises Regions 5, 6, and 7 (snake-like, but contiguous), and District 3 includes Regions 1, 2, and 4 (contiguous, but a weird shape). Due to the additional constraints, the sizes of the districts are now 136, 130, and 124, up from 10 to 12.

Adding conditional constraints that require that if Regions 3 and 8 are in a district, then Region 5 (or, alternatively, Regions 5 and 7 or Regions 4, 6, and 7) must be in the same district, we add constraints

$$x_{31} + x_{81} - x_{51} \leq 1,$$
$$x_{32} + x_{82} - x_{52} \leq 1, \text{ and}$$
$$x_{33} + x_{83} - x_{53} \leq 1.$$

The solution now has District 1 contain Regions 1, 4, and 8 (not contiguous), District 2 has Regions 2 and 3, and District 3 includes 5, 6, and 7 (which is long and

5.3 Solution Methods for Integer Programming Problems

not very wide). The number of voters in the districts are now 136, 124, and 130, so that the total deviation is still 12.

This process then continues until a contiguous solution has been found. It is also possible to continue the process in order to avoid unpleasant shapes, such as District 3 (Regions 5, 6, and 7) in the present solution. We could accomplish this by requiring that "if Regions 5 and 7 are in one district, then Region 8 cannot be in the same district," or similar constraints.

5.3 Solution Methods for Integer Programming Problems

This section will examine some of the techniques that can and are used to solve integer programming problems. The first subsection will briefly describe the ideas behind cutting plane methods. The second subsection thoroughly discusses the basic features of branch-and-bound techniques, and the third subsection illustrates some heuristic methods and how they may apply to some integer programming problems.

5.3.1 Cutting Plane Methods

As mentioned in the introduction to this chapter, the first exact techniques for the solution of integer programming problems were the so-called *cutting plane techniques*. Their general idea may be explained as follows. Recall that some linear programming problems always have optimal integer solutions, such as the assignment problem and the transportation problem (the latter only if all supplies and demands are integer-valued). Given Dantzig's corner point theorem (see Sect. 2.3 of this book), this means that the feasible set of those two problems (and others like them) has corner points, all of which are integer-valued. Clearly, very few formulations have this property. If it were now feasible to find a restricted feasible set that includes all-integer solutions of the original feasible set but has all of its extreme points integer-valued (something usually called a *convex hull*), then we could simply solve the linear programming relaxation of the problem with this restricted feasible set and then automatically obtain an integer solution. Unfortunately, obtaining the convex hull of a feasible set is very difficult and the suggested approach is not computationally viable.

However, it is not necessary to have the entire convex hull at our disposal. And this is where cutting plane methods come in. The idea of cutting plane methods is to locally approximate the convex hull. This is done as follows. We first solve the linear programming relaxation of the given integer programming problem. If the optimal solution happens to be integer, we are done. Suppose now that it is not. We then formulate a cutting plane, i.e., an additional constraint that does two things: (1) it must cut off (i.e., make infeasible) the present optimal solution, while (2) it cannot cut off any feasible integer point.

As a numerical illustration, consider the following all-integer programming problem:

Fig. 5.5 Illustration of the cutting plane method

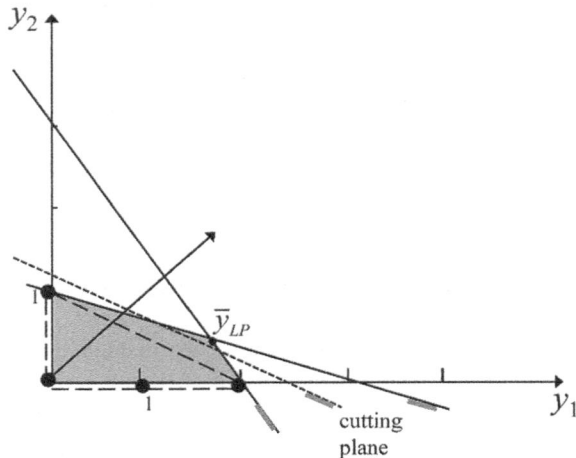

$$P : \text{Max } z = y_1 + y_2$$
$$\text{s.t.} \quad 3y_1 + 2y_2 \leq 6$$
$$y_1 + 3y_2 \leq 3$$
$$y_1, \quad y_2 \geq 0 \text{ and integer.}$$

The shaded area in Fig. 5.5 shows the feasible set of the linear programming relaxation, and the point shown as $\overline{y}_{LP} = (12/7, 3/7)$ is the optimal solution of the linear programming relaxation. The triangle shown by the broken lines that connect the points (0, 0), (2, 0), and (0, 1) is the convex hull of the feasible set for the integer programming problem. The dotted line shows the boundary of the cutting plane $5y_1 + 10y_2 \leq 12$. Plugging in the coordinates of $\overline{y}_{LP}$, we see that this point violates the condition of the cutting plane, as $90/7 = 12\frac{6}{7} \not\leq 12$. This is also apparent in the graph. On the other hand, all four feasible integer points satisfy the condition, so that the condition is indeed a cutting plane.

While cutting planes appear to be a very good idea, their computational performance for general integer programming problems has been disappointing. In particular, the slices that are cut off tend to become very small, requiring a very large number of additional constraints to be added in the process, which adds to the size of the problem and the degree of difficulty.

Below we will demonstrate this fact for one specific type of cut, a so-called *Dantzig cut*. We have chosen this particular type of cutting plane, as it does not require any knowledge beyond the solution that is typically provided by a solver. Our discussion is designed to provide a general idea about cutting planes, it is not our intention to suggest the practical use of this particular type of cutting plane which turned out to be rather inefficient. Other, more efficient, cutting planes work on the same principle.

5.3 Solution Methods for Integer Programming Problems

Again, let us start with the general idea. Suppose that we have an all-integer linear programming problem in which the set of variables includes all slack and excess variables, so that all constraints are written as equations. Let there be n nonnegative variables (including the slack and excess variables) and m structural equation constraints. Furthermore, assume that the present optimal solution (starting with the linear programming relaxation) has at least one variable assume a noninteger value. At this point, we can separate the variables into two disjoint sets B and N, where B includes all variables that presently have a positive value, while N includes all variables that are presently zero. If the solution is nondegenerate, the set B includes exactly m variables, and the set N includes exactly $(n-m)$ variables. In case of primal degeneracy, the set N will include more than $(n-m)$ variables, in which case we define N as any $(n-m)$ variables that are presently zero.

A *Dantzig cut* is now an additional constraint that requires that the sum of all variables in the set N is at least "1." The validity of this constraint as a cutting plane (i.e., a constraint that makes the present solution infeasible, but that does not invalidate any feasible integer points) can be shown as follows. First of all, as in the present solution all variables in the set N equal zero, this solution does not satisfy the requirement of the new constraint, so that the cutting plane does cut off the present solution. On the other hand, any feasible solution to the original integer problem will need to have at least one variable in N assume a positive value, which, since this is an all-integer optimization problem, must be at least one. Hence, the sum of all the variables that are presently at the zero level, must be at least one, hence Dantzig's cut.

Once the cutting plane has been formulated and is added to the formulation, the problem is solved again. Note that in practice, it is not required to use a "cold start," i.e. start the optimization from scratch. Much rather, it is typically a good idea to use a "warm start," in which the cutting plane is added to the formulation and the solver continues from the previously optimal solution. At this point, we do not have to concern ourselves with such technical details. Note that each time a cutting plane is added to the system, the value of the objective function either stays the same (in case of dual degeneracy) or decreases (in maximization problems, it increases in minimization problems). This is simply due to the fact that the feasible set shrinks each time a cut is added. An important part of the development of a cutting plane is the proof that eventually, the sequential addition of cutting plane results in an integer optimal solution. Such a proof is not available for Dantzig cuts, meaning that it is not guaranteed that the process will eventually terminate with an integer solution.

We will now explain the procedure by means of a numerical

Example Consider the following all-integer programming problem:

$$P : \text{Max } z = 3y_1 + 2y_2$$
$$\begin{align} \text{s.t. } 3y_1 + 7y_2 &\leq 22 & (I) \\ 5y_1 + 3y_2 &\leq 17 & (II) \\ y_1 &\geq 2 & (III) \\ y_1, y_2 &\geq 0 \text{ and integer.} \end{align}$$

Adding slack variables S_1 and S_2 to the left-hand sides of the first two constraints and subtracting an excess variable E_3 from the left-hand side of the third constraint, we obtain the following formulation with $n = 5$ variables and $m = 3$ structural constraints:

$$P : \text{Max } z = 3y_1 + 2y_2$$
$$\begin{align} \text{s.t. } 3y_1 + 7y_2 + S_1 &= 22 \\ 5y_1 + 3y_2 + S_2 &= 17 \\ y_1 - E_3 &= 2 \\ y_1, y_2, S_1, S_2, E_3 &\geq 0 \text{ and integer.} \end{align}$$

Solving the linear programming relaxation of this problem, we obtain the optimal solutions $\bar{y}_1 = 2.0385$, $\bar{y}_2 = 2.2692$, $\bar{S}_1 = \bar{S}_2 = 0$, and $\bar{E}_3 = 0.0385$ with an objective value $\bar{z} = 10.65385$. The solution is nondegenerate with the variables S_1 and S_2 being included in the set N, so that the first Dantzig cut is

$$S_1 + S_2 \geq 1$$

(or, given that the original constraints imply that $S_1 = 22 - 3y_1 - 7y_2$ and $S_2 = 17 - 5y_1 - 3y_3$, the cut could also be written as $8y_1 + 10y_2 \leq 38$, but we will continue with the original version of the cut). Subtracting a new excess variable E_4 from the left-hand side of this cut, we obtain $S_1 + S_2 - E_4 = 1$. Adding this cut to the problem and solving it again, we obtain the new solution $\bar{y}_1 = 2.1538$, $\bar{y}_2 = 2.0769$, $\bar{S}_1 = 1$, $\bar{S}_2 = \bar{E}_4 = 0$, and $\bar{E}_3 = 0.1538$ with an objective value $\bar{z} = 10.61539$. Clearly, another cut is required. The sequence of cutting planes that are generated in the process is shown in Table 5.12.

It is apparent that even in this very small problem, a large number of cutting planes are required to solve the problem. Computational experience with Dantzig's cutting planes confirms this. Figure 5.6 shows that the cutting planes in this example only slice off tiny parts of the feasible set. Different cuts, particularly the so-called deep cuts perform much better, but can still not compete with branch-and-bound methods, at least not for general (unstructured) integer programming problems.

One possible modification in the above example could have used the objective function to derive a cut. The idea is this. Given the fact that all variables must be integer, the value of the objective function $z = 3y_1 + 2y_2$ must also be integer. Given

5.3 Solution Methods for Integer Programming Problems

Table 5.12 Cutting planes generated in the example

Optimal solution	Cutting plane
$\bar{y}_1 = 2.0385, \bar{y}_2 = 2.2692, \bar{S}_1 = 0, \bar{S}_2 = 0, \bar{E}_3 = 0.0385$ with $\bar{z} = 10.65385$ (optimal solution of the linear programming relaxation).	$S_1 + S_2 \geq 1$ or $S_1 + S_2 - E_4 = 1$
$\bar{y}_1 = 2.1538, \bar{y}_2 = 2.0769, \bar{S}_1 = 1, \bar{S}_2 = 0, \bar{E}_3 = 0.1538, \bar{E}_4 = 0$ with $\bar{z} = 10.61539$.	$S_2 + E_4 \geq 1$ or $S_2 + E_4 - E_5 = 1$
$\bar{y}_1 = 2.2692, \bar{y}_2 = 1.8846, \bar{S}_1 = 2, \bar{S}_2 = 0, \bar{E}_3 = 0.2692, \bar{E}_4 = 1, \bar{E}_5 = 0$ with $\bar{z} = 10.5769$.	$S_2 + E_5 \geq 1$ or $S_2 + E_5 - E_6 = 1$
$\bar{y}_1 = 2.3846, \bar{y}_2 = 1.6923, \bar{S}_1 = 3, \bar{S}_2 = 0, \bar{E}_3 = 0.3846, \bar{E}_4 = 2, \bar{E}_5 = 1, \bar{E}_6 = 0$ with $\bar{z} = 10.53846$.	$S_2 + E_6 \geq 1$ or $S_2 + E_6 - E_7 = 1$
$\bar{y}_1 = 2.5, \bar{y}_2 = 1.5, \bar{S}_1 = 4, \bar{S}_2 = 0, \bar{E}_3 = 0.5, \bar{E}_4 = 3, \bar{E}_5 = 2, \bar{E}_6 = 1, \bar{E}_7 = 0$ with $\bar{z} = 10.5$.	$S_2 + E_7 \geq 1$ or $S_2 + E_7 - E_8 = 1$
$\bar{y}_1 = 2.6154, \bar{y}_2 = 1.3077, \bar{S}_1 = 5, \bar{S}_2 = 0, \bar{E}_3 = 0.6154, \bar{E}_4 = 4, \bar{E}_5 = 3, \bar{E}_6 = 2, \bar{E}_7 = 1, \bar{E}_8 = 0$ with $\bar{z} = 10.46154$.	$S_2 + E_8 \geq 1$ or $S_2 + E_8 - E_9 = 1$
$\bar{y}_1 = 2.7308, \bar{y}_2 = 1.1154, \bar{S}_1 = 6, \bar{S}_2 = 0, \bar{E}_3 = 0.7308, \bar{E}_4 = 5, \bar{E}_5 = 4, \bar{E}_6 = 3, \bar{E}_7 = 2, \bar{E}_8 = 1, \bar{E}_9 = 0$ with $\bar{z} = 10.42308$.	$S_2 + E_9 \geq 1$ or $S_2 + E_9 - E_{10} = 1$
$\vdots$	$\vdots$
$\bar{y}_1 = 2, \bar{y}_2 = 2, \bar{S}_1 = 2, \bar{S}_2 = 1, \bar{E}_3 = 0$ with $\bar{z} = 10$ (optimal all-integer solution)	

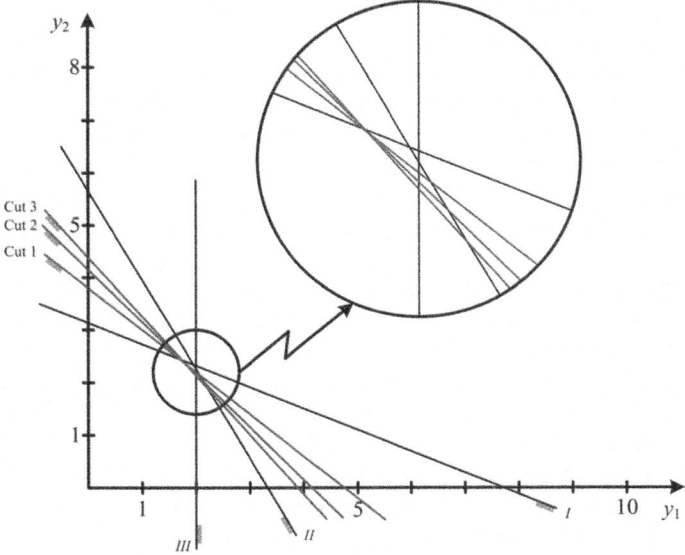

Fig. 5.6 Cuts with Dantzig's method

that we have just solved the linear programming relaxation of the problem that has an objective value of $\bar{z} = 10.65385$, we know that $\bar{z} \leq 10$ must hold. We can then define the cutting plane $3y_1 + 2y_2 + S_4 = 10$. Solving the problem with this added constraint results in the solution $\bar{y}_1 = 3.3333$, $\bar{y}_2 = 0$, $\bar{S}_1 = 12$, $\bar{S}_2 = 0.3333$, $\bar{E}_3 = 1.3333$ and $\bar{S}_4 = 0$ with an objective value of $\bar{z} = 10$. (Since the objective value has not changed, we presently encounter dual degeneracy.)

The next cutting plane is then $y_2 + S_4 \geq 1$, or, in its proper form, $3y_1 + y_2 + S_5 = 9$. The resulting solution has $\bar{y}_1 = 2.6667$, $\bar{y}_2 = 1$, $\bar{S}_1 = 7$, $\bar{S}_2 = 0.6667$, $\bar{E}_3 = 0.667$, $\bar{S}_4 = \bar{S}_5 = 0$, again with an objective value of $\bar{z} = 10$.

The next cut is $S_4 + S_5 \geq 1$, or, rewritten in terms of the original variables and the new slack variable S_6, it is written as $6y_1 + 3y_2 + S_6 = 18$. The optimal solution is then $\bar{y}_1 = \bar{y}_2 = 2$, $\bar{S}_1 = 2$, $\bar{S}_2 = 1$, $\bar{E}_3 = \bar{S}_4 = 0$, $\bar{S}_5 = 1$, $\bar{S}_6 = 0$, with a value of the objective function of $\bar{z} = 10$. This solution is an integer optimum.

5.3.2 Branch-and-Bound Methods

In contrast to cutting plane methods, the concept of branch-and-bound in its many variations is nowadays accepted as the universal solver for mixed- and all-integer linear programming problems. The basic idea is simple. We first solve again the linear programming relaxation. Again, if its coordinates of the optimal solution point satisfy the required integrality conditions, the process terminates. Suppose now that this is not the case. We then select any variable that is required to be integer but, at the present solution, is not. As an example, let the variable $y_7 = 3.4$ at the present solution. Given that y_7 is required to be integer, the present solution does not satisfy all integrality constraints. We will then subdivide the given problem (often called the "parent") into two new problems (the "children"), a process usually referred to as "divide and conquer" (or, by some critics, "double your trouble"). Each of the two children contains exactly the same constraints as its parent plus one single additional constraint. One child (often called the "left child" based on the graphical depiction) has the additional constraint $y_7 \leq 3$, while the other child (the "right child") has the additional constraint $y_7 \geq 4$. The union of the feasible sets of the two problems includes all feasible points of the original problem, except all of those, for which $3 < y_7 < 4$. The reason that we can actually cut out such a "corridor" without deleting important parts of the feasible set is that none of the solutions that we eliminate from consideration is feasible, in that none of them contains any point that has an integer value for the variable y_7.

Continuing in this way, we will build up what is known as a *solution tree*. Each "node" of the tree, shown as a small box, represents a problem formulation and a solution. Once we decide to work on a node or solution, we "branch" from it, meaning we generate its children by adding a constraint as explained above. Once we have branched from a node, it is no longer *active*. Also, nodes that represent integer solutions and those that represent problem formulations that do not have feasible solutions are also not active. This is best explained in terms of a small numerical example.

5.3 Solution Methods for Integer Programming Problems

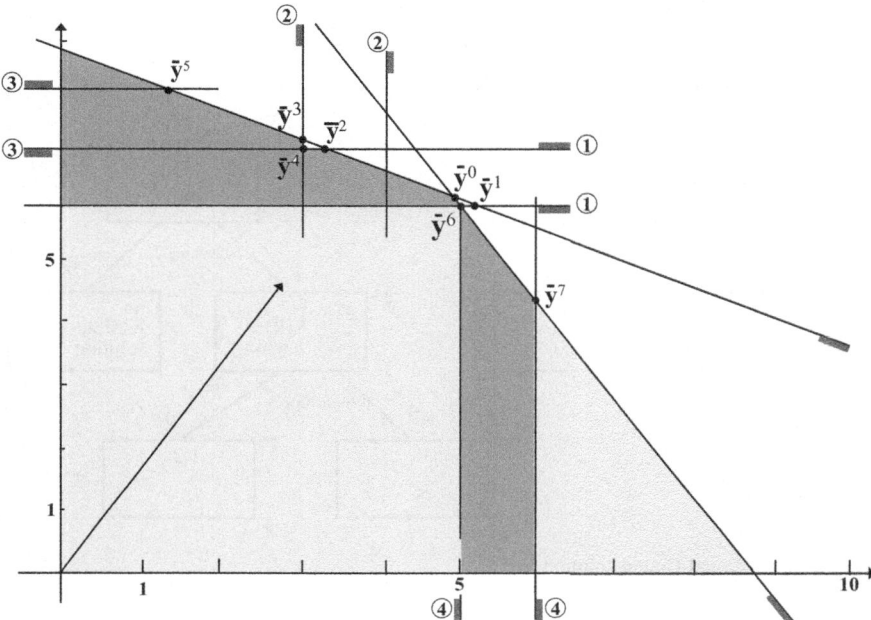

Fig. 5.7 Graphical representation of the Branch-and-Bound Method

Consider the following all-integer optimization problem:

$$P : \text{Max } z = 5y_1 + 9y_2$$
$$\text{s.t.} \quad 5y_1 + 11y_2 \leq 94 \quad \text{Constraint } I$$
$$10y_1 + 6y_2 \leq 87 \quad \text{Constraint } II$$
$$y_1, \quad y_2 \geq 0 \text{ and integer.}$$

The graphical representation of the problem is shown in Fig. 5.7, where the feasible set of the linear programming relaxation is the shaded area. The optimal solution of the linear programming relaxation is depicted as the point $\bar{\mathbf{y}}_{LP}$. The intermediate results of our computations will be displayed and collected in the solution tree shown in Fig. 5.8. At present, only the top node, known as the *root of the tree*, is known. At this point, this is also the only active node.

The first step is to examine the solution. In this example, both variables are supposed to be integer and neither of them is. This means that we can start working on either variable. Here, we choose to start working on y_2. (Alternatively, we could have started working on y_1. This would have resulted in a completely different solution tree—which one is smaller and easier cannot be said in advance—which is shown in Fig. 5.9.) At present, $y_2 = 6.3125$, so that we will eliminate the corridor $6 < y_2 < 7$, so that our first branching, shown in the solution tree in Fig. 5.8 branches from the root tree to the two children, the one on the left with the additional constraint $y_2 \leq 6$, and the one on the right with the additional constraint $y_2 \geq 7$. In

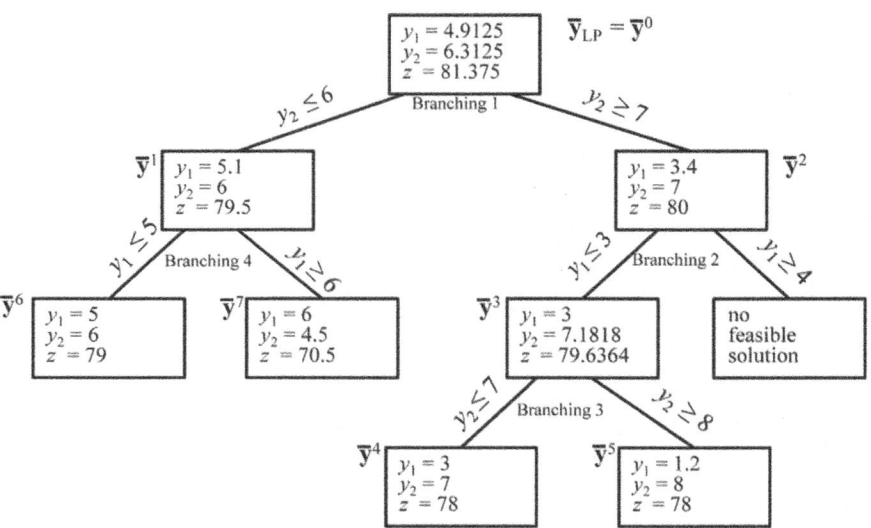

Fig. 5.8 Branch-and-Bound solution tree

Fig. 5.7, the feasible set of the right child is now the triangle with vertices at $(0, 7)$, $(0, 8.5455)$, and $\bar{y}^2 = (3.4, 7)$. Using the objective function, the optimal solution of this subproblem is $\bar{y}^2$. On the other hand, the feasible set of the left child is the trapezoid with corner points $(0, 0)$, $(0, 6)$, $\bar{y}^1 = (5.1, 6)$, and $(8.7, 0)$. The optimal solution of this subproblem is $\bar{y}^1$.

At this point we have two active nodes with solutions $\bar{y}^1$ and $\bar{y}^2$. Among all active nodes, we always choose the one with the best value of the objective function (i.e., highest for maximization problems and lowest for minimization problems) to examine and further work on, if necessary. In this example, it is the node with the solution $\bar{y}^2$ (now considered the parent). The solution $\bar{y}^2$ has values of 3.4 and 7 for the two variables, meaning that while the variable y_2 is integer as required, the variable y_1 is not and we will have to work on it. This means that from this solution, we will perform our second branching that adds the constraint $y_1 \leq 3$ to the left child, and $y_1 \geq 4$ to the right child. The additional constraints are shown in Fig. 5.7 as short lines that will have to be considered in conjunction with the triangle (the feasible set of the parent) derived earlier. This results in a feasible set of the left child that is the trapezoid with corner points $(0, 6)$, $(0, 8.5455)$, $\bar{y}^3 = (3, 7.1818)$, and $(3, 6)$, the optimal solution of which is $\bar{y}^3$. On the other hand, the right child has an empty feasible set, as the intersection of the triangle of its parent and the constraint $y_1 \geq 4$ is empty. This is now indicated in the solution tree in Fig. 5.8. At this point we have two active nodes: the nodes labeled with solutions $\bar{y}^1$ and $\bar{y}^3$. All other nodes developed so far have either already been branched from or have no feasible solution and thus need no longer to be considered.

5.3 Solution Methods for Integer Programming Problems

The node with the best (here: highest) value of the objective functions value is $\overline{y}^3$, so that our work continues here. Now the variable $y_2 = 7.1818$ is again noninteger and we will branch on it. Considering this node as the present parent, we branch to the children by using the additional constraints $y_2 \leq 7$ and $y_2 \geq 8$, respectively. The feasible set of the left child is the line segment with end points $(0, 7)$ and $(3, 7)$. The optimal solution of this child is $\overline{y}^4 = (3, 7)$ with a value of the objective function of 78. This is our first integer solution so far in the tree. The feasible set of the right child is the triangle with vertices $(0, 8)$, $(0, 8.5455)$, and $\overline{y}^5$, with $\overline{y}^5$ being its optimal solution, whose objective value is also 78.

At this point, the active nodes are $\overline{y}^1$, $\overline{y}^4$, and $\overline{y}^5$. Inspection reveals that $\overline{y}^1$ has the highest value of the objective function, so that work will continue here, regardless of the fact that we have already found an integer solution (that solution may not be the best). This node is now temporarily considered the parent. In this solution, $y_1 = 5.1$, which violates the required integrality condition.

Recall that the feasible solution of this parent is the shaded area below $y_2 \leq 6$. Branching $y_1 \leq 5$ and $y_1 \geq 6$, we obtain the two lightly shaded feasible sets: the rectangle with vertices $(0, 0)$, $(0, 6)$, $\overline{y}^6 = (5, 6)$, and $(5, 0)$ for the left child, and the triangle with vertices $(6, 0)$, $\overline{y}^7 = (6, 4.5)$, and $(8.7, 0)$ for the right child. The respective optimal solutions are $\overline{y}^6$ and $\overline{y}^7$, whose values of the objective function are 79 and 70.5, respectively.

At this point, we have four active nodes with solutions $\overline{y}^4$, $\overline{y}^5$, $\overline{y}^6$, and $\overline{y}^7$. The node with the best value of the objective function is $\overline{y}^6$, whose solution is $\overline{y}^6 = (5, 6)$ and objective value 79. Work will continue on this node. Inspection reveals that all integrality conditions are now satisfied, and no other active node has an objective function value this large, so that $\overline{y}^6$ represents an optimal solution. Since no other active node has an equal objective value, this optimal solution is also unique, so that the problem is fully solved.

We have seen that the branch-and-bound method solves integer programming problems by a sequence of linear programming problems. It is not at all uncommon that one integer programming problem requires the solution of hundreds of thousands of linear programming problems, all similar, but differing by a few constraints. Consider again the solution tree in Fig. 5.8 for some additional considerations. For instance, the additional constraints to be considered at a node are the constraints at all branchings from the root of the tree to the node under consideration. For instance, the node with optimal solution $\overline{y}^5$ represents a problem with the given objective function, all constraints of the linear programming relaxation (here: constraints *I* and *II*), as well as the additional constraints $y_2 \geq 7$, $y_1 \leq 3$, and $y_2 \geq 8$. An immediate consequence of this principle leads to the observation that as you move on some path down the tree, the values of the objective function of the problems either stay the same or get worse (i.e., go down for maximization problems and go up for minimization problems). This occurs as by adding constraints, the feasible set gets smaller, which cannot result in better objective values.

It is also apparent that even a very small problem with only two variables can have sizeable solution trees. When using a computer with linear programming software installed, it is possible to practice by using the computer to solve the linear programming problems at the nodes of the solution tree, while constructing the solution tree manually. In order to solve the integer programming problem in such "computer-assisted" fashion, it is necessary to learn how to edit the linear programming problems. For instance, return from the problem with the solution $\bar{\mathbf{y}}^5$ to $\bar{\mathbf{y}}^1$ requires moving up to the root of the tree (thus removing the constraints $y_2 \geq 8$, $y_1 \leq 3$, and $y_2 \geq 7$, and then moving down to the node labeled $\bar{\mathbf{y}}^1$ by adding the constraint $y_2 \leq 6$).

For practice, readers may either use the graphical approach or the computer-assisted approach to solve the same problem but start branching at the root node on the variable y_1 rather than y_2. This will result in the solution tree shown in Fig. 5.9. Note that this tree requires six branchings and the optimal solution is found at the node labeled $\bar{\mathbf{y}}^9$.

It is also important to realize that the very same procedure can be used to solve mixed-integer linear programming problems. Whenever the best active node has been chosen, we have to consider what is required to be integer and what is not. Branching is then done only on variables that are required to be integer. As an illustration, use the same example as above, except that $y_1 \geq 0$, while $y_2 \geq 0$ and integer. Again, we first solve the linear programming relaxation and find the root node of the solution tree. Now as $y_1 = 4.9125$, it satisfies the nonnegativity constraint, which is all that is required of this variable. On the other hand, $y_2 = 6.3125$, which satisfies the nonnegativity constraint, but not the integrality constraint, so we must branch on y_2 (there is no choice as in the all-integer problem). The branching is identical to the first branching shown in Fig. 5.8. At this point, $\bar{\mathbf{y}}^2$ is the better solution with $y_1 = 3.4$, $y_2 = 7$ with an objective value of 80. At this point,

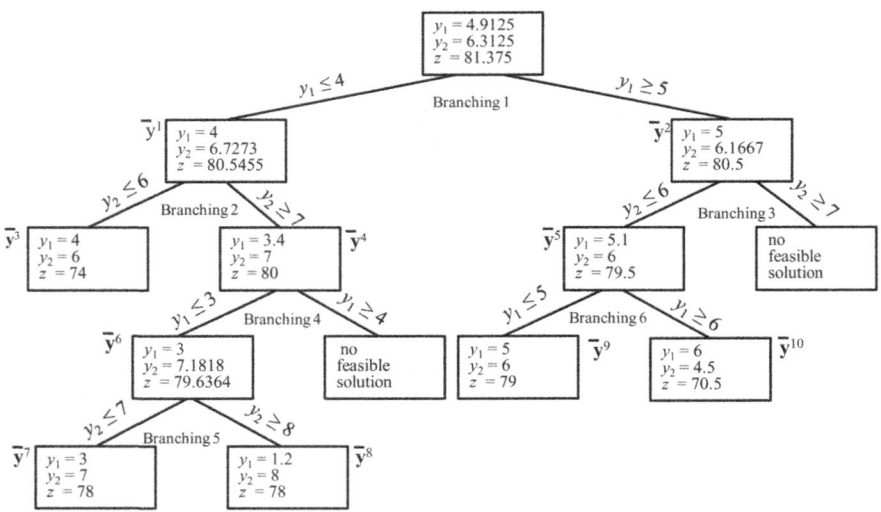

Fig. 5.9 Alternative Branch-and-Bound solution tree

5.3 Solution Methods for Integer Programming Problems

the procedure terminates, since y_1 is nonnegative as required, and y_2 is also nonnegative and integer, as required; hence $\bar{\mathbf{y}}^2$ is the optimal solution for this problem.

This is also the time to demonstrate what the term "bound" in branch-and-bound actually means. The idea is that during the procedure, the upper and lower bounds on the optimal objective value of the all- or mixed-integer programming problem are constantly tightened. Consider the tree shown in Fig. 5.9. Initially, as shown in the first section of this chapter, we only know that $\bar{z}_{IP} \leq \bar{z}_{LP}$, i.e. the optimal objective value of the integer problem is no better than 81.375. For maximization problems, the upper bound on $\bar{z}_{IP}$ is always the objective value of the best-known active node, while the lower bound is the objective value of the best-known integer solution. Hence, initially $\bar{z}_{IP} - \infty \leq 81.375$. After the first branching, the upper bound has decreased to 80.5455. After the second branching, the upper bound is reduced to 80.5, and the lower bound is now 74, as the integer solution $\bar{\mathbf{y}}^3$ is now known. Branching 3 further reduces the upper bound to 80, branching 4 reduces the upper bound to 79.6364, branching 5 reduces the upper bound to 79.5, while the lower bound is now increased to 78, as the integer solution $\bar{\mathbf{y}}^7$ has become known. Finally, after branching 6 the upper and lower bounds coincide at 79, which terminates the process.

Another aspect of the bounding procedure is that any active node with an objective function value worse (i.e., smaller for maximization and larger for minimization problems) than the best integer solution found so far can be removed from the set of active nodes. Branching from such a node could obviously never produce a child with an integer solution that might be optimal.

We will now demonstrate how to deal with cases in which the mixed- or all-integer programming problem has no feasible solution. As an illustration, consider the all-integer programming problem

$$P : \text{Max } z = y_1 + 4y_2$$
$$\text{s.t. } 28y_1 + 7y_2 \leq 49$$
$$30y_1 - 6y_2 \geq 36$$
$$y_1, \quad y_2 \geq 0 \text{ and integer.}$$

The solution tree for this problem, given that branching commences with y_2, is shown in Fig. 5.10. Branching 1 results in one child of the root having no feasible solution, while the other has a noninteger solution. Branching on this sole active node, we obtain two children, both of whom have no feasible solutions. At this point, there are no further active nodes, and no solution has been found. The tree that is obtained if we were to start branching on y_1 is even smaller: the two children of the root of the tree both have no feasible solution.

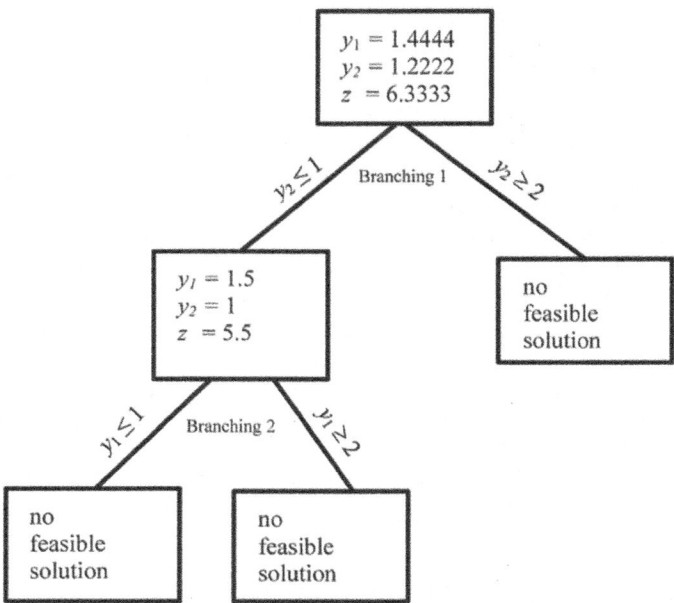

Fig. 5.10 Solution tree for problem with no feasible integer solution

5.3.3 Heuristic Methods

It will have become clear in the above discussion that for very large problems, exact methods may not be able to solve a given integer linear programming problem within a reasonable time frame. Surely, whenever an exact method such as the branch-and-bound technique described in the previous section has found an integer solution, one could terminate computations, even though the integer solution may not be optimal. As a matter of fact, while such a procedure will result in a feasible solution, it may be far from optimal. Instead, users often use heuristic techniques to find (hopefully reasonably good) solutions quickly. This section will describe such a technique.

In order to illustrate the technique, consider the following knapsack problem:

$$P: \text{Max } z = 12y_1 + 20y_2 + 31y_3 + 17y_4 + 24y_5 + 29y_6$$
$$\text{s.t.} \quad 2y_1 + 4y_2 + 6y_3 + 3y_4 + 5y_5 + 5y_6 \leq 19$$
$$y_1, \quad y_2, \quad y_3, \quad y_4, \quad y_5, \quad y_6 = 0 \text{ or } 1.$$

In order to employ the *Greedy Method* (sometimes colloquially referred to as "best bang for the buck" method), we first need to rank the variables in nonincreasing order of their "value" to us. Rather than simply using the coefficients in the objective function to rank the variables, we compute the "value per weight" ratios for each product by dividing the contribution to the objective function by the coefficient

5.3 Solution Methods for Integer Programming Problems

Table 5.13 "Value per weight" of the individual items

Variable	y_1	y_2	y_3	y_4	y_5	y_6
Value per weight	12/2 = 6	20/4 = 5	31/6 = 5.1667	17/3 = 5.6667	24/5 = 4.8	29/5 = 5.8
Rank	1	5	4	3	6	2

Table 5.14 First swap move

Leaving variable	Entering variable	New solution	ΔR	Δz
y_1	y_2	0, 1, 1, 1, 0, 1	$-2 + 4 = +2$	$-12 + 20 = +8$

in the constraint. We then rank the variables in nonincreasing order of these ratios as shown in Table 5.13.

Starting with all variables set to zero, the Greedy algorithm will now increase the values of variables one by one, starting with the highest rank, as long as resources are available.

In Step 1, we set $y_1 := 1$, which consumes 2 resource units and contributes 12 units to the objective,

in Step 2, we set $y_6 := 1$, which consumes 5 resource units and contributes 29 units to the objective,

in Step 3, we set $y_4 := 1$, which consumes 3 resource units and contributes 17 units to the objective,

in Step 4, we set $y_3 := 1$, which consumes 6 resource units and contributes 31 units to the objective,

in Step 5, we set $y_2 := 1$, which consumes 4 resource units. Stop and backtrack, i.e., re-set the variable $y_2 := 0$, as this latest assignment exceeds the availability of resources.

In summary, the solution at the termination of the Greedy algorithm is

$$\mathbf{y} = [y_1, y_2, y_3, y_4, y_5, y_6] = [1, 0, 1, 1, 0, 1]$$

which uses 16 resource units and has a value of the objective function of $z(\mathbf{y}) = 89$.

In the second phase, we use a simple improvement heuristic of the "interchange" or "swap" type. This heuristic method raises one arbitrarily chosen variable from a level of zero to one, while decreasing another also arbitrarily chosen variable from one to zero. Whenever such an exchange is feasible and increases the value of the objective function, it becomes our new starting point. The procedure is repeated until no further improvements are possible. Note that given the present solution, we have an extra 3 resource units available. In the tables below, we list the variable that leaves the solution (i.e., the variable whose value is reduced from its present value of one to zero), the entering variable (i.e., the variable whose value is increased from its present value of zero to one), the resulting new solution, the marginal resource usage ΔR, and the change Δz of the objective function that results from the swap.

Table 5.15 Second swap move

Leaving variable	Entering variable	New solution	ΔR	Δz
y_2	y_1	1, 0, 1, 1, 0, 1	$-4+2=-2$	$-20+12=-8$
y_2	y_5	0, 0, 1, 1, 1, 1	$-4+5=1$	$-20+24=+4$

Table 5.16 Third swap move

Leaving variable	Entering variable	New solution	ΔR	Δz
y_3	y_1	1, 0, 0, 1, 1, 1	$-6+2=-4$	$-31+12=-19$
y_3	y_2	0, 1, 0, 1, 1, 1	$-6+4=-2$	$-31+20=-11$
y_4	y_1	1, 0, 1, 0, 1, 1	$-3+2=-1$	$-17+12=-5$
y_4	y_2	0, 1, 1, 0, 1, 1	$-3+4=+1$: infeasible	
y_5	y_1	1, 0, 1, 1, 0, 1	$-5+2=-3$	$-24+12=-12$
y_5	y_2	0, 1, 1, 1, 0, 1	$-5+4=-1$	$-24+20=-4$
y_6	y_1	1, 0, 1, 1, 1, 0	$-5+2=-3$	$-29+12=-17$
y_6	y_2	0, 1, 1, 1, 1, 0	$-5+4=-1$	$-29+20=-9$

It is apparent that the first swap move in Table 5.14 results in an improvement, so that the solution $\mathbf{y} = [0, 1, 1, 1, 0, 1]$ becomes the new basis with resource consumption of 18 (so that we could use one additional resource unit) and objective value of $z(\mathbf{y}) = 97$. Starting with this solution, we perform again pairwise exchanges shown in Table 5.15.

Again, it was possible to improve the solution. The new solution is $\mathbf{y} = [0, 0, 1, 1, 1, 1]$, which uses all 19 resource units that are available, and has a value of the objective function of $z(\mathbf{y}) = 101$. With this new benchmark solution, we start another series of potential improvements, shown in Table 5.16.

At this point, all feasible pairwise exchanges result in decreases of the value of the objective function, so that the procedure terminates. Note that the fact that all resource units happen to be used is a coincidence.

It is also worth pointing out that the Greedy algorithm alone without the second-phase Swap procedure may result in very poor solutions. As an example, consider the problem

$$P : \text{Max } z = 10y_1 + 8y_2 + 7y_3$$
$$\text{s.t.} \quad 54y_1 + 48y_2 + 47y_3 \leq 100$$
$$y_1, \quad y_2, \quad y_3 = 0 \text{ or } 1.$$

5.3 Solution Methods for Integer Programming Problems

The ranking of the variables is y_1, y_2, and y_3 in that order. Setting the highest-ranked variable y_1 to 1 consumes 54 resource units and achieves an objective value of $z = 10$. No other variables can be set to one, as only 46 resource units remain. However, the solution $\mathbf{y} = [0, 1, 1]$ has an objective value of $z = 15$, much superior to the solution obtained by the Greedy algorithm without subsequent Swap procedure. To see that Greedy alone without any improvement follow-up may reach an objective value of only half the optimum, consider the problem

$$P : \text{Max } z = 1.01 y_1 + 2 y_2$$
$$\text{s.t.} \quad 1.001 y_1 + 2 y_2 \leq 2$$
$$y_1, \quad y_2 = 0 \text{ or } 1.$$

The Greedy algorithm will find the solution $\mathbf{y} = [1, 0]$ with the objective value of $z = 1.01$, while the optimal solution is $\mathbf{y} = [0, 1]$ with the value of the objective function $z = 2$.

Another type of heuristic method can sometimes be advantageous, using Lagrangean relaxation, which we described in Sect. 5.1. Since we are maximizing the Lagrangean objective function z_{Lagr}, the optimization process tends to favor solutions that are feasible for P_{IP}, and this tendency will be stronger with larger values of $\hat{u}_i \geq 0$; for very large values of $\hat{u}_i$, the resulting solutions will therefore be feasible as long as feasible solutions to P_{IP} exist. Note that very high values of the Lagrangean multipliers will put a very high emphasis on feasibility, while the original objective function will be largely ignored. The art is to find a balance between feasibility and the objective value.

A heuristic procedure that exploits this feature commences with some arbitrary value of $\hat{u}_i \geq 0$, and the corresponding relaxation P_{Lagr} is solved. If the solution point is feasible for P_{IP}, the value of $\hat{u}_i \geq 0$ will be reduced and P_{Lagr} is solved again. If, on the other hand, the solution point is not feasible for P_{IP}, the value of $\hat{u}_i \geq 0$ will be increased, and P_{Lagr} is solved again.

Although we have described the procedure based on dualizing a single constraint, it is not difficult to generalize the technique so as to dualize any number of constraints.

As a numerical illustration of the method, we will use the assignment problem described in Sect. 2.2.5 but change it to a maximization problem. For convenience, we restate the problem here as a maximization problem. It is then

P : Max $z = 4x_{11} + 3x_{12} + 1x_{13} + 8x_{21} + 5x_{22} + 3x_{23} + 2x_{31} + 6x_{32} + 2x_{33}$
s.t.
$$x_{11} + x_{12} + x_{13} = 1$$
$$x_{21} + x_{22} + x_{23} = 1$$
$$x_{31} + x_{32} + x_{33} = 1$$
$$x_{11} + x_{21} + x_{31} = 1$$
$$x_{12} + x_{22} + x_{32} = 1$$
$$x_{13} + x_{23} + x_{33} = 1$$
$$x_{11}, x_{12}, x_{13}, x_{21}, x_{22}, x_{23}, x_{31}, x_{32}, x_{33} \geq 0 \text{ and integer.}$$

In addition to the usual constraints, we add the constraint
$$2x_{11} + 3x_{12} + x_{13} + 4x_{21} + 6x_{23} + 5x_{32} + 2x_{33} \leq 8.$$

In many practical applications of the assignment problem, an additional constraint of this type may occur, expressing some kind of capacity, weight, or budget restriction. Without it, the problem can very easily be solved either by using a specialized algorithm, or simply using some all-purpose software. However, in the presence of additional constraints, many nice features of the problem may be lost, and these features made the problem easy to solve. With this in mind, we will refer to such additional constraints as the *complicating constraints*. An obvious way to deal with this complication is to dualize the complicating constraint and solve the resulting assignment problem.

Consider again our numerical example. We first solve the reduced problem P_{red}, i.e., the assignment problem without the complicating constraint, and obtain the unique optimal solution $\widetilde{x}_{13} = \widetilde{x}_{21} = \widetilde{x}_{32} = 1$, and $\widetilde{x}_{ij} = 0$ otherwise. The associated value of the objective function is $\widetilde{z} = 15$. (Note that this solution does *not* satisfy the complicating constraint.) If we dualize the additional constraint so as to write $-\widehat{u}(2x_{11} + 3x_{12} + x_{13} + 4x_{21} + 6x_{23} + 5x_{32} + 2x_{33} - 8)$, we can consider this as a *penalty term* that is added to the objective function of the original problem. The reason is that the expression in brackets will be positive for solution points that violate the additional constraint, so that the penalty term will make a negative contribution to the objective function to the relaxed problem, and the penalty will be bigger with larger values of the multiplier $\widehat{u}$.

Using the dual variable $\widehat{u} = \frac{1}{2}$ and solving the Lagrangean relaxation, we obtain the solution $\widehat{x}_{13} = \widehat{x}_{21} = \widehat{x}_{32} = 1$, and $\widehat{x}_{ij} = 0$ otherwise. This is the same solution we have obtained for the assignment problem without the complicating constraint, so that the additional constraint is still violated.

We take this as an indication for the fact that the dual variable must be set at a higher value than $\widehat{u} = \frac{1}{2}$, so we arbitrarily set $\widehat{u} := 5$. The unique optimal solution to the relaxed problem is then $\widehat{x}_{13} = \widehat{x}_{22} = \widehat{x}_{31} = 1$ and $\widehat{x}_{ij} = 0$ otherwise. This solution satisfies the additional constraint and is therefore feasible to the original problem with an objective value of $z = 8$.

5.3 Solution Methods for Integer Programming Problems

At this point we may investigate whether or not to reduce the penalty, as by decreasing the value of the penalty term $\hat{u} = 5$, we will improve (i.e., increase for this maximization problem) the value of the objective function, while keeping the solution feasible for the original problem. Since $\hat{u} = \frac{1}{2}$ was too small and $\hat{u} = 5$ may have been too big, we can try a value in between such as $\hat{u} = 1$. The resulting relaxed problem then has the unique optimal solution $\hat{x}_{11} = \hat{x}_{22} = \hat{x}_{33} = 1$, and $\hat{x}_{ij} = 0$ otherwise. This solution is feasible for the original problem with an objective value of $z = 11$. We could now choose penalty values between $\frac{1}{2}$ and 1 and continue. (Actually, the solution we found here happens to be optimal.)

Alternatively, we could have dualized not the additional constraint and solved the relaxation (an assignment problem), but we could have dualized all six assignment problem constraints and solved the Lagrangean relaxation, which would be a knapsack problem. Since we are maximizing and since all coefficients are nonnegative, we may write the six assignment constraints as less-than-or-equal constraints. If we were to proceed in this fashion, our numerical example would read as follows:

P_{Lagr} : Max $z_{Lagr} = 4x_{11} + 3x_{12} + 1x_{13} + 8x_{21} + 5x_{22} + 3x_{23} + 2x_{31} + 6x_{32} + 2x_{33}$
$- \hat{u}_1(x_{11} + x_{12} + x_{13} - 1) - \hat{u}_2(x_{21} + x_{22} + x_{23} - 1) - \hat{u}_3(x_{31} + x_{32} + x_{33} - 1)$
$- \hat{u}_4(x_{11} + x_{21} + x_{31} - 1) - \hat{u}_5(x_{12} + x_{22} + x_{32} - 1) - \hat{u}_6(x_{13} + x_{23} + x_{33} - 1)$

s.t. $2x_{11} + 3x_{12} + x_{13} + 4x_{21} + 6x_{23} + 5x_{32} + 2x_{33} \leq 8$

$x_{ij} = 0$ or 1 for all i, j.

Searching for suitable values for the dual variables $\hat{u}_1, \hat{u}_2, \ldots, \hat{u}_6$ is now considerably more difficult since values of the six variables must be chosen simultaneously. Furthermore, the relaxation is an integer problem that, albeit it has only a single constraint, may not be all that easy to solve. On the other hand, dualizing only the complicating constraint as done earlier leaves us with an assignment problem that automatically has integer solution, so that all we have to do is solve a linear programming problem.

Continuing with this approach, we may choose $\hat{u}_1 = \hat{u}_2 = \ldots = \hat{u}_6 = \frac{1}{2}$ and obtain the objective function $z_{Lagr} = 3x_{11} + 2x_{12} + 7x_{21} + 4x_{22} + 2x_{23} + x_{31} + 5x_{32} + x_{33} + 3$. Maximizing this objective function subject to the budget constraint and the zero-one specifications of the variables results in the solution $\hat{x}_{11} = \hat{x}_{21} = \hat{x}_{22} = \hat{x}_{31} = \hat{x}_{33} = 1$, and $\hat{x}_{ij} = 0$ otherwise. Since this solution violates the second, third, and fourth of the six assignment constraints, we can increase the penalty parameters for the violated constraints to, say, $\hat{u}_2 = \hat{u}_3 = \hat{u}_4 = 5$. In the resulting knapsack problem, most objective function coefficients are negative, leading to the solution $\hat{x}_{12} = \hat{x}_{32} = 1$ and $\hat{x}_{ij} = 0$ otherwise. This solution violates only the fifth constraint in the original problem. At this point, we could just increase the value of $\hat{u}_5$ in an attempt to satisfy the fifth constraint and the process will continue.

The above example demonstrates that (1) it is not always apparent beforehand which approach (i.e., dualizing which constraints) is preferable, and (2) which values of the dual variables will result in quick convergence. Actually, choosing good values for the dual variable is more of an art than a science.

Fig. 5.11 Arrangement and numbering of parcels for Problem 1

1	2	3	4
5	6	7	8
9	10	11	12
13	14	15	16

Table 5.17 Decision table

y_1	y_2	OK?
0	0	Yes
0	1	Yes
1	0	Yes
1	1	No

We should also point out that the practical benefit of using Lagrangean relaxation for solving an assignment problem with a complicating constraint will only be present if the assignment problem is of a reasonable size. For instance, for $n = 15$, the assignment problem has more than 10^{12} solutions, making enumeration of all feasible solutions impractical, so that a specialized algorithm is needed.

Exercises

Problem 1 (neighborhood constraints in forestry modeling): This model concerns planning in forestry. In particular, suppose that a landowner has a number of parcels, which, for the sake of simplicity, can be thought of as a regular grid. The idea is now to plan which of the parcels should be harvested. Once it has been decided to harvest a certain parcel, it will be clearcut. One restriction is to ensure that neighboring parcels should not be harvested, so as to ensure avoiding huge clearcut areas that foster erosion and do not sustain wildlife. Suppose that the parcels are arranged and numbered as shown in Fig. 5.11.

Describe how to model the neighborhood constraints to ensure that neighboring parcels are not harvested.

Solution: Define binary variables y_j that assume a value of 1, if parcel j is harvested and 0 otherwise. The constraints we are interested in can then be expressed as conditional constraints of the type "if parcel j is harvested, then its neighbor k cannot be harvested," which will then have to be expressed for all neighbors of each parcel. As an example, consider the neighboring parcels 1 and 2. Table 5.17 shows the four cases.

It is apparent that the only case that is prohibited is when both neighboring parcels are harvested at the same time. As a result, the constraint that prevents the undesirable case from happening, we formulate as $y_1 + y_2 \leq 1$. This type of constraint has to be formulated for *all* pairs of neighbors, e.g. parcels 1 and 2 (as we already did), parcels 1 and 5 (as $y_1 + y_5 \leq 1$), parcels 2 and 3 (as $y_2 + y_3 \leq 1$), parcels 2 and 6 (as $y_2 + y_6 \leq 1$), and so forth. For the above problem, there are no less than 24 such neighborhood constraints. If parcels that only share a corner are also considered neighbors (such as parcels 1 and 6 or parcels 8 and 11), that will add another 18 constraints, as $y_1 + y_6 \leq 1$, $y_8 + y_{11} \leq 1$, and so forth.

5.3 Solution Methods for Integer Programming Problems

Table 5.18 Unit transportation costs

	Customer 1	Customer 2	Customer 3
Warehouse 1	3	7	4
Warehouse 2	2	6	8
Warehouse 3	9	3	4
New warehouse 1	5	6	4
New warehouse 2	7	3	9

Problem 2 (a warehouse distribution problem): A decision maker has presently three warehouses with capacities of 30, 10, and 50 units, respectively. Operating the three warehouses incurs fixed operating costs of $25, $50, and $45, respectively. The first warehouse could be expanded by up to 20 units for a cost of $1 per-unit capacity. It is also possible to close any of the existing warehouses, in which case no operating costs are incurred at that site. Furthermore, two additional sites have been identified, where new warehouses might be opened. The costs to open the two new warehouses are $15 and $25, and their respective capacities are 20 and 30 units.

As far as demand goes, there are three customers with demands of 20, 60, and 40 units. These demands have to be satisfied exactly. The unit transportation costs between the existing and potential warehouses and the customers are shown in Table 5.18, where the first three supply points refer to the existing warehouses, while the last two supply points symbolize the potential new warehouses. Formulate and solve an integer programming problem that minimizes the total costs.

Solution: In order to formulate the problem, we need to define a number of variables. First of all, we define binary variables y_1, y_2, y_3, y_4, and y_5, which assume a value of one, if the warehouse is kept open (for the first three existing warehouses) or is newly opened in case of the last two warehouses. Furthermore, we define the continuous variable w, which denotes the number of capacity units, by which the first warehouse is expanded. In addition, we will define the usual continuous variables x_{ij} as the quantity that is shipped from warehouse i to customer j.

The objective function is then the sum of facility costs (operating costs for the existing warehouses and opening costs for the planned new warehouses), expansion costs for the first warehouse, and transportation costs. The facility costs are, of course, only incurred if the facilities are actually opened, so that we have $25y_1 + 50y_2 + 45y_3$ for the existing warehouses and $15y_4 + 25y_5$ for the new warehouses. The expansion costs for the first warehouse are $1w$. Finally, the transportation costs are $3x_{11} + 7x_{12} + 4x_{13} + 2x_{21} + \ldots + 9x_{53}$.

Consider now the constraints. First there are the capacity constraints of the warehouses. For instance, the second warehouse has a capacity of 10, provided we keep it open. The variable that determines this is y_2, so that the constraint states that the flow out of warehouse 2 cannot exceed the capacity of the warehouse, or $x_{21} + x_{22} + x_{23} \leq 10y_2$. The constraints for the other existing or planned warehouses, except for the first, are similar. The capacity of the first warehouse equals $30y_1$ plus the capacity of the expansion (if any), which is w. We also have to specify that the expansion will only be undertaken if the warehouse is kept open, as it is meaningless to decide to close warehouse 1 and then expand its capacity. This is written as $w \leq 20y_1$. The reason is that if the warehouse is kept open, we have $y_1 = 1$, so that the constraint states that we can expand its capacity of 20 units. On the other hand, if warehouse 1 is closed, $y_1 = 0$ and the constraint states that $w \leq 0$, meaning that no expansion is possible. The demand constraints then require the inflow into the customer sites to be equal to the demand at the site. The complete formulation is then as follows:

$$P: \text{Min } z = 25y_1 + 50y_2 + 45y_3 + 15y_4 + 25y_5 + 1w$$
$$+3x_{11} + 7x_{12} + 4x_{13} + 2x_{21} + 6x_{22} + 8x_{23} + 9x_{31} + 3x_{32} + 4x_{33}$$
$$+5x_{41} + 6x_{42} + 4x_{43} + 7x_{51} + 3x_{52} + 9x_{53}$$

$$\text{s.t. } x_{11} + x_{12} + x_{13} \leq 30y_1 + w$$
$$x_{21} + x_{22} + x_{23} \leq 10y_2$$
$$x_{31} + x_{32} + x_{33} \leq 50y_3$$
$$x_{41} + x_{42} + x_{43} \leq 20y_4$$
$$x_{51} + x_{52} + x_{53} \leq 30y_5$$

$$w \leq 20y_1$$

$$x_{11} + x_{21} + x_{31} + x_{41} + x_{51} = 20$$
$$x_{12} + x_{22} + x_{32} + x_{42} + x_{52} = 60$$
$$x_{13} + x_{23} + x_{33} + x_{43} + x_{53} = 40$$

$$x_{ij} \geq 0 \text{ for all } i, j; w \geq 0$$
$$y_1, \ldots, y_5 = 0 \text{ or } 1.$$

Solving the problem results in a decision to keep warehouses 1 and 3 open, while closing warehouse 2. In addition, warehouse 5 will be opened, while warehouse 4 is not. It was also decided to expand the capacity of warehouse 1 by 10 units. The shipments are as follows: from warehouse 1, we send 20 units to the first and 20 units to the third customer, from warehouse 3 we send 30 units to customer 2 and 20 units to customer 3, and from the newly opened warehouse 5 we ship 30 units to customer 2. The total cost of the operations is $505.

Problem 3 (choosing vehicles for a display): An entrepreneur wants to put up an exhibit featuring some antique cars. The vehicles potentially available are a Bugatti, Cadillac, Cobra, Corvette, Pierce Arrow, and Studebaker. The impact of displaying the individual vehicles has been estimated in terms of the number of people who would make a special trip to see a vehicle as 58, 37, 42, 40, 55, and 33, respectively. The budget of the organizer is $15,000, and the costs to transport the automobiles to the venue (they are presently located at different sites) and the costs of their insurance (depending on the vehicles' estimated value) are $6000, $4000, $3800,

5.3 Solution Methods for Integer Programming Problems

Table 5.19 Decision table for the second requirement in Problem 5

y_4	y_3	OK?
0	0	Yes
0	1	Yes
1	0	No
1	1	Yes

Table 5.20 Decision table for the third requirement in Problem 5

y_1	y_2	OK?
0	0	No
0	1	Yes
1	0	Yes
1	1	Yes

$4200, $5500, and $3200, respectively. The obvious idea is to choose vehicles for the exhibit, so as to maximize the total impact, while staying within the budget. In addition, there are some further requirements.

- Choose at least three vehicles for the exhibit.
- If a Corvette is included in the exhibit, then a Cobra must also be included.
- If a Bugatti is not included in the show, then a Cadillac must be included.

Solution: We first we define binary variables y_1, y_2, y_3, y_4, y_5, and y_6 that assume a value of one, if the first, second, ..., sixth vehicle is included in the exhibit, and 0 otherwise. The formulation of the objective function and the budget constraint is straightforward. Consider now the additional requirements. The number of vehicles included in the exhibit is expressed as the sum of all variables, so that we can write $y_1 + y_2 + y_3 + y_4 + y_5 + y_6 \geq 3$.

The next step is the conditional constraint "if Corvette, then Cobra," or, more formally, "if $y_4 = 1$, then $y_3 = 1$ as well." The possible solutions and their acceptability are shown in Table 5.19.

The undesirable solution $y_4 = 1$ and $y_3 = 0$ can be avoided by including the constraint $y_4 \leq y_3$, which violates the undesirable solution, while it is valid in the other three solutions.

Consider now the last requirement. The conditional constraint is "if not Bugatti, then Cadillac," or, equivalently, "if $y_1 = 0$, then $y_2 = 1$." Again, consider the decision table shown in Table 5.20.

The only solution that violates the condition is the one that has neither of the two vehicles in the exhibit. In other words, at least one of the two vehicles must be in the exhibit, so we can formulate $y_1 + y_2 \geq 1$.

The formulation of the entire problem is then

Table 5.21 Solutions of the car exhibit problem for different budgets

Budget (in $10,000)	11	12	13	14	15	16	17	18	19
Cars included	2, 3, 6	2, 3, 4	1, 3, 6	1, 3, 4	1, 5, 6	1, 3, 5	1, 2, 3, 6	1, 2, 3, 4	1, 3, 5, 6
z-value	112	119	133	140	146	155	170	177	188
Budget (in $10,000)	20	21	22	23	24	25	26	27	
Cars included	1, 3, 4, 5	2, 3, 4, 5, 6	1, 2, 3, 4, 6	1, 3, 4, 5, 6	1, 2, 3, 4, 5	1, 2, 3, 4, 5	1, 2, 3, 4, 5	1, 2, 3, 4, 5, 6	
z-value	195	207	210	228	232	232	232	265	

$$P: \text{Max } z = 58y_1 + 37y_2 + 42y_3 + 40y_4 + 55y_5 + 33y_6$$
$$\text{s.t.} \quad 6{,}000y_1 + 4{,}000y_2 + 3{,}800y_3 + 4{,}200y_4 + 5{,}500y_5 + 3{,}200y_6 \leq 15{,}000$$
$$y_1 + y_2 + y_3 + y_4 + y_5 + y_6 \geq 3$$
$$y_4 \leq y_3$$
$$y_1 + y_2 \geq 1$$
$$y_1, \quad y_2, \quad y_3, \quad y_4, \quad y_5, \quad y_6 = 0 \text{ or } 1.$$

Table 5.21 displays solutions and objective values for a large variety of budgets. It is apparent that there is no feasible solution for any budget strictly less than 11, as we have to include at least three vehicles in the exhibit, and the three least expensive cars, the sixth, third, and second vehicle, cost $3200 + 3800 + 4000 = \$11{,}000$. On the other hand, exhibiting all vehicles costs $26,700, so that any budget at or above this level will enable the organizer to exhibit all vehicles. Also notice the "granularity" of the solutions: an increase in the budget by $1000 results in an increase in the objective value by 7 (if increasing the budget from $11,000 to $12,000), by 14 (if the budget is increased from $12,000 to $13,000), by 7 if the budget is increased from $13,000 to $14,000, and so forth.

Problem 4 (cutting plane method): Consider the following all-integer linear programming problem:

$$P: \text{Min } z = 3y_1 + 5y_2$$
$$\text{s.t.} \quad 9y_1 + 12y_2 \geq 122$$
$$2y_1 + y_2 \geq 6$$
$$y_1 \leq 4$$
$$y_1, \quad y_2 \geq 0 \text{ and integer.}$$

Solve the linear programming relaxation of the problem and then sequentially introduce cutting planes to solve the problem for eight steps or to optimality, whichever comes first.

5.3 Solution Methods for Integer Programming Problems

Table 5.22 Cutting planes for Problem 6

#	Solution	Cutting plane
1	Optimal solution of the LP relaxation: $\bar{y}_1 = 4, \bar{y}_2 = 7.1667, \bar{E}_1 = 0, \bar{E}_2 = 9.1667, \bar{S}_3 = 0$ with $\bar{z} = 47.8333$	$E_1 + S_3 \geq 1$, or, equivalently, $8y_1 + 12y_2 - E_4 = 119$
2	$\bar{y}_1 = 4, \bar{y}_2 = 7.25, \bar{E}_1 = 1, \bar{E}_2 = 9.25, \bar{S}_3 = 0, \bar{E}_4 = 0$ with $\bar{z} = 48.25$	$S_3 + E_4 \geq 1$, or, equivalently, $7y_1 + 12y_2 - E_5 = 116$
3	$\bar{y}_1 = 3, \bar{y}_2 = 7.9167, \bar{E}_1 = 0, \bar{E}_2 = 7.9167, \bar{S}_3 = 1, \bar{E}_4 = 0, \bar{E}_5 = 0$ with $\bar{z} = 48.5833$. Note primal degeneracy! Arbitrarily choose E_1 and E_4.	$E_1 + E_4 \geq 1$, or, equivalently, $17y_1 + 24y_2 - E_6 = 242$
4	$\bar{y}_1 = 3.3333, \bar{y}_2 = 7.7222, \bar{E}_1 = 0.6667, \bar{E}_2 = 8.3889, \bar{S}_3 = 0.6667, \bar{E}_4 = 0.3333, \bar{E}_5 = 0, \bar{E}_6 = 0$ with $\bar{z} = 48.6111$.	$E_5 + E_6 \geq 1$, or, equivalently, $31y_1 + 48y_2 - E_7 = 475$
5	$\bar{y}_1 = 3.6667, \bar{y}_2 = 7.5278, \bar{E}_1 = 1.3333, \bar{E}_2 = 8.8611, \bar{S}_3 = 0.3333, \bar{E}_4 = 0.6667, \bar{E}_5 = 0, \bar{E}_6 = 1, \bar{E}_7 = 0$ with $\bar{z} = 48.6389$.	$E_5 + E_7 \geq 1$, or, equivalently, $38y_1 + 60y_2 - E_8 = 592$
6	$\bar{y}_1 = 4, \bar{y}_2 = 7.3333, \bar{E}_1 = 1.1250, \bar{E}_2 = 9.3333, \bar{S}_3 = 0, \bar{E}_4 = 1, \bar{E}_5 = 0, \bar{E}_6 = 2, \bar{E}_7 = 1, \bar{E}_8 = 0$ with $\bar{z} = 48.6667$. Note primal degeneracy! Arbitrarily choose S_3 and E_5.	$S_3 + E_5 \geq 1$, or, equivalently, $6y_1 + 12y_2 - E_9 = 113$
7	$\bar{y}_1 = 3.3750, \bar{y}_2 = 7.7292, \bar{E}_1 = 2, \bar{E}_2 = 8.4792, \bar{S}_3 = 0.625, \bar{E}_4 = 0.75, \bar{E}_5 = 0.375, \bar{E}_6 = 0.875, \bar{E}_7 = 0.625, \bar{E}_8 = 0, \bar{E}_9 = 0$ with $\bar{z} = 48.7708$.	$E_8 + E_9 \geq 1$, or, equivalently, $44y_1 + 72y_2 - E_{10} = 706$
8	$\bar{y}_1 = 3.5, \bar{y}_2 = 7.6667, \bar{E}_1 = 1.5, \bar{E}_2 = 8.6667, \bar{S}_3 = 0.5, \bar{E}_4 = 1, \bar{E}_5 = 0.5, \bar{E}_6 = 1.5, \bar{E}_7 = 1.5, \bar{E}_8 = 1, \bar{E}_9 = 0, \bar{E}_{10} = 0$ with $\bar{z} = 48.8333$.	
...	Optimal all-integer solution: $\bar{y}_1 = 3, \bar{y}_2 = 8, \bar{E}_1 = 1, \bar{E}_2 = 8, \bar{S}_3 = 1$ with $\bar{z} = 49$.	

Solution: When we reformulate the cuts, we write them as equations with the added excess variables in the rightmost column of Table 5.22. This is to indicate which excess variable was used in which cut, which is useful when formulating cuts later in the process. However, when adding the cut to the problem, we write the constraint without the excess variable as an inequality, since the solver will automatically add slack and excess variables.

The successive cutting planes generated for this problem are shown in Table 5.22.

Problem 5 (solving a problem via branch-and-bound): Consider the following all-integer optimization problem:

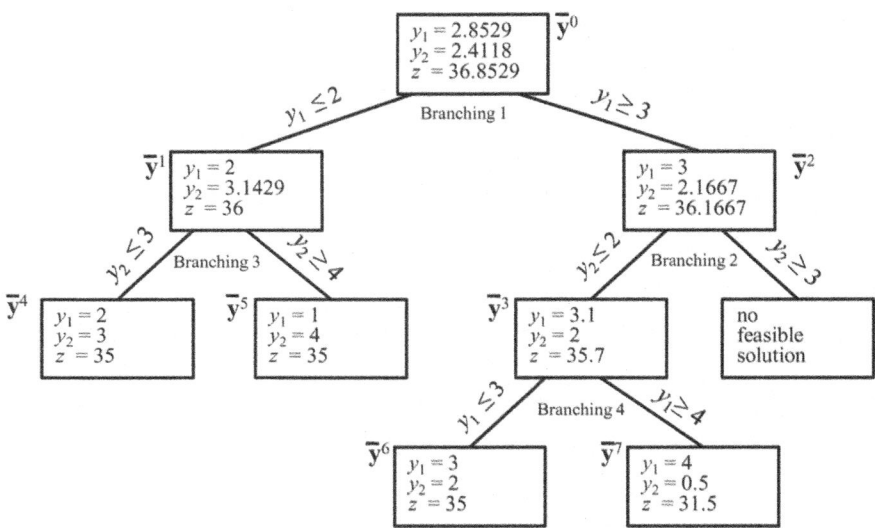

Fig. 5.12 Branching on y_1 first

$$P: \text{Max } z = 7y_1 + 7y_2$$
$$\text{s.t. } 6y_1 + 7y_2 \leq 34$$
$$10y_1 + 6y_2 \leq 43$$
$$y_1, y_2 \geq 0 \text{ and integer.}$$

(a) Produce the solution trees for branching, starting with y_1 and y_2, respectively.
(b) What would the optimal solution be, if the integrality requirement for y_2 had been dropped?
(c) What are the additional constraints that were introduced between the root of the tree and the last integer solution found in the tree, assuming that branching starts with y_1?
(d) What would have happened, if the left child that resulted from the first branching in either tree had an objective value of 34.8?

Solution:

(a) The solution that starts branching on y_1 is shown in Fig. 5.12.
The solution tree that starts branching on y_2 is shown in Fig. 5.13.
The problem has actually three alternative optimal solutions. Note that in both trees, branching 4 is necessary to complete the solution tree. The optimal solutions are: $\bar{y}_1 = 3$ and $\bar{y}_2 = 2$, $\bar{y}_1 = 2$ and $\bar{y}_2 = 3$, as well as $\bar{y}_1 = 1$ and $\bar{y}_2 = 4$, all with a value of the objective function of $\bar{z} = 35$.

5.3 Solution Methods for Integer Programming Problems

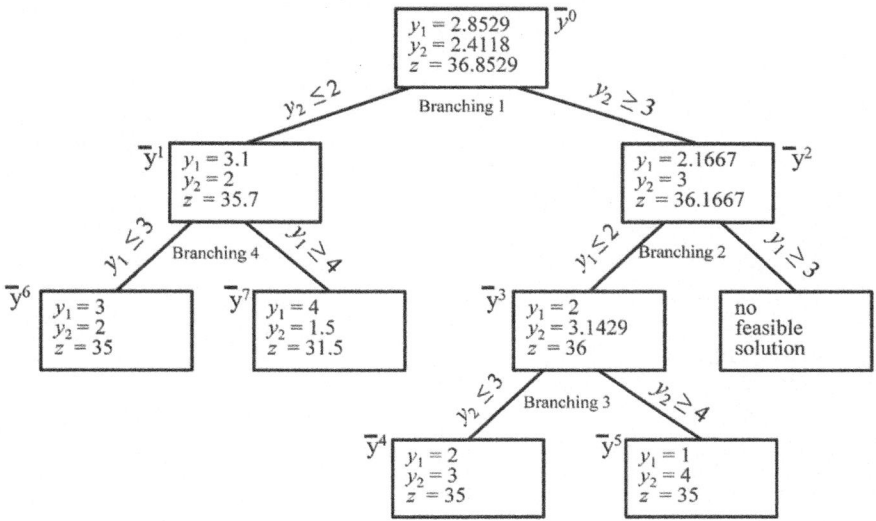

Fig. 5.13 Branching on y_2 first

(b) In this case we would have had to start branching with y_1, so that Fig. 5.12 applies. The optimal solution would be the right child of the root of the tree, i.e., solution $\bar{y}^2$, which has $\bar{y}_1 = 3$ and $\bar{y}_2 = 2.1667$ with an objective value of $\bar{z} = 36.1667$.

(c) The additional constraints are $y_1 \geq 3$, $y_2 \leq 2$, and $y_1 \leq 3$.

(d) In both trees, we would never have branched from the left child of the first branching.

Problem 6 (choosing the correct branch-and-bound tree): Consider the following all-integer programming problem:

$$P : \text{Max } z = 2y_1 + y_2$$
$$\text{s.t.} \quad -3y_1 + 15y_2 \leq 45$$
$$3y_1 - 4y_2 \leq 9$$
$$y_1, \ y_2 \leq 0 \text{ and integer}$$

Four solution trees have been developed by four different individuals, each claiming that their tree is correct. The trees are shown in Fig. 5.14a–d. However, only one of the trees is correct. Which one? For each solution tree, write one sentence that explains why this is or is not the correct tree.

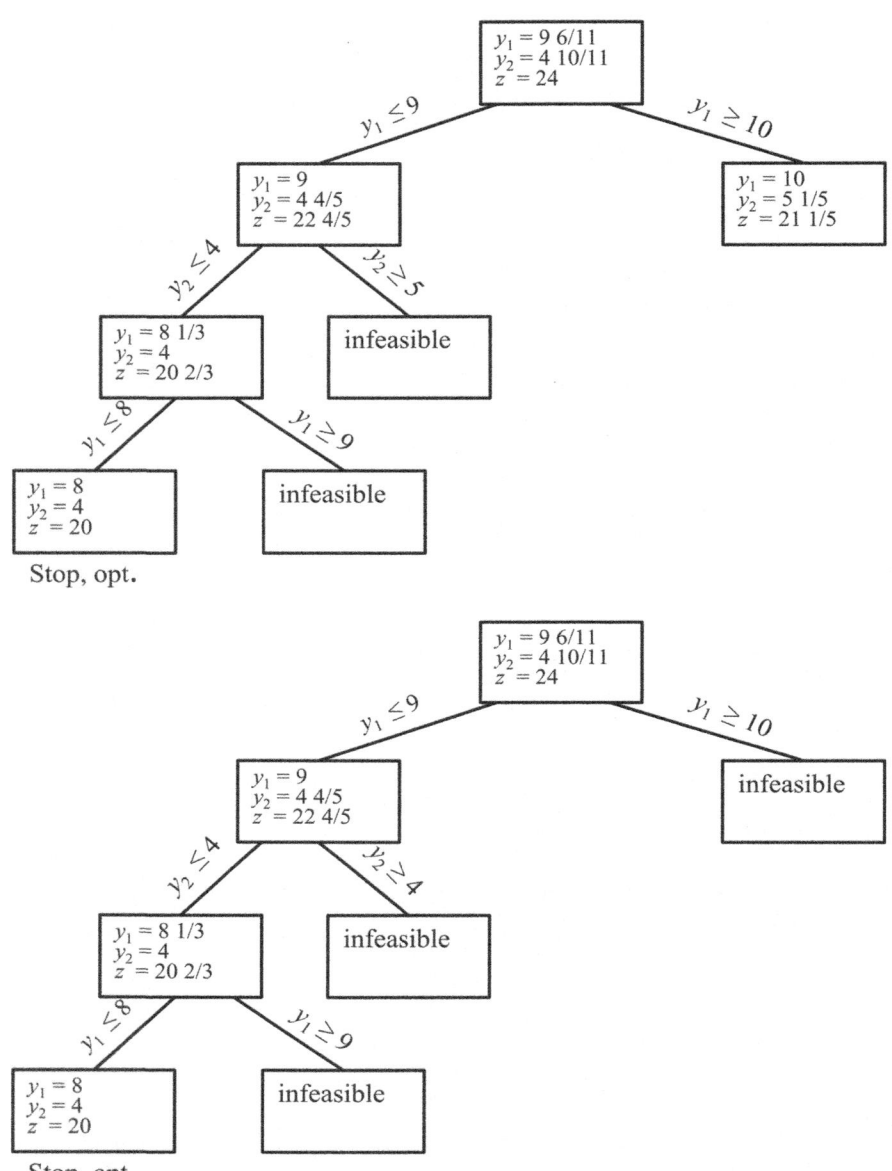

Fig. 5.14 Four different solution trees

5.3 Solution Methods for Integer Programming Problems

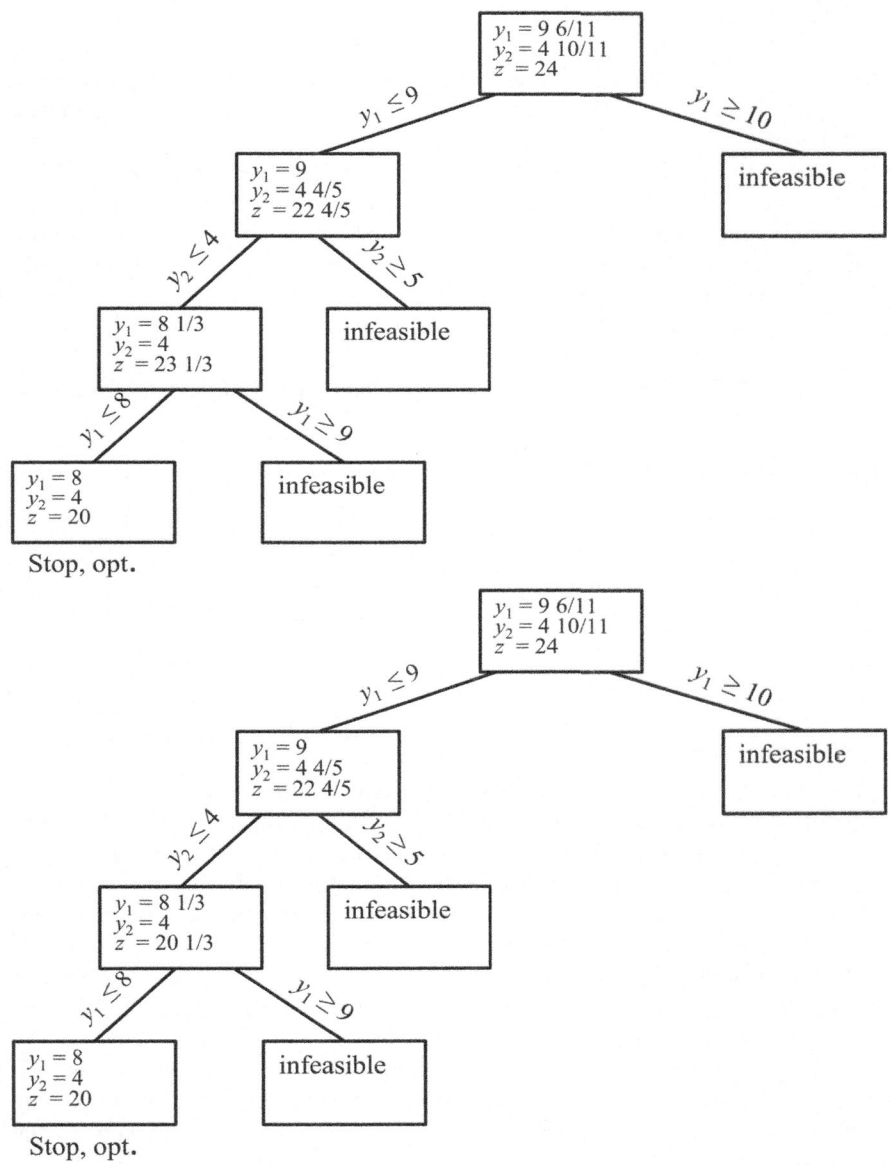

Fig. 5.14 (continued)

210　　　　　　　　　　　　　　　　　　　　5 Integer Linear Programming

Solution:

The solution tree in Fig. 5.14a is false. The right child of the root of the tree is not feasible. If the solution there had been correct, branching should have continued at that node.

The solution tree in Fig. 5.14b is false. The branching on the second level should be $y_2 \leq 4$ (as is), but the branch to the right child should be $y_2 \geq 5$, not $y_2 \geq 4$.

The solution tree in Fig. 5.14c is false. The second branching leading to the left child has the objective value increase from $z = 22^4/_5$ to $z = 23^1/_3$, which cannot happen in a maximization problem.

The solution tree in Fig. 5.14d is correct.

Problem 7 (heuristics: Greedy and Swap method): Consider the following integer programming problem.

$$P : \text{Max } z = 21y_1 + 11y_2 + 65y_3 + 58y_4 + 122y_5$$
$$\text{s.t.} \quad 21y_1 + 10y_2 + 42y_3 + 37y_4 + 64y_5 \leq 640$$
$$y_2 \leq 2$$
$$y_3 \leq 1$$
$$y_4 \geq 1$$
$$y_4 \leq 2$$
$$y_5 \leq 2$$
$$y_1, \quad y_2, \quad y_3, \quad y_4, \quad y_5 \geq 0 \text{ and integer.}$$

Use the Greedy algorithm and a Swap interchange to find a solution.

Solution: Ordering the variables with respect to their objective function contribution per resource unit, the order is y_5, followed by y_4, y_3, y_2, and y_1. Before we start increasing the values of the variables, we need to set the variables to their minimum values, i.e., $y_3 = 1$ and $y_4 = 1$, so that we do not have to worry about lower bounds anymore. This leaves us with $640 - 42 - 37 = 561$ resource units.

We now start the allocation with y_5. The upper bound is 2, so we set $y_5 = 2$, which leaves us with $561 - 2(64) = 433$ resource units. The next best variable is y_4. As its upper bound equals 2 and its value already equals 1, we can only increase y_4 by 1. This leaves $433 - 1(37) = 396$ resource units. The next valuable variable is y_3. It does not have an upper bound, so that we increase its value as much as the remaining resource units allow. We have 396 units left, each unit of y_3 requires 42 units, so that the largest value of $y_3 = 9$. Increasing y_3 by that value leaves us $396 - 9(42) = 18$ resource units left. The next most valuable variable is y_2, whose upper bound is 2. However, the remaining resource units are only good for an increase of 1. This leaves us 8 resource units, which are not sufficient for any other increase. In summary, we have the solution $\mathbf{y} = [0, 1, 10, 2, 2]$, for which the objective value $z = 1021$ may be calculated.

In the Swap procedure we will decrease the value of a variable by one, thus freeing some resources, which we then try to use by increasing the value of some other variable. For instance, decreasing the value of y_2 by one frees 10 units for a total of 18, which is not sufficient to increase any other variable by an integer amount.

5.3 Solution Methods for Integer Programming Problems

Reducing the variable y_3 by one frees up 42 units, so that $42 + 8 = 50$ resource units are now available. Note that it also reduces the objective value by 65. Those resource units may be used to increase the value of y_1 by 2, which increases the objective value by 42, not enough to make up for the loss of 65. Alternatively, we may increase y_4 by one, which increases the objective value by 58, also not sufficient to make up for the loss.

We may now try to reduce the value of y_4 by 1, freeing 37 resource units for a total of 45. Note that the objective value decreases by 58 in the process. The resource may now be used to increase y_3 by one unit, which increases the value of the objective function by 65. This represents a net gain of +7, so that we make this change. The new solution is now $\mathbf{y} = [0, 1, 11, 1, 2]$ with an objective value of $z = 1028$. Three resource units remain available.

The process would continue here. We terminate the procedure at this point. It so happens that the solution found here is optimal.

Problem 8 (Lagrangean relaxation): Consider a maximization assignment problem with coefficients as shown in the matrix

$$\mathbf{C} = \begin{bmatrix} 9 & 2 & 4 \\ 3 & 1 & 5 \\ 6 & 8 & 7 \end{bmatrix},$$

and assume that the following complicating constraint must also be satisfied:

$$5x_{11} + x_{12} + 5x_{13} + 2x_{21} + 4x_{23} + 3x_{31} + 6x_{32} + 5x_{33} \leq 9$$

(a) Solve the assignment problem (e.g., as a linear programming problem), ignoring the complicating constraint. Is the solution unique? Show that it violates the complicating constraint.
(b) Dualize the complicating constraint and solve the resulting Lagrangean relaxation as an assignment problem, using a value of $\widehat{u} = 1$ for the dual variable.
(c) Same as (b) above, but with $\widehat{u} = 2$.
(d) Same as (b) above, but with $\widehat{u} = 1.5$.
(e) Compare and discuss the results in (b), (c), and (d) above.

Solution:

(a) Solving the assignment problem without the complicating constraint, i.e., the reduced problem P_{red}, we find the unique optimal solution $\widetilde{x}_{11} = \widetilde{x}_{23} = \widetilde{x}_{32} = 1$ and $\widetilde{x}_{ij} = 0$ otherwise. The associated maximal value of the objective function ignoring the complicating constraint is $\widetilde{z} = 22$. This solution will yield a left-hand side value of 15 in the complicating constraint, and since the right-hand side value of this constraint is 9, the constraint is violated.

(b) Dualizing the complicating constraint with a value $\widehat{u} = 1$ for the dual variable transfers the problem into a regular assignment problem with maximizing objective and the coefficient matrix

$$\mathbf{C} = \begin{bmatrix} 4 & 1 & -1 \\ 1 & 1 & 1 \\ 3 & 2 & 2 \end{bmatrix}.$$

The solution to this problem is the same is that shown in (a), i.e., $\widehat{x}_{11} = \widehat{x}_{23} = \widehat{x}_{32} = 1$ and $\widehat{x}_{ij} = 0$ otherwise (alternatively $\widehat{x}_{11} = \widehat{x}_{22} = \widehat{x}_{33} = 1$ and $\widehat{x}_{ij} = 0$ otherwise) with objective function value $\widehat{z}_{Lagr} = 16$. The complicating constraint is again violated. The alternative solution also violates the complicating constraint.

(c) With $\widehat{u} = 2$, we obtain

$$\mathbf{C} = \begin{bmatrix} -1 & 0 & -6 \\ -1 & 1 & -3 \\ 0 & -4 & -3 \end{bmatrix},$$

which has optimal solutions $\widehat{x}_{12} = \widehat{x}_{23} = \widehat{x}_{31} = 1$ and $\widehat{x}_{ij} = 0$ otherwise, as well as $\widehat{x}_{11} = \widehat{x}_{22} = \widehat{x}_{33} = 1$ and $\widehat{x}_{ij} = 0$ otherwise and $\widehat{z}_{Lagr} = 15$. Now the left-hand side of the complicating constraint equals 8 in the first solution and 10 in the second, so that in case of the first solution, this constraint is satisfied. The first solution has a value of the objective function of the original problem of $\widehat{z} = 13$.

(d) With $\widehat{u} = 1.5$, we obtain

$$\mathbf{C} = \begin{bmatrix} 1.5 & 0.5 & -3.5 \\ 0 & 1 & -1 \\ 1.5 & -1 & -0.5 \end{bmatrix},$$

which has the unique optimal solution $\widehat{x}_{11} = \widehat{x}_{22} = \widehat{x}_{33} = 1$ and $\widehat{x}_{ij} = 0$ otherwise, which has an optimal value of the objective function $\widehat{z}_{Lagr} = 17$. Now the left-hand side of the complicating constraint equals 10, so that this constraint is again violated. The value of the objective function of the original problem is $\widehat{z} = 17$.

5.3 Solution Methods for Integer Programming Problems

Table 5.23 Summary of results for problem 10

	Problem	Dual variable $\hat{u}$	Optimal solution	Objective function value z_{Lagr}	z	Feasible for P?
(a)	P_{red} (P_{Lagr})	0	$\tilde{x}_{11} = \tilde{x}_{23} = \tilde{x}_{32} = 1$	22	22	No
(b)	P_{Lagr}	1	$\hat{x}_{11} = \hat{x}_{23} = \hat{x}_{32} = 1$	16	22	No
	P_{Lagr}	1	$\hat{x}_{11} = \hat{x}_{22} = \hat{x}_{33} = 1$	16	17	No
(c)	P_{Lagr}	2	$\hat{x}_{12} = \hat{x}_{23} = \hat{x}_{31} = 1$	15	13	Yes
	P_{Lagr}	2	$\hat{x}_{11} = \hat{x}_{22} = \hat{x}_{33} = 1$	15	17	No
(d)	P_{Lagr}	1.5	$\hat{x}_{11} = \hat{x}_{22} = \hat{x}_{33} = 1$	15.5	17	No
–	P	–	$\bar{x}_{12} = \bar{x}_{23} = \bar{x}_{31} = 1$	–	13	Yes

(e) With $\hat{u} = 1$, as in (b), the complicating constraint is violated, indicating that the value of $\hat{u}$ should be increased. With $\hat{u} = 2$ as in (c), the complicating constraint is satisfied, demonstrating that $\hat{u}$ is sufficiently large. Finding an intermediate value such as $\hat{u} = 1.5$ as in (d), the complicating constraint is again violated, showing that the value of 1.5 is too small for the Lagrangean multiplier. Enumerating all solutions of the original assignment problem without additional constraints and then checking which solutions satisfy the complicating constraint, we can see that $\hat{u} = 2$ actually results in the unique optimal solution to the original problem with the complicating constraint, i.e., $\bar{x}_{12} = \bar{x}_{23} = \bar{x}_{31} = 1$ and $\bar{x}_{ij} = 0$ with $\bar{z} = 13$. Note that the relationship $\hat{z}_{Lagr} \geq \hat{z}$ developed in Sect. 4.1 is not violated, even though we obtained a z-value of 22 in (a) which is larger than the objective function values in (b), (c), and (d). This is because $\hat{z}_{Lagr} < \hat{z}$ for solutions that are infeasible, which is the case in (a), (b), and (d). A summary of the results is presented in Table 5.23.

Network Models 6

Graph theory, the subject at the root of this chapter, dates back to 1736, when the Swiss mathematician Leonhard Euler considered the now famed "Königsberg bridge problem." At that time, there were seven bridges across the River Pregel that ran through the city of Königsberg on the Baltic Sea, and Euler wondered whether or not it would be possible to start somewhere in the city, walk across each of the bridges exactly once, and return to where he came from. (It was not.) We will return to Euler's problem in Sect. 6.5. Two hundred years later in 1936, the Hungarian mathematician Denès König wrote the seminal book "The Theory of Finite and Infinite Graphs," that laid the foundations of modern graph theory. The subject was advanced by operations researchers in the 1950s, most prominently by L.R. Ford and D.R. Fulkerson to deal with path and network flows. For a more comprehensive treatment of the topics covered in this chapter, readers are referred to the pertinent literature, e.g. Eiselt and Sandblom (2000), or Murty (2006).

6.1 Definitions and Conventions

The models discussed in this section are optimization problems on a structure commonly known as a graph. A *graph* (for simplicity, we will refer to graphs also as *networks*, even though many graph-theorists will disagree) consists of *nodes* (sometimes referred to as *vertices*) and *arcs* (or *edges*). Many authors refer to undirected connections as edges and directed connections as arcs. A node and an arc leading to or from it are said to be *incident*. The graph in Fig. 6.1 is an example with the nodes n_1, n_2, n_3, n_4, and n_5 represented by circles and the arcs represented by directed or undirected lines. Arcs are written as either a_{ij} or (n_i, n_j), whatever is more convenient. Similarly in an undirected graph, an edge that connects the nodes n_i and n_j is either written (n_i, n_j) or e_{ij}. A graph that contains only undirected edges is called an *undirected graph*, one with only directed arcs is a *directed graph* (frequently referred to as a network), and a graph that includes directed and undirected arcs is called a *mixed graph*.

Fig. 6.1 Mixed graph

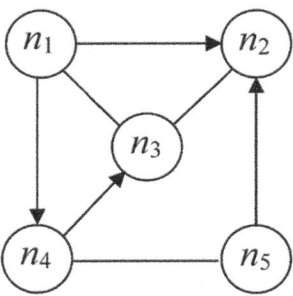

Graphs can be stored in the form of an *adjacency matrix* $\mathbf{A} = (a_{ij})$, in which a_{ij} equals the value of one, if an edge between nodes n_i and n_j or an arc from node n_i to node n_j exists, and 0 otherwise. For the graph in Fig. 6.1, we have the adjacency matrix

$$\mathbf{A} = \begin{bmatrix} 0 & 1 & 1 & 1 & 0 \\ 0 & 0 & 1 & 0 & 0 \\ 1 & 1 & 0 & 0 & 0 \\ 0 & 0 & 1 & 0 & 1 \\ 0 & 1 & 0 & 1 & 0 \end{bmatrix}.$$

If the graph is undirected, its adjacency matrix will be symmetric (i.e., $a_{ij} = a_{ji}$ for all i and j), while this is not the case in directed or mixed graphs (such as the one in our example). It is also possible to indicate not only that a connection exists, but some other feature about this connection, e.g. its distance, cost, or other measures. In such a case, we replace the "1" for existing arcs or edges by the value assigned to the arc or edge. Depending on the meaning of the values, such a matrix could then be a distance matrix, cost matrix, or similar. It is typically assumed that all the values associated with edges and arcs are integer. If this is not originally the case, we can use appropriate rounding or scaling techniques.

A *path* is defined as a sequence of nodes and arcs that starts at a node and ends at some node (possibly, but not necessarily, where it started). In case a path includes directed arcs, they must be traversed in the direction indicated by its arrow. In the graph in Fig. 6.1, the sequence n_5—a_{52}—n_2—a_{23}—n_3 is a path. A *circuit* in a graph is defined as a path that begins and starts at the same node. In the graph in Fig. 6.1, the sequence n_1—a_{14}—n_4—a_{43}—n_3—a_{31}—n_1 is a circuit. A node n_j is said to be *reachable* from another node n_i, if there exists at least one path from n_i to n_j. A *connected graph* is an undirected graph, in which each node is reachable from each other node. The graph in Fig. 6.1 is connected. A directed graph is *strongly connected*, if there exists at least one path from each node n_i to each other node n_j.

A *tree* (or *tree graph*) is defined as a connected undirected graph, with the property that the removal of any arc or edge from it will render the graph disconnected. In that sense, a tree graph is the minimalist connected structure. An example

Fig. 6.2 Tree

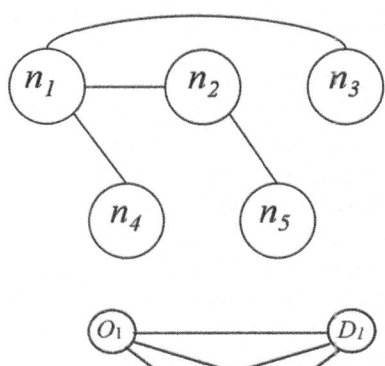

Fig. 6.3 Bipartite graph

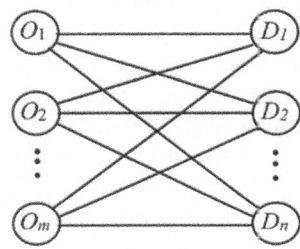

for a tree is shown in Fig. 6.2. Also note that there exists exactly one path between any pair of nodes in a tree.

Another specific type of graph is a so-called *bipartite graph*. It is characterized by the fact that its set of nodes is subdivided into two mutually exclusive and collectively exhaustive subsets. Call these two subsets O and D for reasons to become clear below. Arcs or edges exist in a bipartite graph exclusively between the sets O and D, i.e. each arc or edge has one end in O and the other in D. An example is shown in Fig. 6.3.

We have encountered such graphs in linear programming when discussing the transportation and assignment problems (Sect. 2.2.5) in this book. This also explains the choice of name of the sets: We are shipping from origin O to destination D.

Any other problem-specific notation will be introduced whenever it is needed. We should mention that while all problems discussed in this chapter can be formulated as integer programming problems and solved with any of the pertinent solvers, such a process is typically inefficient and may hide internal structures that allow the user to gain insight into the problem. Typically, specific network-based algorithms are very efficient and outperform general-purpose integer programming solvers when applied to the particular class of problems they were designed to solve. However, such specific methods depend on the network structure. Even a single additional constraint may destroy the particular structure and will no longer allow using the special algorithm, so that we have to return to standard integer programming formulations.

6.2 Social Network Analysis

Social network analysis is a relatively young branch of network analysis. It is concerned about the assessment of and relations between individuals in a group and their importance. In order to start our discussion, suppose that each individual in a group is represented by a node and the interactions between the individuals are shown as a directed or undirected arcs. These interactions can indicate friendship (which must be measured by some means), the fact that two individuals had any kind of contact with each other in a specific time frame, or any other relation. To further explain the concept, consider the graph in Fig. 6.4.

If the graph were to represent the members of a class in school and the edges denote existing friendly relationships, we could consider the student who is represented by node n_7 more of a loner with just a single friend, while the student behind node n_2 appears to be rather popular. Another important application is related to terrorist networks. The goal in their analysis is to find the leaders, lieutenants, and foot soldiers. One of the analyses will attempt to determine which of the elements are the key players in this group. In order to do so, it is necessary to outline measures that express the degree of centrality of a node in the graph. One guiding principle in this quest could be P. Bonacich's assertion that "Power comes from being connected to those who are powerless."

The simplest measure of centrality is the so-called *degree centrality*. The degree $d(n_i)$ of a node n_i in an undirected graph is defined as the number of edges incident to that node. A high degree of a node indicates that the element is connected to many people, so that it may be considered important. In our example, the degrees of the nodes are 3, 5, 4, 3, 3, 2, 1, 4, 3, and 4, respectively, so that we may conclude that the individual behind node n_2 is a central figure in the network.

Another measure is *betweenness centrality*. This measure requires the determination of the shortest paths between all nodes. An efficient algorithm for this task will be provided in Sect. 6.4. In our example in which there are no weights associated with the edges in the graph that could indicate the strength of the link between elements (such as the degree of friendship, the number of phone calls per week

Fig. 6.4 A social network

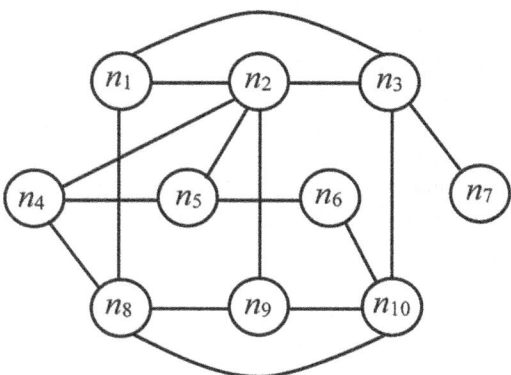

6.2 Social Network Analysis

between the individuals, or a similar measure), we consider the distance between any two nodes simply as the number of arcs or edges on the shortest path between the nodes. In other words, if Mary wants to contact Robert and this is only possible if she talks to Patricia first, who then talks to Ben, who, in turn, talks to Robert (which sounds a bit like Chinese whispers), then we will say that the shortest path between Mary and Robert is 3, as 3 contacts (edges) are necessary to transmit the message.

In our example, let $\mathbf{D} = (d_{ij})$ denote the matrix of shortest paths, in which the element d_{ij} denotes the number of edges on the shortest path from n_i to n_j.

$$\mathbf{D} = \begin{bmatrix} 0 & 1 & 1 & 2 & 2 & 3 & 2 & 1 & 2 & 2 \\ 1 & 0 & 1 & 1 & 1 & 2 & 2 & 2 & 1 & 2 \\ 1 & 1 & 0 & 2 & 2 & 2 & 1 & 2 & 2 & 1 \\ 2 & 1 & 2 & 0 & 1 & 2 & 3 & 1 & 2 & 2 \\ 2 & 1 & 2 & 1 & 0 & 1 & 3 & 2 & 2 & 2 \\ 3 & 2 & 2 & 2 & 1 & 0 & 3 & 2 & 2 & 1 \\ 2 & 2 & 1 & 3 & 3 & 3 & 0 & 3 & 3 & 2 \\ 1 & 2 & 2 & 1 & 2 & 2 & 3 & 0 & 1 & 1 \\ 2 & 1 & 2 & 2 & 2 & 2 & 3 & 1 & 0 & 1 \\ 2 & 2 & 1 & 2 & 2 & 1 & 2 & 1 & 1 & 0 \end{bmatrix}.$$

While the matrix $\mathbf{D}$ above shows the lengths of the shortest paths, the matrix $\mathbf{E}$ indicates where these paths are by listing the nodes on the shortest path(s). Multiple shortest paths are separated by a comma, e.g., between nodes n_1 and n_6 there are three shortest paths, one via nodes n_2 and n_5, one via nodes n_3 and n_{10}, and one via nodes n_8 and n_{10}.

$$\mathbf{E} = \begin{bmatrix} - & - & - & 2,8 & 2 & 25,310,810 & 3 & - & 2,8 & 3,8 \\ - & - & - & - & - & 5 & 3 & 1,4,9 & - & 3,9 \\ - & - & - & 2 & 2 & 10 & - & 1,10 & 2,10 & - \\ 2,8 & - & 2 & - & - & 5 & 23 & - & 2,8 & 8 \\ 2 & - & 2 & - & - & - & 23 & 4 & 2 & 6 \\ 25,310,810 & 5 & 10 & 5 & - & - & 310 & 10 & 10 & - \\ 3 & 3 & - & 23 & 23 & 310 & - & 13,310 & 23,310 & 3 \\ - & 1,4,9 & 1,10 & - & 4 & 10 & 13,310 & - & - & - \\ 2,8 & - & 2,10 & 2,8 & 2 & 10 & 23,310 & - & - & - \\ 3,8 & 3,9 & - & 8 & 6 & - & 3 & - & - & - \end{bmatrix}$$

In order to determine the importance of a node as a detour node, we determine the number of times that a node is a detour node on a shortest path. In case there are multiple shortest paths, say, r shortest paths, each detour node is counted as $1/r$. In our example, n_2 and n_8 are counted ½ each for the shortest path between n_1 and n_4, the detour nodes n_2, n_3, n_5, and n_8 are counted ⅓ each as they each appear once on one of the three shortest paths between n_1 and n_6, while n_{10} is counted ⅔ as it appears on two of the shortest paths between n_1 and n_6. Finally, n_{10} is counted as 1, as it appears on the unique shortest path between n_6 and n_3. Adding up the number of times a node appears as a detour node results in counts of 1⅔, 17⅔, 18⅔, 2⅔, 4⅔, 2, 0, 6⅔, 1⅔, and 12⅓, respectively. This indicates that node n_3 is the node that has the highest betweenness centrality, followed by node n_2. These two elements take a central place in the network.

Yet another measure of centrality is *closeness centrality*. This measure is based on the lengths of shortest paths from a node to all other nodes. Note that again, we need to determine shortest paths for this measure. Formally, closeness centrality is defined as the inverse value of the sum of lengths of shortest paths from the node under consideration to all other nodes. In our example, we can simply add the values in each of the columns of the distance matrix **D** to obtain the sum of lengths of shortest paths, which are 16, 13, 14, 16, 16, 18, 22, 15, 16, and 14. The inverse values are the closeness centrality. It appears that node n_7 is performing worst on this measure (which is also visible in the graph in Fig. 6.4), while n_2 performs best, albeit not by a wide margin.

Other, somewhat more advanced, concepts have been devised, such as *eigenvalue centrality* or one of its modifications, called *Page Rank* (named after one of the co-founders of Google), which, among other uses, is used by search engines to determine the ranks of files as they appear on the screen.

6.3   Network Flow Problems

In its simplest form, a network flow problem can be described as follows. It starts with a graph, in which an unlimited (or, at least, sufficiently large) quantity of a homogeneous good is available at a node called the *source*, while as many units as possible are to be shipped from the source through the network to another node, named the *sink*, where these units are in demand. (Some authors call this an $O - D$ *flow*, which refers to the flow from origin to destination.) Units can be shipped through as many nodes as necessary, as long as they do not exceed the *capacity*, i.e. an upper bound on the flow specified at each of the arcs and/or the nodes in the network. Typically, we assume that there are no losses anywhere in the network. Once the problem has been solved, the optimal *flow pattern* indicates how many units are shipped on which arcs, and the optimal *flow value* indicates how many flow units are shipped through the network.

One of the many examples deals with the evacuation of people from buildings, convention halls, or even cities. As an example, consider the situation shown in Fig. 6.5.

The floor plan depicts a large exhibition hall with the squares denoting kiosks. An escape plan is needed for the auditorium that has only one entrance labeled by n_s at the bottom left of Fig. 6.5, and one exit on the right labeled by n_t. People are assumed to enter the hall through n_s, from where they have to be evacuated through the exhibit

6.3 Network Flow Problems

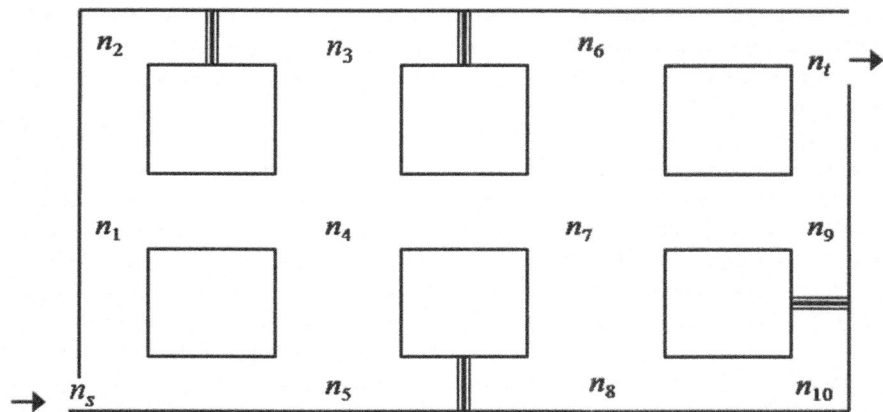

Fig. 6.5 Floor plan

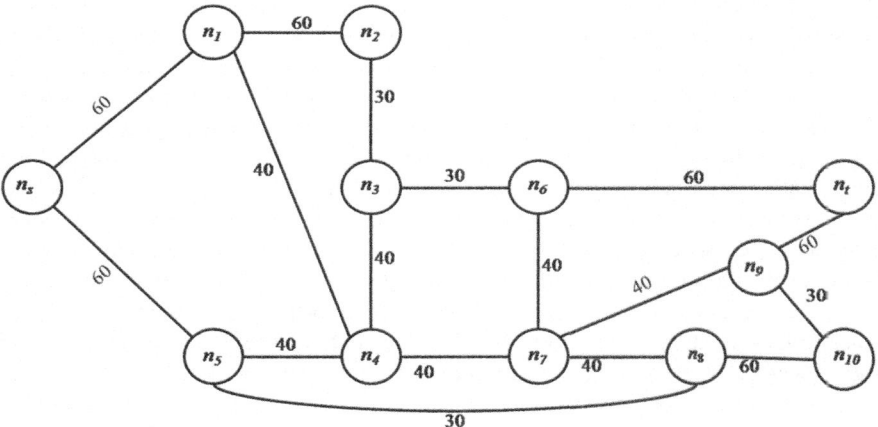

Fig. 6.6 Arc capacities for evacuation

hall to the door at n_t and on to the outside. The walkways along the walls of the exhibit hall are fairly wide, and each of their segments (from one corner to another) was shown to let up to 60 people per minute through in an emergency situation. The walkways in the center are narrower, so that only 40 people per minute can pass along each segment. The steps on some of the outside segments shown by lines across the walkway, present additional obstacles, so that these segments only allow 30 people per minute to pass through them. The question is now how many people can pass through the exhibition hall in each minute. Given the capacity of the auditorium, it can then be computed how long an evacuation of the auditorium through the exhibit hall (from n_s to n_t) will take. The graph in Fig. 6.6 shows an image of the situation with the appropriate arc capacities. The quest is now to find a pattern with the largest possible flow from n_s to n_t.

A number of additional applications of flow problems can be found in Eiselt and Sandblom (2000).

In order to formulate the problem, we first need to define variables. Here, the continuous flow variables are x_{ij}, defined as the flow along the arc a_{ij} from n_i directly to node n_j. Consider the example of Fig. 6.6. Since all edges are undirected, we need to define two variables for each edge, one for each direction. Given that there are 17 edges, we will need 34 variables. (Some of the variables can be ignored, though: we do not have to consider flows into the source, as the entire purpose is to get as much flow out of there as possible. In other words, we do not need to define x_{1s} and x_{5s}. Similarly, we do not need to consider flows out of the sink.) The objective will then maximize the flow through the network. This flow can be expressed as the number of flow units that are sent out of the source. In the example of Fig. 6.6, this is $x_{s1} + x_{s5}$. Alternatively, we can consider the number of flow units that are sent into the sink, i.e., $x_{6t} + x_{9t}$.

As far as constraints are concerned, we have to consider two types. The first set of constraints will have to ensure that no flow is lost or created at any of the nodes. Such constraints are commonly called *conservation equations*, *(flow) balancing equations*, or *Kirchhoff node equations* in reference to their counterpart in electrical networks. They are formulated by requiring that the number of units that flow into a node equals the number of units that flow out of a node. In the above example, the conservation equation for the node n_3 is $x_{23} + x_{43} + x_{63} = x_{32} + x_{34} + x_{36}$. Conservation equations have to be formulated for each node except for the source and the sink. The second set of constraints formulates the capacity constraints that require that the flow along each segment does not exceed the upper limit. Again, in this example we have $x_{s1} \leq 60$, $x_{s5} \leq 60$, $x_{12} \leq 60$, $x_{21} \leq 60$, and so forth. Adding the nonnegativity constraints, a mathematical formulation has been obtained and the problem could be solved with any off-the-shelf linear programming software. Whenever available, specialized algorithms are, however, much faster, as they can exploit the special structure of the problem.

Ford and Fulkerson were the first to describe a method to solve the maximal flow problem. Their technique belongs to the large class of the so-called *labeling methods*. Labeling methods have been developed for many network models, and they all have in common that they are very efficient. The idea of Ford and Fulkerson's incremental method is to (incrementally) increase the flow in forward arcs, and to decrease the flow in backward arcs. Both steps are necessary in order to reach an optimal solution. In order to illustrate, consider a "forward arc," in which we would like to increase the flow. If the capacity of such an arc is, say, 7, and the present flow in the arc is 4, then we can increase the flow on this arc by the present slack of $7-4 = 3$ units. On the other hand, if a "backward arc" has a capacity of 7 and a present flow of 4, we can decrease the flow on this arc by no more than its present flow, i.e., by 4 units. This is the fundamental ideal of the labeling technique. Starting with the source, the method attempts to label nodes along arcs. The first part of a node's label indicates the neighboring node it was labeled from, while the second

6.3 Network Flow Problems

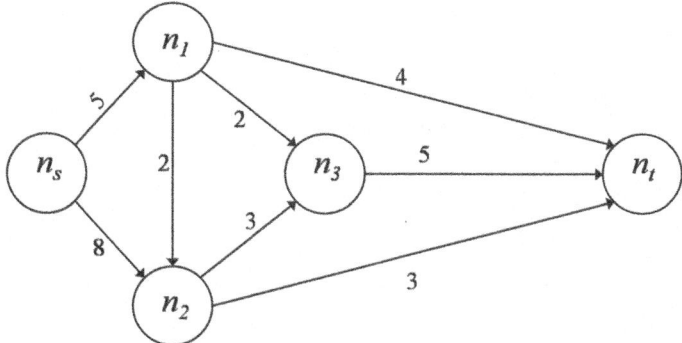

Fig. 6.7 Network for maximal flow example

part of the label indicates the possible flow change on the arc along which the labeling took place.

The objective value—the second part of the label—is determined by the objective value achieved so far plus whatever contribution to the objective function has to be accounted for in this step. (This is a standard procedure used in what is called *dynamic programming*.) More specifically, labeling a node n_j from a node n_i along an arc a_{ij} means that the incremental flow that can be squeezed from the source to n_j equals the minimum of what incremental flow can be sent from the source to n_i plus whatever slack capacity we have along the arc a_{ij}.

Once all nodes that can be labeled have been labeled, the iteration comes to an end in one of two states. We either have been able to label the sink, in which case a *breakthrough* has occurred, or we were not able to label the sink, which is referred to as a *nonbreakthrough*. If a breakthrough has occurred, we are able to increase the flow through a network, while in case of a nonbreakthrough, the current flow is maximal, and the procedure terminates.

The algorithm is best explained by means of a numerical.

Example Consider the network in Fig. 6.7, whose numbers next to the arcs indicate the capacity of the arc.

Suppose that the source n_s is now labeled with (n_s, ∞), indicating that we start at the node n_s, and, given the present zero flow through the network, we can change the flow by any amount we choose. At this point we could label either n_1 and/or n_2. Suppose we choose n_1. Since we label the node n_1 from n_s, the arc (n_s, n_1) is a forward arc and we attempt to increase its flow. The capacity of the arc is 5, while its present flow is zero, meaning that we can increase the flow in this arc by 5 units. This is indicated in the label of n_1, which is then $(n_s, 5)$. Now the nodes n_s and n_1 are labeled, and the process continues. At this point, we have a large number of choices: we can now either label n_2 from the source n_s, or label n_2, n_3, or n_t from n_1. Any technique that labels all nodes from one node before moving on is referred to as

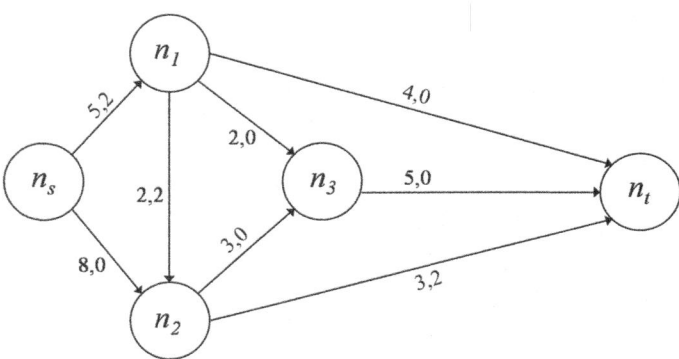

Fig. 6.8 First flow pattern

breadth-first-search, while a labeling strategy that attempts to move on as deeply into the network as possible is called *depth-first-search*. Suppose that we choose to label n_2 from n_1. Again, we are moving with the direction of the arrow, so that this is another forward labeling step. The label of the node n_2 will then be $(n_1, \min \{5, 2-0\}) = (n_1, 2)$. The reason is that while we could increase the flow from the source on some path (we do not have to know at this point which path) to n_1 by 5 units, we can only ship two more units from n_1 to n_2. At this point, n_2 has been labeled and one of the many choices to continue labeling is to label the sink n_t from n_2. The label of the sink n_t is then $(n_2, \min \{2, 3-0\}) = (n_2, 2)$. Now the sink has been labeled, and we have achieved a breakthrough.

Whenever a breakthrough occurs, we are able to increase the flow through the network by at least one unit. To do so, we now have to retrieve the path on which the flow change is possible. This is where the first part of the labels comes in. In a backward recursion, we start at the sink n_t. Its label indicates that its predecessor is n_2. Now the label of n_2 shows that its predecessor was n_1, whose predecessor, in turn, was n_s. This means that we have successfully retrieved the path n_s—n_1—n_2—n_t. This is the path on which the flow will be increased by 2 units, which is the second part of the label of the sink. The resulting flow pattern is shown in Fig. 6.8, where the arcs have two values: its capacity and its present flow.

We now delete all labels except the label of the source and start anew. We can again label n_1 from the source n_s, this time the label of n_1 is $(n_s, \min \{\infty, 5-2\}) = (n_s, 3)$. From n_1, it is possible to label the sink, whose label is then $(n_1, \min \{3, 4-0\}) = (n_1, 3)$. Again, we have obtained a breakthrough and the flow can be increased by 3 units. The path on which this increase takes place is determined by following the labels backward from the sinks, which results in n_s—n_1—n_t. The resulting flow pattern is shown in Fig. 6.9.

Again, resetting the labels, the process begins again. At this point, it is no longer possible to label the node n_1 from the source, as the flow has reached the capacity. However, we can still label the node n_2 with $(n_s, \min \{\infty, 8-0\}) = (n_s, 8)$. From n_2, we can label the sink n_t with the label $(n_2, \min \{8, 3-2\}) = (n_2, 1)$, and we have

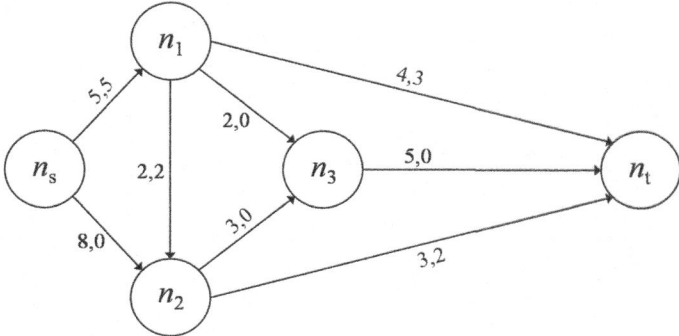

Fig. 6.9 Second flow pattern

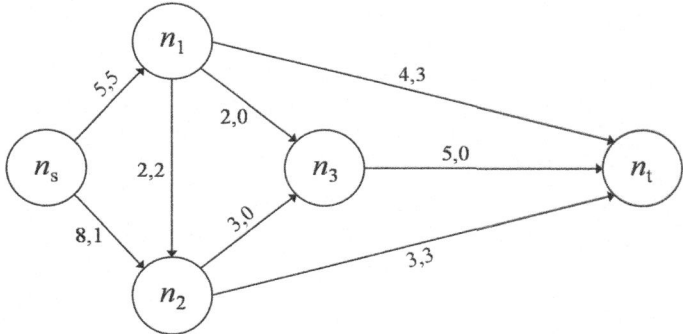

Fig. 6.10 Third flow pattern

achieved another breakthrough, on which the flow can be changed by one unit. The path can be retrieved as n_s—n_2—n_t. The flow pattern is then shown in Fig. 6.10.

The next iteration again starts by labeling the node n_2 from the sink, which is again our only choice. The label of the node n_2 is now $(n_s, \min\{\infty, 8-1\}) = (n_s, 7)$. We can now label n_3 from n_2, so that the label on node n_3 is $(n_2, \min\{7, 3-0\}) = (n_2, 3)$. From n_3, we can then label the sink with $(n_3, \min\{3, 5-0\}) = (n_3, 3)$, and another breakthrough has occurred. The path on which the flow is changed can be retrieved as n_s—n_2—n_3—n_t, and on all arcs along that path the flow is increased by 3 units. The resulting flow pattern is then shown in Fig. 6.11.

Notice that so far, we have only used the forward labeling in the incremental method. The next iteration commences by resetting the labels and restarting the process. Given that the source is labeled as usual with (n_s, ∞), the only choice is now to label the node n_2 with $(n_s, \min\{\infty, 8-4\}) = (n_s, 4)$. From the node n_2, no forward labeling is possible. However, we can follow the arc from n_1 to n_2 against the direction of the arc in a backward labeling step. This results in node n_1 receiving the label $(n_2, \min\{4, 2\}) = (n_2, 2)$. From node n_1, we can then either label n_t or n_3. We arbitrarily choose n_3, label it with $(n_1, \min\{2, 2-0\}) = (n_1, 2)$, and from n_3, we

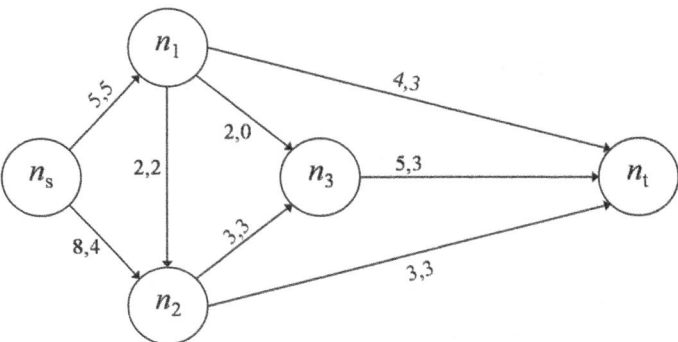

Fig. 6.11 Fourth flow pattern

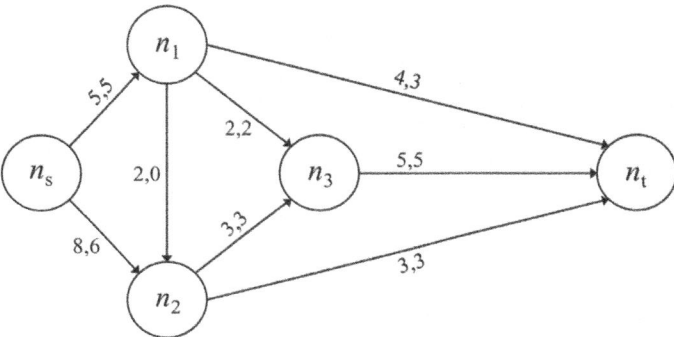

Fig. 6.12 Fifth and optimal flow pattern

can then label the sink n_t with $(n_3, \min\{2, 5-3\}) = (n_3, 2)$. Another breakthrough has occurred, and the flow through the network can be increased by 2 units. The flow along which the flow will be changed is retrieved through backward recursion as n_s—n_2—n_1—n_3—n_t, where the arc from n_1 to n_2 is used in reverse direction. The new flow pattern is determined by increasing the flow in all forward arcs on that path, while decreasing the flow in the solitary backward arc on the path. The resulting flow pattern is shown in Fig. 6.12.

In the next step, we start again by labeling the node n_2 from the source, which receives the label $(n_s, \min\{\infty, 8-6\}) = (n_s, 2)$. At this point, further progress is blocked. The only unlabeled node adjacent to the source is n_1, and the arc (n_s, n_1) is filled to capacity. From the node n_2, further (forward) flow to n_3 and n_t cannot be sent, as both arc flows are at capacity. Also, labeling node n_1 from n_2 is not possible, as the arc flow on that (backward) arc is already at the lower bound of zero. At this point we have labeled all nodes that can be labeled, and we were not able to label the sink. This is a nonbreakthrough. This indicates that the present flow pattern is indeed a *maximal flow* with a total flow of 11 units, the number of flow units that leave the source and, since no flow is lost along the way, the number of units that arrive at the

Fig. 6.13 Cuts in a network

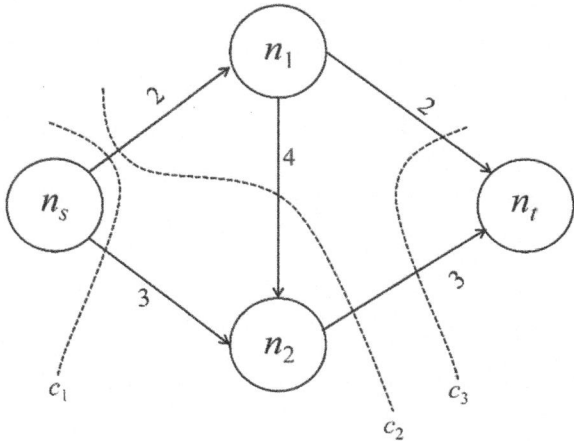

sink. We would like to point out that this maximal flow pattern is not unique: sending one less unit from n_1 to n_3, and on to n_t and shipping it instead directly from n_1 to n_t results in a different flow pattern with the same flow value.

Assessing the situation, we find that we now have the set $N_s = \{n_s, n_2\}$ of labeled nodes, and its complement $N_t = \{n_1, n_3, n_t\}$ of unlabeled nodes. Observe that the flows of all arcs that lead from a node in N_s to a node in N_t (here: the arcs (n_s, n_1), (n_2, n_3), and (n_2, n_t)) are at capacity, while the flows of all arcs leading from a node in N_t to a node in N_s (here: the arc (n_1, n_2)) are at the zero level. All arcs that lead from a node in N_s to a node in N_t are said to be included in the *minimal cut C*. In our example, $C = \{(n_s, n_1), (n_2, n_3), (n_2, n_t)\}$. Adding the capacities (*not* flows) of all arcs in the minimal cut results in the value (or capacity) of the minimal cut, which, in or example, equals $5 + 3 + 3 = 11$. This leads to the famous

Theorem (Ford and Fulkerson) The value of a maximal flow equals the value/capacity of a minimal cut.

The minimal cut constitutes a bottleneck in the network. If we want to increase the capacities of some arcs in the network so as to be able to increase the flow through the network, we have to increase the capacities of arcs that are in the minimal cut(s). Note that the minimal cut is not necessarily unique. As an example, consider the network in Fig. 6.13 with capacities next to the arcs. The broken lines refer to the cuts $C_1 = \{(n_s, n_1), (n_s, n_2)\}$, $C_2 = \{(n_s, n_1), (n_2, n_t)\}$, and $C_3 = \{(n_1, n_t), (n_2, n_t)\}$. All cuts have a capacity equal to 5. Note that in case of multiple cuts, the method described above will find only the minimal cut that is closest to the source.

A variety of extensions of the maximal flow problem exists. One of the most popular generalizations is the *min-cost feasible flow problem*. Again, the idea is very simple. In a network with a designated source and sink, each arc a_{ij} has a lower bound λ_{ij} and an upper bound κ_{ij} on the flow. (Recall that in the above max flow problem all lower bounds were assumed to be zero.) In addition, it is assumed to cost

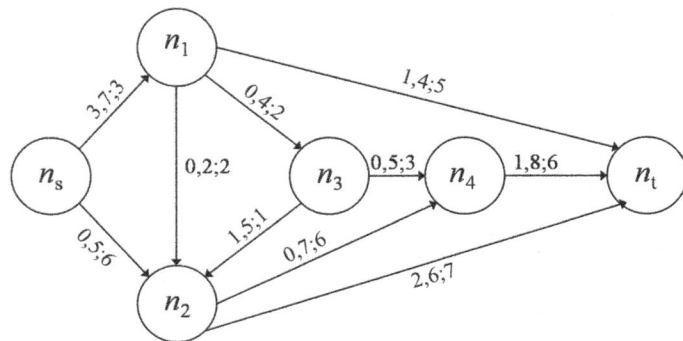

Fig. 6.14 Network for min-cost feasible flow problem

c_{ij} dollars to send one unit of flow on the arc a_{ij}. The problem is now to find a flow from source to sink that respects all lower and upper bounds on the flows and that minimizes the total shipping costs.

As an example, consider the graph in Fig. 6.14, where the numbers next to the arcs symbolize the lower bounds λ_{ij}, the upper bounds κ_{ij}, and the unit costs c_{ij}, respectively.

The cost-minimizing flow problem has three types of constraints: the usual conservation equations, and the lower and upper bounds on each arc flow, respectively. This specific problem can be formulated as follows:

P : Min $z = 3x_{s1} + 6x_{s2} + 2x_{12} + 2x_{13} + 5x_{1t} + 6x_{24} + 7x_{2t} + 1x_{32} + 3x_{34} + 6x_{4t}$

s.t. $x_{s1} - x_{12} - x_{13} - x_{1t} = 0$

$x_{s2} + x_{12} + x_{32} - x_{24} - x_{2t} = 0$

$x_{13} - x_{32} - x_{34} = 0$

$x_{34} + x_{24} - x_{4t} = 0$

$x_{s1} \geq 3 \qquad x_{13} \leq 4$

$x_{32} \geq 1 \qquad x_{32} \leq 5$

$x_{1t} \geq 1 \qquad x_{1t} \leq 4$

$x_{2t} \geq 2 \qquad x_{34} \leq 5$

$x_{4t} \geq 1 \qquad x_{24} \leq 7$

$x_{s1} \geq 7 \qquad x_{2t} \leq 6$

$x_{s2} \geq 5 \qquad x_{4t} \leq 8$

$x_{12} \geq 2$

$x_{ij} \geq 0$ for all arcs i,j.

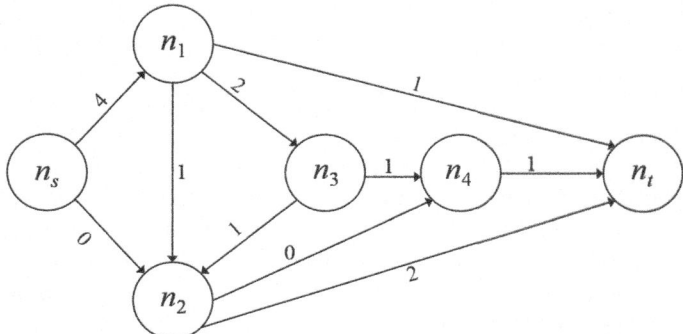

Fig. 6.15 Optimal flow pattern

The solution can be obtained by any standard optimization package or specialized algorithm. The optimal flow pattern is shown in Fig. 6.15, the total flow from source to sink is 4, and the associated minimal total transportation costs are 47.

Further extensions are possible. One such extension has a required flow value $\bar{f}$. A typical network modification that accomplishes this requirement will simply add a return arc (n_t, n_s) with zero costs and lower and upper bound equal to $\bar{f}$. The original flow problem can then be reformulated as a circulation, in which the conservation equations have to hold for *all* nodes, including the source and the sink. Working with the mathematical formulation, all we have to do is add a single constraint, which requires that the flow value equals $\bar{f}$. If $\bar{f} = 7$ in the above example, we either add the constraint $x_{s1} + x_{s2} = 7$, or, alternatively, $x_{1t} + x_{2t} + x_{4t} = 7$ and resolve. (In the above numerical example, the new solution will be the same as that shown in Fig. 6.15, except that an additional 3 flow units are shipped from n_s to n_1 and on to n_t.)

It is now also possible to demonstrate how to cast the standard transportation problem (see Sect. 2.2.5) into the mold of cost-minimal network flow problems. In terms of the network, we will make the following modifications. In addition to the existing origins and destinations, we first create an artificial source n_s and artificial sink n_t. We then connect the source with all origins and connect all destinations with the sink. (This type of problem reformulation can be used for all problems with multiple sources and sinks, provided that the goods are homogeneous and do not need to be distinguished.) The lower bounds of all (source, origin) connections are zero, while the upper bounds equal the supplies available at the respective origin. Similarly, all (destination, sink) arcs have a zero lower bound and an upper bound that equals the demand at the respective destination. Both types of arcs have zero costs. The existing arcs that connect the origins and the destinations have zero lower bounds and arbitrarily large upper bounds (except in cases, in which capacities need to be considered), and carry the costs specified for the original problem. Note that so far, the zero flow would be optimal, as none of the arcs requires a flow greater than zero, and any flow from source to sink costs money. In order to force flow through the network, we connect the sink with the source by means of an artificial arc (n_t, n_s) that has zero costs, and an upper and lower bound that are both equal to the minimum of the total supply and the total demand at the sources and destinations, respectively. This way, the (sink, source) arc forces as many flow units through the network as are in demand or are needed, whatever is less.

Another possible extension includes capacity constraints at the nodes. Two approaches are possible. The first uses the given network and modifies it so that it includes node capacities. This can be accomplished by "splitting" all of the nodes with node capacities. In particular, a node n_i with node capacity κ_i is then replaced by an "in-node" n'_i into which all arcs lead that led into the original node n_i, an "out-node" n''_i, out of which all arcs lead, that lead out of the original node n_i. Finally, the two new nodes n'_i and n''_i are connected by an arc (n'_i, n''_i), whose arc capacity is the original node capacity κ_i. What we have done in this approach is simply to replace the node capacity by an arc capacity.

An alternative approach simply uses a mathematical programming formulation. The number of flow units that flow through a node equals the number of units that enter (and, as nothing is lost, leave) a node, and an appropriate additional constraint is added. In the min-cost flow example of Fig. 6.14 a capacity of, say, 5 units through node n_2 can be written as $x_{s2} + x_{12} + x_{32} \leq 5$, or, equivalently, by using the outflow, as $x_{24} + x_{2t} \leq 5$.

6.4 Shortest Path Problems

Similar to the maximal flow problem discussed in the previous section, shortest path problems are easily described. Given a network with a prespecified source and sink node as well as arc values c_{ij} that denote the cost (or distance, fuel, or any other disutility) of traveling from node n_i directly to node n_j along arc a_{ij}, the task is to find the shortest path from the source to the sink. The literature typically distinguishes between one-to-one shortest path problems (those that search the shortest path between source and sink), one-to-all shortest path problems (in which the task is to find the shortest paths between the source and all other nodes in the network), and the all-to-all shortest path problems (where the task is to determine the shortest paths between all pairs of nodes). Clearly, it would be possible—albeit inefficient—to use an algorithm for the one-to-one shortest path problem and apply it repeatedly so as to solve the one-to-all and all-to-all shortest path problems.

This section first describes a way to reformulate the one-to-one shortest path problem, so that it fits into the mold of the cost-min feasible flow problem. It then describes the workings of an all-to-all shortest path algorithm that is not only very efficient, but also needed in areas such as location models, where all shortest paths have to be known before any location algorithm can even start.

First, we will discuss how to reformulate a shortest path problem as a cost-minimizing flow problem. The idea is to force one flow unit through the network, and let the optimizer find the cost-minimal, i.e., shortest, path. This is done by having lower bounds of zero and upper bounds of one for all arcs, coupled with the actual costs or distances specified for all arcs. So far, the optimal solution would be the zero

6.4 Shortest Path Problems

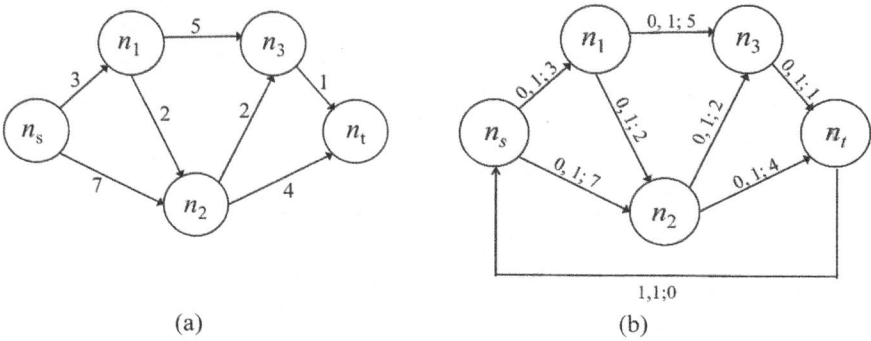

Fig. 6.16 Transformation of a shortest path problem into a min-cost flow problem. (**a**) Original graph, (**b**) Transformed graph

flow, as it is feasible, and no cheaper solution can exist (at least not as long as the arc distances are nonnegative). We then add the circulatory arc (n_t, n_s) with lower and upper bound equal to one. This forces a single unit through the network and will result in the desired solution. As an example, the graph transformation is shown in Fig. 6.16, where Fig. 6.16a shows the original graph with the distances or costs next to the arcs, while Fig. 6.16b has the lower bounds, the upper bounds, and the cost/distances at the arcs.

As far as the mathematical formulation is concerned, we can use the usual min-cost objective function, coupled with two sets of constraints. The first single constraint ensures that exactly one unit leaves the source. The second set of constraints contains the usual conservation equations, which ensure that the flow unit that leaves the source has only one place to go: the sink. In the example of Fig. 6.16, the formulation would be as follows:

$$P : \text{Min } z = 3x_{s1} + 7x_{s2} + 5x_{13} + 2x_{12} + 2x_{23} + 1x_{3t} + 4x_{2t}$$
$$\text{s.t. } x_{s1} + x_{s2} = 1$$
$$x_{s1} - x_{12} - x_{13} = 0$$
$$x_{s2} + x_{12} - x_{23} - x_{2t} = 0$$
$$x_{13} + x_{23} - x_{3t} = 0$$
$$x_{ij} \geq 0 \text{ for all } i,j.$$

Alternatively, the first constraint $x_{s1} + x_{s2} = 1$ could be replaced by the constraint $x_{2t} + x_{3t} = 1$, representing the flow into the sink.

Shortest path problems have many real-world applications. In addition to the obvious applications, in which the shortest path in a road network is to be found (e.g., for GPS-based navigation systems), shortest path problems occur in scenarios that seemingly have nothing to do with shortest paths. As an example, assume that a process has been "discretized," i.e. is subdivided into a finite number of states that describe the system at that point in time. Each node of the network symbolizes a state of the system, and an arc indicates a possible transition from one state to another. The

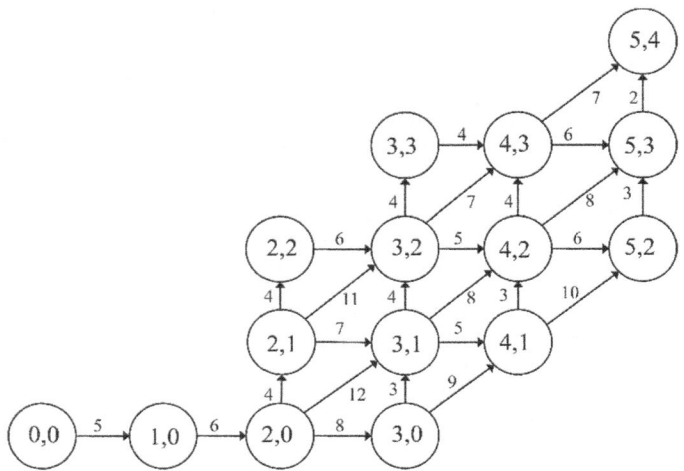

Fig. 6.17 Graph for fighter aircraft problem

arc values show the amount of resources, such as time, money, or fuel, that a transition takes.

A good example of such a problem deals with the problem of getting an aircraft from a standstill position to a certain speed and altitude in the fasted possible way. This problem is particularly relevant for fighter aircraft.

As a numerical example, consider the situation shown in Fig. 6.17. The two numbers in each node indicate the state the aircraft is in: the first component is the ground speed in 100 mph, and the second number shows the aircraft's altitude in 1000 ft. Initially, the aircraft is at (0, 0), i.e. standing still on the ground. Each node describes a state the aircraft can be in with respect to speed and altitude, and the arcs between these states indicate the possible transitions from one state to another. The values next to the arcs show the time (in seconds) that is required to make the transition from one node to another. For example, it takes five seconds to bring the aircraft from the standstill position at (0, 0) to (1, 0), i.e. a speed of 100 mph at zero altitude (meaning on the runway). Suppose that it is desired to bring the aircraft from a standstill position to a speed of 500 mph and an altitude of 4000 ft. as quickly as possible.

At each state in this example, the pilot has three options: either stay at the same altitude and speed up, remain at the same speed and climb, or speed up and climb simultaneously. The shortest path in the above example includes the nodes (0, 0), (1, 0), (2, 0), (2, 1), (2, 2), (3, 2), (4, 3), and (5, 4). In other words, the instructions to the pilot would state to bring the aircraft from a standstill position to a speed of 200 mph, then remain at that speed and climb 2000 ft., then stay at that altitude and speed up to 300 mph, and then accelerate and climb simultaneously to the desired speed of 500 mph and altitude of 4000 ft. This way, it will take 39 sec, the length of the shortest path, to reach the desired state.

6.4 Shortest Path Problems

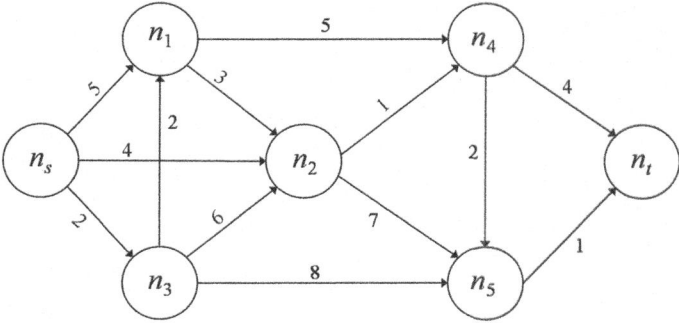

Fig. 6.18 Example for Dijkstra's technique

One of the most popular special-purpose techniques for the determination of shortest paths from one node to all other nodes is Dijkstra's technique that was first published in the late 1950s. It is a highly efficient method that belongs to the class of the so-called *label setting techniques* (as opposed to other *label-correcting techniques*). The main idea is to label the nodes n_s, n_1, ..., with labels $L(n_s)$, $L(n_1)$, ..., each consisting of two parts. The first is the immediate predecessor of the node on the shortest path known so far, and the second part is the length of the shortest path known so far. Throughout the procedure, we distinguish between nodes that have a *temporary label* and those with a *permanent label*. The label of a permanently labeled node indicates the actual length of the shortest path from the source to this node as well as the immediate predecessor on that path, while a temporary label comprises the length of the presently shortest known path and the node's immediate predecessor on it. In each step of the algorithm, one node with a temporary label is chosen and its label made permanent. The upper bounds on the estimates of the shortest paths to all direct successors of this node are revised, and the process is repeated until the labels of all nodes are made permanent. It is important to realize that Dijkstra's method is only applicable to networks with nonnegative arc lengths.

To initialize the method, assume that the source n_s is labeled $L(n_s) = (n_s, 0)$, while all other nodes n_j are labeled $L(n_j) = (n_j, \infty)$. In the beginning, all nodes are assumed to be temporary. The method now proceeds as follows. We choose the temporarily labeled node whose (second part of the) label is minimal among all temporarily labeled nodes. Ties are broken arbitrarily. The label of this node is then made permanent. Suppose this node is n_i. All of this node's direct successors are then investigated by comparing their second part of the label with the label of n_i plus the arc length of the arc a_{ij}. If the present label of n_j is smaller, we leave it unchanged; if it is larger, we replace it by setting $L(n_j) = (n_i, L(n_i) + c_{ij})$, i.e., by the label of n_i plus the length of the arc that connects n_i and n_j.

As an example of the procedure, consider the network in Fig. 6.18.

As indicated above, in the initialization step ("Step 0"), we label the source as $L(n_s) = (n_s, 0)$ and all other nodes n_j with $L(n_j) = (n_j, \infty)$, and let all node labels be

Table 6.1 Permanent and temporary labels during the Dijkstra method

Step #	$L(n_s)$	$L(n_1)$	$L(n_2)$	$L(n_3)$	$L(n_4)$	$L(n_5)$	$L(n_t)$
0	$(n_s, 0)^*$	(n_1, ∞)	(n_2, ∞)	(n_3, ∞)	(n_4, ∞)	(n_5, ∞)	(n_t, ∞)
1		$(n_s, 5)$	$(n_s, 4)$	$(n_s, 2)^*$	(n_4, ∞)	(n_5, ∞)	(n_t, ∞)
2		$(n_3, 4)^*$	$(n_s, 4)$		(n_4, ∞)	$(n_3, 10)$	(n_t, ∞)
3			$(n_s, 4)^*$		$(n_1, 9)$	$(n_3, 10)$	(n_t, ∞)
4					$(n_2, 5)^*$	$(n_3, 10)$	(n_t, ∞)
5						$(n_4, 7)^*$	$(n_4, 9)$
6							$(n_5, 8)^*$

temporary. The node with the lowest temporary label is the sink, so that it is chosen, and its label is made permanent. All computations discussed here are summarized in Table 5.1, where the first time a label has been made permanent, it receives a "*" and due to its permanent status, it is not listed again below. Choosing n_s to receive a permanent label means that the labels of its direct successors n_1, n_2, and n_3 may have to be revised (the labels of all other nodes remain unchanged). Their present labels are compared with the label from n_s, which is the source's label (here: 0) plus the length of the arc from n_s to the node in question. For n_1, the comparison is between ∞ and $0 + 5$, which is 5, so that n_1 is now labeled from n_s, which is indicated in its new temporary label $L(n_1) = (n_s, 5)$. Similarly, the labels of the nodes n_2 and n_3 are $(n_s, 4)$ and $(n_s, 2)$, respectively.

We are now at the end of Step 1 in Table 6.1, where we choose the node with the lowest temporary label. In this example, the node is n_3, as it has the smallest label with "2." We now make this label permanent, indicate this by a star, and revise the labels of its successors n_1, n_2, and n_5 in Step 2. The present label of the node n_1 indicates that a path of length 5 is already known from the source. If we were to label n_1 from n_3, its label would be $2 + 2 = 4$, which is shorter, so that the new label of $L(n_1) = (n_3, 4)$. For n_2, we find that the presently shortest known path is of length 4 (its present label), while labeling the node from n_3 would lead us to a path of length of $2 + 6 = 8$. Since this new path is longer, we ignore it and leave the label of n_2 unchanged. Finally in this step, the present label of n_5 indicates a path of length ∞ is known, which is compared to the label the node would receive if labeled from n_3, which is $2 + 8 = 10$. This new label is shorter, so that node n_5 receives the new label $L(n_5) = (n_3, 10)$.

This process continues until all nodes have been permanently labeled. The results are shown in Table 6.1.

The results in Table 6.1 can now be used to determine the *tree of shortest paths rooted at* n_s. This is done by choosing all permanent labels and connecting the node with its direct predecessor as specified in the label. In our example, the node n_t has n_5 as its direct predecessor, so we introduce the arc a_{5t}. The node n_5 has n_4 in its label, so we introduce the arc a_{45}, and so forth. The resulting arborescence (a directed tree) is shown in Fig. 6.19.

The numbers next to the nodes are the lengths of the shortest paths from the source n_s to all nodes in the network.

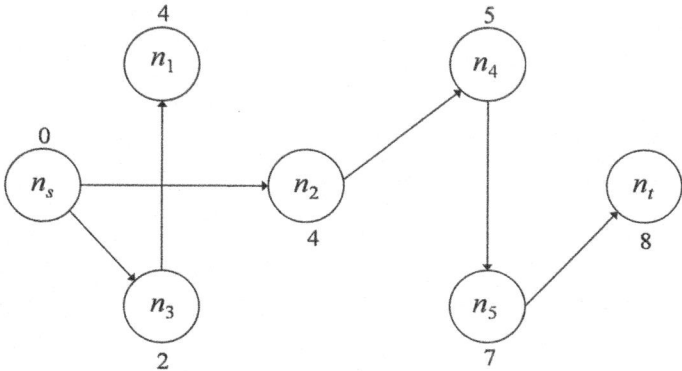

Fig. 6.19 Arborescence for shortest paths

Sometimes (as, for instance, in location models), it is required to determine the paths between all pairs of nodes. Clearly, Dijkstra's technique could be used by considering one node as a source, determine the arborescence of all shortest paths rooted at that node, and then repeat the process with all nodes as roots. This is somewhat tedious, and there is a very efficient technique, called the *Floyd–Warshall method*, that performs this task directly. All it requires are some matrix operations. The method starts and works with the direct distance matrix $\mathbf{C}^0$, which includes all node-to-node distances of the original problem. An iterative step in iteration k can then be described as follows. First define row k and column k as the *key row* and *column*. We then compare the shortest presently known distance between node n_i and node n_j with a detour that uses node n_k. The shorter of the two distances is then used as the new shortest known distance. This process is repeated for all pairs of nodes. Formally, creating matrix $\mathbf{C}^{k+1}$ from matrix $\mathbf{C}^k$, we use the relation

$$c_{ij}^{k+1} = \min\left\{c_{ij}^k, c_{ik}^k + c_{kj}^k\right\}.$$

Starting with the original matrix $\mathbf{C}^0$, we compute the matrices $\mathbf{C}^1, \mathbf{C}^2, \ldots, \mathbf{C}n$. The matrix $\mathbf{C}^n$ (given that the graph has n nodes) is then the matrix of shortest paths.

We will illustrate this procedure by means of an example. The graph and its distances of the example are shown in Fig. 6.20.

The direct distance matrix of the mixed graph in the example is

Fig. 6.20 Example for the Floyd–Warshall method

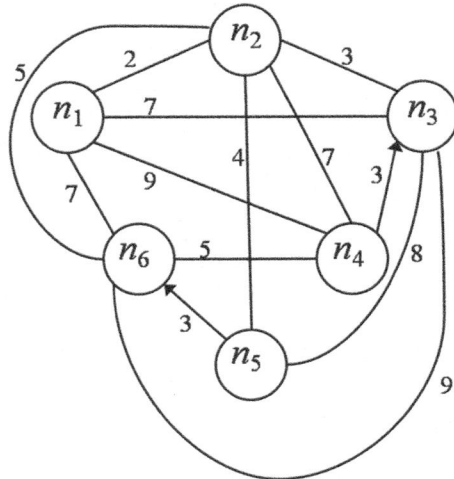

$$\mathbf{C}^0 = \begin{bmatrix} 0 & 2 & 7 & 9 & \infty & 7 \\ 2 & 0 & 3 & 7 & 4 & 5 \\ 7 & 3 & 0 & \infty & 8 & 9 \\ 9 & 7 & 3 & 0 & \infty & 5 \\ \infty & 4 & 8 & \infty & 0 & 3 \\ 7 & 5 & 9 & 5 & \infty & 0 \end{bmatrix}.$$

Note that in case of undirected graphs, the direct distance matrix is symmetric, which is not the case here. In the first iteration $k = 1$, we will consider the node n_1 as a possible detour node. This is done by using the first row and column as the key row and column. This row and column is copied without a change into the next matrix. For all other elements, we compare the present distance with the one that uses n_1 as a detour node. As an example, the lowest known distance from n_2 to n_3 that is presently known is $c_{23} = 3$. This is compared with the detour that uses node 1, i.e., the path that uses the arcs (n_2, n_1) and (n_1, n_3), whose lengths are 2 and 7, respectively. In other words, we now compare 3 with $2 + 7 = 9$. Since the original length is shorter, we keep it and continue with the next pair of nodes. Next, we compare the distances from n_2 to n_4, without and with detour via n_1. We find that $c_{24} = 7 < 2 + 9 = c_{21} + c_{14}$, so that again, we make no change. We continue this way and find no changes, until we reach the connection from n_3 to n_4. At present, $c_{34} = \infty$, i.e., there is no direct connection. We compare this distance with the detour that uses n_1, i.e. the arcs (n_3, n_1) and (n_1, n_4). Those distances are 7 and 9, so that we are now able to reach n_4 from n_3 on a path of length $7 + 9 = 16$. This turns out to be the only change from matrix $\mathbf{C}^0$ to matrix $\mathbf{C}^1$, which is indicated in $\mathbf{C}^1$ by an asterisk.

6.4 Shortest Path Problems

$$\mathbf{C}^1 = \begin{bmatrix} 0 & 2 & 7 & 9 & \infty & 7 \\ 2 & 0 & 3 & 7 & 4 & 5 \\ 7 & 3 & 0 & 16* & 8 & 9 \\ 9 & 7 & 3 & 0 & \infty & 5 \\ \infty & 4 & 8 & \infty & 0 & 3 \\ 7 & 5 & 9 & 5 & \infty & 0 \end{bmatrix}$$

Starting now with $\mathbf{C}^1$, the process is repeated by using the second row and column as key row (column) and comparing all known distances in $\mathbf{C}^1$ with the detour that uses n_2 as detour node. Again, we copy the key row and column without any changes to $\mathbf{C}^2$. Here are some of the computations. For the connection from n_1 to n_3, we presently have a length of 7, which we compare with the detour via node n_2, which has a length of $2 + 3 = 5$, so that the new distance is shorter. Similarly, the distance from n_1 to n_4 without the detour via n_2 is 9, with the detour it is $2 + 7 = 9$, a tie. The distance from n_1 to n_5 without the detour is ∞, with the detour it is $2 + 4 = 6$. The results are shown in the matrix $\mathbf{C}^2$, again with all changes indicated by an asterisk.

$$\mathbf{C}^2 = \begin{bmatrix} 0 & 2 & 5* & 9 & 6* & 7 \\ 2 & 0 & 3 & 7 & 4 & 5 \\ 5* & 3 & 0 & 10* & 7* & 8* \\ 9 & 7 & 3 & 0 & 11* & 5 \\ 6* & 4 & 7* & 11* & 0 & 3 \\ 7 & 5 & 8* & 5 & 9* & 0 \end{bmatrix}$$

The remaining iterations are shown without further comment.

$$\mathbf{C}^3 = \begin{bmatrix} 0 & 2 & 5 & 9 & 6 & 7 \\ 2 & 0 & 3 & 7 & 4 & 5 \\ 5 & 3 & 0 & 10 & 7 & 8 \\ 8* & 6* & 3 & 0 & 10* & 5 \\ 6 & 4 & 7 & 11 & 0 & 3 \\ 7 & 5 & 8 & 5 & 9 & 0 \end{bmatrix}, \mathbf{C}^4 = \begin{bmatrix} 0 & 2 & 5 & 9 & 6 & 7 \\ 2 & 0 & 3 & 7 & 4 & 5 \\ 5 & 3 & 0 & 10 & 7 & 8 \\ 8 & 6 & 3 & 0 & 10 & 5 \\ 6 & 4 & 7 & 11 & 0 & 3 \\ 7 & 5 & 8 & 5 & 9 & 0 \end{bmatrix},$$

$$\mathbf{C}^5 = \begin{bmatrix} 0 & 2 & 5 & 9 & 6 & 7 \\ 2 & 0 & 3 & 7 & 4 & 5 \\ 5 & 3 & 0 & 10 & 7 & 8 \\ 8 & 6 & 3 & 0 & 10 & 5 \\ 6 & 4 & 7 & 11 & 0 & 3 \\ 7 & 5 & 8 & 5 & 9 & 0 \end{bmatrix}, \text{ and } \mathbf{C}^6 = \begin{bmatrix} 0 & 2 & 5 & 9 & 6 & 7 \\ 2 & 0 & 3 & 7 & 4 & 5 \\ 5 & 3 & 0 & 10 & 7 & 8 \\ 8 & 6 & 3 & 0 & 10 & 5 \\ 6 & 4 & 7 & 8* & 0 & 3 \\ 7 & 5 & 8 & 5 & 9 & 0 \end{bmatrix}$$

The matrix $\mathbf{C}^6$ now includes the lengths of the shortest paths between all pairs of nodes. We should note that it is also possible to construct a second set of matrices parallel to the computations made above, so as to keep track not only of the lengths of the shortest paths, but also the paths themselves. This is, however, beyond the scope of this volume. Interested readers are referred to books such as Eiselt and Sandblom (2000).

6.5 Spanning Tree Problems

Similar to the models in the previous sections in this chapter, the problem behind spanning trees is easily explained. Suppose there is an undirected graph that includes the potential edges that connect the nodes of the graph. The values associated with the edges indicate their costs. The problem is now to choose some of the edges and include them in the solution, so that the total costs are minimized, while the network remains connected. This may result in a graph, in which the path from one node to another leads through many intermediate nodes.

It is apparent that what we are looking for a tree, as any graph with fewer edges than a tree has will no longer be connected, while any graph that has more edges than a tree will include at least one cycle, which is unnecessary.

Problems of this nature occur whenever it is very expensive to establish edges. Typical examples are road networks, networks of power lines, or networks of sewer tunnels. The example shown in Fig. 6.21 shows all possible connections that may be established, coupled with their respective costs.

The task is now to find a subset of existing edges (the edges that belong to connections that will actually be built), so as to minimize the total amount of money necessary to connect all nodes with each other. The resulting connected subgraph of

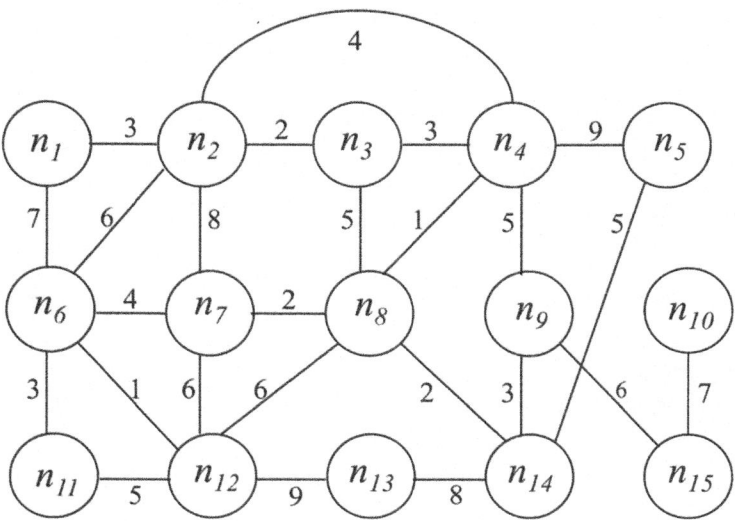

Fig. 6.21 Example for the minimal spanning tree problem

the original graph with n nodes is a *minimal spanning tree* of the given graph. Given that the result will be a tree, one can show that the optimal solution will include exactly $(n-1)$ edges.

A variety of solution methods exist for this type of problem. Here, we describe the *Kruskal technique*, which was first published in 1956. Actually, the method is nothing but a Greedy technique (see, e.g., Sect. 5.3.3 and Chap. 17). However, while the Greedy heuristic finds only approximate solutions in general, it is guaranteed to find an optimal solution for this problem. It can be described as follows. We first sort the edges in order of nondecreasing arc values, where ties are broken arbitrarily. Starting with the edge that has the lowest arc value, we introduce one arc at a time, provided it does not form a cycle with the already existing edges. The procedure continues until $(n-1)$ arcs are included in the solution.

In the example of Fig. 6.21, we order the edges, which results in Table 6.2. The table shows the edges and their costs in nondecreasing order.

Table 6.2 indicates which edges are introduced and which are not. In order to visualize the process, consider Fig. 6.22. Note that after edge $a_{13,14}$ is introduced, we have introduced 14 edges, and the process terminates.

6.6 Routing Problems

Routing problems are among the most frequently used network models in practice. We distinguish between two classes of routing models: *arc routing* and *node routing* models. The first example of arc routing was provided by Euler and his "Königsberg

Table 6.2 Edges of Fig. 6.21 in order of their value

Arc	a_{48}	$a_{6,12}$	a_{23}	a_{78}	$a_{8,14}$	a_{12}	a_{34}	$a_{6,11}$	$a_{9,14}$	a_{24}
Cost	1	1	2	2	2	3	3	3	3	4
Inserted?	Yes	Yes	Yes	Yes	Yes	Yes	Yes	Yes	Yes	No
Arc	a_{67}	a_{38}	a_{49}	$a_{5,14}$	$a_{11,12}$	a_{26}	$a_{7,12}$	$a_{8,12}$	$a_{9,15}$	a_{16}
Cost	4	5	5	5	5	6	6	6	6	7
Inserted?	Yes	No	No	Yes	No	No	No	No	Yes	No
Arc	$a_{10,15}$	a_{27}	$a_{13,14}$							
Cost	7	8	8							
Inserted?	Yes	No	Yes							

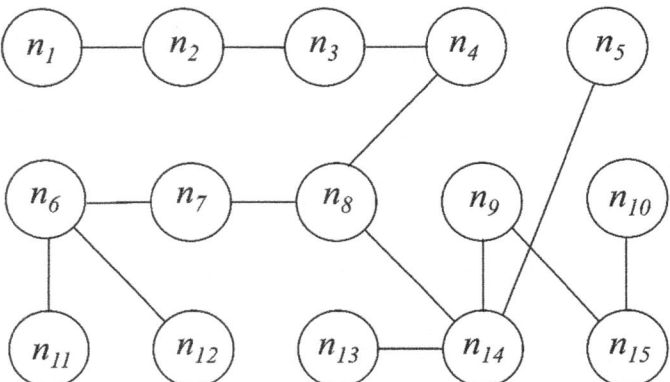

Fig. 6.22 Minimal spanning tree

Bridge Problem" described at the beginning of this chapter. The idea common to all arc routing problems is to find a path in a given graph, so that each arc is used on the path exactly or at least once. Similarly, in node routing problems, the idea is to find a path that starts at some node and returns to it, while using each node exactly or at least once in the process. Combinations of the two classes are vehicle routing problems, which belong to the most difficult routing problems.

The best-known arc routing problem is the *Chinese Postman Problem*. The name is due to Meigu Guan, who in the course of the "cultural" revolution in China was assigned to the position of a postal worker in the early 1960s. There, he considered the problem of a letter carrier, who would pick up the mail at some point (a node in the street network) and deliver it to the individual households by walking along each street at least once. The objective of the model is to minimize total distance walked, and the constraints ensure that mail is delivered to the houses on all streets of the network. While some versions of the model are easy to solve, others remain difficult. There are many important and popular applications of Chinese Postman Problems, including street cleaning and snow removal. Clearly, those problems are more

6.6 Routing Problems

difficult, as they will include hierarchies of streets, e.g. highways are typically plowed after a snowstorm before small neighborhood streets.

Similar to arc routing, the field of node routing has a long history. It starts with Hamilton's "trip around the world" developed in 1856, a game in which players have to find a tour that visits all desirable places (the nodes in a graph) exactly once. The best-known version of node routing is the famed *traveling salesman problem*, surely one of the most popular models in all of operations research. The story (not a real application, but a scenario that gave the model its name) is that a traveling salesman attempts to sell his goods in a number of cities in his region. He must visit each city exactly once and return to the place he started from. In order to have as much time as possible with the customers, the objective is to minimize the total travel time (or distance) of the tour. Applications of traveling salesman problems abound, many of them seemingly unrelated. One such example is the drilling of holes into sheet metal with the use of an automated drill press. Drilling the hole takes the same amount of time regardless of the sequence, in which the holes are drilled, so that the objective is to minimize the amount of time it takes to move the metal into the position, in which the next hole is to be drilled. This is nothing but a traveling salesman problem.

6.6.1 Arc Routing Problems

This section will deal with existence conditions for Euler tours, an extension of the Chinese Postman Problem, in directed and undirected graphs. We continue describing a technique that allows us to find such a tour, given that it exists. Finally, should an Euler tour not exist, we demonstrate how the Chinese Postman Problem can be solved, given that the underlying graph is directed or undirected (but not mixed).

In order to formalize our discussion, we first need to define some graph properties. For simplicity, we will consider only undirected and directed graphs in this section; we leave the discussion of mixed graphs to the advanced literature. We first need to define some properties of graph in addition to those already discussed at the beginning of this chapter. For undirected graphs, recall that we defined the *degree of a node* as the number of edges incident to it. For instance, node n_1 in Fig. 6.2 in Sect. 6.1 has a degree of 3, node n_2 has a degree of 2, and each of the nodes n_3, n_4, and n_5 has a degree of 1. In the case of directed graphs, we define the *indegree of a node* as the number of arcs leading into that node, while the *outdegree of a node* measures the number of arcs leading out of that node. We then call a graph *Eulerian* or an *Euler graph*, if it has a circuit that uses each arc or edge exactly once. A circuit with this property is called an Euler tour and sometimes referred to as "unicursal."

At this point, there are three issues that have to be resolved. Firstly, we need a tool to determine whether or not any given graph is Eulerian. Secondly, if this is the case, then we need an algorithm that actually determines an Euler tour. Thirdly, and finally, in case a graph is not Eulerian, we will have to find a circuit that uses all edges at least once, so that the total length of the tour is minimized. This is the

Fig. 6.23 Example for Fleury's algorithm

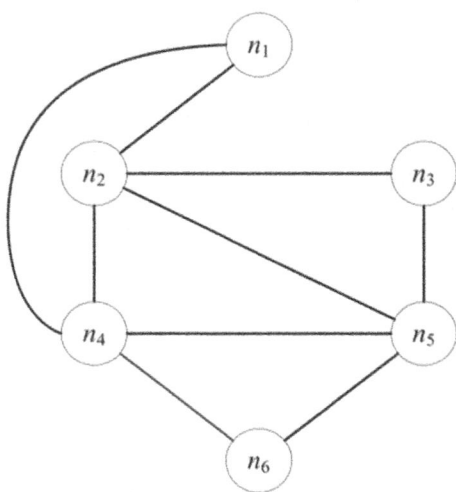

Chinese Postman Problem. The remainder of this section will discuss these three issues for undirected and directed graphs.

First, consider undirected graphs. There is a very simply existence condition that indicates whether or not any given graph is Eulerian. It states that a graph is Eulerian if and only if it is connected and all of its nodes have even degrees. We will justify this condition later after describing a method that allows us to find an Euler circuit in case one exists.

First, we need to define a *bridge* as an edge, whose removal from the graph will disconnect the network into two connected components, each containing at least one edge. Next, we remove this edge from the graph and repeat the step. In order to construct an Euler circuit, we start at any arbitrary node. This method was first described in the late 1880s and is sometimes called *Fleury's algorithm* in honor of its inventor.

In order to explain the procedure, consider the following:

Example Consider the graph in Fig. 6.23.

It is readily apparent that the graph in the figure is Eulerian as all degrees of its nodes are even (the nodes n_1, n_3, and n_6 have a degree of 2, while the nodes n_2, n_4, and n_5 have degrees of 4). Suppose that we start our tour at node n_1. From here, we can either move to n_2 or n_4, arbitrarily choose n_4. Once that move is made, we can remove the edge e_{14} from the graph, as we have used it and we cannot use it again on our tour. From n_4, we can now either move to n_2, n_5, or n_6, as none of these connections represents a bridge. Arbitrarily choose n_6, from where we can only move to n_5. Our choices at this point are e_{52}, e_{53}, and e_{54}, and since none of them represents a bridge, we arbitrarily choose e_{52} and move to n_2. Given that we have

6.6 Routing Problems

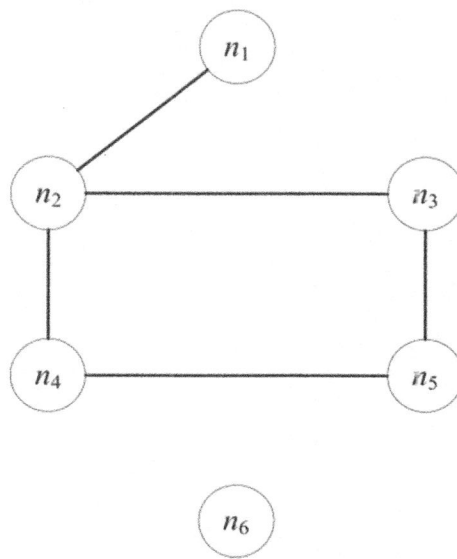

Fig. 6.24 Example with edges removed

removed all edges that were used on the tour so far, we obtain the graph shown in Fig. 6.24.

The node n_2 has three connections left, but e_{21} represents a bridge—meaning that if we were to cross it, we would be stuck at n_1, unable to return and move along the remaining edges. We thus take either of the other two edges, say, e_{23}. From n_3 we continue on to n_5, on to n_4, n_2, and we then finally return to n_1 and complete the circuit.

We can use this process to explain the existence condition for Eulerian circuits that we stated earlier. When we move out of the starting node and remove the first edge after having used it, the degree of the starting node decreases by one. From that point on, we enter all nodes on the tour once and leave it again (maybe repeating it multiple times as in the above example for nodes n_4, n_5, and n_2), so that its degree decreases by 2 each time we enter and leave a node. This means that passing through a node and removing the edges by which we enter and leave the node decreases its degree by an even number. Finally, when we complete the tour and enter the starting node for the last time, we remove the edge by which we enter it, thus reducing its degree by one. In summary, the degrees of all nodes have now been reduced to zero by an even number. This justifies the existence condition. (By the way, in case we do not require the tour to begin and end at the same node, the existence condition for a tour is easily modified—all we have to do is require that the graph is connected and that all but two of its nodes have even degrees. The nodes with odd degrees will then be the starting and the ending points of the tour.)

Consider now directed graphs. Again, we first state the condition that guarantees the existence of a Eulerian tour in the graph. For such a tour to exist, it is required

Fig. 6.25 Directed graph with an Euler circuit

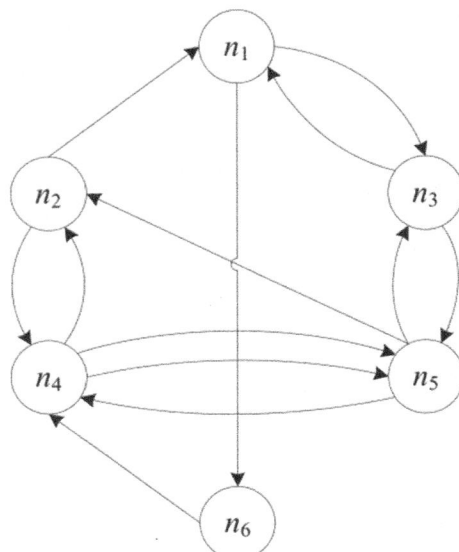

that the graph is strongly connected and that the indegree and outdegree are equal for each node in the network. As an example, consider the graph in Fig. 6.25.

At node n_6 the in- and outdegrees equal 1, at nodes n_1, n_2, and n_3, in- and outdegrees equal 2, and at nodes n_4 and n_5, in- and outdegrees equal 3. Furthermore, between each pair of nodes there is a path in either direction, so that the graph is strongly connected. This means that an Euler circuit exists in this graph. We can determine it again with Fleury's algorithm, minding the rule never to enter a subgraph that we cannot leave again. One such tour, starting at, say, n_4, will be $(n_4, n_5, n_2, n_4, n_5, n_4, n_2, n_1)$ (at this point we cannot move to n_6, as this would lead us to n_4 without having covered all edges), so that we continue to $(n_3, n_5, n_3, n_1, n_6, n_4)$, which completes the tour.

So far, we have discussed the first two issues related to arc routing—demonstrate the existence of Eulerian tours in undirected and directed graphs (or the lack thereof), and, if such tours exist, how to actually find them. Assume now that a Eulerian tour does not exist in the given graph. This will require that we use some of the edges or arcs more than once. Given that each edge or arc has now a nonnegative length associated with it, our task is now to add copies of the arcs or edges to the graph, so that the *augmented graph* is Eulerian and an Euler tour can be found. This task is to be accomplished so as to minimize the total length of the tour (which is identical to stating that we minimize the sum of the arc or edge values of all arcs and edges that have been added to the graph).

In order to explain the formulation of the problem, consider the following numerical example shown in Fig. 6.26, in which the numbers next to the edges indicate their lengths.

Fig. 6.26 Graph with no Eulerian tour

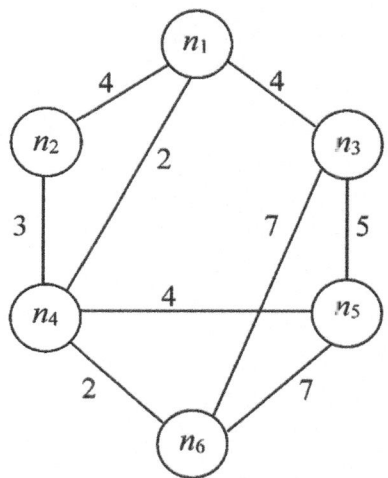

Clearly, the graph in Fig. 6.26 is not Eulerian, as the degrees of nodes n_1, n_3, n_5, and n_6 are odd. The degrees of these nodes must each be increased by one or any other odd number, so as to make the graph Eulerian. However, since we are trying to find a cost-minimal expansion of the given graph, we only need to increase the degree by one each. This is done by adding edges of the shortest paths that connect nodes with odd degree. In order to formulate the problem, we need to define zero-one variables x_{ij}, which assume a value of one, if a new connection between nodes n_i and n_j is introduced, and zero otherwise. In order to determine the distance between the pairs of nodes with odd degrees, we need to find the shortest paths between all of them. For this purpose, we can use any of the techniques described in Sect. 6.3.

The objective function is then to minimize the sum of distances of all connections introduced in addition to the existing edges (which must be traversed anyway). The constraints require that the number of edges that are added to the original graph to make up the augmented graph equals one for each node with an odd degree.

To solve the numerical problem in our example, we first determine the shortest paths between all pairs of nodes with odd degrees. Here, the distance matrix for the nodes n_1, n_3, n_5, and n_6 is

$$C = \begin{bmatrix} 0 & 4 & 6 & 4 \\ 4 & 0 & 5 & 7 \\ 6 & 5 & 0 & 6 \\ 4 & 7 & 6 & 0 \end{bmatrix}.$$

Defining variables x_{ij}, which assume a value of one, if a copy of the shortest path between n_i and n_j is to be added to the original non-Eulerian graph, and zero otherwise, we can then formulate our numerical example as

Fig. 6.27 Augmented graph for shortest path example

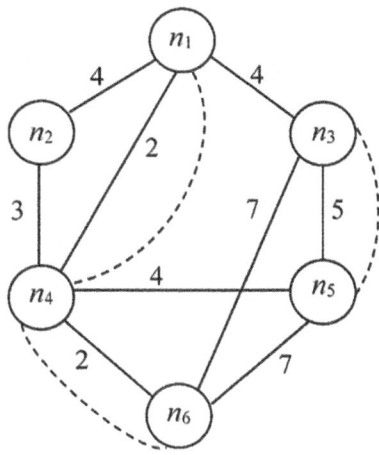

$$P : \text{Min } z = 4x_{13} + 6x_{15} + 4x_{16} + 5x_{35} + 7x_{36} + 6x_{56}$$
$$\text{s.t. } x_{13} + x_{15} + x_{16} = 1$$
$$x_{13} + x_{35} + x_{36} = 1$$
$$x_{15} + x_{35} + x_{56} = 1$$
$$x_{16} + x_{36} + x_{56} = 1$$
$$x_{13}, x_{15}, x_{16}, x_{35}, x_{36}, x_{56} = 0 \text{ or } 1.$$

Note that the formulation of this type of problem is nothing but an assignment problem (see Sect. 2.2.5) with

- the elements on the main diagonal being ineligible for assignments, and
- a symmetry condition $x_{ij} = x_{ji}$ for all i and j.

Solving the problem results in $\bar{x}_{16} = \bar{x}_{35} = 1$ and $\bar{x}_{ij} = 0$ otherwise. In other words, the connection between n_1 to n_6 (i.e., the edges e_{14} and e_{46}) as well as the connection between n_3 and n_5 (the edge e_{35}) has to be included in the solution. The resulting augmented graph is shown in Fig. 6.27.

Note that now all nodes in the augmented graph have even degree, so that the graph satisfies the Euler condition. Fleury's algorithm can now be used to find an Euler tour. Arbitrarily starting at n_1, one such tour is $(n_1, n_2, n_4, n_1, n_4, n_6, n_5, n_4, n_6, n_3, n_5, n_3, n_1)$, or, equivalently, $(e_{12}, e_{24}, e_{41}, e_{14}, e_{46}, e_{65}, e_{54}, e_{46}, e_{63}, e_{35}, e_{53}, e_{31})$.

Finally, consider the Chinese Postman Problem in directed graphs. In order to explain the formulation, consider the following numerical

Example Consider the directed graph shown in Fig. 6.28, whose numbers next to the arcs denote the time (or distance or cost) required moving along it.

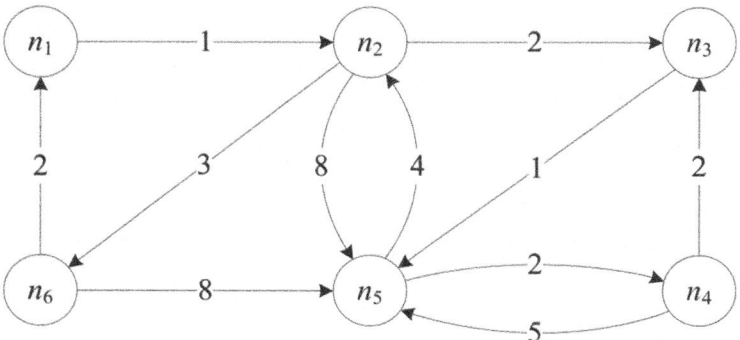

Fig. 6.28 Graph for the Chinese Postman Problem

In this graph, only at node n_1 in- and outdegrees are equal. Node n_2 has an indegree of 2 and an outdegree of 3, node n_3 has an indegree of 2 and an outdegree of 1, node n_4 has an indegree of 1 and an outdegree of 2, node n_5 has an indegree of 4 and an outdegree of 2, and node n_6 has an indegree of 1 and an outdegree of 2. It is evident that a Eulerian tour does not exist in this graph. In order to solve the Chinese Postman Problem, we first define sets of nodes N^- and N^+, so that N^- includes all nodes, whose outdegrees are higher than their indegrees, while N^+ includes all nodes whose indegrees are higher than their respective outdegrees. In this example, we have $N^+ = \{n_3, n_5\}$ and $N^- = \{n_2, n_4, n_6\}$. We now determine which connections have to be added in order to make the graph Eulerian. This requires that we define variables x_{ij} for all $n_i \in N^+$ and $n_j \in N^-$, which indicate how many copies of the connection from n_i to n_j must be added so as to make the graph Eulerian. For instance, if $x_{36} = 2$, this would indicate that we must introduce two additional copies of the connection from n_3 to n_6 in the network. However, this does not mean that a copy of the arc a_{36} should be introduced (which, in this case, would refer to an arc that does not even exist). Instead, we need to determine the shortest paths from all nodes in N^+ to all nodes in N^-, which can be accomplished by any of the pertinent algorithms discussed earlier in this chapter.

Associated with each variable x_{ij} is the cost that is incurred if we actually introduce the connection from node n_i to node n_j. This cost is nothing but the length of the shortest path from n_i to n_j. In this example, the shortest paths from the nodes in N^+, i.e. n_3 and n_5, to the nodes in N^-, viz., n_2, n_4, and n_6, are collected in the distance matrix

$$\mathbf{C} = \begin{bmatrix} 5 & 3 & 8 \\ 4 & 2 & 7 \end{bmatrix}.$$

The objective function will then minimize the total length of the additional arcs that are added to the original graph, so that the resulting *augmented graph* is Eulerian. The constraints ensure that the sum of additional arcs that lead out of each node in N^+ equals the difference between the indegree and the outdegree of that

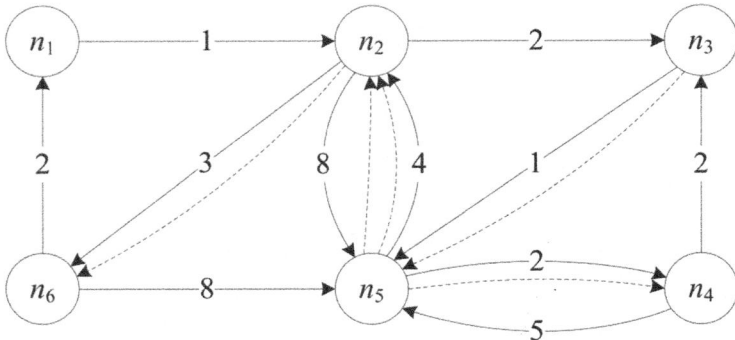

Fig. 6.29 Augmented graph for the Chinese Postman Problem

node. Similarly, for each node in N^- we require that the additional arcs leading into that node equal the difference of its outdegree and its indegree. The formulation of the problem in our example is then

$$P : \text{Min } z = 5x_{32} + 3x_{34} + 8x_{36} + 4x_{52} + 2x_{54} + 7x_{56}$$
$$\text{s.t. } x_{32} + x_{34} + x_{36} = 1$$
$$x_{52} + x_{54} + x_{56} = 2$$
$$x_{32} + x_{52} = 1$$
$$x_{34} + x_{54} = 1$$
$$x_{36} + x_{56} = 1$$
$$x_{32}, x_{34}, x_{36}, x_{52}, x_{54}, x_{56} \geq 0 \text{ and integer.}$$

It is apparent that the formulation is nothing but a transportation problem, see Sect. 2.2.5. Our problem has multiple optimal solutions. One such solution is $\bar{x}_{36} = \bar{x}_{52} = \bar{x}_{54} = 1$, and $\bar{x}_{ij} = 0$ otherwise. The value $\bar{x}_{36} = 1$ means that one copy of all arcs on the shortest path from n_3 to n_6 is included in the augmented graph, i.e. the arcs a_{35}, a_{52}, and a_{26}. Similarly, the value $\bar{x}_{52} = 1$ indicates that all edges on the shortest path from n_5 to n_2 are included (which is a_{52}, the direct connection), and $\bar{x}_{54} = 1$ means that all arcs on the shortest path from n_5 to n_4 are included (which is only the single arc a_{54}). The augmented graph with the additional arcs shown as broken lines is shown in Fig. 6.29.

It is also worth pointing out that an alternative optimal solution has $\bar{x}_{32} = \bar{x}_{54} = \bar{x}_{56} = 1$, and $\bar{x}_{ij} = 0$ otherwise. Although the solution to the transportation problem is completely different, it turns out that this solution adds the exact same arcs to the original graph, so that again the augmented graph of Fig. 6.29 results.

In either case, the augmented graph is now Eulerian and a tour can be found with Fleury's algorithm. Arbitrarily starting at n_1, one such tour is $(n_1, n_2, n_3, n_5, n_4, n_3, n_5, n_4, n_5, n_2, n_6, n_5, n_2, n_5, n_2, n_6, n_1)$, or, equivalently, $(a_{12}, a_{23}, a_{35}, a_{54}, a_{43}, a_{35},$

a_{54}, a_{45}, a_{52}, a_{26}, a_{65}, a_{52}, a_{25}, a_{52}, a_{26}, a_{61}). The total length of the tour is found by adding up the lengths of all arcs, it is 52.

6.6.2 Node Routing Problems

In this section, we first formulate a traveling salesman problem. We then show ways to deal with some of the difficult constraints. Finally, we describe a simple heuristic method that allows us finding hopefully good, but not necessarily optimal solutions quickly.

At first glance, formulating a traveling salesman problem appears easy. We need to formulate zero-one variables x_{ij} that assume a value of one, if the arc a_{ij} is part of the tour, and 0 otherwise. The objective function is then simply to minimize the sum of arc values, each multiplied with their respective binary variable. As far as constraints go, we have to ensure that the traveling salesman tour enters each node exactly once and that it leaves each node exactly once. In order to explain the formulation, consider as an example the graph in Fig. 6.30.

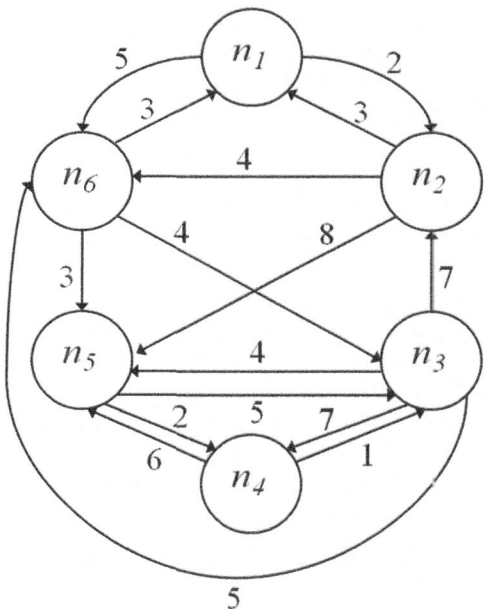

Fig. 6.30 Graph for the traveling salesman problem

The formulation as described above is then as follows:

$$\text{Min } z = 2x_{12} + 5x_{16} + 3x_{21} + 8x_{25} + 4x_{26} + 7x_{32} + 7x_{34} + 5x_{36} + 4x_{35}$$
$$+ 1x_{43} + 6x_{45} + 5x_{53} + 2x_{54} + 3x_{61} + 4x_{63} + 3x_{65}$$

$$\text{s.t. } x_{12} + x_{16} = 1 \qquad x_{21} + x_{61} = 1$$

$$x_{21} + x_{25} + x_{26} = 1 \qquad x_{12} + x_{32} = 1$$

$$x_{32} + x_{34} + x_{35} + x_{36} = 1 \qquad x_{43} + x_{53} + x_{63} = 1$$

$$x_{43} + x_{45} = 1 \qquad x_{34} + x_{54} = 1$$

$$x_{53} + x_{54} = 1 \qquad x_{25} + x_{35} + x_{45} + x_{65} = 1$$

$$x_{61} + x_{63} + x_{65} = 1 \qquad x_{16} + x_{26} + x_{36} = 1$$

$$x_{ij} = 0 \text{ or } 1 \text{ for all } i,j.$$

The first six constraints in the above formulation force the outflow of each node to be equal to one (meaning that the traveling salesman tour will leave each node exactly once), while the second set of six constraints requires that the inflow into each node equals one (meaning that the tour enters each node exactly once). If the original graph would have direct connections between all pairs of nodes, the above formulation would actually be identical to that of an assignment problem. We know that for this type of problem, the zero-one constraints are satisfied without us requiring them, so that the problem can be solved as a standard linear programming problem. The result we find includes the following nonzero variables: $\bar{x}_{12} = \bar{x}_{21} = \bar{x}_{36} = \bar{x}_{43} = \bar{x}_{54} = \bar{x}_{65} = 1$, and all other variables equal zero. The value of the objective function for this solution equals $\bar{z} = 16$.

It is easily apparent that this solution does satisfy the constraints but is not what we were looking for: rather than one tour, the solution includes two subtours (n_1—n_2—n_1) and (n_3—n_6—n_5—n_4—n_3). What we will have to add are the so-called *subtour elimination constraints*. There are different ways of doing this, but the most efficient sets of such constraints require a huge number of constraints. To be precise, for a graph of n nodes, there are 2^n such subtour elimination constraints. As a result, the modeler will refrain from including all of these constraints from the beginning, but rather solve the problem without them. Then, if the solution has no subtours, we are done. Otherwise, a single relevant subtour elimination constraint is introduced, the problem is solved again, and the process is repeated until a single tour emerges. This is the process we follow here.

The idea is now this. First, we select some subtour, define N_s as the set of nodes in the chosen subtour, and let $\overline{N}_s$ denote the complement of this set. Note that the present solution has no arc leading out of N_s to any node in $\overline{N}_s$. Therefore, we define a constraint that requires at least one arc in the solution to lead out of a node in N_s to a node in $\overline{N}_s$. In our example, choose the subtour ($n_1 - n_2 - n_1$), so that $N_s = \{n_1, n_2\}$

6.6 Routing Problems

and $\overline{N}_s = \{n_3, n_4, n_5, n_6\}$. The set $\{a_{16}, a_{25}, a_{26}\}$ includes all arcs that lead from N_s to $\overline{N}_s$, so that we can formulate the constraint

$$x_{16} + x_{25} + x_{26} \geq 1.$$

We now solve the problem again with this additional constraint. Note that due to the additional constraint the "assignment structure" of the problem is lost, so that it is necessary to explicitly include the zero-one requirements for all variables, which makes the problem considerably more difficult. The optimal solution of the problem is $\bar{x}_{12} = \bar{x}_{26} = \bar{x}_{35} = \bar{x}_{43} = \bar{x}_{54} = \bar{x}_{61} = 1$ with an objective value of $\bar{z} = 16$ (the same as before, so apparently, there were alternative optimal solutions to the problem in the first step). This means that our tour is (n_1—n_2—n_6—n_1) and (n_3—n_5—n_4—n_3), meaning that we have successfully eliminated the previous subtour, but now have a solution with another subtour, so that another subtour elimination constraint must be added.

Given the subtour (n_1—n_2—n_6—n_1), our sets are $N_s = \{n_1, n_2, n_6\}$ and $\overline{N}_s = \{n_3, n_4, n_5\}$, so that the set of arcs from N_s to $\overline{N}_s$ is $\{a_{25}, a_{63}, a_{65}\}$. The additional constraint can then be written as

$$x_{25} + x_{63} + x_{65} \geq 1.$$

Solving the problem again results in the solution $\bar{x}_{16} = \bar{x}_{21} = \bar{x}_{32} = \bar{x}_{43} = \bar{x}_{54} = \bar{x}_{65} = 1$ with a value of the objective function $\bar{z} = 21$. This solution includes the tour (n_1—n_6—n_5—n_4—n_3—n_2—n_1), which no longer includes a subtour. This is the optimal solution.

While this procedure may be feasible for small and medium-sized problems, it is not suitable for large-scale applications. Here, we may resort to heuristic algorithms. In the simplest case, we may use the Greedy algorithm (also referred to as *the nearest neighbor method*) to find a tour. In this application, we would start the Greedy algorithm with some node, find the nearest neighbor (provided it does not result in a subtour), move on to the next neighbor, again avoiding subtours, and so forth. Note that the number of degrees of freedom is constantly decreasing while we make choices. (Not that there is anything new in that: whenever you make a choice such as spending money on some item, you will have less choices, i.e., money, for future decisions.)

In order to explain the procedure, consider the distance matrix

$$\mathbf{D} = \begin{bmatrix} 0 & 4 & 7 & 3 & 2 & 6 \\ 3 & 0 & 4 & 9 & 6 & 8 \\ 5 & 4 & 0 & 8 & 3 & 4 \\ 6 & 8 & 3 & 0 & 2 & 5 \\ 1 & 7 & 2 & 3 & 0 & 4 \\ 8 & 7 & 4 & 9 & 4 & 0 \end{bmatrix}.$$

Arbitrarily starting with the node n_1, the nearest neighbor, i.e., the smallest element in the first row of **D** (other than the element $d_{11} = 0$, which would just lead from n_1 back to itself), is the connection to n_5 at a distance of 2. From n_5, i.e., in row 5 of the matrix, the nearest neighbor is n_1 at a distance of 1. However, this connection would create a subtour, so we look for the next-shortest distance. It is the link from n_5 to n_3 with a distance of 2. From n_3, the nearest neighbor is n_5 with a distance of 3, but going back to n_5 would create a subtour. The next-shortest distances are those to n_2 and n_6, both with a distance of 4. Any tie-breaking rule can be used, here, we choose n_2. From n_2, the nearest neighbor is n_1, which cannot be chosen, as it would create a subtour. The next-nearest neighbor is n_3, which cannot be chosen for the same reason. The next-nearest neighbor is n_5, which is also not eligible. The next-nearest neighbor is n_6 at a distance of 8, which must be chosen. Note how the lack of degrees of freedom forces us to choose undesirable links at the later stages of the algorithm. Actually, at this point there is no degree of freedom left, and we have to continue on to n_4, the last remaining node, and from there return to n_1. The two distances are 9 and 6, respectively, so that the entire tour is of length $l = 31$. In summary, the tour is n_1—n_5—n_3—n_2—n_6—n_4—n_1. Clearly, this procedure will always result in a traveling salesman tour. However, we can show that the tour found in this manner can be arbitrarily bad.

In a "Phase 2" procedure, we can now attempt to improve the solution found earlier. One way to do this is a "pairwise exchange" or "swap" method. It very simply exchanges two (usually, but not necessarily, adjacent) nodes on the tour. In the above tour, we may try to avoid the long distance from n_6 to n_4 by switching the order, in which these two nodes are visited on the tour. Doing so results in the tour n_1—n_5—n_3—n_2—n_4—n_6—n_1, which has length $l = 30$, which is better than the present solution, so that this new tour becomes the starting point for further improvements.

Starting with the new tour, we may now attempt to avoid the direct connection from n_2 to n_4, which is also a very long leg of the tour. Switching the order of these two nodes results in the tour n_1—n_5—n_3—n_4—n_2—n_6—n_1, which is of length $l = 36$. This is higher than the previous (best known) solution, so that the switch is not made. This swap process can be continued, until no swap change is able to further decrease the length of the tour. Again, this being a heuristic method, an optimal solution may not be found.

It is sometimes useful to use what is known as a *multistart procedure*. This means that rather than starting with some random node (n_1 in the above example), finding a solution and then trying to improve it, we may try out some or all different nodes as potential starting points. With each such node, we obtain a tour. We would then choose the best tour, and try to find improvements from there.

In the above example, we could use the Greedy algorithm to find tours starting with each of the six nodes. In addition to the tour that starts with n_1, which has already been determined above, we obtain the tours n_2—n_1—n_5—n_3—n_6—n_4—n_2 of length $l = 28$, n_3—n_5—n_1—n_4—n_6—n_2—n_3 of length $l = 23$, n_4—n_5—n_1—n_2—n_3—n_6—n_4 of length $l = 24$, n_5—n_1—n_4—n_3—n_2—n_6—n_5 of length $l = 23$, and n_6—n_3—n_5—n_1—n_4—n_2—n_6 of length $l = 27$. The tours that start with n_3 and n_5

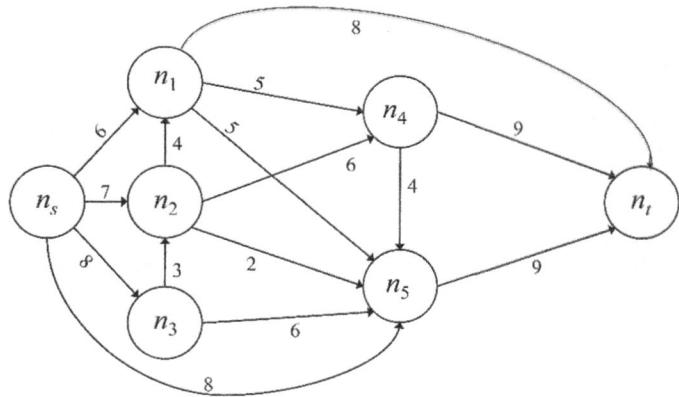

Fig. 6.31 Network for Problem 1

are best, and the swap process would start with them. Any of these heuristic procedures is computationally cheap and, once some tour has been obtained, the process can be terminated at any point in time.

Exercises

Problem 1 (maximal flow algorithm, minimal cut): Consider the network in Fig. 6.31, in which the numbers next to the directed arcs denote the capacity of the arcs.

Use the method by Ford and Fulkerson to determine a maximal flow. Show the flow pattern. What is the value of the maximal flow? Which arcs are in the minimal cut you have found?

Solution: Starting with a flow of zero at all arcs, we first increase the flow on the following paths:

path (n_s, n_1, n_t): 6 units,
path (n_s, n_2, n_4, n_t): 6 units,
path (n_s, n_3, n_5, n_t): 6 units, and
path (n_s, n_5, n_t): 3 units.

The flow pattern is then shown in the network in Fig. 6.32, where the two numbers next to the arcs indicate the arc's capacity and its present flow, respectively.

The next step consists of an increase of the flow on the following two paths:

path $(n_s, n_3, n_2, n_1, n_t)$: 2 units, and
path $(n_s, n_2, n_1, n_4, n_t)$: 1 unit.

The resulting flow pattern is shown in Fig. 6.33.

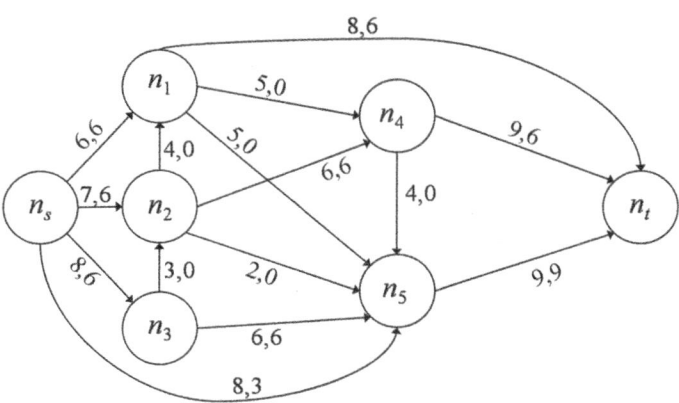

Fig. 6.32 First flow pattern for Problem 1

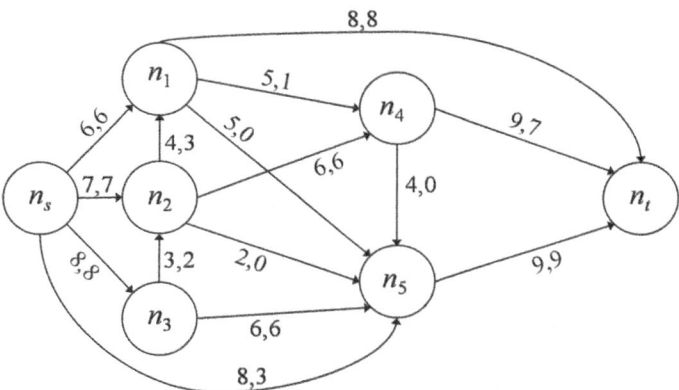

Fig. 6.33 Second flow pattern for Problem 1

At this point, there are no more degrees of freedom, and labeling can only be done on the path (n_s, n_5, n_3 (backward labeling at this point), n_2, n_1, n_4, n_t). The flow on this path can be changed by one unit, resulting in the flow pattern shown in the network in Fig. 6.34.

Starting with the pattern in Fig. 6.34, we can label the nodes n_s, n_3, and n_5. At this point, a nonbreakthrough occurs, and we can conclude that the flow pattern in Fig. 6.34 is maximal. The corresponding flow value is 25. The minimal cut includes the capacities of all arcs that lead from labeled to unlabeled nodes. Here the minimal cut includes the arcs (n_s, n_1), (n_s, n_2), (n_3, n_2), and (n_5, n_t). Note that the Ford and Fulkerson method only finds the cut that is closest to the source. In this example, another minimal cut exists with arcs (n_s, n_1), (n_2, n_1), (n_2, n_4), and (n_5, n_t).

6.6 Routing Problems

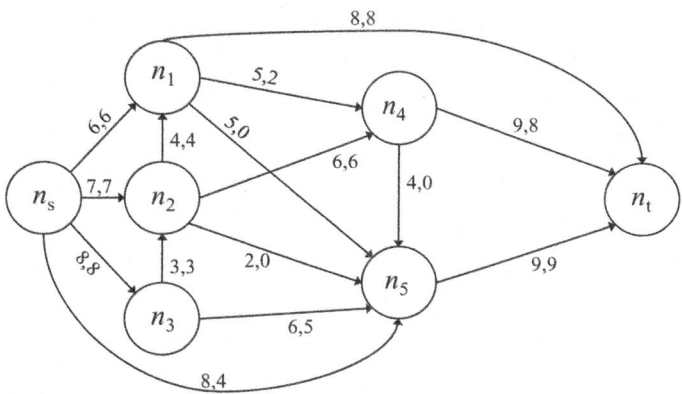

Fig. 6.34 Third and optimal flow pattern for Problem 1

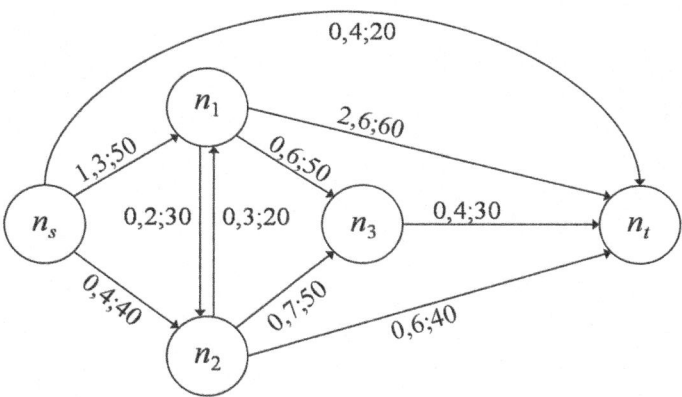

Fig. 6.35 Network for Problem 2

Problem 2 (formulation of a feasible flow problem with node constraints):
Consider the network shown in Fig. 6.35, where the numbers next to the arcs consist of the lower and upper bound on the flow in the arc, while the double-digit numbers next to the arcs indicate the per-unit cost of the flow through the arc.

In addition, we want to ensure that a total of 5 units flow from the source to the sink and exactly two units flow through the node n_3. Formulate a cost-minimizing feasible flow model for this problem.

Solution: Defining variables x_{ij} as the flow from node i to node j on the arc connecting the two nodes, we can formulate the problem as follows:

P : Min $z = 20x_{st} + 50x_{s1} + 40x_{s2} + 20x_{12} + 50x_{13} + 60x_{1t} + 20x_{21} + 50x_{23}$
$+ 40x_{2t} + 30x_{3t}$

s.t. $x_{st} + x_{s1} + x_{s2} = 5$ (forcing 5 units through the network)

$x_{s1} + x_{21} - x_{12} - x_{13} - x_{1t} = 0$ (conservation equation for node n_1)

$x_{s2} + x_{12} - x_{21} - x_{23} - x_{2t} = 0$ (conservation equation for node n_2)

$x_{13} + x_{23} - x_{3t} = 0$ (conservation equation for node n_3)

$x_{13} + x_{23} = 2$ $\Big($or, equivalently, $x_{3t} = 2$: forces a flow of 2 through node $n_3\Big)$

$$\left.\begin{array}{l} x_{st} \leq 4 \\ x_{s1} \leq 3 \\ x_{s2} \leq 4 \\ x_{12} \leq 2 \\ x_{13} \leq 6 \\ x_{1t} \leq 6 \\ x_{21} \leq 3 \\ x_{23} \leq 7 \\ x_{2t} \leq 6 \\ x_{3t} \leq 4 \end{array}\right\} \text{(upper limits on arc flows)}$$

$$\left.\begin{array}{l} x_{s1} \geq 1 \\ x_{1t} \geq 2 \end{array}\right\} \text{(lower limits on arc flows)}$$

$x_{ij} \geq 0$ for all i, j.

The optimal flow pattern is shown in Fig. 6.36, where the numbers next to the arcs are the arc flows. The total cost for the flow pattern is $480.

Changing the requirement regarding the throughput of node n_3 from "exactly 2 units" to "at most 2 units" results in a flow of 3 units on the arc from n_s to n_t, and 2 units on the path from n_s to n_1 and n_t, i.e. zero flow through node n_3. The total costs of this solution are significantly less, at $280.

Problem 3 (shortest path, Dijkstra method): Consider the network in Fig. 6.37 and determine the shortest paths from n_s to all other nodes.

Solution: Table 6.3 shows the labeling process.

The tree with the shortest distances is shown in Fig. 6.38. Note that in this example, the tree is a simple path.

Problem 4 (shortest path, Floyd–Warshall method): Consider the network shown in Fig. 6.39.

6.6 Routing Problems

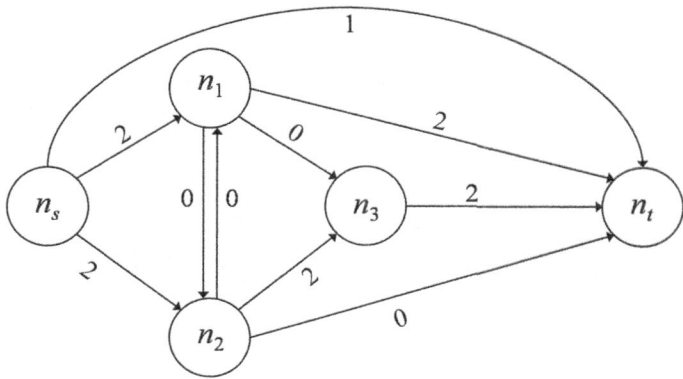

Fig. 6.36 Optimal flow pattern for Problem 2

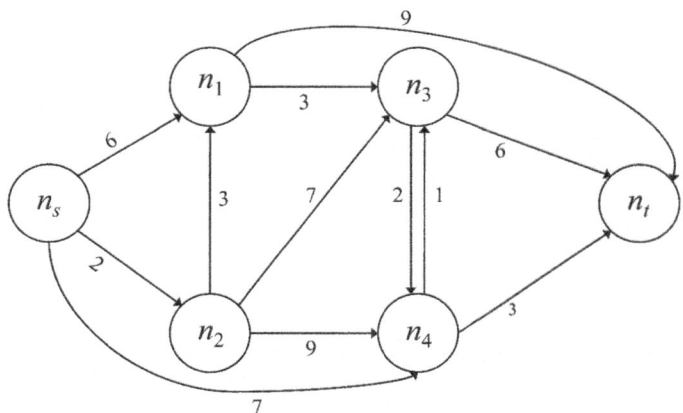

Fig. 6.37 Network for Problem 3

Table 6.3 Labels of the nodes in the shortest path example

Step #	$L(n_s)$	$L(n_1)$	$L(n_2)$	$L(n_3)$	$L(n_4)$	$L(n_t)$
0	$(n_s, 0)^*$	(n_1, ∞)	(n_2, ∞)	(n_3, ∞)	(n_4, ∞)	(n_t, ∞)
1		$(n_s, 6)$	$(n_s, 2)^*$	(n_3, ∞)	$(n_s, 10)$	(n_t, ∞)
2		$(n_2, 5)^*$		$(n_2, 9)$	$(n_s, 10)$	(n_t, ∞)
3				$(n_1, 8)^*$	$(n_s, 10)$	$(n_1, 14)$
4					$(n_3, 10)^*$	$(n_1, 14)$, or $(n_3, 14)$
5						$(n_4, 13)^*$

Fig. 6.38 Optimal solution for Problem 3

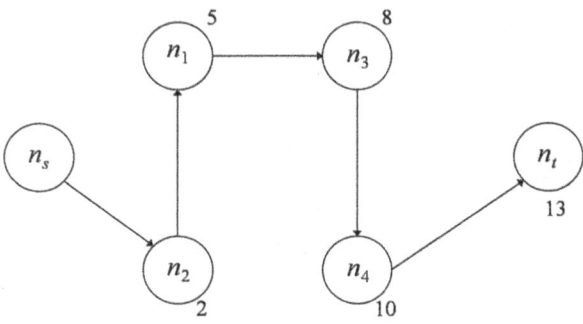

Fig. 6.39 Network for Problem 4

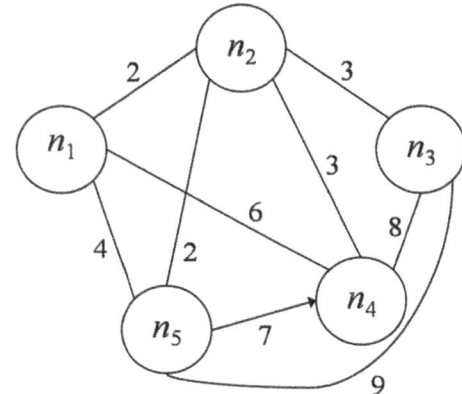

Use the Floyd–Warshall algorithm to determine the shortest paths between all pairs of nodes.

Solution: The direct distance matrix is

$$\begin{bmatrix} 0 & 2 & \infty & 6 & 4 \\ 2 & 0 & 3 & 3 & 2 \\ \infty & 3 & 0 & 8 & 9 \\ 6 & 3 & 8 & 0 & \infty \\ 4 & 2 & 9 & 7 & 0 \end{bmatrix}.$$

The first two steps result in the following approximations (where the elements with a "*" indicate changes in the step that led to the new solution):

$$\begin{bmatrix} 0 & 2 & \infty & 6 & 4 \\ 2 & 0 & 3 & 3 & 2 \\ \infty & 3 & 0 & 8 & 9 \\ 6 & 3 & 8 & 0 & \infty \\ 4 & 2 & 9 & 7 & 0 \end{bmatrix} \begin{bmatrix} 0 & 2 & \infty & 6 & 4 \\ 2 & 0 & 3 & 3 & 2 \\ \infty & 3 & 0 & 8 & 9 \\ 6 & 3 & 8 & 0 & 10* \\ 4 & 2 & 9 & 7 & 0 \end{bmatrix} \begin{bmatrix} 0 & 2 & 5* & 5* & 4 \\ 2 & 0 & 3 & 3 & 2 \\ 5* & 3 & 0 & 6* & 5* \\ 5* & 3 & 6* & 0 & 5* \\ 4 & 2 & 5* & 5* & 0 \end{bmatrix}$$

After this, there are no further changes, so that the last matrix is indeed the shortest path matrix.

Problem 5 (minimal spanning tree): A manufacturing firm wants to connect its nine plants by rail. The cost of establishing the line segments is shown in the symmetric table below, where a number in row i and column j indicates the fixed costs to connect plant i with plant j.

$$\begin{bmatrix} 0 & 23 & 64 & 15 & 39 & 66 & 18 & 84 & 93 \\ 23 & 0 & 35 & 27 & 56 & 63 & 29 & 91 & 37 \\ 64 & 35 & 0 & 96 & 62 & 45 & 33 & 72 & 48 \\ 15 & 27 & 96 & 0 & 38 & 22 & 74 & 61 & 88 \\ 39 & 56 & 62 & 38 & 0 & 42 & 95 & 54 & 67 \\ 66 & 63 & 45 & 22 & 42 & 0 & 25 & 43 & 47 \\ 18 & 29 & 33 & 74 & 95 & 25 & 0 & 51 & 32 \\ 84 & 91 & 72 & 61 & 54 & 43 & 51 & 0 & 26 \\ 93 & 37 & 48 & 88 & 67 & 47 & 32 & 26 & 0 \end{bmatrix}$$

Solution: Putting the costs in nondecreasing order (ties are broken arbitrarily) results in the sequence 15, 18, 22, 23, 25, 26, 27, 29, 32, 33, 35, 37, 38, 39, 42, 43, 45, 47, 48, 51, 54, 56, 61, 62, 63, 64, 66, 67, 72, 74, 84, 88, 91, 93, 95, and 96. Starting at the beginning, we introduce connections (1, 4), (1, 7), (4, 6), and (1, 2) with costs of 15, 18, 22, and 23. The next cheapest connection is (6, 7) with costs of 25. However, plants 6 and 7 are already connected. The next connection is (8, 9) with costs of 26, which we introduce. The next connection on the list is (2, 4) with costs of 27. Plants 2 and 4 are already connected, so that we reject this connection and continue. The next connection is (2, 7) with costs of 29, which is also rejected. The next connection is (7, 9) with costs of 32, which is introduced. This is followed by connection (3, 7) with costs of 33, which is introduced. This is followed by the connections (2, 3) with costs of 35 and (2, 9) with costs of 37; both of which are rejected. The connection (4, 5) with costs 38 connects two previously unconnected nodes, and thus it is introduced. At this point, eight arcs have been introduced and all plants are connected. Thus, an optimal solution has been determined. Its costs are 207.

Problem 6 (Chinese Postman Problem in undirected graphs): A local post office plans the route of its letter carrier. The region has eight street junctions and

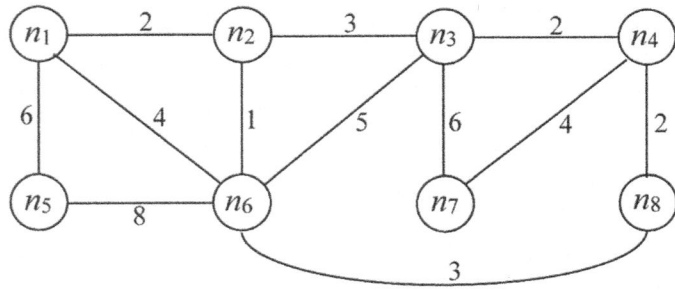

Fig. 6.40 Network for Problem 6

Table 6.4 Shortest paths and path lengths for Problem 6

	n_1	n_2	n_4	n_6
n_1	–	2 (e_{12})	7 (e_{12}, e_{23}, e_{34})	3 (e_{12}, e_{26})
n_2	2 (e_{21})	–	5 (e_{23}, e_{34})	1 (e_{26})
n_4	7 (e_{43}, e_{32}, e_{21})	5 (e_{43}, e_{32})	–	5 (e_{48}, e_{86})
n_6	3 (e_{62}, e_{21})	1 (e_{62})	5 (e_{68}, e_{84})	–

twelve street segments, which the carrier has to serve. Figure 6.40 shows the streets and the junctions as well as the distances (in hundreds of feet).

(a) Is it possible for the post office to find a tour for the postman, so that the letter carrier has to traverse each street exactly once?
(b) In case a tour as sought under (a) does not exist, determine a tour that minimizes the total distance that the letter carrier has to walk. What is the length of such a tour?
(c) The agreement with the letter carriers' union stipulates that the post office has to pay 20¢ per ft. and year to the letter carrier. It has now become possible to hire an outside firm for the service. However, there are two individuals who agree to do the job, but only if they can split the tour among themselves in a specific way, as each has a preference for certain neighborhoods. In particular, Joe, the first potential carrier, would like to cover the streets $e_{12}, e_{15}, e_{23}, e_{26}, e_{34}, e_{37}$, and e_{48}, while Jim, our second carrier, prefers to deliver the mail along the remaining streets. How much can the post office offer them, if they do not want to pay more than they do for a single carrier under the present system?

Solution:

(a) The question is whether or not the graph in Fig. 6.40 is Eulerian. Since the degrees of the nodes are 3, 3, 4, 3, 2, 5, 2, and 2, respectively, not all degrees are even numbers, so that such a tour does not exist.
(b) The nodes with odd degrees are n_1, n_2, n_4, and n_6. The distances between these nodes and the actual paths are shown in Table 6.4.

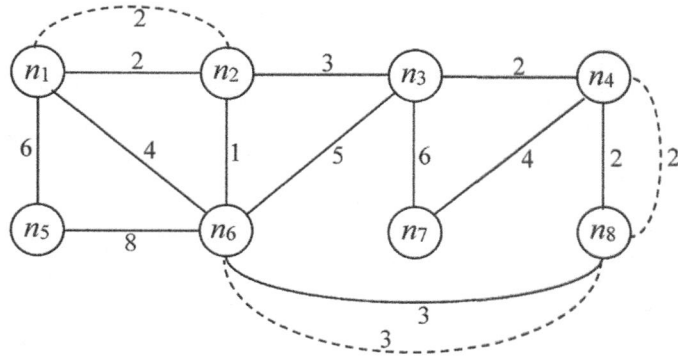

Fig. 6.41 Network with added edges for Problem 6

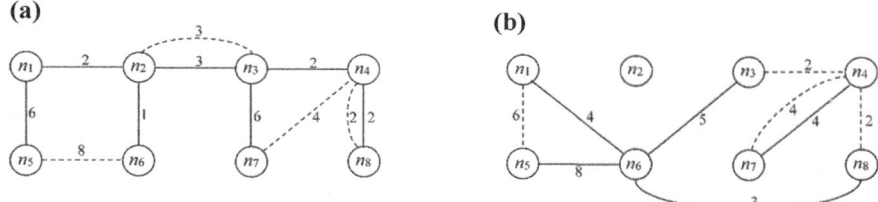

Fig. 6.42 Solution for Problem 6 (c). (a) Joe's preference, (b) Jim's preference

Solving the symmetric assignment problem, we find the unique optimal solution $\bar{x}_{12} = \bar{x}_{21} = \bar{x}_{46} = \bar{x}_{64} = 1$ and $\bar{x}_{ij} = 0$ otherwise. The value assignment $\bar{x}_{12} = 1$ means that we add one copy of the shortest path from n_1 to n_2 to the graph in Fig. 6.40 (meaning one copy of the edge e_{12}), and the assignment $\bar{x}_{46} = 1$ means that we add one copy of the shortest path from n_4 to n_6 to the graph, i.e. copies of the edges e_{48} and e_{86}. The resulting graph (with the added edges shown as broken lines) is depicted in Fig. 6.41.

One of the many possible tours in the graph is n_1—n_2—n_3—n_4—n_8—n_6—n_2—n_1—n_5—n_6—n_3—n_7—n_4—n_8—n_6—n_1 with a total length of 53, i.e. 5300 ft.

(c) At present, the length of the tour is 5300 ft. as determined under (b), meaning that the costs to the post office are $1060 per year. Given Joe and Jim's offer, we first determine their respective tours. The streets they want to cover are shown in Figs. 6.42a, b, respectively, where the solid lines indicate the streets they desire to cover, while the broken lines show the streets they have to walk more than once.

The total lengths of their tours are 3900 ft. and 3800 ft. for a total of 7700 ft. Denoting by p the money the post office can pay them per foot covered, the break-

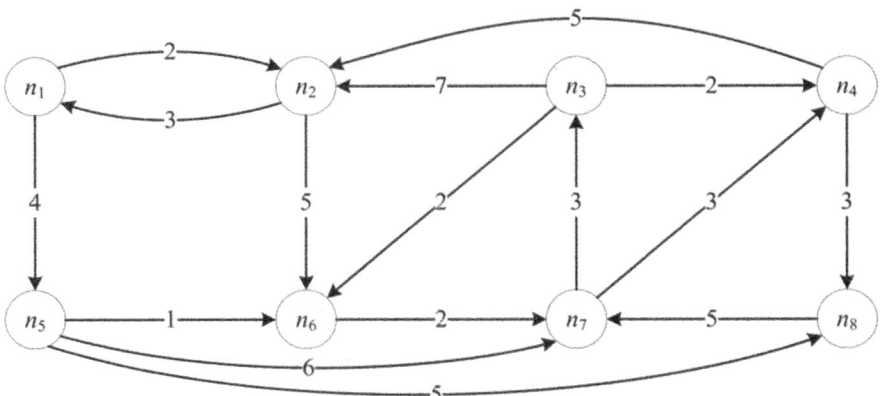

Fig. 6.43 Street network for Problem 7

even point is obtained for $1060 = 7700p$ or $p = 13.7662\cent$. This will result in payments of $536.88 to Joe and 523.12 to Jim.

Problem 7 (Chinese Postman Problem in directed graphs): City Council plans the snow plowing strategies for next winter. The graph in Fig. 6.43 shows the street network, which includes only one-way streets. The numbers next to the arcs denote the amount of time (in minutes) that is required to plow the streets.

(a) Without doing any computations, what is the shortest possible time in which the streets may be plowed? Is it actually possible to plow the streets in this amount of time? Explain.

(b) If the answer in (a) is negative, find the shortest tour in which all streets can be plowed.

(c) Suppose that each minute of a snowplow costs $5. A private contractor has now offered to plow streets a_{57} and a_{58}. What is the maximal amount of money that the municipality can pay the contractor, so that they do not pay more in total than if they do the job themselves?

Solution:

(a) The shortest possible plowing time is the sum of all of the arc values of the graph in Fig. 6.43. In our example, this shortest possible time is 58 min. However, this time can only be realized if there were a tour, in which the plow traverses each street exactly once, i.e., if an Euler tour exists. Given that the indegree does not equal the outdegree for each node, such a tour does not exist, so that 58 min can only be used as a lower bound on the shortest amount of time, in which all streets can be plowed.

(b) Given that the original graph does not include an Euler tour, we have to determine the augmented graph. In our example, the indegrees of the nodes are 1, 3, 1, 2, 1, 3, 3, and 2, while the outdegrees are 2, 2, 3, 2, 3, 1, 2, and 1, so that the sets are $N^+ = \{n_2, n_6, n_7, n_8\}$ $N^- = \{n_1, n_3, n_5\}$. The shortest paths and their lengths for all pairs from a node in N^+ to a node in N^- are shown in Table 6.5.

6.6 Routing Problems

Table 6.5 Shortest paths and path lengths for Problem 7

	n_1	n_3	n_5
n_2	3 (a_{21})	10 (a_{26}, a_{67}, a_{73})	7 (a_{21}, a_{15})
n_6	13 ($a_{67}, a_{74}, a_{42}, a_{21}$)	5 (a_{67}, a_{73})	17 ($a_{67}, a_{74}, a_{42}, a_{21}, a_{15}$)
n_7	11 (a_{74}, a_{42}, a_{21})	3 (a_{73})	15 ($a_{74}, a_{42}, a_{21}, a_{15}$)
n_8	16 ($a_{87}, a_{74}, a_{42}, a_{21}$)	8 (a_{87}, a_{73})	20 ($a_{87}, a_{74}, a_{42}, a_{21}, a_{15}$)

The pertinent transportation problem can then be written as follows:

$$P : \text{Min } z = 3x_{21} + 10x_{23} + 7x_{25} + 13x_{61} + 5x_{63} + 17x_{65} + 11x_{71} + 3x_{73}$$
$$+ 15x_{75} + 16x_{81} + 8x_{83} + 20x_{85}$$

s.t. $x_{21} + x_{23} + x_{25} = 1$

$x_{61} + x_{63} + x_{65} = 2$

$x_{71} + x_{73} + x_{75} = 1$

$x_{81} + x_{83} + x_{85} = 1$

$x_{21} + x_{61} + x_{71} + x_{81} = 1$

$x_{23} + x_{63} + x_{73} + x_{83} = 2$

$x_{25} + x_{65} + x_{75} + x_{85} = 2$

$x_{ij} \geq 0$ and integer for all i, j.

Solving the problem, we find many alternative optimal solutions. One of them has $\bar{x}_{21} = 1$, $\bar{x}_{63} = 2$, $\bar{x}_{75} = 1$, and $\bar{x}_{85} = 1$, while all other $\bar{x}_{ij} = 0$ for all other variables. This means that one copy of the arcs on the shortest path from n_2 to n_1 is added (the single arc a_{21}), two copies of all arcs on the shortest path from n_6 to n_3 are added (the arcs a_{67} and a_{73}), one copy of all arcs on the shortest path from n_7 to n_5 are added (the arcs a_{74}, a_{42}, a_{21}, and a_{15}), and one copy of all arcs on the shortest path from n_8 to n_5 are added (the arcs $a_{87}, a_{74}, a_{42}, a_{21}$, and a_{15}). The resulting graph is shown in Fig. 6.44.

This graph is now Eulerian, and we can easily determine an Euler tour, such as n_1, $n_2, n_1, n_5, n_6, n_7, n_4, n_8, n_7, n_3, n_6, n_7, n_4, n_2, n_1, n_5, n_7, n_3, n_4, n_2, n_1, n_5, n_8, n_7, n_4$, n_2, n_6, n_7, n_3, n_2, and n_1, and the length of the tour is 106.

(c) Given the length of the tour of 106, the costs for the snow plow are $106(5) = \$530$. Tentatively deleting the arcs a_{57} and a_{58} results in the network shown in Fig. 6.45.

	n_1	n_3
n_2	3	10
n_6	13	5

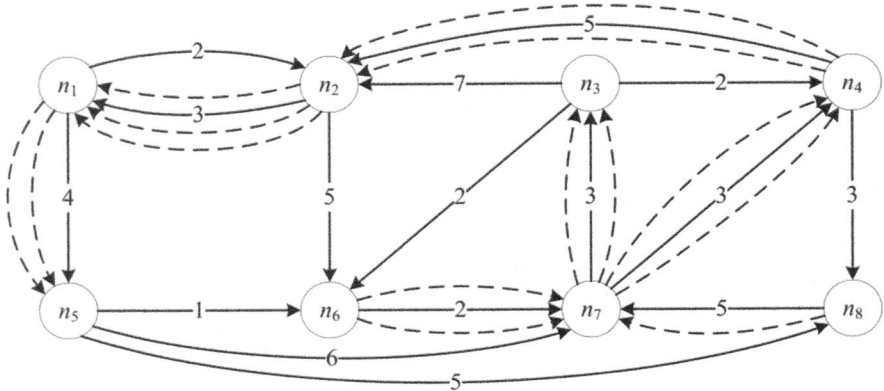

Fig. 6.44 Augmented graph for Problem 7

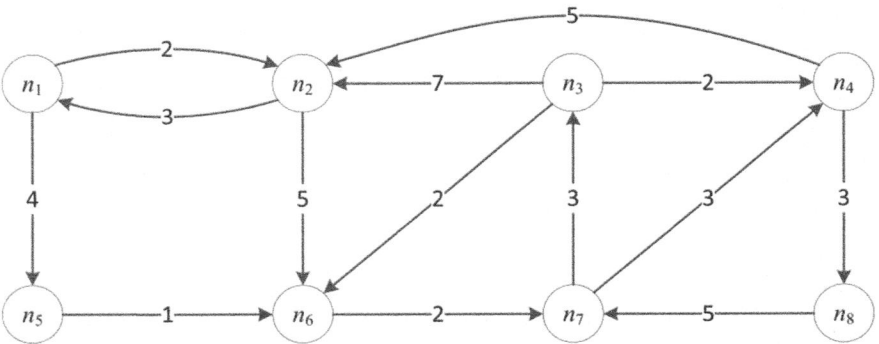

Fig. 6.45 Network for Problem 7 (c)

The problem can again be formulated as a transportation problem with cost matrix and the shortest paths happen to be the same as under (b). The "supplies" (i.e., the indegrees minus the outdegrees of the two source nodes) are 1 and 2, while the "demands" (i.e., the outdegrees minus the indegrees of the two sink nodes) are 1 and 2. The optimal solution includes $\bar{x}_{21} = 1$ and $\bar{x}_{63} = 2$. The augmented Eulerian graph is shown in Fig. 6.46.

The length of the tour is 60, so that the costs are \$300. This means savings of $530 - 300 = \$230$, which is the maximal amount that the municipality can spend for the plowing of the two additional arcs. It is apparent that the two arcs the contractors offered to plow are causing significant costs if the municipality does the plowing. This raises the issue of the apportionment of costs.

Problem 8 (heuristics applied to a traveling salesman problem): A pharmaceutical company uses glass containers to store their chemicals. Over time, they

6.6 Routing Problems

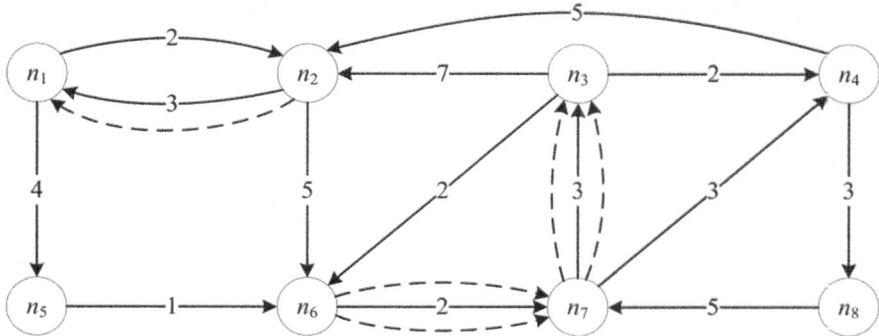

Fig. 6.46 Augmented network for Problem 7 (c)

reuse the containers, but if they do, they are required to clean them. The cleaning costs depend on what was stored in the container before, and what will be stored next. The cleaning costs of the six chemicals A, B, C, D, E, and F are shown in the following matrix:

$$\begin{bmatrix} 0 & 6 & 5 & 3 & 5 & 4 \\ 2 & 0 & 7 & 9 & 3 & 8 \\ 5 & 5 & 1 & 6 & 4 & 2 \\ 4 & 2 & 3 & 2 & 1 & 3 \\ 9 & 6 & 4 & 3 & 0 & 7 \\ 2 & 3 & 8 & 1 & 6 & 0 \end{bmatrix}.$$

For example, if chemical C is stored in a glass that was used for F before, then the cleaning costs are 8 (from F to C, i.e., element (F, C), in the sixth row and third column).

(a) Assume that presently, chemical D is stored in a container. Use the Greedy algorithm to determine a sequence that has each of the six chemicals stored in a container exactly once. (Ties are broken arbitrarily.) Clearly specify the sequence that results from the application of the Greedy algorithm. What are the costs associated with the sequence?
(b) Use the pairwise exchange method to improve the solution determined under (a). Examine all adjacent pairs until an exchange results in an improvement. Make this improvement and stop (even though additional improvements may be possible).

Solution:

(a) Starting with D, we obtain the sequence D—E—C—F—A—B—D. The length of the tour (here: the cost) is $l = 1 + 4 + 2 + 2 + 6 + 9 = 24$.

(b) In this application, the starting point at D is fixed. The following swap steps can be made:

Swap E and C. Result: D—C—E—F—A—B—D, length $l = 31$, reject.
Swap C and F. Result: D—E—F—C—A—B—D, length $l = 36$, reject.
Swap F and A. Result: $D - E - C - A - F - B - D$, length $l = 26$, reject.
Swap A and B. Result: $D - E - C - F - B - A - D$, length $l = 15$, accept.
This solution becomes the new benchmark from which the swap steps continue.
If there is no fixed starting point, it would make sense to start swapping with the most expensive direct connection, in our case $B - D$.

References

Eiselt HA, Sandblom C-L (2000) Integer programming and network models. Springer, Berlin
Murty K (2006). http://www-personal.umich.edu/~murty/books/network_programming/

Location Models 7

This chapter introduces the basic ideas of location models. We first provide a short introduction to the subject and enumerate some of its major components. This is followed by a detailed discussion of the major classes of location models. For an in-depth treatment of the models described in this chapter, we refer readers to Eiselt and Sandblom (2004), Eiselt and Marianov (2011), Daskin (2013) and Laporte et al. (2019).

7.1 The Major Elements of Location Problems

The origins of location theory are shrouded in history. The first to discuss location models (and here, we use the term in the widest possible sense) were mathematicians. One of the famous location-based puzzles (regarding the point in the triangle from which the sum of distances to the triangles' vertices is minimal) were investigated and solved by Torricelli and Fermat in the seventeenth century. The geographer von Thünen wrote about his famed "von Thünen circles" regarding the location of economic activities around a central place, and the German geographer Weber wrote a treatise concerning location models early in the twentieth century. Hakimi (1964) introduced location models to the field of operations research. We will encounter his famous theorem below. Since then, thousands of contributions have been made by researchers from fields as diverse as mathematics, computer science, geography, business administration, and economics.

While many of us have a pretty good idea what location problems are, let us take a step back and examine its major components. The three main components of every location model are space, customers, and facilities. Supplies exist at the facilities, demand occurs at the customer sites, and the goods are "somehow" transported from the facilities to the customers. Let us now look at these components in some detail.

We distinguish between two major classes of location models, those that occur in the plane (or sometimes in three-dimensional space), and those that occur in transportation networks. While not altogether correct, location models in the plane tend to look at the problem from a macro point of view, while those models in transportation networks investigate the scenario from a micro perspective. Location problems in the plane are called *continuous location models* (as the facilities that are to be located can be sited anywhere in the space under consideration) in contrast to *discrete location models*, many of which occur in networks. In discrete location models, facilities can be located only at a finite number of points.

These two classes of problems are also different from the way the variables are defined: determining the location of the variables in continuous models in the two-dimensional plane requires the definition of variables $(x_1, y_1), (x_2, y_2), \ldots$ that symbolize the coordinates of the facilities 1, 2, ... On the other hand, in network models, we need a variable y_j for each potential location that will assume a value of one, if we actually do locate at site j, and 0 if we do not. This places continuous location models into the field of linear or nonlinear optimization, while network location models are typically formulated and solved as integer programming problems.

As far as parameters go, we assume that the locations of our customers are known, and so are their demands, which are commonly referred to as weights. Formally, customer i (or, equivalently, customers at demand point i) are assumed to have a weight of w_i. In the 2-dimensional (Euclidean) plane, customer i is assumed to be located at a point with coordinates (a_i, b_i), while in a network, customer i can be found at node n_i. Note that the points at which customers are located typically represent either census units, towns, or other customer agglomerations.

Assume now that transportation takes place in the plane. While there exist many different metrics and gauges, most authors use either rectilinear (or Manhattan) distances, Euclidean or straight-line distances, or, sometimes, squared Euclidean distances. Suppose that a customer is located at a point with coordinates (a, b), while the facility is located at (x, y). The *rectilinear* or *Manhattan distance* between the customer and the facility is then defined as $|a - x| + |b - y|$, i.e., it is assumed that all movements take place parallel to the axes of the system of coordinates. The *Euclidean distance* between the customer and the facility is defined as $\sqrt{(a-x)^2 + (b-y)^2}$, and the *squared Euclidean distance* between these two points is $(a - x)^2 + (b - y)^2$. As an example, the Manhattan, Euclidean, and squared Euclidean distances between the points (2, 3) and (7, 1) are 7, $\sqrt{29} \approx 5.3852$, and 29, respectively.

In network models, it is common practice to assume that movements take place on the shortest path between the customer and the facility. This means that the shortest path algorithms (see Sect. 6.3) will typically have to be applied when building a location model. In general, we will use the term "distance," by which we may mean the actual mileage between points or any other disutility such as time or costs.

Other components of location problems include the following:

- The number of facilities (which may be fixed by the decision maker or may be endogenous to the model).

7.1 The Major Elements of Location Problems

- The magnitude of the demand (which may be fixed, e.g., for essential goods, or may depend on the proximity of the facility to the customer).
- The way customers are assigned to facilities (they either choose themselves, as is the case in the context of retail facilities, which are customer choice models, or firms allocate customers to their facilities in allocation models, as is the case in deliveries from warehouses).
- The type of deterministic or probabilistic parameters we are dealing with.
- Single-level or hierarchical models (hierarchical models often exist in the context of health care, with levels such as doctor, local clinic, and regional hospital), where a higher-level facility can accomplish all tasks a lower-level facility can. A clearly defined referral system is crucial in multi-level models.
- Competitive or noncompetitive models (most competitive scenarios use game theory as a tool and have firms compete in location, prices, and quantities).

One main component of any location model is the objective function pursued by the decision maker. Until the mid-1970s, location models assumed that the facilities to be located by the decision maker would be "attractive" to the customers in some sense. That way, proximity to the customers was the main part of the objective. Many facilities, however, do not fall into that mold. Consider, for instance, a grocery store, a facility that customers normally would consider desirable to have in their proximity. However, as noisy deliveries are made very early in the morning, the same facility may very well be considered (at least partially) undesirable. Clearly, power plants and landfills will be considered undesirable by most residents in the vicinity. The NIMBY syndrome ("not in my back yard") attests to that. While a customer would attempt to "pull" a desirable facility towards himself, he will attempt to "push" away an undesirable facility. In addition to push and pull models, there are also "balancing" models, in which the decision maker attempts to locate facilities so as to balance the interactions between individual facilities and the customers. This is often the case in the location of automobile dealerships, motels of a chain, and fast-food franchises, which are located so as to avoid cannibalizing demand from its own (other) facilities.

Among the pull objectives, three classes have received most of the attention. The first of these classes of models deals with *covering objectives*. The main idea is to locate facilities, so that as many customers as possible are covered, i.e., within a given distance of any of the facilities. Such objectives are often used for the location of emergency equipment, i.e., ambulances, police stations, and fire stations. The other two classes are *center and median locations*. In both of them, the decision maker locates facilities so as to minimize a measure of distance. In center problems, the planner will attempt to locate facilities, so that the largest distance to customers is as small as possible. This objective was justified (in some sense) by Rawls's "Theory of Justice," according to which the quality of a solution is determined by the lowest quality of any of its components. While such objectives may be justified in the context of reliability systems, it takes an extreme view and is strongly biased towards outliers. Median problems, on the other hand, are probably the largest class in location models. Their objective is the minimization of the weighted total distance.

In other words, the distance between a customer and his closest facility is determined, which is then multiplied by the magnitude of the customer's demand. Considering the distance as a proxy for the cost of shipping one truckload from the facility to the customer (or vice versa), the customer's demand then denotes the number of truckloads to be shipped. The sum of all of these weighted distances is then a proxy expression for the costs of the transportation of the goods to all customers. Similarly, from a customer's point of view, a median solution minimizes the average customer—facility distance, thus benefitting customers as well.

Applications of location problems range from the location of schools to warehouses and church camps, equipment for the removal of oil spills, warning sirens in towns in case of floods, hurricanes, or tsunamis, bottling plants, landfills, newspaper transfer points, sewage treatment plants, and many other facilities. Other applications include nonphysical spaces. An example are brand positioning models, in which each dimension of the space represents a feature that is deemed relevant to customers. The features of each product will then determine its location in this feature space. Similarly, each (potential) customer can be represented in the same space by his ideal point, i.e., the point that has the features most preferred by the customer. Assuming that each customer purchases the brand closest to his ideal point, it is then possible to determine the estimated sales of each product. Then it is also possible to relocate, i.e., redesign, a brand so as to maximize the amount of demand it captures. The model has also been used to study the positions of political candidates, who attempt to maximize their share of the vote.

7.2 Covering Problems

The idea behind covering models is to locate facilities that provide some service required by customers. If customers are positioned within a certain predefined critical distance $\bar{d}$ from any of the facilities, then they are considered served or "covered." Two objectives for the location of facilities are to either cover all customers in the network with the smallest number of facilities or, alternatively, to cover as many customers as possible with a given number of facilities. Typical examples of applications of covering models are found when emergency facilities are to be located. Whether the facilities in question are fire stations, ambulances, police cruisers, or any similar "facilities," the objective is to maximize protection. However, measuring protection is very difficult. In order to find an expression for "protection," we would need to know the value of responding to an emergency from different distances or times. We can safely assume that an increase in the time to respond to an area for fire protection will mean that the fire has a greater chance to spread and it may reduce the chance to save property or a life if one is in jeopardy. A similar argument applies to other types of protection. Unfortunately, measurements of the value of protection are nearly impossible to make. In the case of fire protection, standards of service have been suggested by the Insurance Services Office that, when met, virtually guarantee an adequate level of protection or, at least, the lowest premiums for fire protection. For fire services, the most commonly

7.2.1 The Location Set Covering Problem

One of the first models that was developed to site emergency service facilities that incorporates a maximum service distance standard is the *Location Set Covering Problem* (*LSCP*) introduced in the early 1970s. The objective of the *LSCP* is to locate the smallest number of facilities such that each demand node is "covered" by one or more facilities. A demand node is said to be covered, if there is a facility within a prespecified distance from the demand node. In the context of the location of a fire hall, the decision maker could specify that a building is covered (or sufficiently protected), if it is within 5 miles (or 7 min or some similar measure) of the fire hall.

If we assume that the cost to purchase land and build a facility is roughly the same for all nodes (or is small in comparison to maintaining each of the needed fire crews), then the objective of using the least number of facilities is equivalent to minimizing the cost of providing service to all demand.

Throughout this chapter, we will use the subscript i to denote customers, while the subscript j is employed to denote facilities. For simplicity, we also assume that all customers and facilities will be located at the nodes of the network. We can then define d_{ij} as the shortest distance (or time, cost, or any other disutility) on a path between the demand node n_i and the facility site at node n_j. In addition, let the service standard $\overline{d}$ denote the maximal service distance or time as specified by the decision maker. We can then define a coverage matrix $\mathbf{C}(\overline{d}) = (c_{ij})$, so that $c_{ij} = 1$, if customers at node n_i can be covered from a facility located at node n_j, and 0 otherwise. As an example, consider the network in Fig. 7.1.

The symmetric matrix of shortest distances is $\mathbf{D}$.

$\mathbf{D}$:

	n_1	n_2	n_3	n_4	n_5	n_6	n_7	n_8	n_9
n_1	0	5	11	14	6	12	14	20	18
n_2	5	0	6	9	7	11	15	19	13
n_3	11	6	0	9	11	5	16	13	7
n_4	14	9	9	0	8	4	16	12	16
n_5	6	7	11	8	0	6	8	14	17
n_6	12	11	5	4	6	0	14	8	12
n_7	14	15	16	16	8	14	0	7	9
n_8	20	19	13	12	14	8	7	0	16
n_9	18	13	7	16	17	12	9	16	0

Given a covering distance of $\overline{d} = 10$, the covering matrix $\mathbf{C}(10)$ is then.

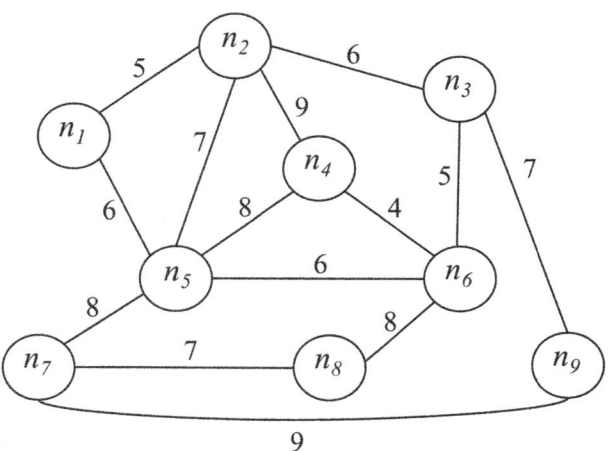

Fig. 7.1 Network example

		n_1	n_2	n_3	n_4	n_5	n_6	n_7	n_8	n_9
	n_1	1	1	0	0	1	0	0	0	0
	n_2	1	1	1	1	1	0	0	0	0
	n_3	0	1	1	1	0	1	0	0	1
	n_4	0	1	1	1	1	1	0	0	0
C(10):	n_5	1	1	0	1	1	1	1	0	0
	n_6	0	0	1	1	1	1	0	1	0
	n_7	0	0	0	0	1	0	1	1	1
	n_8	0	0	0	0	0	1	1	1	0
	n_9	0	0	1	0	0	0	1	0	1

The problem can then be formulated as

$LSCP$: Min $z = y_1 + y_2 + y_3 + y_4 + y_5 + y_6 + y_7 + y_8 + y_9$

s.t. $y_1 + y_2 + y_5 \geq 1$

$y_1 + y_2 + y_3 + y_4 + y_5 \geq 1$

$y_2 + y_3 + y_4 + y_6 + y_9 \geq 1$

$y_2 + y_3 + y_4 + y_5 + y_6 \geq 1$

$y_1 + y_2 + y_4 + y_5 + y_6 + y_7 \geq 1$

$y_3 + y_4 + y_5 + y_6 + y_8 \geq 1$

$y_5 + y_7 + y_8 + y_9 \geq 1$

$y_6 + y_7 + y_8 \geq 1$

$y_3 + y_7 + y_9 \geq 1$

$y_1, y_2, \ldots, y_9 = 0$ or 1.

7.2 Covering Problems

The objective function minimizes the number of facilities that are located, and the constraints ensure that at least one facility is located within reach of every customer.

We first solve the problem as a linear programming problem, as often the solution is integer without us requiring it, a property sometimes referred to as "integer friendly." In this example, the solution of the linear programming problem is $\bar{y}_1 = 0$, $\bar{y}_2 = \frac{1}{2}$, $\bar{y}_3 = \frac{1}{4}$, $\bar{y}_4 = 0$, $\bar{y}_5 = \frac{1}{2}$, $\bar{y}_6 = \frac{1}{4}$, $\bar{y}_7 = \frac{3}{4}$, and $\bar{y}_8 = \bar{y}_9 = 0$. The value of the objective function for this solution is $z = 2\frac{1}{4}$. Including zero-one conditions for all variables results in the optimal solution $\bar{y}_5 = \bar{y}_6 = \bar{y}_7 = 1$ and all other variables zero, i.e., facilities should be located at the nodes n_5, n_6, and n_7. However, this solution is not unique: other optimal solutions with three facilities locate them at nodes n_2, n_3, and n_7, or at n_3, n_5 and n_8, or at n_2, n_3 and n_8, at n_2, n_5, n_7, or at n_2, n_6, and n_7.

In order to further explore the problem, assume now that the distance standard is changed to $\bar{d} = 9$. The resulting covering matrix $\mathbf{C}(9)$ is actually identical to that found for $\bar{d} = 10$, which is not surprising, as there is no distance of magnitude 10 in the matrix of shortest distances. Suppose now that $\bar{d} = 8$. The coverage matrix $\mathbf{C}(8)$ is then

		n_1	n_2	n_3	n_4	n_5	n_6	n_7	n_8	n_9
	n_1	1	1	0	0	1	0	0	0	0
	n_2	1	1	1	0	1	0	0	0	0
	n_3	0	1	1	0	0	1	0	0	1
	n_4	0	0	0	1	1	1	0	0	0
$\mathbf{C}(8)$:	n_5	1	1	0	1	1	1	1	0	0
	n_6	0	0	1	1	1	1	0	1	0
	n_7	0	0	0	0	1	0	1	1	0
	n_8	0	0	0	0	0	1	1	1	0
	n_9	0	0	1	0	0	0	0	0	1

Solving the problem, we obtain a solution that locates facilities at n_3, n_5, and at one of n_6, n_7, or n_8. In other words, even with a distance standard of $\bar{d} = 8$, three facilities are sufficient to cover all customers.

If the distance standard is reduced further to $\bar{d} = 7$, four facilities are necessary to cover all customers. These facilities will have to be located at n_2, n_3, n_6, and n_7, or at n_3, n_5, n_6, n_7, or at n_2, n_3, n_6, n_8, or at n_3, n_5, n_6, n_8. A further reduction of the distance to $\bar{d} = 6$ results in five facilities being required to cover all customers. The facilities will be located at n_6, n_7, n_8, n_9, and either n_1 or n_2. In case the distance standard is set to $\bar{d} = 5$, we need six facilities to cover all customers. The facilities will be located at n_1, n_5, n_6, n_7, n_8, and n_9.

For a distance standard of $\bar{d} = 4$, we need no less than eight facilities. In particular, there will be a facility at each node, except for either node n_4 or n_6. Once the distance standard is $D < 4$, there must be a facility at each node.

On the other hand, increasing the distance standard from the original $\bar{d} = 10$ to $\bar{d} = 11$, we can determine that only two facilities are needed to cover all customers. These facilities will be located at n_5 and n_7.

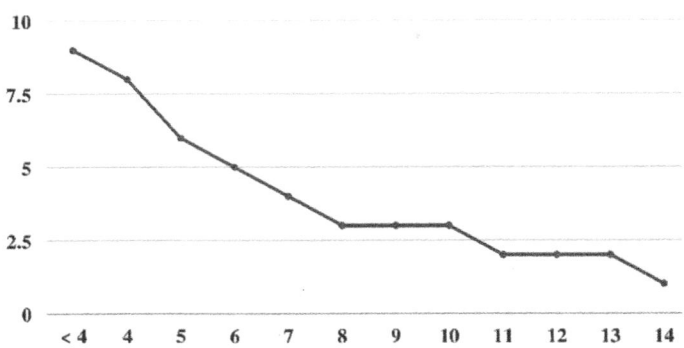

Fig. 7.2 Number of facilities versus distance standard

Finally, for any distance standard $\bar{d} \geq 14$, only a single facility is needed to cover all customers. This facility will have to be located at the node n_6. Note that this solution is unique.

The results are summarized in Fig. 7.2, which has the distance standard on the ordinate and the number of facilities required to cover all nodes at the abscissa.

One observation we can make in these computations is that there typically are multiple optimal solutions for location set covering problems. Furthermore, we would like to mention that there exist rules that may allow users to reduce the size of the problem, see, e.g., Eiselt and Sandblom (2004). However, given today's computational power, such rules are of limited usefulness.

7.2.2 The Maximal Covering Location Problem

In contrast to the location set covering problem discussed above, the *Maximal Covering Location Problem* (*MCLP*) does not attempt the task to cover *all* customers. Given a fixed number of p facilities, the task is to locate these facilities so as to cover the largest possible number of customers (or the largest total demand). In addition to the parameters defined in the previous section, we also need w_i, which denotes the number of customers (or the magnitude of the demand) at node n_i.

As an illustration, consider again the example of the previous section with the assumption that exactly two facilities are to be located. Furthermore, let the covering distance be $\bar{d} = 10$. The weights (i.e., the number of customers at a node) are given as follows:

n_1	n_2	n_3	n_4	n_5	n_6	n_7	n_8	n_9
120	160	100	110	130	140	190	220	200

In order to formulate the problem, we not only need the binary location variables y_j, but we also need coverage variables x_i, $i = 1, \ldots, n$. A coverage variable x_i assumes a value of one, if a customer at node n_i is covered by at least one facility, and

7.2 Covering Problems

zero otherwise. The main reason for these additional variables is to avoid double counting. The formulation of our problem is then

$$MCLP : \text{Max } z = 120x_1 + 160x_2 + 100x_3 + 110x_4 + 130x_5 + 140x_6$$
$$+ 190x_7 + 220x_8 + 200x_9$$
$$\text{s.t. } y_1 + y_2 + y_3 + y_4 + y_5 + y_6 + y_7 + y_8 + y_9 = 2$$
$$x_1 \leq y_1 + y_2 + y_5$$
$$x_2 \leq y_1 + y_2 + y_3 + y_4 + y_5$$
$$x_3 \leq y_2 + y_3 + y_4 + y_6 + y_9$$
$$x_4 \leq y_2 + y_3 + y_4 + y_5 + y_6$$
$$x_5 \leq y_1 + y_2 + y_4 + y_5 + y_6 + y_7$$
$$x_6 \leq y_3 + y_4 + y_5 + y_6 + y_8$$
$$x_7 \leq y_5 + y_7 + y_8 + y_9$$
$$x_8 \leq y_6 + y_7 + y_8$$
$$x_9 \leq y_3 + y_7 + y_9$$
$$y_j = 0 \text{ or } 1 \text{ for all } j \text{ and}$$
$$x_i = 0 \text{ or } 1 \text{ for all } i.$$

Each term in the objective function will count all customers at a node being covered, if and only if the node is covered by a facility. The main reason for the use of the covering variables x_i is to ensure that customers who are covered by more than one facility will not be counted more than once. The first constraint ensures that exactly two facilities will be located. The remaining constraints each define the coverage of a node. In particular, a node is considered covered, if there is at least one facility within covering distance. In this example, consider the node n_1. This node could potentially be covered from a facility at node n_1, n_2, or n_5. Thus, its covering constraint is $x_1 \leq y_1 + y_2 + y_5$, which ensures that if there is no facility at either n_1, n_2, or n_5, then the right-hand side value of the constraint equals zero, which forces the variable x_1 to assume a value of zero as well. On the other hand, if there exists at least one variable at any one of the three nodes, the right-hand side value of the inequality is at least one, which renders it redundant, as x_1 is defined as a zero-one variable anyway. However, while x_1 could assume a value of either zero or one in such a case, the objective function includes the term $120x_1$ which is part of what is to be maximized. This pushes the value of x_1 to as large a value as possible, so that it will assume a value of "1" whenever possible.

The optimal solution of the problem is $\bar{y}_5 = \bar{y}_7 = 1$, meaning that we will locate facilities at the nodes n_5 and n_7. All coverage variables except x_3 equal one, and the value of the objective function equals $\bar{z} = 1270$. In case of very large networks, it may be difficult to solve the problem exactly, so, rather than attempting to find an optimal solution, we may resort to a heuristic algorithm. The heuristic we use below is of the Greedy type in that it locates one facility at a time and in each step it does so by locating the next facility, so as to maximize the number of additional customers that are covered in that step. Being a heuristic, the myopic Greedy procedure may therefore not necessarily find an optimal solution.

The number of customers that are covered by a single facility located at one of the nodes can then be determined by multiplying the coverage matrix by the vector of weights from the left, i.e., by computing $\mathbf{wC}\left(\overline{d}\right)$. The jth component of the resulting vector indicates the number of customers that will be covered if a facility were to be located at the node n_j. In our example, we obtain

$$[120, 160, 100, 110, 130, 140, 190, 220, 200] \begin{bmatrix} 1 & 1 & 0 & 0 & 1 & 0 & 0 & 0 & 0 \\ 1 & 1 & 1 & 1 & 1 & 0 & 0 & 0 & 0 \\ 0 & 1 & 1 & 1 & 0 & 1 & 0 & 0 & 1 \\ 0 & 1 & 1 & 1 & 1 & 1 & 0 & 0 & 0 \\ 1 & 1 & 0 & 1 & 1 & 1 & 1 & 0 & 0 \\ 0 & 0 & 1 & 1 & 1 & 1 & 0 & 1 & 0 \\ 0 & 0 & 0 & 0 & 1 & 0 & 1 & 1 & 1 \\ 0 & 0 & 0 & 0 & 0 & 1 & 1 & 1 & 0 \\ 0 & 0 & 1 & 0 & 0 & 0 & 1 & 0 & 1 \end{bmatrix}$$

$= [410, 620, 710, 640, 850, 700, 740, 550, 490]$.

The facility that serves most customers is one that is located at n_5, from where it covers 850 customers, so we locate a facility there. Note that this first facility is located optimally. In order to facilitate the computations, we can most easily avoid double counting by deleting the rows that belong to nodes that are already covered by a facility at n_5. In our example, these are the nodes $n_1, n_2, n_4, n_5, n_6, n_7$. Deleting column n_5 as well (as we have already located a facility there), the reduced coverage matrix is

	n_1	n_2	n_3	n_4	n_6	n_7	n_8	n_9
n_3	0	1	1	1	1	0	0	1
n_8	0	0	0	0	1	1	1	0
n_9	0	0	1	0	0	1	0	1

Multiplying this matrix with the remaining vector of weights [100, 220, 200] from the left, we obtain [0, 100, 300, 100, 320, 420, 220, 300], whose unique maximum is in the position that belongs to n_7, and we locate a facility at that site. That way, we cover an additional 420 customers, and we can now delete the rows that belong to nodes n_8 and n_9 as well as the column of n_7. Since we have now exhausted our resources, having located two facilities as required, customers at n_3 will remain unserved and we have succeeded covering a total of 850 + 420 = 1270 customers by locating the two facilities at nodes n_5 and n_7. This, as we happen to know, is actually the optimal solution.

Typically, a construction heuristic such as Greedy will be followed by an improvement heuristic. One such improvement heuristic is the swap-interchange improvement heuristic. The idea is simple: in each step, one facility is removed from its present location and relocated to a site, at which there presently is no facility. If such a move increases coverage, it is accepted, otherwise, the search for better solution continues until some stop criterion is satisfied.

We will illustrate this procedure by using again the example shown above. However, since the Greedy heuristic already found an optimal solution, we start our demonstration with some nonoptimal solution. Suppose that we have located facilities at nodes n_3 and n_6. The two facilities cover customers at the nodes n_2, n_3, n_4, n_6, and n_9 and n_3, n_4, n_5, n_6, and n_8, respectively, so that all customers except those at nodes n_1 and n_7 are covered.

We first try to remove a facility from its present location at n_3 and move it to n_1 instead. Such a move means that we lose coverage of customers at n_2 and n_9 for a combined loss of 360 demand units, while the facility at the new location will cover customers at n_1, n_2, and n_5 (of whom customers at n_5 were already covered by the facility at n_6), so that we gain 280 new customers. This indicates that our move results in a net loss of 80 customers, so that the move is rejected.

Instead, we may try to relocate a facility from n_3 to n_2. A similar argument will reveal the same loss of 80 customers, so this move is also rejected. Yet another possibility is to relocate the facility from n_3 to n_4. This move results in a net loss of 200 customers, so again, the move is rejected.

Consider now the relocation of a facility from n_3 to n_5. Again, the loss of customers due to the removal of a facility from n_3 equals 360 customers, while the relocation of the facility at n_5 will cover the previously uncovered customers at n_1, n_2, and n_7 for a gain of 470 customers, so that there is a total net gain of 110 customers. This means that the move is accepted, so that the new solution has facilities located at n_5 and n_6. This process is repeated until no further improvements are possible. Again, the swap method is a heuristic and as such is not guaranteed to find an optimal solution.

We would like to conclude this section by mentioning a number of possible extensions of the basic covering models discussed above. One possibility to deal with outliers (whose coverage is very expensive) is to attempt to cover at least a certain proportion of the population within a prespecified distance or time. For instance, one could attempt to locate fire stations that have 90% of the potential customers within 8 min of the station.

Another important issue concerns congestion. Especially when resources are scarce (e.g., ambulances), an important issue is what happens if a service call arrives while the unit is presently busy. A possible way to deal with such situations is the introduction of backup coverage. In other words, the decision maker may try not just to cover each potential customer once but try to cover a certain proportion of customers more than once. The obvious question then concerns the tradeoff between primary coverage for some people versus backup coverage for others.

7.3 Center Problems

Another classical problem type in location analysis comprises the so-called *center problems*. The basic idea of all center problems is that they locate facilities so as to minimize the longest distance between any customer and his closest facility. One of the problems associated with the concept of centers is their exclusive focus on the customer with the longest facility–customer distance. This can lead to highly undesirable outcomes. In case of problems that involve multiple facilities, the center objective minimizes the maximal distance between any customer and its nearest facility.

Throughout this section, we assume again that all demand is clustered at the nodes of the network (or at given points in the plane). We can then distinguish between *node centers*, which are facility locations for which only the nodes are considered, and *absolute centers*, in which case facilities may be located anywhere on the network (including on arcs and not necessarily at nodes) or in the plane.

The need for this distinction is apparent by considering a graph with only a single arc, at whose edges we have the nodes n_1 and n_2. Either node would serve as a single node center (whose critical distance, i.e., the distance from the facility to the farthest customer) equals the length of the arc. However, the point on the graph that minimizes the maximal distance is at the center of the arc, from where the longest distance to either customer is only half the length of the arc.

7.3.1 1-Center Problems

The simplest center problem is the 1-node center problem on a network. The problem is to find a facility location, so as to minimize the maximal distance between the facility and any of the customers. As an illustration, consider again the graph in Fig. 7.1 in the previous section of this chapter.

Recall that the matrix of shortest distances was

	n_1	n_2	n_3	n_4	n_5	n_6	n_7	n_8	n_9
n_1	0	5	11	14	6	12	14	20	18
n_2	5	0	6	9	7	11	15	19	13
n_3	11	6	0	9	11	5	16	13	7
n_4	14	9	9	0	8	4	16	12	16
n_5	6	7	11	8	0	6	8	14	17
n_6	12	11	5	4	6	0	14	8	12
n_7	14	15	16	16	8	14	0	7	9
n_8	20	19	13	12	14	8	7	0	16
n_9	18	13	7	16	17	12	9	16	0

Again, the rows symbolize the customers, while the columns stand for the facilities. If we were to position a facility at node n_1, then the distances between the customers and the facility would be found in the first column. The longest such distance is then the largest number in the first column, in our case 20.

A simple brute-force enumeration procedure now suggests itself for the determination of a 1-node center on graphs: tentatively locate a facility at a node, determine the distance to the farthest customers, repeat for all potential facility locations, and choose the location with the shortest distance. In other words, determine all column maxima in the matrix, and chose the 1-center facility location as the column with the minimal such distance. This will determine an optimal solution. In our example, the

7.3 Center Problems

column maxima are [20, 19, 16, 16, 17, 14, 16, 20, 18], so that the 1-node center is located at node n_6 and the longest facility–customer distance is 14.

Consider now the 1-absolute center problem in the plane with Manhattan distances. This problem has a very simple closed-form solution. Assume that the n customers are located at the points $P_1, P_2, \ldots, P_n$ with coordinates $(a_1, b_1), (a_2, b_2)$, $\ldots, (a_n, b_n)$, and that the facility will be located at a point with the coordinates (x, y). Note that regardless of the number of customers, the problem has only two variables, viz., x and y.

In order to solve the problem, we need to define five auxiliary variables. They are

$$\alpha_1 = \max_i \{a_i + b_i\}$$

$$\alpha_2 = \max_i \{-a_i + b_i\}$$

$$\alpha_3 = \max_i \{a_i - b_i\}$$

$$\alpha_4 = \max_i \{-a_i - b_i\}$$

$$\alpha_5 = \max \{(\alpha_1 + \alpha_4), (\alpha_2 + \alpha_3)\}.$$

A 1-absolute center is then located at (x, y) with coordinates.
$\bar{x} = \frac{1}{2}(\alpha_3 - \alpha_4)$ and $\bar{y} = \frac{1}{2}(-\alpha_3 - \alpha_4 + \alpha_5)$ with the longest distance $\bar{z} = \frac{1}{2}\alpha_5$, and, alternatively,
$\bar{x} = \frac{1}{2}(\alpha_1 - \alpha_2)$ and $\bar{y} = \frac{1}{2}(\alpha_1 + \alpha_2 - \alpha_5)$ with longest distance $\bar{z} = \frac{1}{2}\alpha_5$, as well as on every point on the line segment between these two solutions. For a simple proof, see, e.g., Eiselt and Sandblom (2004).

As a numerical example, consider a problem with six customers, who are located at the points P_1: (0, 0), P_2: (0, 3), P_3: (1, 6), P_4: (4, 5), P_5: (4, 2), and P_6: (5, 0). We can then calculate

$$\alpha_1 = \max \{0+0, 0+3, 1+6, 4+5, 4+2, 5+0\} = 9,$$
$$\alpha_2 = \max \{-0+0, -0+3, -1+6, -4+5, -4+2, -5+0\} = 5,$$
$$\alpha_3 = \max \{0-0, 0-3, 1-6, 4-5, 4-2, 5-0\} = 5,$$
$$\alpha_4 = \max \{-0-0, -0-3, -1-6, -4-5, -4-2, -5-0\} = 0, \text{ and}$$
$$\alpha_5 = \max \{(9+0), (5+5)\} = 10.$$

The 1-absolute center facility location is then at $\bar{x} = 2\frac{1}{2}$ and $\bar{y} = 2\frac{1}{2}$, and, alternatively, $\bar{x} = 2, \bar{y} = 2$, both with $\bar{z} = 5$.

The Manhattan distances between the individual customers and the facility are 5, 3, 5, 4, 2, and 5, respectively, in case the facility is located at (2½, 2½), and 4, 3, 5, 5, 2, 5, respectively, in case the facility is located at (2, 2). It is apparent that the longest customer-facility distance equals 5 in both cases.

In case Euclidean distances are used, the 1-center in the plane has an interesting geometric interpretation: it is at the center of the smallest enclosing circle. In Fig. 7.3 customers are shown as dots, while the center of the smallest enclosing circle is shown as a star.

Fig. 7.3 Smallest enclosing circle for given customer locations

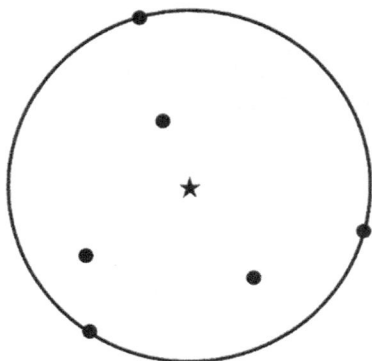

7.3.2 p-Center Problems

This subsection considers node p-center problems in networks. The objective of p-center problems is to locate a given number p of facilities, so that the longest distance between any customer and his closest facility is as small as possible. While, from a theoretical point of view, p-center problems are difficult, a simple bisection method can be employed to solve p-center problems as a sequence of covering problems.

The method commences with an upper bound $\overline{D}$, a lower bound $\underline{D}$, and an initial guess of the covering distance $\overline{d}$ between the lower and the upper bound. We then determine the smallest number of facilities required to cover all nodes given $\overline{d}$. The model to do so is, of course, the location set covering problem which we dealt with in the previous section. Suppose that the number of facilities required to cover all customers is $p(\overline{d})$. If $p(\overline{d}) > p$, then $\underline{D}$ has to be revised upward; if $p(\overline{d}) \leq p$, then $\overline{D}$ might be decreased. This process continues until a covering distance is found that cannot be reduced any further without requiring additional facilities.

The initial interval $[\underline{D}, \overline{D}]$ must be sufficiently large to include the optimal solution. An obvious choice of the lower bound is $\underline{D} = 0$, while $\overline{D} = (n-1) \max \{d_{ij}\}$ with the maximum taken over all edge lengths d_{ij} in the network, is an upper bound on the length of the longest path in the network. The reason is that no path in the network can have more than $(n-1)$ edges (otherwise it will include a cycle), and the length of each edge is no more than that of the longest edge. Once lower and upper bounds on the covering distance are determined, we apply a bisection search. This process can be described as follows.

First, we bisect the present interval $[\underline{D}, \overline{D}]$ and choose a covering distance $\overline{d}$ in the center of this interval. In particular, we choose $\overline{d} := \lfloor \frac{1}{2}(\overline{D} + \underline{D}) \rfloor$. We then solve a location set covering problem with the service standard $\overline{d}$. The resulting number of facilities required to cover all nodes is denoted by $p(\overline{d})$. If $p(\overline{d}) \leq p$, then we reduce the upper bound of the interval $[\underline{D}, \overline{D}]$ to $\overline{D} := \overline{d}$, otherwise we move up the lower part of the interval to $\underline{D} := \overline{d} + 1$. Whenever the lower and upper bounds of the interval are equal, we have found the optimal solution, in which case the last-solved covering problem provides the location pattern of the facilities. As long as $\overline{D} \neq \underline{D}$, we

Fig. 7.4 Network for 3-center problem

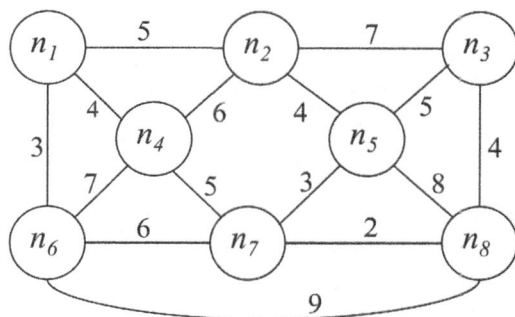

determine a new covering distance $\bar{d}$ on the basis of the interval $[\underline{D}, \overline{D}]$ with a new lower or upper bound. Then a new covering distance $\bar{d}$ is computed as the center point of this interval, and the procedure is repeated.

As a numerical illustration, consider the network in Fig. 7.4 and assume that $p = 3$ facilities are to be located.

We initialize the computations with a lower bound of $\underline{D} = 0$ and an upper bound of $\overline{D} = 7(9) = 63$. Bisecting the interval results in the trial value of $\bar{d} = 31$, for which the set covering problem has a solution of $p(\bar{d}) = p(30) = 1$, i.e., a single facility, located anywhere, will cover all the nodes. As $p(31) = 1 < 3 = p$, we set $\overline{D} := 31$.

The results of the subsequent iterations are summarized in Table 7.1.

7.4 Median Problems

As opposed to the center problems with their minimax objective discussed in the previous section, this section is devoted to median problems which have minisum objectives. In other words, they will locate facilities, so as to minimize the sum of distances to the customers. This feature makes this type of objective amenable to applications in the public and the private sector. Consider, for instance, the location of a public facility such as a library. The municipal planner will attempt to make the library as accessible as possible to all of its potential patrons. This may be done by minimizing the average distance between the library and its customers. It is not difficult to demonstrate that as long as the magnitude of the demand remains constant, minimizing the sum of facility-customer distances is the same as minimizing the average facility-customer distance.

7.4.1 Minisum Problems in the Plane

Throughout this section, we assume that the task is to locate a single new facility anywhere in the plane. The n customers are assumed to be located at points $P_1, P_2,$

Table 7.1 Bisection search to find a p-center solution in the example

Iteration #	$\underline{D}$	$\overline{D}$	$\overline{d}$	$p(\overline{d})$	Location
1	0	63	31	1	Anywhere
2	0	31	15	1	Anywhere
3	0	15	7	2	e.g., (n_1,n_8), or (n_2,n_7), or (n_3, n_6)
4	0	7	3	5	n_2, n_3, n_4, then n_7, and one of n_1 and n_6
5	4	7	5	2	(n_1, n_5), or (n_1, n_8)
6	4	5	4	3	e.g., n_1 and n_5, and one of n_3 and n_8
7	4	4	4		Stop, optimal.

..., P_n with coordinates (a_i, b_i) and their demand is denoted by the weight w_i. The task at hand is now to locate a facility, whose coordinates are the variables (x, y). Note that regardless of the number of customers in the problem, the model has no more than two variables. As discussed above, the objective is to minimize the weighted sum of customer-facility distances, a proxy of the total cost of the transportation. This type of problem (in the plane) is frequently referred to as a *Weber problem* in reference to the work by the German economist-turned-sociologist Weber (1868–1958) that culminated in the publication of his book in 1909.

The simplest problem occurs when rectilinear (Manhattan) distances are used. The objective function is then separable, meaning that it is possible to optimize one variable at a time without affecting the other. It turns out that the actual distances are irrelevant, it is only important how the customers are located in relation to each other, but not how far from each other. Actually, the procedure is very simple. We first scan the customers from left to right (or right to left) along the abscissa and add their weights, until the sum of weights for the first time matches or exceeds half of the total weight. We then repeat the process by scanning the facilities and adding their weights from top to bottom (or bottom to top), until again the reach or exceed half the total weight for the first time. The combination of the two coordinates is the location that minimizes the sum of weighted rectilinear distances.

In order to illustrate the procedure, consider twelve customers $P_1, P_2, ..., P_{12}$, whose locations are $(a_1, b_1), (a_2, b_2), ..., (a_{12}, b_{12})$, respectively, and their weights are $w_1, w_2, ..., w_{12}$, respectively. The numerical values of the coordinates and weights are shown in Table 7.2.

The arrangement of customers can be visualized in Fig. 7.5.

First scan the abscissa. From left to right, customers can be rearranged as P_2, then P_1 and P_3, then P_7, followed by P_4 and P_6, then P_5 and P_8, and so forth. Adding up weights in that order results in $20 + (20 + 10) + 40 + (10 + 30) + (50 + 60) + 60 + ... + 80$. Given that the total weight, i.e., the total demand of all customers, equals 500, we need to add up these values until, for the first time, they reach or exceed the value of $500/2 = 250$. This happens for customer P_{12} with a total of 300, at the value $x = 7$. (The same result is achieved if we add up values starting from the right). This procedure is now repeated for the ordinate by adding up weights from either bottom to top or top to bottom. From the bottom, we add $20 + (10 + 60) + (20 + 50) + (10 +$

7.4 Median Problems

Table 7.2 Locations and weights of customers in the example

Point P_j	P_1	P_2	P_3	P_4	P_5	P_6	P_7	P_8	P_9	P_{10}	P_{11}	P_{12}
a_i	2	1	2	4	5	4	3	5	8	10	9	7
b_i	1	4	5	2	4	6	7	9	9	10	5	2
w_i	20	20	10	10	50	30	40	60	70	80	50	60

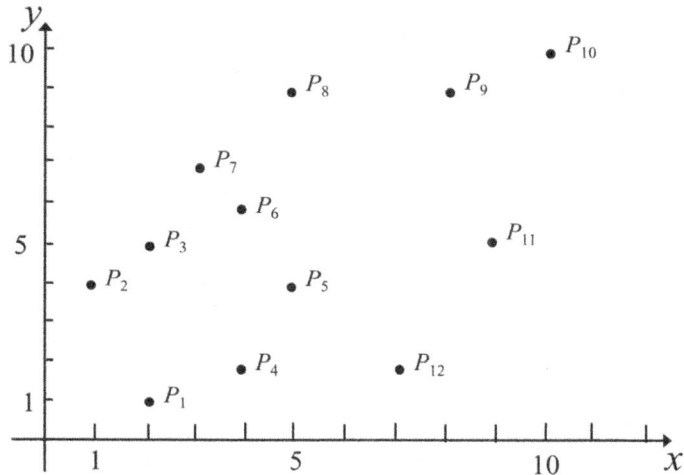

Fig. 7.5 Location of customers

50) + 30, at which point we have reached the value of 250. This occurs for customer P_6 at $y = 6$, so that this is the optimal y-coordinate.

However, since the critical value of 250 was achieved exactly, there are alternative optimal solutions. If we had added up weights from top to bottom, we would have achieved the value of 250 for customer P_7 at the coordinate $y = 7$, which is also optimal. As a matter of fact, not only are the two points $(x, y) = (7, 6)$ and $(7, 7)$ optimal, all points with $x = 7$ and y anywhere between 6 and 7 are optimal. The value of the objective function can then be determined by computing the customer-facility distances for all customers, multiplying them by the appropriate weights, and adding them up. In our example, for the optimal location at (7, 6) we obtain the total weighted distance $20(10) + 20(8) + 10(6) + 10(7) + 50(4) + 30(3) + 40(5) + 60(5) + 70(4) + 80(7) + 50(3) + 60(4) = 2510$.

Consider now the same example as before, but with the weights of P_5 and P_{12} interchanged, i.e., $w_5 = 60$ and $w_{12} = 50$. Now the optimized location is between 5 and 7 on the abscissa, and between 6 and 7 on the ordinate, so that *all* locations in the rectangle with the corners at (5, 6), (5, 7), (7, 7), and (7, 6) are optimal.

Another model works with the same scenario and also minimizes the sum of weighted distances. However, it uses squared Euclidean distances instead of rectilinear distances. Again, the objective function turns out to be separable, so that we are able to optimize for our variables x and y separately. Since these variables are continuous, we can take partial derivatives, set them equal to zero, and solve for the variables x and y. This process results in an optimal solution in which the x-coordinate of the facility equals the sum of weighted x-coordinates of the customers divided by the total weight, and similar for the y-coordinates. The solution obtained by using squared Euclidean distances is the *center-of-gravity* of the given points. As a numerical illustration, consider again the above example with data shown in Table 7.2. Here, we obtain the optimal facility location at $\bar{x} = \frac{2(20)+1(20)+\cdots+7(60)}{20+20\cdots+60} = 3140/500 = 6.28$. The optimal y-coordinate is determined in a similar fashion as $\bar{y} = 3170/500 = 6.34$. Note the difference between this solution and that obtained by using rectilinear distances: despite the strong pull of the customers with large weights in the northeast corner, the solution that uses squared Euclidean distances locates the facility more towards the west than the solution that uses rectilinear distances. This is due to the fact that by squaring distances, long distances receive a heavy emphasis, and the minimization function will try to avoid them. In that sense, a minisum objective with squared Euclidean distances will find a solution that includes features of the usual minisum, but also those of the minimax objective.

Using Euclidean distances, we can apply a similar approach. However, the objective function is no longer separable. Setting partial derivatives to zero, we obtain the optimality conditions

$$x = \frac{\sum_{i=1}^{n} \frac{w_i a_i}{\sqrt{(a_i-x)^2+(b_i-y)^2}}}{\sum_{i=1}^{n} \frac{w_i}{\sqrt{(a_i-x)^2+(b_i-y)^2}}}, \quad \text{and} \quad y = \frac{\sum_{i=1}^{n} \frac{w_i b_i}{\sqrt{(a_i-x)^2+(b_i-y)^2}}}{\sum_{i=1}^{n} \frac{w_i}{\sqrt{(a_i-x)^2+(b_i-y)^2}}}.$$

Notice that in these two relations, the variables x and y appear on the left-hand sides as well as on the right-hand sides. A "trick" first devised by Weiszfeld in 1937 is to use these relations in an iterative procedure. It starts with a guess (x, y), inserts it on the right-hand side of the equations, computes new values for x and y on the left-hand side, uses these values on the right-hand side to compute new values on the left, and so forth.

As a starting point, it may be a good idea to use the center-of-gravity, or the solution given rectilinear distance. In order to demonstrate the *Weiszfeld method*, consider the following numerical

Example There are three customers at the points P_1, P_2, and P_3 with coordinates $(0, 0)$, $(3, 0)$, and $(0, 5)$, respectively. The demands of the customers are $w_1 = 30$, $w_2 = 20$, and $w_3 = 40$. As a starting point, we use $(x, y) = (5, 5)$, even though this is quite an unreasonable initial point. The reason is that the optimal solution will always be located in the triangle formed by the customers, and our starting point is not. We have chosen this point to demonstrate that normally—there are counterexamples, though—this technique converges very quickly. Before starting, note that using rectilinear distances in this example results in an optimal facility location at (0, 0), while squared Euclidean distances will locate the facility at (⅔, 2²⁄₉).

Given a starting point with coordinates $x = 5$ and $y = 5$, we compute the next trial point at

$$x = \frac{\dfrac{0(30)}{\sqrt{(0-5)^2+(0-5)^2}} + \dfrac{3(20)}{\sqrt{(3-5)^2+(0-5)^2}} + \dfrac{0(40)}{\sqrt{(0-5)^2+(5-5)^2}}}{\dfrac{30}{\sqrt{(0-5)^2+(0-5)^2}} + \dfrac{20}{\sqrt{(3-5)^2+(0-5)^2}} + \dfrac{40}{\sqrt{(0-5)^2+(5-5)^2}}} \approx .6982,$$

and

$$y = \frac{\dfrac{0(30)}{\sqrt{(0-5)^2+(0-5)^2}} + \dfrac{0(20)}{\sqrt{(3-5)^2+(0-5)^2}} + \dfrac{5(40)}{\sqrt{(0-5)^2+(5-5)^2}}}{\dfrac{30}{\sqrt{(0-5)^2+(0-5)^2}} + \dfrac{20}{\sqrt{(3-5)^2+(0-5)^2}} + \dfrac{40}{\sqrt{(0-5)^2+(5-5)^2}}} \approx 2.5068.$$

Given the new trial point $(x, y) = (0.6982, 2.5068)$—which, incidentally, is quite similar to the center-of-gravity determined earlier—we can then compute the next trial point in similar fashion. It turns out to be $(x, y) = (0.5366, 2.3512)$, which is quite close to the previous solution. This procedure terminates whenever some stop criterion is satisfied, e.g., there is only a small change in the solution or the value of the objective function.

A good tool that allows us to envisage the forces that determine the solution is the *Varignon frame* named after the French mathematician Varignon, 1654–1722. Imagine a board, in which holes have been drilled at the points at which customers are located. One string is fed through each hole and all strings are tied together in one knot. Below the board, weights are attached to the strings, such that the weights are proportional to the demand of the respective customer. Given only gravity and in the absence of friction, the knot will then settle at the optimal point. A graph of the Varignon frame is shown in Fig. 7.6.

7.4.2 Minisum Problems in Networks

Consider now the location of a single new facility on a network. The facility location that minimizes the sum of weighted customer–facility distances is usually referred to as a 1-*median*. An obvious extension of the 1-median problem is the *p-median problem* that locates p facilities so as to minimize the sum of distances between all customers and their respective closest facilities. Given n customers in the network, we assume that $p < n$; otherwise, the optimal solution locates a facility at each customer site.

The main result we use is due to Hakimi (1964), who proved the following

Fig. 7.6 Varignon frame

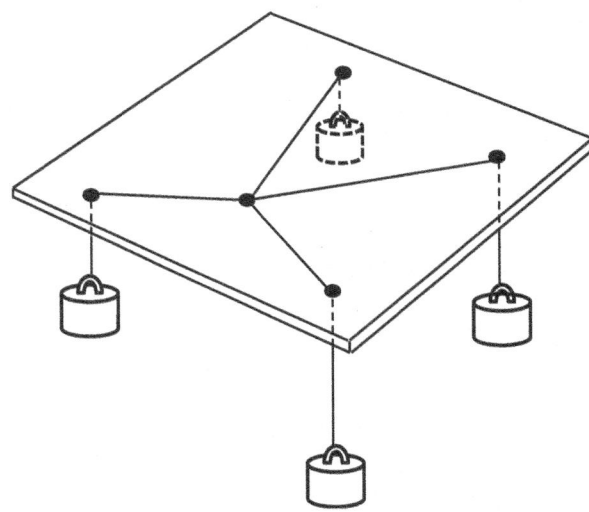

Theorem 6.1 At least one optimal solution of a p-median problem is located at a node.

Given this result, we no longer have to search for an optimal solution anywhere on an arc but can restrict ourselves to the nodes of the network. In that sense, this theorem is reminiscent of Dantzig's corner point theorem for linear programming that also restricted the set of possible optimal locations from an infinite to a finite set.

First consider the case, in which the decision maker's task is to determine a 1-median in a network. We assume that the matrix of shortest paths has already been determined. Assuming that the matrix $\mathbf{D} = (d_{ij})$ includes the lengths of the shortest paths between all pairs of nodes and given the vector $\mathbf{w} = (w_i)$ of customer demands, we now use Hakimi's theorem and compute the costs for all potential facility locations at the nodes of the network. This can be accomplished by vector-matrix multiplication. In other words, determine the vector $\mathbf{wD}$, which includes in its j-th element the total transportation cost (i.e., weighted distance) from all customers to a facility located at node j. As an illustration consider the following

Example The graph in Fig. 7.7 includes the single-digit node-to-node (direct) distances next to its edges and double-digit weights next to its nodes.

The symmetric matrix of shortest path distances is then

7.4 Median Problems

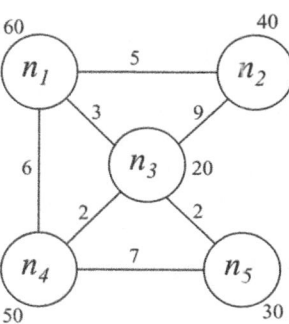

Fig. 7.7 Network for p-median example

D:

	n_1	n_2	n_3	n_4	n_5
n_1	0	5	3	5	5
n_2	5	0	8	10	10
n_3	3	8	0	2	2
n_4	5	10	2	0	4
n_5	5	10	2	4	0

while the vector of demands is $\mathbf{w} = [60, 40, 20, 50, 30]$. Multiplying the vector and the matrix results in $\mathbf{wD} = [660; 1260; 660; 860; 940]$. This means that if we were to locate a facility at, say node n_4, then our total transportation costs would be 860. Choosing the minimum in this cost vector reveals that we should either locate at node n_1 or at node n_3; in both cases, the total transportation costs are 660. In order to choose between the two alternatives, the decision maker may use secondary criteria. Having shown how to locate a single facility, we now demonstrate how to locate a given number p of facilities.

Arguably, p-median problems are among the most researched location models in practice. Since they are difficult to solve, most large-scale problems will be solved by heuristic algorithms. This section will describe one construction heuristic and two improvement heuristics.

As usual, the Greedy heuristic works in sequential fashion. In the first stage, it computes the weighted transportation costs for tentative locations at all nodes, and then permanently locates the first facility at the node that allows the transportation at the lowest cost. This is exactly the same step as that in the exact method that locates a single facility, i.e., the 1-median on a network.

Suppose now that a number less than p of facilities have already been permanently located, and the Greedy heuristic attempts to locate an additional facility. This is accomplished by tentatively locating the new facility at one (presently unoccupied) node at a time. Suppose we tentatively locate the new facility at node n_j. The method will then compute the shortest distance between each customer at node n_i and the closest facility that either exists already or is proposed (at n_j). These distances will be collected in the column vector $\mathbf{d}_j$. The weighted distance is then computed by multiplying the weights and these distances, i.e., $\mathbf{wd}_j$. This number expresses the additional total transportation costs that are incurred if we locate the new facility at node n_j. This process is repeated for all possible tentative locations, and the minimum is chosen. The location of the new facility is then made permanent, and the process continues until the desired number of facilities has been located.

We will illustrate this procedure by a numerical.

Example Consider again the network in Fig. 7.7 and assume that the task is to locate $p = 3$ facilities, so as to minimize the total transportation costs. The matrix of shortest distances **D** was already determined in the previous section, and so were the vector of weights **w** and the cost vector **wD** = [660; 1260; 660; 860; 940]. Again, we choose to locate at either node n_1 or at node n_3. Arbitrarily choose node n_1.

We now tentatively locate the second facility at the nodes n_2, n_3, n_4, and n_5. For the trial location at the node n_2, we now have facilities at n_1 and n_2, so that the distances between a customer and the closest of our facilities are found by taking the minima of columns 1 and 2 in the distance matrix. Here, we obtain $\mathbf{d}_2 = [0, 0, 3, 5, 5]^T$, so that $\mathbf{wd}_2 = 460$. For the trial location at n_3, the shortest customer–facility distances are found by computing the minimum among columns 1 and 3 in the distance matrix **D**, resulting in $\mathbf{d}_3 = [0, 5, 0, 2, 2]^T$, so that $\mathbf{wd}_3 = 360$. Similarly, we compute $\mathbf{d}_4 = [0, 5, 2, 0, 4]^T$ with $\mathbf{wd}_4 = 360$, and $\mathbf{d}_5 = [0, 5, 2, 4, 0]^T$ with $\mathbf{wd}_5 = 440$. Among those trials, the lowest location costs are found at either node n_3 or at node n_4. Again, we arbitrarily break the tie and choose a facility location at n_3.

We now have facilities permanently located at the nodes n_1 and n_3. In this iteration, we tentatively locate facilities at n_2, n_4, and n_5, one at a time. For a trial location at n_2, we have locations at n_1, n_2, and n_3, so that we determine the shortest customer-facility distances by computing the minima in the first three columns. This results in $\mathbf{d}_2 = [0, 0, 0, 2, 2]^T$ and $\mathbf{wd}_2 = 160$. Similarly, the trial facilities at n_4 and n_5 result in distances and total transportation costs of $\mathbf{d}_4 = [0, 5, 0, 0, 2]^T$ with $\mathbf{wd}_4 = 260$ and $\mathbf{d}_5 = [0, 5, 0, 2, 0]^T$ with $\mathbf{wd}_5 = 300$. The lowest costs are found for the tentative location at n_2, which is now made permanent. Since we have now located all available facilities, the process terminates with facilities located at n_1, n_2, and n_3. The total transportation costs are then 160.

If we had broken the tie for the first facility in the same way and located at n_1, but chose n_4 for the second facility, we would have ended up with facilities located at n_1, n_2, and n_4 for total costs of 140. On the other hand, if we had broken the tie for the location of the first facility in favor of n_3, we would have ended up with facilities located at n_1, n_2, and n_3 with total transportation costs of 160, the same location pattern as that determined earlier. Notice that none of the tie-breaking rules has been proved superior.

As usual, any construction heuristic should be followed by an application of an improvement heuristic. In this section, we will describe two techniques. The first heuristic we apply is the so-called *location-allocation heuristic*. Simply put, it alternates between location and allocation steps. In particular, the technique is initialized with any solution. The better the solution the procedure starts with, the better the solution may be expected to turn out (although not necessarily so).

7.4 Median Problems

Suppose now that any facility location with the required p facilities has been determined by eyeballing, an application of the greedy method, or any other technique. The first step of the improvement heuristic is then the allocation phase. In it, we simply assign each customer to its closest facility. This results in p clusters, each with a single facility. In the location phase, we then consider one cluster at a time, remove the facility from it, and determine an optimal facility location in it. This new facility location may or may not be equal to the previous location of the facility. This process is repeated for all clusters.

If there has been any change regarding the facility locations in the last step, the procedure is repeated, until there are no further changes.

Example Consider again the above example and suppose that facilities have been located at n_3, n_4, and n_5. Allocating each customer to its closest facility results in clusters $\{n_1, n_2, n_3\}$ for the facility at n_3, $\{n_4\}$ for the facility at n_4, and $\{n_5\}$ for the facility at n_5. The transportation costs for this initial location pattern are 500. The location phase of the heuristic method begins by considering the first cluster. The weights of the customers at n_1, n_2, and n_3 are $\mathbf{w}^* = [60, 40, 20]$, and the partial distance matrix for the three nodes is

$$\mathbf{D}^* = \begin{array}{c|ccc} & n_1 & n_2 & n_3 \\ \hline n_1 & 0 & 5 & 3 \\ n_2 & 5 & 0 & 8 \\ n_3 & 3 & 8 & 0 \end{array}$$

Computing the total costs for all three potential facility locations in this cluster (i.e., determining a new single facility in this cluster as shown earlier in this section), we obtain 260, 460, and 500, so that we choose the node n_1 as the new facility location in this cluster. The other two clusters include only one node each, so that relocation is not possible. Thus the new location pattern includes facilities at the nodes n_1, n_4, and n_5 (which replaces the previous locations at n_3, n_4, and n_5.)

Given these facility locations, we again assign customers to their closest open facilities. The new clusters have customers $\{n_1, n_2\}$ assigned to the facility at n_1, $\{n_3, n_4\}$ assigned to the facility at n_4, and $\{n_5\}$ assigned to the facility at n_5. Note that alternatively, the node n_3 could have been assigned to the facility at node n_5 instead. The first cluster has weights $\mathbf{w}^* = [w_1, w_2] = (60, 40)$ and with a distance matrix of $\mathbf{D}^* = \begin{bmatrix} 0 & 5 \\ 5 & 0 \end{bmatrix}$, we obtain costs of $\mathbf{w}^*\mathbf{D}^* = [200, 300]$, so that the facility will remain at n_1 with a partial cost of 200. The weight vector and distance matrix for the second cluster are $\mathbf{w}^* = [w_3, w_4] = [20, 50]$, and the distance matrix is $\mathbf{D}^* = \begin{bmatrix} 0 & 2 \\ 2 & 0 \end{bmatrix}$, so that the transportation costs for the two potential facility locations in this cluster are $\mathbf{w}^*\mathbf{D}^* = [100, 40]$, meaning that the facility is again located at node n_4 with a partial cost of 40. The last cluster only includes a single node, so that the facility is located at n_5 in that cluster with partial costs of 0. The total costs of this location arrangement are then $200 + 40 + 0 = 240$.

These locations are the same as those in the previous iteration, indicating that the location—allocation algorithm has converged. The solution has facilities located at n_1, n_4, and n_5 with total transportation costs of 240. Notice again that this solution is not optimal.

The second heuristic we describe in this chapter is the so-called *vertex substitution method*, a technique that employs a simple "swap" step. Again, start with any location arrangement of the required p facilities. In each iteration, the method tentatively moves one of the facilities from its present location to an unassigned location. If the swap reduces the total transportation costs, we have a new solution, and the process continues. Otherwise, we continue with another pair of locations. The process terminates, if no swap reduces the costs any further.

Example Consider again the above example and initialize the method with the facilities again located at the nodes n_3, n_4, and n_5. Moving a facility from n_3 to n_1 results in a cost reduction from 500 to 240, so the move is made, and the new solution has facilities located at n_1, n_4, and n_5. Moving a facility from n_1 to n_2 raises the costs to 340, so the move is not made. Moving a facility from n_1 to n_3 increases the costs to 500, and again, the move is not made. The move of a facility from n_4 to n_2 leaves the cost at 240, so a tie-breaking rule must be used. Here, we keep the facilities at their present locations. Moving a facility from n_4 to n_3 increases the cost to 300, so the move is not made. Moving a facility from n_5 to n_2 decreases the costs to 160, so we make the move and have a new location arrangement with facilities located at nodes n_1, n_2, and n_4. The procedure is repeated until no further improvements are possible.

7.5 Other Location Problems

This section is designed to introduce some additional types of location models that have been discussed in the literature. The first such model deals with *undesirable facilities*. While it appears apparent that facilities such as sewage treatment plants, landfills, power plants, or prisons are undesirable to have in the neighborhood of a residential area, things are more subtle than that. As a matter of fact, just about all facilities have desirable and undesirable features. Take, for instance, a hospital. Few people will argue that it would be great to have such a facility nearby, the wailing sirens of ambulances that can and will be turned on at any time of day or night will surely be considered a nuisance. A similar argument applies to facilities such as prisons: whereas few people would like to have them nearby (other than maybe to visit relatives), the many employees who work in these facilities would not appreciate having them at a great distance from their home.

Modeling of the location of undesirable facilities can be done in two different ways. We either use the standard cost-minimizing objective and define "forbidden regions" in which facilities cannot be located, or we use an objective function that pushes the facilities away from the customers, rather than pulling them towards them as cost-minimizing objectives do. There are a number of problems related to both approaches. Using forbidden regions in networks is easy: if it is not desirable or permitted to locate a facility at a node, we simply do not define a location variable y_j for that node (or, equivalently, set the location variable to zero). The process is much more complicated in the plane, where the forbidden regions must be defined as system of linear or nonlinear inequalities. Using a "push" objective, i.e., an objective that pushes facilities away from customers or population as opposed to pulling facilities towards them, is not a straightforward process, either: first of all, if we optimize on an unconstrained set, such an objective would attempt to push the facilities towards infinity, which is obviously not reasonable. This means that it will be required to define a feasible set, in which all facilities must be located. Again, the objective that maximizes the weighted sum between customers and facilities will tend to have the facilities locate at the border of the feasible set. A good example of this behavior is the location of nuclear power plants in France, many of which have been sited along the Rhine River, i.e., the border with Germany. Another problem is that simply exchanging the "Min" for a "Max" in the objective function will not suffice. The reason is that the usual cost-minimizing objective will automatically assign a customer to his nearest facility, which is reasonable in almost all relevant contexts. Similarly, a maximization objective, the objective will automatically assign a customer to his farthest facility, which does not make sense in most cases: customers are most affected by the nearest undesirable facility (as its effects will be most pronounced on the customer), rather than the farthest facility. In order to ensure that the effects of the closest facility are measured and counted towards the objective (e.g., the overall pollution level on the population), additional constraints are required. They result in fairly large formulations for even small problems, thus often necessitating the use of heuristic algorithms.

Another strand of work deals with location models that have "equity" objectives. The idea is to locate facilities, so as to make them as equally accessible to all (potential) customers. Models of this nature have a variety of difficulties associated with them. First of all, it is not obvious what "equity" is. Dictionaries will define it as "fairness," a concept just as vague. The location models in this category all deal with equality, rather than equity. More precisely, they attempt to locate facilities, so as to make the shortest customer-facility distances (i.e., the usual assignments) as equal as possible. Many measures for equality have been described: the range (i.e., the difference between the shortest and the longest of any customer-facility distances), the variance of these distances, the *Lorenz curve* (a tool that economists use to display income disparities), and the related *Gini index*. It is important to realize that "equity" objectives should always be coupled with an efficiency objective, as otherwise, they tend to result in degenerate solutions. As an example, consider two customers located at the ends of a line segment, and a third customer who is just below the center of that line segment. The optimal solution for any single-facility location problem with equity objective has the facility located at the center of the circle, on whose circumference all three customers lie. This point can be very far away from the customers. Worse yet, as the facility moves closer to the customers

Fig. 7.8 Problem with equality objective

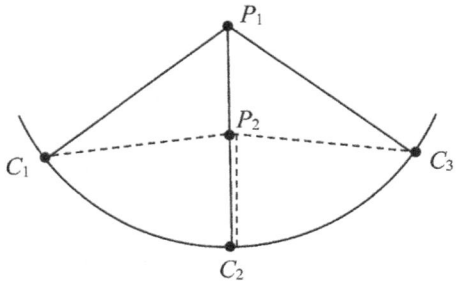

(a move from which all customers would benefit, as they are all closer to the facility), the solution is less equal and thus an equality objective will not make it. This is clearly undesirable and can only be avoided if some efficiency objective is considered as well. The situation is shown in Fig. 7.8, where P_1 is the facility location that is equidistant to the customers C_1, C_2, and C_3 (the solid black lines), while the facility location P_2 is closer to each of the customers, but no longer equidistant from them (the broken lines).

Another active area of research concerns the *location of hubs*. Hubs are an essential concept in a number of industries, most prominently the airline industry. Typically, airline flights between an origin and a destination are routed through one or two hubs. The inconvenience of having to change planes once or twice is acceptable due to the fact that without hubs, many origin-destination pairs would not permit any flights at all between them due to low traffic volume. The flight volumes to the hubs (or concentrators) are, however, often sufficient to justify flights to them. In addition, flights between (remote) origins or destinations and more central hubs are typically done by small commuter planes, whose costs per passenger-mile are typically considerably higher than the costs of larger airplane that operate between hubs. This is the reason for inter-hub discounts, i.e., lower costs between hubs. Hub location problems are difficult to solve exactly. The major reason for the difficulties is the size of the problem. Typically, the formulation will use binary variables of the type y_{iklj}, which assume a value of one, if the traffic from origin i to destination j is routed through hubs k and l. For example, in a network with one hundred nodes, each of which is an origin, a destination, and a potential hub, there would be a hundred million zero-one variables. Even considering today's powerful computing equipment, this is a very large problem.

Competitive location models were introduced by Hotelling in 1929. A mathematical statistician by trade, he considered the simplest competitive location problem one can think of: two profit-maximizing duopolists locate one facility each on a line segment. They offer a homogeneous good and compete in locations and prices. Hotelling concluded that a situation, in which the duopolists would cluster together at the center of the market, is stable. It took 50 years to prove him wrong, and it is known today that there exists no stable solution in his original version of the problem. Today, we consider two versions of competitive location models. In the first class of problems, the main quest is to find, as Hotelling did, stable solutions.

7.5 Other Location Problems

These are called *Nash equilibria*. A Nash equilibrium is said to exist, if none of the participants in the game has an incentive to unilaterally move out of the current situation, which in this case means to change his location or price. The second class of problem concerns *von Stackelberg solutions*. The economist von Stackelberg considered two groups of participants: leaders and followers. The leader will choose his actions (here: his location) first, knowing that a follower will locate later. Note that the leader's planning will require assumptions concerning the follower's objectives and perception. This is typically summarized in a reaction function, which delineates the follower's reaction to each of the leader's courses of action. In contrast, once the leader has made his decision, the follower only has to observe what the leader has done and make his own decisions accordingly. It is apparent that the follower's problem is a conditional optimization problem (finding an optional location for the follower's facilities, given the leader's facility locations), while the leader's problem is very difficult, as even the determination of the reaction function requires the solution of a zero-one integer programming problem for each of the leader's possible courses of action.

Finally, consider *extensive facilities* and facility *layout problems*. In both areas, the facilities can no longer be represented as points on a map, so that the sizes of the facilities are no longer negligible in relation to the space they are located in. Problems of this nature are much more difficult than "simple" location problems. The main reason is that the shape of the facilities must now also be considered. Typical examples of layout problems are the arrangements of workstations in an office, the placement of rooms in a hospital (operating rooms, supply rooms, recovery rooms, etc.), and the allocation of spaces in a shopping mall to stores. The best-known facility layout model is the quadratic assignment problem. Generally speaking, the purpose of this problem is to assign items to empty slots. Depending on the specific application, this may mean work functions to stations, offices to empty rooms, or drill bits to slots on a drill. One way to formulate quadratic assignment problems is to define binary variables y_{ijkl}, which assume a value of one, if item i is assigned to slot j and item k is assigned to slot l, and zero otherwise. (Notice the similarity to hub problems discussed above.) The advantage of this formulation is its linear objective function, while the disadvantage is the very large number of variables. Another formulation uses double-subscripted binary variables y_{ij}, which equal one, if item i is assigned to slot j, and zero otherwise. This formulation has much fewer variables, but its disadvantage is its quadratic objective function (hence the name of the formulation). To this day, exact solutions for quadratic assignment problems with more than about 30 items and slots remain elusive.

Exercises

Problem 1 (a location set covering problem): An administrative district includes 18 small villages. One of the functions of the district officer is to ensure that each community is reasonably well served in case of a fire. It was established that no village should be farther than 8 min from its closest fire hall. The graph with the villages and the distances (in minutes) between them is shown in the Fig. 7.9. All villages must be covered.

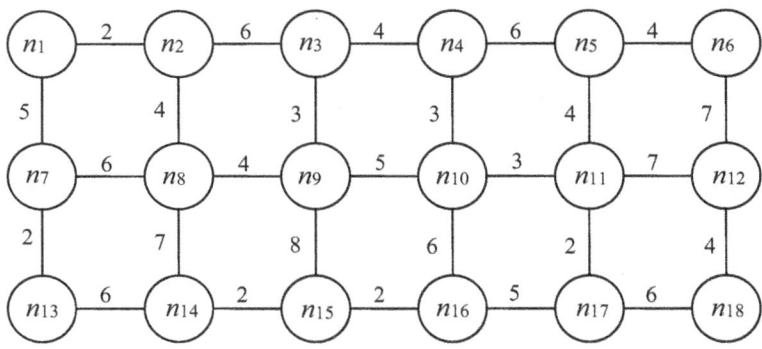

Fig. 7.9 Graph for Problem 1

How many facilities will be needed to cover all customers? The covering matrix is shown in Table 7.3.

Solution:

The formulation is

$$P : \text{Min } z = y_1 + y_2 + y_3 + y_4 + y_5 + y_6 + y_7 + y_8 + y_9 + y_{10} + y_{11} + y_{12}$$
$$+ y_{13} + y_{14} + y_{15} + y_{16} + y_{17} + y_{18}$$

s.t. $y_1 + y_2 + y_3 + y_7 + y_8 + y_{13} \geq 1$

$y_1 + y_2 + y_3 + y_7 + y_8 + y_9 \geq 1$

$y_1 + y_2 + y_3 + y_4 + y_8 + y_9 + y_{10} \geq 1$

$y_3 + y_4 + y_5 + y_9 + y_{10} + y_{11} \geq 1$

$y_4 + y_5 + y_6 + y_{10} + y_{11} + y_{17} \geq 1$

$y_5 + y_6 + y_{11} + y_{12} \geq 1$

$y_1 + y_2 + y_7 + y_8 + y_{13} + y_{14} \geq 1$

$y_1 + y_2 + y_3 + y_7 + y_8 + y_9 + y_{13} + y_{14} \geq 1$

$y_2 + y_3 + y_4 + y_8 + y_9 + y_{10} + y_{11} + y_{15} \geq 1$

$y_3 + y_4 + y_5 + y_9 + y_{10} + y_{11} + y_{15} + y_{16} + y_{17} \geq 1$

$y_4 + y_5 + y_6 + y_9 + y_{10} + y_{11} + y_{12} + y_{16} + y_{17} + y_{18} \geq 1$

$y_6 + y_{11} + y_{12} + y_{18} \geq 1$

$y_1 + y_7 + y_8 + y_{13} + y_{14} + y_{15} \geq 1$

$y_7 + y_8 + y_{13} + y_{14} + y_{15} + y_{16} \geq 1$

$y_9 + y_{10} + y_{13} + y_{14} + y_{15} + y_{16} + y_{17} \geq 1$

$y_{10} + y_{11} + y_{14} + y_{15} + y_{16} + y_{17} \geq 1$

$y_5 + y_{10} + y_{11} + y_{15} + y_{16} + y_{17} + y_{18} \geq 1$

$y_{11} + y_{12} + y_{17} + y_{18} \geq 1$

$y_j = 0 \text{ or } 1 \text{ for all } j$

The optimal solution locates facilities at nodes (n_8, n_9, n_{11}), (n_8, n_{10}, n_{11}), (n_8, n_{10}, n_{12}), or (n_8, n_{11}, n_{13}). In all cases, three facilities are sufficient to cover *all* customers as required.

7.5 Other Location Problems

Table 7.3 Coverage matrix for Problem 1

	n_1	n_2	n_3	n_4	n_5	n_6	n_7	n_8	n_9	n_{10}	n_{11}	n_{12}	n_{13}	n_{14}	n_{15}	n_{16}	n_{17}	n_{18}
n_1	1	1	1	0	0	0	1	1	0	0	0	0	1	0	0	0	0	0
n_2	1	1	1	0	0	0	1	1	1	0	0	0	0	0	0	0	0	0
n_3	1	1	1	1	0	0	0	1	1	1	0	0	0	0	0	0	0	0
n_4	0	0	1	1	1	0	0	0	1	1	1	0	0	0	0	0	0	0
n_5	0	0	0	1	1	1	0	0	0	1	1	1	0	0	0	0	1	0
n_6	0	0	0	0	1	1	0	0	0	0	0	0	0	0	0	0	0	0
n_7	1	1	0	0	0	0	1	1	1	0	0	0	1	1	0	0	0	0
n_8	1	1	1	0	0	0	1	1	1	1	1	0	1	1	1	0	0	0
n_9	0	1	1	1	0	0	1	1	1	1	1	1	0	0	1	0	0	0
n_{10}	0	0	1	1	1	0	0	1	1	1	1	1	0	0	0	0	0	0
n_{11}	0	0	0	1	1	0	0	1	1	1	1	1	0	0	0	0	0	0
n_{12}	0	0	0	0	1	0	0	0	1	1	1	1	0	0	0	0	0	0
n_{13}	1	0	0	0	0	0	1	1	1	1	1	0	1	1	1	1	1	1
n_{14}	0	0	0	0	0	0	1	1	0	1	1	0	1	1	1	1	1	1
n_{15}	0	0	0	0	0	0	0	1	1	0	0	0	1	1	1	1	1	0
n_{16}	0	0	0	0	0	0	0	0	1	0	0	0	1	1	1	1	1	0
n_{17}	0	0	0	0	0	0	0	0	0	0	0	0	1	1	1	1	1	1
n_{18}	0	0	0	0	0	0	0	0	0	0	0	0	0	0	0	0	1	1

Fig. 7.10 Village network for Problem 2

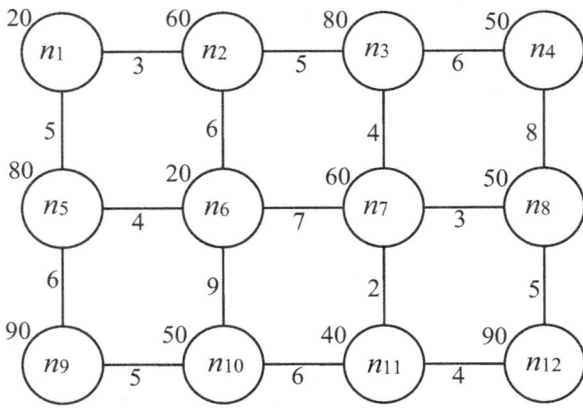

Problem 2 (a maximal covering location problem, heuristics): A small rural district has the task of locating health posts in the country to serve remove villages. A village is said to be covered, if it is within ten miles of the health post. The district administration can pay for no more than two health posts. The geographical layout of the villages is shown in Fig. 7.10, in which the single-digit numbers indicate the distances between the villages, and the double-digit numbers are the populations of the villages.

The covering matrix is then as follows:

	n_1	n_2	n_3	n_4	n_5	n_6	n_7	n_8	n_9	n_{10}	n_{11}	n_{12}
n_1	1	1	1	0	1	1	0	0	0	0	0	0
n_2	1	1	1	0	1	1	1	0	0	0	0	0
n_3	1	1	1	1	0	0	1	1	0	0	1	1
n_4	0	0	1	1	0	0	1	1	0	0	0	0
n_5	1	1	0	0	1	1	0	0	1	0	0	0
n_6	1	1	0	0	1	1	1	1	1	1	1	0
n_7	0	1	1	1	0	1	1	1	0	1	1	1
n_8	0	0	1	1	0	1	1	1	0	0	1	1
n_9	0	0	0	0	1	1	0	0	1	1	0	0
n_{10}	0	0	0	0	0	1	1	0	1	1	1	1
n_{11}	0	0	1	0	0	1	1	1	0	1	1	1
n_{12}	0	0	1	0	0	0	1	1	0	1	1	1

(a) Use the Greedy heuristic to locate the two health posts, so as to maximize the benefit of the health posts to the people, i.e., cover as many people as possible.

(b) Demonstrate the Swap technique by exchanging the facility that was located first with two other facilities, one at a time. (Choose the facilities with the smallest subscripts). What are the new coverages, and would you make either of the swaps permanent?

7.5 Other Location Problems

Solution: (a) Premultiplying the coverage matrix by the vector of weights, we obtain the number of covered customers if a facility were to be located at node n_j as:

n_1	n_2	n_3	n_4	n_5	n_6	n_7	n_8	n_9	n_{10}	n_{11}	n_{12}
260	320	450	240	270	470	500	390	240	350	390	370
						↑max					

Therefore, we locate the first facility at n_7. In the next step, we determine:

n_1	n_2	n_3	n_4	n_5	n_6	n_7	n_8	n_9	n_{10}	n_{11}	n_{12}
100	100	20	0	190	190	–	0	170	90	0	0
				↑max	↑max						

The second facility is then located at either n_5 or n_6. Consequently, the facilities should be located at n_5 and n_7 or at n_6 and n_7. The total coverage is 690.

(b) Temporarily swap n_7 and n_1. The new solution has then facilities at n_1, n_5 (or n_1, n_6) and it will cover 350 (or 550) customers, so that there will be no permanent swap.

Temporarily swap n_7 and n_2, and the new solution has facilities at n_2 and n_5 (or n_2 and n_6). It covers 410 (or 550) customers, so again, there will be no permanent swap.

Problem 3 (1-median and 1-center on a network): Industrial customers have contracted demands for heat pumps. These units are to be delivered from the warehouse of a central supplier to the companies. The supplier is now attempting to locate the warehouse, so as to minimize the transportation cost of the pumps to its customers. The demand is fairly constant throughout the year. The delivery is per pickup truck, one heat pump at a time, resulting in a linear cost function. Figure 7.11 shows the supplier's customers, their double-digit demands, and the single-digit distances between the customers.

A consultant of the supplier had suggested locating the warehouse on the link between the nodes n_2 and n_3 at a distance of 5.4 from n_2. They have based their argument on the large weights of the adjacent nodes n_2 and n_3 that provide a strong pull to locate the warehouse there.

(a) Without any calculations, do you agree with the consultant's recommendation?
(b) Find a location on the network that minimizes the total delivery cost. How much more expensive was the consultant's recommendation?
(c) Ignore now the weights at the nodes and assume that the same graph were to be used by some planner to locate a vertex 1-center. Where would this center be located?

Fig. 7.11 Network for Problem 3

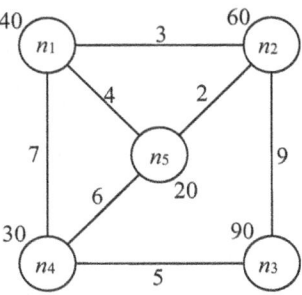

Solution:

(a) No. This is a 1-median problem and there is more to a solution than just weight.

(b)
$$\mathbf{w} = [40, 60, 90, 30, 20], \mathbf{D} = \begin{bmatrix} 0 & 3 & 12 & 7 & 4 \\ 3 & 0 & 9 & 8 & 2 \\ 12 & 9 & 0 & 5 & 11 \\ 7 & 8 & 5 & 0 & 6 \\ 4 & 2 & 11 & 6 & 0 \end{bmatrix}.$$

Then $\mathbf{wD} = [1550; 1210; 1390; 1330; 1450]$, so that node n_2 is the optimal location and the total transportation costs are 1210. The consultant's recommendation is $[40, 60, 90, 30, 20] [8.4, 5.4, 3.6, 8.6, 7.4]^T = 1390$. The costs are about 15% higher than at optimum.

(c) The question is to determine a 1-node center. The column maxima are [12, 9, 12, 8, 11] with 8 being the minimum, so that node 4 will be the optimal location.

Problem 4 (1-node center in the plane with Manhattan distances): A rural district is planning the location of its fire station. The station will have to serve seven villages. The coordinates of the villages are (0, 0), (6, 1), (2, 3), (8, 3), (0, 4), (4, 5), and (3, 7). As the people in the district are fierce defenders of property rights, the roads were constructed parallel to the rectangular fields.

(a) What is the optimal minimax location of the fire station?
(b) Suppose that it takes the fire truck and its crew 3 min to get ready after an emergency call, 2 min to drive one distance unit, and another 3 min to get the hoses and pumps going. Given the solution determined under (a), how much time elapses between the phone call and the beginning of the firefighting action at the site of the most distant customer? How long does it take on average?

7.5 Other Location Problems

Solution:

(a)
$$\alpha_1 = \max\{0, 7, 5, 11, 4, 9, 10\} = 11,$$
$$\alpha_2 = \max\{0, -5, 1, -5, 4, 1, 4\} = 4,$$
$$\alpha_3 = \max\{0, 5, -1, 5, -4, -1, -4\} = 5, \quad \text{and}$$
$$\alpha_4 = \max\{0, -7, -5, -11, -4, -9, -10\} = 0, \quad \text{so that}$$
$$\alpha_5 = \max\{11, 9\} = 11.$$

Hence the optimal coordinates of the facility are
$\bar{x} = \frac{1}{2}(5) = 2\frac{1}{2}$ and $\bar{y} = \frac{1}{2}(6) = 3$, as well as
$\bar{x} = \frac{1}{2}(7) = 3\frac{1}{2}$ and $\bar{y} = \frac{1}{2}(4) = 2$,
both with $\bar{z} = \frac{1}{2}(11) = 5\frac{1}{2}$.

(b) Worst case: $6 + 2(5\frac{1}{2}) = 17$ min. Average case = 15 min in case of the first, and 14.14 min in case of the second optimal solution.

Problem 5 (p-node center in a graph): Consider again Fig. 7.11 and ignore the weights at the nodes. For $p = 2$, determine the vertex p-center with the bisection search algorithm.
Solution:

$$\overline{D} = (n-1)\max\{d_{ij}\} = 4(9) = 36, \underline{D} = 0.$$

$\bar{d} = 18$. Set cover: $p(18) = 1$, locate anywhere. Set $\overline{D} = 18$.
$\bar{d} = 9$. Set cover: $p(9) = 1$ with facility at node n_2 or n_4. Set $\overline{D} = 9$.
$\bar{d} = 4$. Set cover: $p(4) = 3$ with facilities at nodes n_4, n_3, and one of $n_1, n_2,$ and n_5. Set $\underline{D} = 5$
$\bar{d} = 7$. Set cover: $p(7) = 2$, with facilities at, e.g., n_1 and n_3, or n_3 and n_5. Set $\overline{D} = 7$
$\bar{d} = 6$. Set cover: $p(6) = 2$, with facilities at, e.g., n_1 and either n_3 or n_4. Set $\overline{D} = 6$
$\bar{d} = 5$. Set cover: $p(5) = 2$ with facilities at e.g., n_1 and either n_3 or n_4. Set $\overline{D} = 5$. Now $\overline{D} = 5 = \underline{D}$, Stop.

Problem 6 (1-median problem in the plane with Manhattan distances): Customers are located at (0, 0), (7, 0), (6, 3), (5, 2), (5, 7), (3, 4), (6, 6), and (2, 8), respectively, with weights 20, 40, 30, 50, 10, 60, 70, and 80, respectively.
Given Manhattan distances, determine the point that minimizes the sum of distances to a single new facility. What are the total weighted costs from the new facility to the customers?

Fig. 7.12 Graph for Problem 8

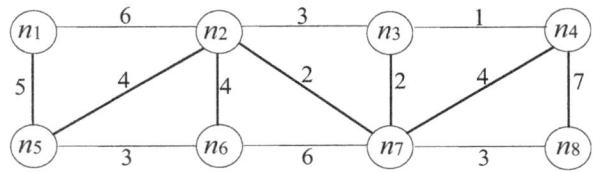

Solution: The sum of weights equals 360, so that the location of the facility should be at (5, 4), with distances to the customers of 9, 6, 2, 2, 3, 2, 3, and 7, so that the weighted sum (total costs) equals 1500.

Problem 7 (center-of-gravity and Weiszfeld method): Customers are located at (0, 0), (6, 0), and (10, 5) with weights 4, 7, and 2, respectively.

(a) Calculate the center-of-gravity, i.e., the optimal solution of the minisum location problem with squared Euclidean distances.
(b) Start with an initial guess of (1, 1) and perform one iteration with Weiszfeld's method to determine the 1-median.

Solution: (a) center-of-gravity is at (62/13, 10/13) ≈ (4.77, 0.77).
(b)

$$x = \frac{\frac{0}{\sqrt{2}} + \frac{42}{\sqrt{26}} + \frac{20}{\sqrt{97}}}{\frac{4}{\sqrt{2}} + \frac{7}{\sqrt{26}} + \frac{2}{\sqrt{97}}} = \frac{10.2676}{4.4043} = 2.3313, y = \frac{\frac{0}{\sqrt{2}} + \frac{0}{\sqrt{26}} + \frac{10}{\sqrt{97}}}{\frac{4}{\sqrt{2}} + \frac{7}{\sqrt{26}} + \frac{2}{\sqrt{97}}} = \frac{1.0153}{4.4043} = 0.2305.$$

Problem 8 (p-median with Greedy and vertex substitution heuristics): Consider the undirected graph shown in Fig. 7.12.

Assume that the customers $n_1, n_2, ..., n_8$ have demands of 38, 25, 13, 18, 15, 21, 32, and 40, respectively. Suppose now that the coverage distance is $\bar{d} = 4$.

(a) Set up the capture table and apply the *Greedy* heuristic to locate three facilities. Where should the facilities be located, and what is the total capture?
(b) Use the vertex substitution heuristic to improve the solution.

Solution:
(a)

Potential facility location	n_1	n_2	n_3	n_4	n_5	n_6	n_7	n_8
Capture	38	124	88	88	61	61	128	72

Locate one facility at n_7. The next captures are then

Potential facility location	n_1	n_2	n_3	n_4	n_5	n_6	n_8
Capture	38	36	0	0	36	36	0

7.5 Other Location Problems

Locate one facility at n_1. The next captures are then.

Potential facility location	n_2	n_3	n_4	n_5	n_6	n_8
Capture	36	0	0	36	36	0

Locate one facility at either n_2, n_5 or at n_6. Hence the solution locates facilities at either (n_7, n_1, n_2), at (n_7, n_1, n_5), or at (n_7, n_1, n_6) with a total capture of 202. The facilities capture everything, so there can be no better solution.

(b) Given the solution determined in (a), an improvement is not possible since all customers are covered by the 3 facilities. Using the vertex substitution procedure for practice anyway, starting with facilities located at, say, (n_1, n_5, n_6), we presently capture customers at nodes n_1, n_2, n_5, and n_6, i.e., a total of 99 customers. Exchanging n_1 and n_3, the new solution has facilities at (n_3, n_5, n_6) and we capture customers at nodes n_2, n_3, n_4, n_5, and n_6, i.e., a total of 92 customers. This solution is worse than its predecessor and it is thus rejected. Now exchange n_5 and n_8. The new solution locates facilities at n_1, n_6, and n_8. These facilities capture customers at nodes n_1, n_2, n_5, n_6, n_7, and n_8, i.e., a total of 171 customers. This is an improvement, and it thus becomes our new solution. The process then continues from here.

Problem 9 (p-median in a network with Greedy, location-allocation heuristic): Consider the graph in Fig. 7.13.
(a) Determine the 2-median by using the Greedy heuristic.
(b) Improve the solution found under (a) by the location-allocation heuristic.

Solution:
(a) The weights are $\mathbf{w} = [60, 20, 40, 70, 40, 10]$ and the distance matrix is

$$\mathbf{D} = \begin{bmatrix} 0 & 2 & 6 & 8 & 5 & 4 \\ 2 & 0 & 4 & 10 & 7 & 3 \\ 6 & 4 & 0 & 8 & 11 & 2 \\ 8 & 10 & 8 & 0 & 3 & 7 \\ 5 & 7 & 11 & 3 & 0 & 9 \\ 4 & 3 & 2 & 7 & 9 & 0 \end{bmatrix}.$$

Locate the first facility: $\mathbf{wD} = [1080, 1290, 1460, 1190, 1180, 1230]$, so that the first facility is located at n_1 at a minimal total cost of 1080.

For the second facility, we obtain costs of $[-, 950, 820, 440, 530, 810]$, so that the second facility will be located at n_4 for a total z-value of 440.

(b) Allocation: To the facility at n_1, we allocated customers at nodes n_1, n_2, n_3, and n_6, and to the facility at n_4, we allocate the customers at n_4 and n_5. The optimization on the two problems then results in cost vectors [320, 310, 460, 380] with the minimum occurring at n_2, and [120, 210] with the minimum at n_4. The next step then allocates to the facility at n_2 the customers at n_1, n_2, n_3, and n_5, to the facility at n_4, the customers at n_4 and n_5, which is the same as before. Hence the

Fig. 7.13 Graph for Problem 9

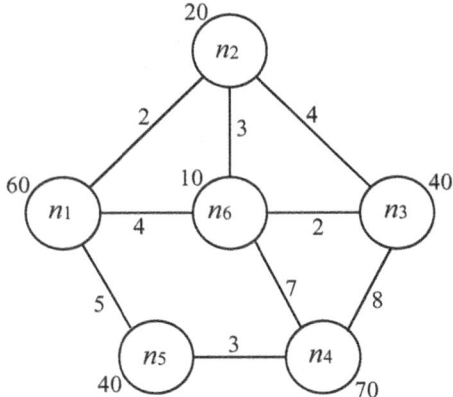

method has converged with an optimal solution with facilities at n_2 and n_4 with a total cost of 310 + 120 = 430.

References

Daskin MS (2013) Network and discrete location: models, algorithms, and applications, 2nd edn. Wiley, Hoboken, NJ

Eiselt HA, Marianov V (eds) (2011) Foundations of location analysis. Springer, New York

Eiselt HA, Sandblom C-L (2004) Decision analysis, location models, and scheduling problems. Springer, Berlin

Hakimi S (1964) Optimum location of switching centers and the absolute centers and medians of a graph. Oper Res 12:450–459

Laporte G, Nickel S, Saldanha da Gama F (eds) (2019) Location science, 2nd edn. Springer Nature, Cham, Switzerland

Project Networks   8

Back in the days when projects were dealt with by a single individual or a group of workers working sequentially, there was no need for project networks. As an example, consider the construction of a house. Somewhat simplistically, assume that a single individual wants to build a log cabin. He will first dig a hole in the ground for the foundation, then pour the cement, then lay the logs one by one, and so forth. Each job is completely finished before the next task begins. This is a sequential plan, and there is very little that can be done as far as planning is concerned. Consider, however, some of the issues that have arisen as a result of the division of labor. Nowadays, the plumbers can work at the same time the electrician does, but not before the walls have been established, which is also required for the roof to be put up. Given these interdependencies, planning is necessary in case time is an issue. Clearly, while it is *possible* that, say, electricians and plumbers can work in the building at the same time, it is *not necessary* to use this parallelism: we can still have the two contractors work one after the other if we so wish. The project will take longer, but it is possible.

Project networks were designed by a number of firms in the 1950s. The so-called *critical path method* (or *CPM* for short) was developed by du Pont de Nemours and the Remington Rand Univac corporations for construction projects. At roughly the same time, the US Navy, in conjunction with the Lockheed Aircraft Corporation and the consulting firm of Booz, Allen, and Hamilton devised the *Program Evaluation and Review Technique* (*PERT*) for their Polaris missile program. Even though *CPM* and *PERT* have completely independent backgrounds, today, we can consider them very close brothers: their underlying ideas are identical, the resulting networks are identical, and the only difference is that *CPM* is a deterministic technique, *PERT* is (partially) stochastic.

This chapter is organized as follows. The first section will introduce the elements of the critical path method, demonstrate its graphical representation, and describe basic planning with the critical path method. The remaining sections of this chapter deal with extensions of the basic concept: the second section allows the acceleration of the project (a process that introduces costs into the model), the third section allows resources to be used (which, with the obvious limitations, results in an optimization model that allocates the scarce resources), and the final section of this chapter discusses the probabilistic *PERT* method.

8.1 The Critical Path Method

Before getting into details, the planner will have to decide on what level the planning will take place. We may be interested in planning on the macro level in order to get the "bigger picture." As an example, when building a house, we may have "foundation work" as a single task, another one is "electrical work," another is "plumbing," another is "roofing," and so forth. Zooming in closer, we may look at each such task as a project in itself. For instance, plumbing may include subtasks such as the installation of pipes, connections to outside lines or septic systems, etc. Zooming in even further, the installation of lavatories may be further subdivided into the mounting of the washbasin, the connection of the faucet to the pipes, and so forth. We can construct project networks on each of these levels.

Once the level has been decided upon, the task in its entirety—the building of the house, design and planting of a public or private garden, an individual's university studies—will be referred to as a *project*. Each project can now be subdivided into individual *tasks* or *activities*. Planting a bed of grape hyacinths, installing the flashing on the roof, or taking a specific course in a university program are typical examples of such activities. Associated with each activity is a *duration* that indicates how long it takes to complete the activity. In *CPM*, the durations of all activities are assumed to be known with certainty (which is the only distinction to *PERT*, where the durations have underlying probability distributions).

In addition to the activities and their durations, we also need to have *precedence relations*. Such relations indicate which activities must be completely finished before another activity can take place. For example, in order to start the activity "drive with the car to grandmother," the activities "gas up the car," "pack the gifts (wine and cheese) for grandmother," and "lock the house" must be completely finished. A complete set of precedence relations will specify the (immediate) predecessors of each activity.

In the analysis in this section, we are only concerned about time. There is no optimization that takes place here, the goal is to determine the earliest time by which the project can be completed. And while making these calculations, we can also find out when each task can and must be started and finished. That allows us to determine bottleneck activities in the project, whose delay will delay the entire project. In subsequent sections, we will include other components in the basic network, including money and other resources.

One of the huge advantages of project networks and one of the reasons for their popularity among users is the ease with which they can be understood and visualized. Following the old adage that "a manager would much rather live with a problem he cannot solve than with a solution he cannot understand," many managers have adopted project networks as a standard tool in their toolkit. This was helped

8.1 The Critical Path Method

Table 8.1 Precedence relations for sample network

Activity	Immediate predecessor	Duration (in weeks)
A	–	5
B	A	3
C	A, B	7
D	B	4
E	B, C	6
F	C, D, E	4
G	D	2
H	F, G	9
I	F, G	6
J	I	2

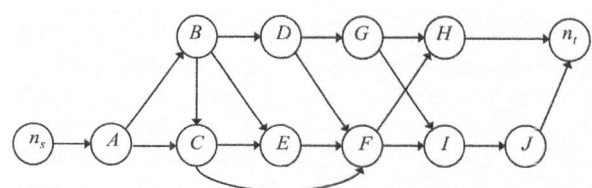

Fig. 8.1 Activity-on-node project network

tremendously by a change in the representation that was made some time during the 1990s. Traditionally, each activity was represented by an arc in a network, while nodes represented events (which are no more than the beginning and/or the end of activities). Such a representation is referred to as an *activity-on-arc (AOA) representation*. While *AOA* networks have some advantages, they are rather difficult to construct and may require a lot of artificial activities, called *dummy activities*. A more modern way to visualize relations between activities is the *activity-on-node (AON) representation* which we will use in this chapter. In it, each node represents an activity, while the arcs represent the precedence relations. In addition to the given activities, each project network contains two artificial nodes n_s and n_t, which are the starting node (the *source*) and the terminal node (the *sink*), which symbolize the beginning and the end of the project. The project start has arcs leading out of it to all activities that do not have any predecessors, while all nodes that represent activities without successors will have arcs leading directly to the terminal node "Project End." For all other activities j, we introduce an arc from node i to node j, if activity i is an immediate predecessor of j.

As a numerical illustration, consider the precedence relations shown in Table 8.1.

The *AON* network that includes all of the activities and precedence relations is shown in Fig. 8.1. Like all *AON* project networks, it has a unique start n_s, a unique end n_t, all arcs are directed, and no cycles can possibly exist, as they would require an activity to be completely finished before it can actually start, which is obviously impossible.

As an example for (nested) precedence relations, consider activity F. It is clear from the project network that the activities C, D, and E must be completely finished before activity F can commence. However, activity C requires that A and B are completely finished, activity D requires that B is finished, and activity E requires that B and C are completely finished. So, in reality activity F cannot commence before the activities A, B, C, D, and E are all finished. This does not present a difficulty, but users should be aware of these implicit precedence relations. Incidentally, while it is not necessary to specify that activity C is a direct predecessor of F since E must precede F anyway, and C precedes E, its inclusion does not introduce an error.

We are now interested in finding the earliest possible time at which the entire project can be completed. For computational convenience, we assume that the project start occurs at time $t = 0$ and the end of the project occurs at time T, which will be determined in the process. Consider now one of the many (14 in this case, to be precise) paths of the network that lead from n_s to n_t, say $n_s - A - C - E - F - H - n_t$. Starting at the end and moving back towards the beginning one step at a time, we notice that n_t can only finish if H is completely finished, which, in turn, requires that F is completely done, which, in turn, requires that E is fully done, etc. The length of any path from source to sink is defined as the sum of durations of all of its activities. In this example, the path we are presently looking at has the length of $5 + 7 + 6 + 4 + 9 = 31$. This also means that while there are other precedence relations, the ones on this path result in a project duration of 31, which is thus a lower bound on the project duration. From this, we can conclude that the project duration is equal to the length of the longest path in the network.

While it would certainly be possible to determine the longest path in a network by enumeration (even though there will be a lot of paths), a different type of computation is preferred as it will provide decision makers with additional and very valuable information. This procedure will be described in the following.

The first part of the procedure uses what is commonly known as a *forward pass*, a *forward sweep*, or a *forward recursion*. In the forward recursion, we compute the *earliest possible starting times* (*ES*) of all activities, as well as their *earliest possible finishing times* (*EF*). We start labeling the nodes in the forward pass with the source node. The earliest starting time of the source node is arbitrarily set to $ES(n_s) = 0$. For any node n_i the earliest possible finishing time EF equals the earliest possible starting time ES plus the duration of the activity n_i, which we will denote by d_i. For the source, this results in $EF(n_s) = 0$. In order to continue labeling (which in the forward pass means assigning values ES and EF to a node), we will use the following

Rule 1: In the forward pass, a node can only be labeled if all of its predecessors have been labeled.

If more than one node satisfies the condition in Rule 1, labeling continues with any of the nodes that satisfy the condition. The label ES of a node is then the maximum among *all EF* labels of its direct predecessors. The reason for this calculation is easily explained. Suppose that an activity has three direct predecessors with the earliest possible finishing times of 4, 7, and 9, respectively. It is apparent that we cannot start before 4, as none of the predecessors has been finished. Between time 4 and 7, the first predecessor is finished, but we will have to wait until all predecessors have been finished, which is the case at time 9.

8.1 The Critical Path Method

Table 8.2 The earliest possible starting and finishing times

Activity	n_s	A	B	C	D	E	F	G	H	I	J	n_t
Earliest possible starting time ES	0	0	5	8	8	15	21	12	25	25	31	34
Earliest possible finishing time EF	0	5	8	15	12	21	25	14	34	31	33	34

Applying this rule to our example means that, once n_s has been labeled, we can now label node A. Node A has only one predecessor, so that $ES(A) = EF(n_s) = 0$. Given that, we can calculate $EF(A) = ES(A) + d_A = 0 + 5 = 5$. Now that node A is labeled, we can continue labeling with nodes B and C. Arbitrarily choose node B. Activity B also has only one predecessor, which is A. Consequently, $ES(B) = EF(A) = 5$, and $EF(B) = ES(B) + d_B = 5 + 3 = 8$. Now that activity B has been labeled, we can continue labeling node C. As activity C has activities A and B as predecessors, we have $ES(C) = \max\{(EF(A), EF(B))\} = \max\{5, 8\} = 8$, so that $EF(C) = ES(C) + d_C = 8 + 7 = 15$. Labeling can now continue with nodes D and E. The results of the forward labeling phase are summarized in Table 8.2.

At this point, the sink n_t has been labeled, and the ES and EF labels of the sink (which are necessary equal, as the artificial activity n_t has zero duration), indicate the minimal total duration of the project. In other words, we now know that the project can be completed at the earliest in $T = 34$ weeks. This terminates the forward sweep.

In the backward sweep, we will compute the latest allowable finishing times LF and the latest allowable starting times LS of all activities (in that order). By "latest allowable," we refer to the time we can start or finish an activity, given that the project has to be completed by time T (in our example $T = 34$). While the forward sweep started with the source being labeled first, the backward sweep commences by labeling the sink. Now we consider a node labeled, if we have computed its LF and LS values. As the project duration T is now known, we will label the sink node $LF(n_t) = LS(n_t) = T$. The rule for labeling other nodes is now.

Rule 2: In the backward pass, a node can only be labeled if all of its successors have been labeled.

With the sink as the only labeled node, only the activities H and J can be labeled. Arbitrarily choose node H. This node has only one successor, so that $LF(H) = LS(n_t) = 34$. Given that activity H must be finished no later than 34 and its duration is $d_H = 9$, we determine that $LS(H) = LF(H) - d_H = 34 - 9 = 25$. The process for node J is similar, and we obtain $LF(J) = LS(n_t) = 34$, and $LS(J) = LF(J) - d_J = 34 - 2 = 32$. At this point, the nodes n_t, H and J are labeled, so that we can continue labeling only with node I (as node G has node I as a successor, which is not yet labeled). Node I has again only one successor, so that its label can easily be calculated as $LF(I) = LS(J) = 32$ and $LS(I) = LF(I) - d_I = 32 - 6 = 26$. At this point, we can continue labeling the nodes G and F. Arbitrarily choose F. Node F has activities H and I as successors. In order to determine its latest finishing time, consider this. As we have just computed, its two successors H and I cannot start any later than 25 and 26, respectively. If activity F were to finish any later than, say,

Table 8.3 The latest allowable starting and finishing times

Activity	n_s	A	B	C	D	E	F	G	H	I	J	n_t
Latest allowable starting time LS	0	0	5	8	17	15	21	23	25	26	32	34
Latest allowable finishing time LF	0	5	8	15	21	21	25	25	34	32	34	34

25½, activity H could not start on time. In other words, in order to avoid delaying the entire project it is necessary that *all* activities can start on time, so that $LF(F) = \min\{LS(H), LS(I)\} = \min\{25, 26\} = 25$. The resulting latest allowable starting time is then $LS(F) = LF(F) - d_F = 25 - 4 = 21$. The next node to be labeled is G. Since its successors are I and H, we have $LF(G) = \min\{LS(I), LS(H)\} = \min\{26, 25\} = 25$, and $LS(G) = LF(G) - d_G = 25 - 2 = 23$. The procedure continues in this fashion until the source is labeled. The labels of all nodes are shown in Table 8.3.

A good test for correctness is to examine $LF(n_s)$ in the backward sweep, this value must be zero. If it is not, an error has been made. However, the converse is not true: even if $LF(n_s) = 0$, it does not mean that the backward recursion has been done correctly. Furthermore, we realize that for every activity we have $LF - LS = EF - ES$.

Now that all earliest possible and latest allowable starting and finishing times have been computed, we are able to determine which of the activities are critical when scheduled and which are not. As an example, consider activity G. So far, we have determined that the earliest possible time that we can schedule the activity is at $ES(G) = 12$, while the latest possible finishing time of G is $LF(G) = 25$. This gives us a time window of $25 - 12 = 13$ weeks during which the activity has to be scheduled. That is not a problem, since the duration of the activity is only two weeks, so that there is plenty of leeway. An expression of the magnitude of this leeway is the *total float* of the activity G, which we will abbreviate here as $TF(G)$. The total float of an activity is the magnitude of the time window for its schedule minus the duration of the activity. Formally for activity G, we have $TF(G) = LF(G) - ES(G) - d_G = 25 - 12 - 2 = 11$. Decision makers can use the information provided by the total float as the amount of time by which the duration of an activity can increase without delaying the entire project. Thus, activities with a large float are safe in that their duration can increase significantly without delaying the entire project. On the other hand, an activity with a large float indicates that there might be too many resources allocated to this activity. Redirecting some of these resources elsewhere may result in possible decreases of the durations of other activities.

As another example, consider activity H. Its time window ranges from $ES(H) = 25$ to $LF(H) = 34$. With its duration of $d_H = 9$, we can compute the activity's total float as $TF(H) = LF(H) - ES(H) - d_H = 34 - 25 - 9 = 0$. This indicates that there is absolutely no leeway when scheduling activity H. Activities that have no leeway in the schedule are referred to as *critical activities*. A critical activity has the property

8.1 The Critical Path Method

Fig. 8.2 Node representation by a box

$ES(D)$	d_D	$EF(D)$
	D	$TF(D)$
$LS(D)$	d_D	$LF(D)$

that as soon as its duration increases, the entire project will be delayed, regardless of how small the time increase actually is.

Critical activities are very similar to those resources in optimization problems whose constraints are satisfied as equations at optimum and thus represent bottlenecks in the problem. On the other hand, noncritical activities in project networks can be compared to resources or constraints that have positive slacks or excess variables at optimum. Note that for the ease of computations, we can calculate the total float also as $TF = LF - EF = LS - ES$.

In order to have all information available at a glance, it is useful to draw the network in a slightly different way. Rather than representing each node by a circle with its name in it, we suggest representing a node by a box with nine different fields, as shown in Fig. 8.2. We will refer to the different fields by the geographic directions they are found in, e.g., the field in the north, the southwest, etc. The center of the node is reserved for the name of the activity (here activity D). The fields in the north and in the south both show the duration d_D of the activity.

Recall that during the forward sweep, we compute the earliest possible starting time $ES(D)$ and the earliest possible finishing time $EF(D)$ of the activity under consideration. They are found in the northwest and the northeastern fields, respectively. In the backward sweep, we determine the latest allowable starting time $LS(D)$ and the latest allowable finishing time $LF(D)$ of the activity. This information is put into the fields in the southwest and southeast, respectively. Finally, the field in the east will include the total float TF, which is computed after the forward and backward passes have been completed. The field in the west will remain empty for now. It is designed for the resource consumption of the activity, which is not considered in the basic model.

The three fields in the top row of each node read from left to right symbolize the relation $ES + d = EF$, while the three fields at the bottom of a node read from right to left show the relation $LF - d = LS$. This is why we have chosen to include the duration of the activity twice, once in the field in the north and again in the south.

Rather than using tables for the display of the information, we can work directly on the graph, which is much easier, as it provides all required information at a glance. The project network for our example is shown in Fig. 8.3.

Having all of this information at hand, we are now able to determine what gave the method its name, *viz.*, the *critical path*. Formally, the critical path is a path from

Fig. 8.3 Project network, for example, in text

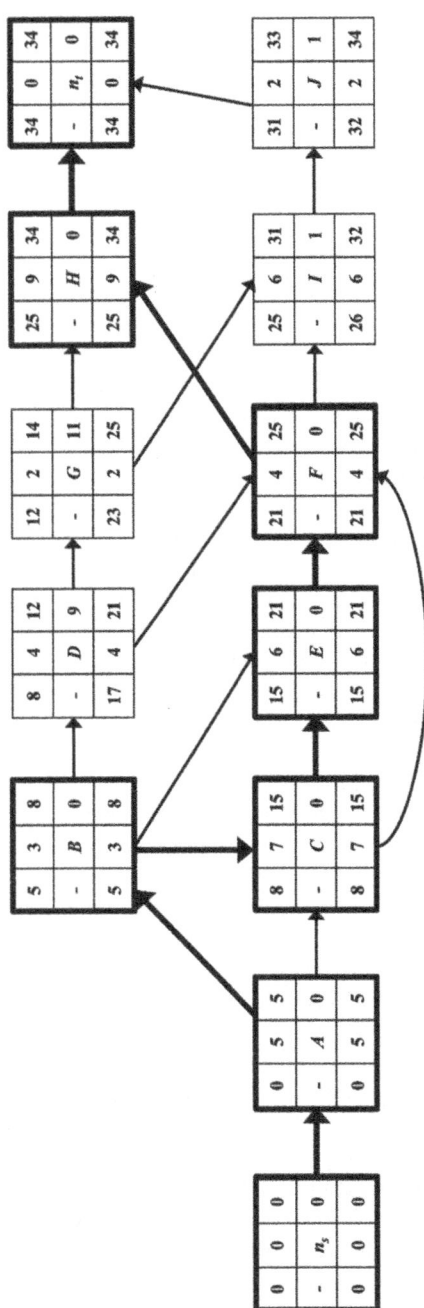

8.1 The Critical Path Method

the source n_s to the sink n_t that includes only critical activities, such that node j can directly succeed node i on the critical path, only if $ES(j) = LF(i)$. The critical path in our example is shown in bold lines and it includes the activities $n_s - A - B - C - E - F - H - n_t$. Clearly, the length of that path, obtained by adding its activity durations, equals $T = 34$. Note that the critical path does *not* include the link from node B to node E, even though E directly follows B, but $ES(E) = 15 > 8 = LF(B)$. This clearly indicates that it is not sufficient to simply connect all neighboring nodes that have zero total float. Furthermore, it may happen that a project network has multiple critical paths. The next section will have examples of that case.

We conclude this section by summarizing the procedure that determines the critical path:

1. Use the forward pass to calculate the earliest possible starting and finishing times of all activities.
2. Use the backward pass to calculate the latest allowable starting and finishing times of all activities.
3. Calculate the total floats of all activities.
4. Determine the critical path.

While the above procedure is highly visual, it is also possible to formulate the problem as an integer programming problem and solve it with any of the pertinent methods. In order to do so, define zero-one variables x_{ij}, which assume a value of 1, if the activities n_i and n_j (in that order) are located on the critical path, and 0 if they are not. Recall that the critical path is the longest path in the network, we can then formulate the objective function as the sum of the products of the durations d_j and the variables x_{ij}, while the constraints are the usual node equations, along with a constraint that requires that exactly one unit leaves the source.

In our example, we obtain the following formulation:

P : Max $z = 5x_{sA} + 3x_{AB} + 7x_{AC} + 7x_{BC} + 4x_{BD} + 6x_{BE} + 6x_{CE} + 4x_{CF} + 4x_{DF} + 2x_{DG} + 4x_{EF} + 9x_{FH} + 6x_{FI} + 9x_{GH} + 6x_{GI} + 2x_{IJ}$

s.t. $x_{sA} = 1$

$x_{sA} = x_{AB} + x_{AC}$

$x_{AB} = x_{BC} + x_{BD} + x_{BE}$

$x_{AC} + x_{BC} = x_{CE} + x_{CF}$

$x_{BD} = x_{DG} + x_{DF}$

$x_{BE} + x_{CE} = x_{EF}$

$x_{CF} + x_{DF} + x_{EF} = x_{FH} + x_{FI}$

$x_{DG} = x_{GH} + x_{GI}$

$x_{GH} + x_{FH} = x_{Ht}$

$x_{GI} + x_{FI} = x_{IJ}$

$x_{IJ} = x_{Jt}$

$x_{ij} = 0$ or 1 for all i, j.

The solution indicates that $x_{sA} = x_{AB} = x_{BC} = x_{CE} = x_{EF} = x_{FH} = x_{Ht} = 1$ and $x_{ij} = 0$ otherwise. This is the same critical path found earlier with the graphical method.

8.2 Project Acceleration

So far, we have considered time as the only criterion in project networks. Also note that the technique described in the previous section did not involve any optimization, all we have done is determined when the project can be finished by finding the length of the longest path and which of the activities are bottlenecks in the system. In this section, we will return to the basic model, but allow the possibility to accelerate, at a cost, individual activities, so as to be able to finish the project earlier. The result will be a list that shows possible finishing times of the project and the amounts that will have to be paid to reach them. This will enable the planner to decide what combination of money spent and project duration best fits the specific situation.

In order to describe the situation, consider a single activity. As before, the activity will have what we now call a *normal duration*. Since we will engage in a marginal analysis, the cost of the activity at its normal duration is immaterial (we will have to engage in the activity in any case), and the only costs we consider are those that are incurred due to the acceleration of the activity. Suppose that the normal duration of our activity is 7 hrs. It is now possible to use more resources (e.g., more manpower, more tools, contracting out part of the activity, or any similar measure) to accelerate this activity. Suppose that it costs \$20 to reduce the duration of the activity to 6 hrs. Using more resources still, additional money can reduce the duration of the activity further. For simplicity, we assume that the cost function of the acceleration is linear, meaning that reducing the duration by another hour to 5 hrs costs another \$20 for a total of \$40. Note that normally the cost function is superlinear, meaning that reducing the duration by 1 hr costs, say \$$x$, reducing it by another hour costs more than \$$x$, yet another reduction is more expensive still, and so forth.

It is quite apparent that the reduction has some limitations, below which we cannot reduce the duration of the activity any further. The shortest activity duration of an activity that can be achieved is customarily referred to as *crash time*, and the process of acceleration is sometimes called *crashing*. Our task is now to determine which activities should be accelerated or crashed, so as to achieve the desired result at the lowest possible cost.

As an illustration of the concept, consider the numerical example shown in Fig. 8.4.

The project has four activities A, B, C, and D, whose normal times, crash times, and unit acceleration costs are shown next to the nodes. For example, activity D normally takes 8 hrs (there are no costs incurred at this duration), but we can reduce the duration down to 7, 6, or 5 hrs. Each hour of acceleration costs \$200. Note that activity B cannot be accelerated.

8.2 Project Acceleration

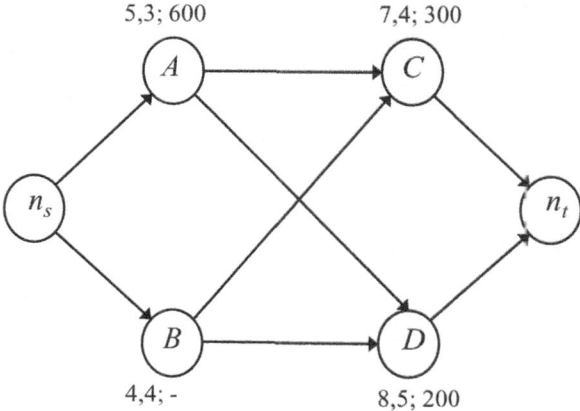

Fig. 8.4 Project network for acceleration example

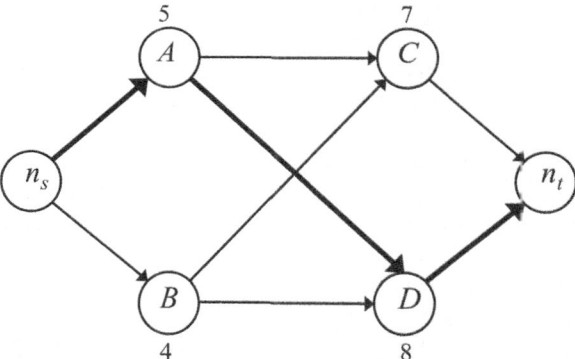

Fig. 8.5 Critical path with normal activity durations

In the following, we first describe a procedure based on this network, as it provides some insight so as to what happens during the process.

Figure 8.5 shows the project network under consideration along with the normal durations of the activities (in hrs) and the critical path, which is shown in bold arcs. (Note that we normally have to use the forward sweep/backward sweep procedure in each step, but since the project network here is so small, we may enumerate the four paths $A-C, A-D, B-C$, and $B-D$, determine their respective lengths and choose the longest path; it is the critical path). The present duration of the project is 13 hrs.

We now have to determine which activities to accelerate. Recall that the project duration is determined by the length of the longest path in the network. This means that as long as we are not accelerating an activity on the longest, i.e., the critical path, the project duration will not be reduced. This leads to the important realization that we must accelerate an activity on the critical path. And, among those activities, we will choose the one that minimizes our marginal, i.e., additional costs. In our example, we have a choice between either accelerating activity A at a cost of \$600, or activity D at a cost of \$200. Since it is less expensive to accelerate activity

Fig. 8.6 Critical paths with activity D accelerated

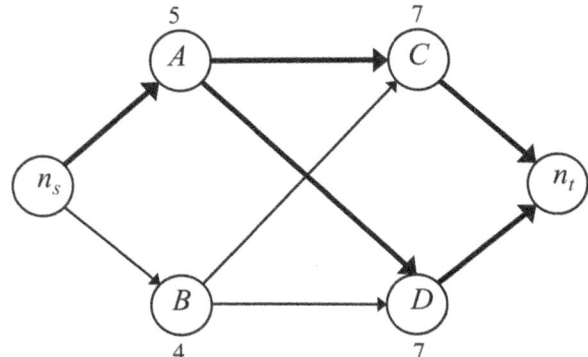

D, we reduce its duration by a single hr to 7 hrs. We now have a new network (even though the network structure has not changed and never will during the computations, only one activity duration has changed), and we have to determine the critical path and the project duration anew. The result is shown in Fig. 8.6.

We notice that there are now two critical, i.e., longest paths in the network. They are n_s–A–C–n_t and n_s–A–D–n_t, both having a length of 12. Accelerating the project further will pose some additional difficulties. In order to demonstrate these problems, suppose that we were to again accelerate activity D. This would cause the path n_s–A–D–n_t to be only 11 hrs long, while the path n_s–A–C–n_t would still be 12 hrs long and as such, would now be the only critical path. In other words, we would have spent another $200 and still would have to wait 12 hrs to finish the project. This means that we have to refine the rule somewhat to indicate which activities must be accelerated in order to speed up the project. In fact, we will have to accelerate a set of activities, so that at least one of the activities in this set is on each of the critical paths. In our small network, it is possible to examine the network and enumerate the possibilities. In the network in Fig. 7.6, we can either accelerate activity A (at a cost of $600), or the activities C and D (at a cost of $300 + $200 = $500). Before making the actual decision, we have to ascertain that all of these accelerations are actually possible, i.e., that the present durations are all above the crash times. In our case, the activities A, C, and D have present durations of 5, 7, and 7, while their crash times are 3, 4, and 5, so that all activities can actually be accelerated. Since the cheapest option is to accelerate activities C and D, we accelerate each of these activities by one unit each. Based on the new activity durations, we also determine the new critical path(s). The results are shown in Fig. 8.7.

If any further acceleration is required, the options are the same as before. Checking the possibility to accelerate, we find that the present activity durations of the nodes on the critical path A, C, and D are 5, 6, and 6, while their crash times are 3, 4, and 5, so durations of all of these activities can be reduced further. The least expensive option again involves accelerating activities C and D by 1 hr each, resulting in the situation shown in Fig. 8.8.

8.2 Project Acceleration

Fig. 8.7 Critical paths with activities C and D accelerated

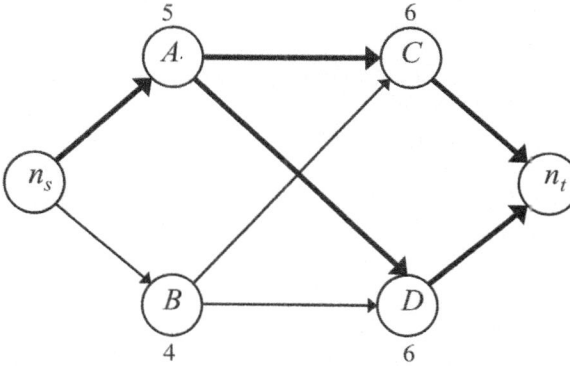

Fig. 8.8 Critical paths with activities C and D accelerated

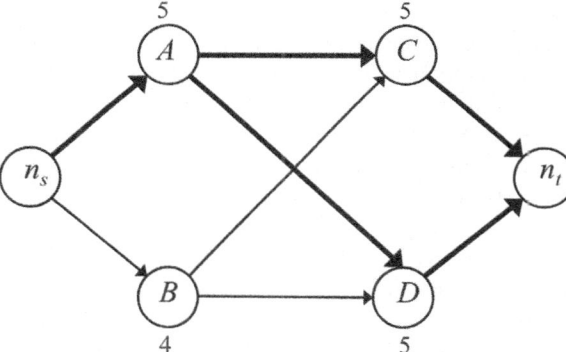

Fig. 8.9 Critical paths with activity A accelerated

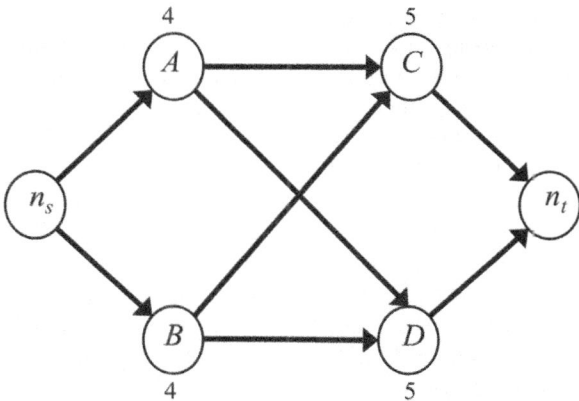

While the critical paths are still the same, the situation has changed. Activity D has now reached its crash time and no longer can be accelerated. This means that the only way to accelerate both critical paths simultaneously is to speed up activity A at a cost of $600. The resulting situation is shown in Fig. 8.9.

Table 8.4 Summary of the project acceleration process

Accelerate activity	Total acceleration costs ($)	Critical path	Project duration (hrs)
–	0	A–D	13
D	0 + 200 = 200	A–C and A–D	12
C and D	200 + 500 = 700	A–C and A–D	11
C and D	700 + 500 = 1200	A–C and A–D	10
C and D	1200 + 600 = 1800	A–C, A–D, B–C, and A–D	9

Notice that at this point in time, all four paths from the source to the sink are critical. This means that the next acceleration will have to ensure that at least one activity on each path in the network is accelerated. However, in this case, at least one path, viz., the path n_s–B–D–n_t can no longer be accelerated at all: both activities B and D are at their respective crash times. This means that the shortest possible project duration is $T = 9$ and it can be achieved at a cost of $1800. The results achieved in the above computations are shown in Table 8.4.

This is where the decision maker comes in. He can now determine what it costs to accelerate the project and whether or not it is worth it. This is a good example of what operations research does best: prepare alternative decisions (rather than actually make them).

A few concluding comments are in order. It became clear that as the process moved forward, more and more paths became critical, implying that more and more activities had to be accelerated in order to reduce the project duration further. This does, of course, imply increasing costs from each unit of acceleration. This was to be expected, of course: at first, it is easy to accelerate a project, but as the timeframe becomes tighter and tighter, the costs skyrocket.

If it appears too cumbersome to go through the entire process, a quick idea of how short the project duration can actually be is to crash all activities and determine the critical path on that basis. While this will certainly result in the shortest possible project duration, it is usually not necessary to crash all activities to reach the same overall duration. A good illustration is the above example. The optimized durations of 4, 5, 4, and 5 of the activities A, B, C, and D were sufficient to reduce the project duration to 9, while the crash durations of the activities are 3, 4, 4, and 5. Their use would result in the same overall project duration of 9.

Finally, we will delineate a linear programming formulation that finds the results for arbitrary project durations. To formulate the problem, define variables d_j and t_j with d_j denoting the actual duration of activity j, and t_j symbolizing t_j as the earliest possible finishing times of node/activity j for all activities including the source and the sink. The objective then is the sum of products of the unit acceleration costs and the number of units by which an activity has been accelerated, viz., the difference between normal time and the actual time d_j. Then there are four types of constraints. First, there is the single constraint t_t equals the desired project duration. Secondly, all activities without predecessors have their variables t_j set equal to their actual durations. The third set of constraints has a constraint $t_j \geq t_i + d_j$ for each activity j (including the sink t), for which a predecessor i has been specified. The fourth and last set of constraints requires that the activity durations each are between crash time and normal time. In our example, we formulate:

$$P : \text{Min } z = 600(5 - d_A) + 300(7 - d_C) + 200(8 - d_D)$$

$$\begin{aligned}
\text{s.t. } & t_t = 13 & & t_t \geq t_D \\
& t_A = d_A & & d_A \geq 3 \\
& t_B = 4 & & d_A \leq 5 \\
& t_C \geq t_A + d_C & & d_C \geq 4 \\
& t_C \geq t_B + d_C & & d_C \leq 7 \\
& t_D \geq t_A + d_D & & d_D \geq 5 \\
& t_D \geq t_B + d_D & & d_D \leq 8 \\
& t_t \geq t_C & & d_j, t_j \geq 0 \text{ for all } j.
\end{aligned}$$

By successively changing the value of t_t in the first constraint, we can (re-)create the results in Table 8.4.

8.3 Project Planning with Resources

So far, our discussion has focused on time planning. In the process, we have assumed that sufficient resources are available to perform the activities in the time specified for the individual activities. In this section, we will extend the basic model by adding a resource requirement. For simplicity, we will use only a single resource, such as manpower, backhoes, machinery, or any other resource relevant to the project. For simplicity, we will refer to the resource as employees throughout this section. When we associate a resource requirement of, say, 30 units to an activity, we mean that 30 employees are required throughout the duration of the activity. As an illustration, consider again the project network in Fig. 8.3 in Sect. 8.1. Furthermore, assume that the resource consumptions of the individual activities are as shown in Table 8.5.

In order to schedule the activities of the project network, it is useful to employ a so-called *Gantt chart*. In essence, it is a horizontal bar chart, which features the individual activities on the ordinate, while the abscissa is a time axis. Clearly, the activities on the critical path are scheduled from their earliest possible (or, equally,

Table 8.5 Resource consumption of the activities in the example

Activity	A	B	C	D	E	F	G	H	I	J
Employees required	10	20	40	20	25	30	10	25	20	25

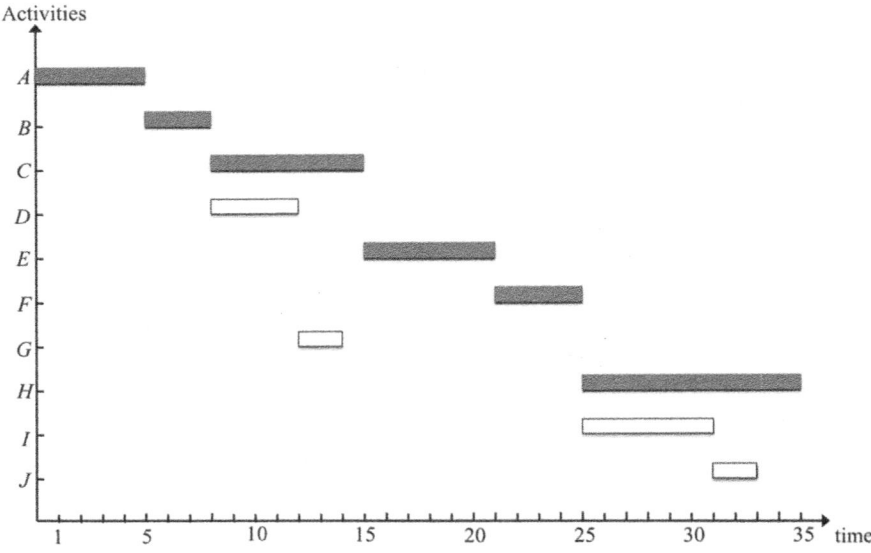

Fig. 8.10 Gantt chart

latest allowable) start times, so that they form a non-overlapping sequence of bars that has no gaps. The bold bars in Fig. 8.10 belong to critical activities and their position in the graph cannot be changed.

The matter is different with the noncritical activities, which have some leeway for their schedule. Suppose now that we use a heuristic method for scheduling them, which includes a rule that states that all noncritical activities should be scheduled as early as possible. The regular bars that belong to the activities D, G, I, and J in Fig. 8.10 show how these activities are scheduled. This schedule now has very clear resource implications. From time $t = 0$ to $t = 5$, we only perform activity A, so that we need 10 employees. From $t = 5$ to $t = 8$, we perform only activity B, which requires 20 employees. Starting at $t = 8$ to time $t = 12$, we perform the activities C and D simultaneously. This requires $40 + 20 = 60$ employees. At time $t = 12$, activity D is finished, while C is still going on. However, at $t = 12$, activity G is also scheduled, so that 40 employees for C and 10 employees for activity G are needed. This process continues until the project is finished. The resource requirements are shown in the *resource requirement graph* in Fig. 8.11.

It is apparent that the resource requirement is very low in the beginning, then peaks, drops and increases again towards the end. If we were to be able to employ only casual labor that we need to pay only when needed, then the total resource requirement is the shaded area in Fig. 8.11. Calculating the size of the area from the beginning, we have 10 employees needed for 5 weeks, 20 employees needed for 3 weeks, 60 employees needed for 4 weeks, and so forth, for a total of 1155 employee weeks. Assuming that we pay employees \$15 per hour for 8 hrs a day and 5 days a week, each employee will cost us \$600 (plus fringe benefits, which we

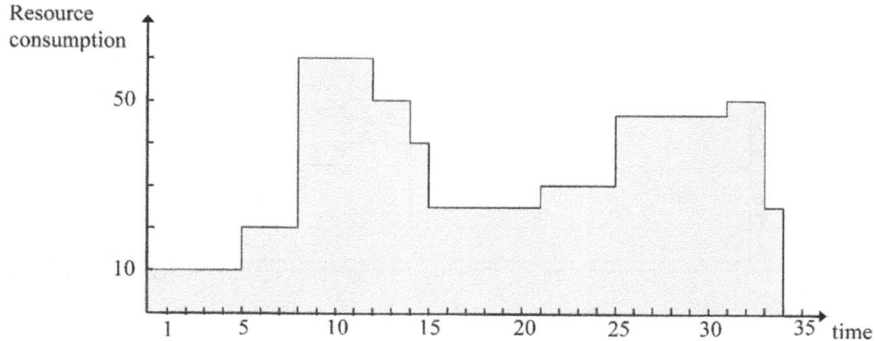

Fig. 8.11 Resource requirement graph

ignore here for simplicity). This means that the resource costs will be $693,000 for the entire project. Note that these costs for casual labor will remain the same, regardless of how we schedule.

The situation changes dramatically if we have to hire all needed employees for the entire duration of the project. As the highest manpower requirement at any point in time is 60, this is the smallest number of permanent employees required for the duration of the project. Employing 60 employees for the total of 34 weeks at a cost of $600 per week costs $1,224,000, more than 76% more than the costs for casual labor. This is caused by the fact that more than 43% of the time the employees are paid, they are actually idle.

This calls for a different schedule whose maximal resource requirement is as low as possible. This type of objective is of the minimax type, where we search to minimize the maximum resource required at any point in time. Rather than using the heuristic that schedules all activities as early as possible (not a bad choice in general, as it allows for some noncritical activities to increase in duration without jeopardizing the finishing of the project on time), we will use another heuristic that schedules all activities as late as possible. The Gantt diagram that belongs to that schedule and the associated resource consumption graph are shown in Fig. 8.12.

It turns out that while this schedule uses the exact same number of employee weeks of casual labor—1155—the highest employee requirement at any one point is only 50. This means that the costs for employees working throughout the project is $1,020,000, which is 16.67% less than with the "earliest possible" schedule. The idle time here is still 32%, though.

Other heuristics exist for the scheduling of noncritical activities. Depending on the problem, they may be able to reduce the number of required resources further. The problem can also be solved by exact methods, but the integer programming problem that must be formulated for that purpose is quite difficult. For details, see, e.g., Eiselt and Sandblom (2004).

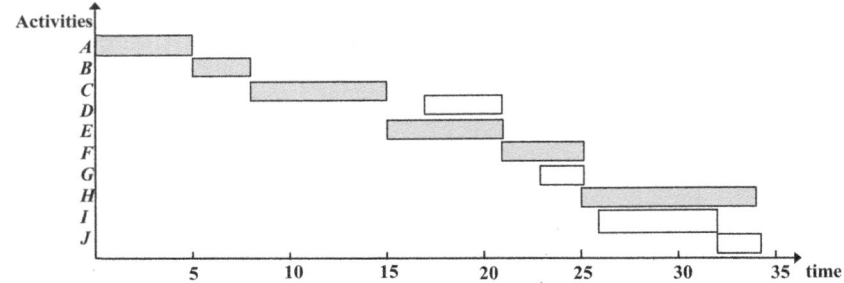

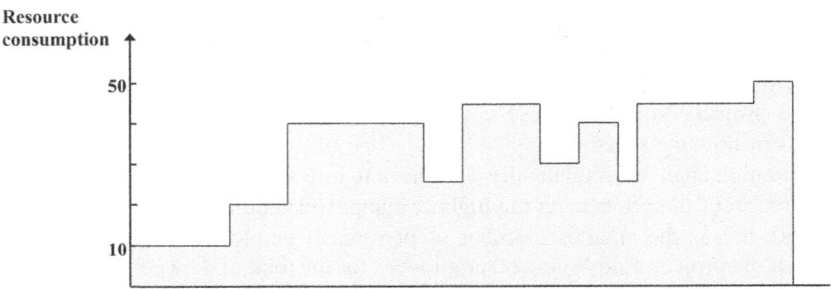

Fig. 8.12 Gantt chart and resource requirement graph for alternative schedule

8.4 The *PERT* Method

All project planning models discussed so far have in common that they are deterministic. More specifically, they have assumed that all components of the network—the activities, their durations, and the precedence structure—are known with certainty. This section will change that. In particular, we assume here that the activity durations are no longer known with certainty. It is important to realize that this is only one component that can be probabilistic: fully stochastic networks are dealt with by very sophisticated project network tools such as *GERT* (*graphical review and evaluation technique*), which are beyond the scope of this book.

The *PERT* method discussed in this section assumes that the duration of the activities are random variables with known underlying probability distributions. We do assume that the durations of activities are independent of each other. This is a fairly strong assumption, not justified in cases such as construction, where occurrences such as bad weather will affect many of the activities to take longer than they normally would. As usual in all of operations research, it is necessary to check the assumptions carefully, and if the assumptions do not fit the scenario under consideration, don't use the model.

Traditionally, it has been assumed that the duration of a single activity follows Euler's beta distribution. This assumption has been much criticized in the literature. However, we can derive the same formulas without making such a strict and

Fig. 8.13 Empirical rule for activity durations

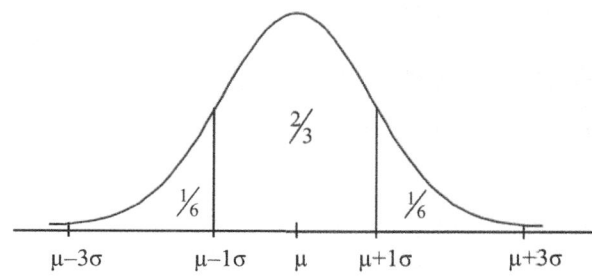

controversial assumption. Our assumption is that the activity durations follow some symmetric bell-shaped distribution. For symmetric bell-shaped distributions, the *empirical rule* in statistics is known to apply; see Appendix C. It states that about all observations are within three standard deviations about the mean, while about two-thirds of all observations are within one standard deviation about the mean. This leaves 1/6 of the total mass for each of the two tails of the distribution. This situation is shown in Fig. 8.13.

We can then define three time estimates for the duration of each activity: a *most likely* time estimate t_m (the mode of the distribution), a *pessimistic estimate* t_p, and an *optimistic estimate* t_o. The most likely time t_m, set at μ, is associated with the central part of the distribution, the pessimistic estimate t_p, set at $\mu + 1\sigma$, belongs to the right tail of the distribution, and the optimistic estimate t_o, set at $\mu - 1\sigma$, is on the left tail of the distributions. Their weights are 2/3, 1/6, and 1/6 as shown in Fig. 8.13. Based on these estimates, we can then compute a (weighted) mean for the duration of an activity as $t = -\frac{1}{6}t_o + \frac{2}{3}t_m + \frac{1}{6}t_p = \frac{t_o + 4t_m + t_p}{6}$, which is exactly what was obtained by using the much stronger assumption of the beta distribution. Similarly, with $t_p = \mu + 3\sigma$ and $t_o = \mu - 3\sigma$, we can determine the variance of the activity duration, which is then $\sigma^2 = \frac{1}{36}(t_p - t_o)^2$.

Armed with this information, we can then determine the mean duration and its variance for all of the given activities from the three time estimates, specified for each of them. Given the mean activity durations, we will use them in exactly the same way we dealt with time estimates in the critical path method. Going through the regular procedure—forward sweep, backward sweep, computation of floats, critical path—we can determine the critical path on the basis of the mean durations of the activities. In addition to the usual information about project duration, critical and noncritical activities, and sensitivity analyses on the basis of floats, we can use the variances of the durations on the critical path to make probability statements. In particular, we can provide the decision maker with an estimate concerning the probability with which the project can be completed within a certain time.

In order to explain the concept, consider a numerical example, whose numerical information is provided in Table 8.6. The graph for this project is the same as that for the examples in Sects. 8.1 and 8.3, but the time estimates are obviously different.

Our first task is to calculate the mean activity durations for the ten activities. They are 6, 3, 7, 5, 6, 4, 2, 9, 6, and 2, respectively. Those time estimates are then used in

Table 8.6 Numerical information for example

Activity	Immediate predecessor	Time estimates (in hrs)		
		Optimistic	Most likely	Pessimistic
A	–	5	6	7
B	A	3	3	3
C	A, B	5	7	9
D	B	3	4	11
E	B, C	3	6	9
F	C, D, E	4	4	4
G	D	1	2	3
H	F, G	4	10	10
I	F, G	4	6	8
J	I	2	2	2

the standard procedure discussed in the first section of this chapter. The results are shown in Fig. 8.14.

Consider now the critical path $n_s - A - B - C - E - F - H - n_t$. The mean project duration of $\mu = 35$ has already been computed in the procedure. We now have to calculate the variance on this path. The variances of the individual activities on the path (in order of their appearance) are $\frac{4}{36}, 0, \frac{16}{36}, \frac{36}{36}, 0$, and $\frac{36}{36}$. The sum of these variances equals $\sigma^2 = \frac{92}{36}$, so that the standard deviation equals $\sigma = \sqrt{\frac{92}{36}} \cong 1.5986$. Furthermore, invoking the central limit theorem (probably *not* justifiably so, as the number of arcs on the critical path is fairly small), the project duration is approximately normal with mean μ and standard deviation σ.

Given this information, we are now able to provide the decision maker with some rough estimate about the probability with which the project can be finished within a prespecified time frame T. Suppose that the decision maker wants to know what the probability is that the project is finished within $T = 36$ hrs. In order to be able to use the standard normal distribution (see Appendix C), we calculate the z-score as $z = \frac{T-\mu}{\sigma} = \frac{36-35}{1.5986} \cong 0.6255$, we find that the probability $P(X \leq 36) = 73.42\%$. When calculating these probabilities, it is always useful to draw the normal distribution function and indicate which area we are looking for. The area relevant to this question is shown in Fig. 8.15a, where it constitutes the shaded area plus the entire mass to the left of the mean which, by definition, equals 0.5.

Similarly, we could compute the probability that the project takes more than 37 hrs. Such information may be needed by the decision maker, as the late completion of the project may carry a penalty with it. The z-score for the completion time of 37 is $z = \frac{37-35}{1.5986} \cong 1.2511$, from which we obtain a probability of $P(X \geq 37) = 10.54\%$. The area of interest under the normal distribution function is shown in Fig. 8.15b.

Finally, we compute the probability that the project is completed between 33 and 36 hrs. The area of interest is shown in Fig. 8.15c, and, due to the symmetry of the normal distribution, it can be computed as $P(33 \leq X \leq 36) = P(33 \leq X \leq 35) + P(35 \leq X \leq 36) = P(35 \leq X \leq 37) + P(35 \leq X \leq 36) = .3946 + .2342 = .6288$. In

8.4 The PERT Method

Fig. 8.14 Project network, for example, in text

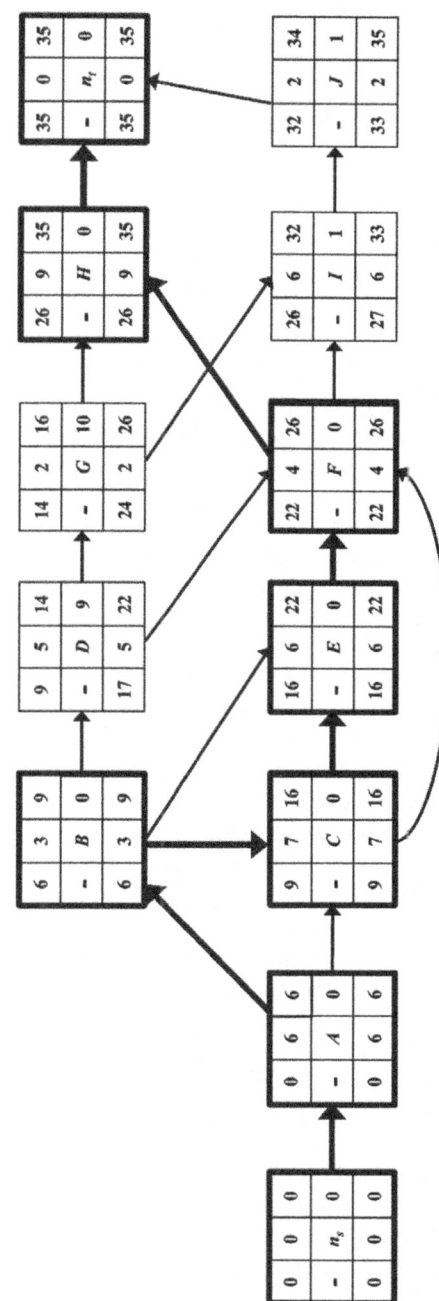

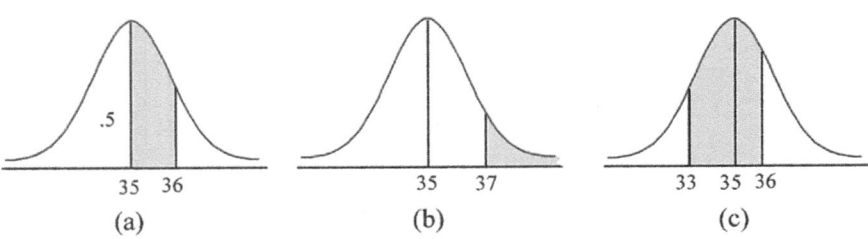

Fig. 8.15 Probabilities for various events

other words, chances are about 63% that our project will be completed within the specified time window.

A few comments are in order. First of all, much care should be taken with the probability statements. They have been derived with a lot of assumptions, so that it makes absolutely no sense to report them to the decision maker with two, four, or even more digits to the right of the decimal point, implying great accuracy. These probabilities are rough estimates, and this should be emphasized throughout.

Secondly, what we have used is what is called the time-critical path. In other words, we have determined the critical path on the basis of estimated activity durations only, and then computed the probabilities. This does not necessarily provide the planner with the true critical path. As an example, consider two paths in some network, one with a mean duration $\mu = 100$ and a standard deviation of $\sigma = 10$, while a second, obviously noncritical, path exists in the network with $\mu = 99$ and $\sigma = 100$. Note that the noncritical path is shorter, but has a much higher standard deviation. The probability to finish the project within $T = 110$ hrs, computed on the basis of the critical path as we do, is then 84.13%. However, computing the same probability on the basis of the noncritical path is only 54.38%, meaning that the former result (that we would obtain with our procedure), grossly overestimates the likelihood to finish the project within the specified time frame. This problem persists in a somewhat different guise even in our example: all probability statements assume that the critical path remains critical. Note, however, that the noncritical activities I and J in our example have a very small float, making them almost critical. If their durations were to increase by fairly small, insignificant amounts, they would become critical, and their standard deviations, which have been completely ignored so far, would suddenly have to be counted. This is yet another reason to treat the probability statements we have calculated with the utmost caution.

Exercises

Problem 1 (acceleration of a project): Consider a project network with four activities, their normal durations, their shortest possible durations (which can be achieved at extra cost), and the acceleration cost per time unit. Details are shown in Table 8.7.

8.4 The PERT Method

Table 8.7 Details of the network in Problem 1

Activity	Immediate predecessor(s)	Normal duration (in days)	Shortest possible duration (in days)	Unit cost of acceleration ($)
A	–	4	3	80
B	A	5	3	30
C	A	3	2	50
D	A, C	2	1	40

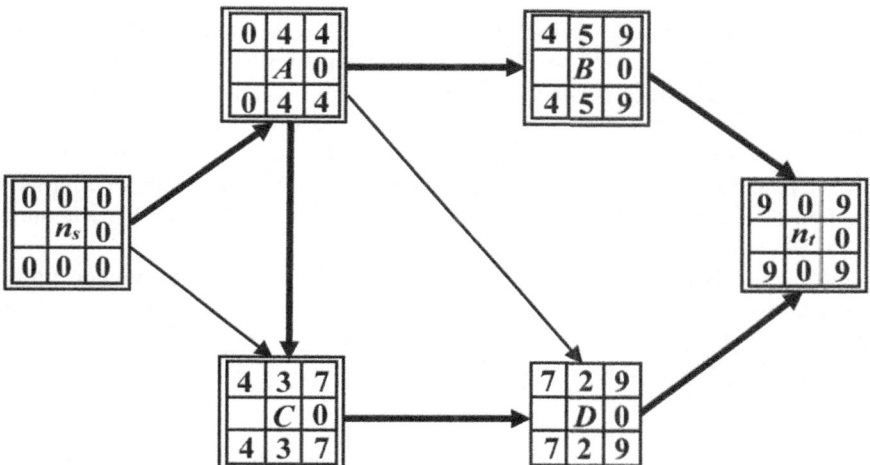

Fig. 8.16 Project network for Problem 1

(a) Draw the project network.
(b) Use the forward recursion, the backward recursion, calculate the floats, and determine the critical path(s). What is the duration of the project?
(c) Accelerate the project by one day. Clearly indicate which (combinations of) activities could be accelerated in order to speed up the project, and which activity or activities should be accelerated. What are the associated costs?
(d) What if the decision maker needs some further project acceleration? What is the shortest project duration, and what are the associated costs?

Solution:
(a), (b) See Fig. 8.16.
The project duration is 9 days.

(c) Given the two critical paths A–B and A–C–D, respectively, we can either accelerate activity A (cost: $80) or the activities B and C (cost: $30 + $50 = $80), or the activities B and D (cost: $30 + $40 = $70). Given its lowest cost, we accelerate activities B and D for a total cost of $70. Now the activity durations are 4, 4, 3, and 1, respectively, and the project duration is $T = 8$. Note that at this point, the activity duration of activity D has reached crash time and can no longer be accelerated. Also, note that the system of critical paths has not changed.

Table 8.8 Details about the network in Problem 2

Activity	Immediate Predecessor	Duration (in days)	Resource consumption
A	–	3	40
B	–	7	40
C	A	5	30
D	A, B, C	2	60
E	B, D	5	70

(d) After the first acceleration in (c), we can either accelerate A (cost $80) or activities B and C (cost $80). Arbitrarily choosing activity A, the activity durations are then 3, 4, 3, and 1, respectively, total acceleration costs are $150, and the project length is $T = 7$. The two critical paths are still the only critical paths in the network. At this point, we can only accelerate activities B and C, as A and D are both at their respective crash times. Accelerating B and C costs $80. This leads to a project duration of $T = 6$, which cannot be accelerated any further. The total acceleration cost to get to this point is $230.

Problem 2 (scheduling with resources, Gantt chart and resource consumption graph): A project has been subdivided into five activities. Their immediate predecessors, activity durations, and resource consumption (e.g., the number of employees required) are shown in Table 8.8.

(a) Draw the project network.
(b) Calculate all earliest and latest starting and finishing times as well as the total floats of all activities. What is the critical path? What is its duration?
(c) Draw the Gantt chart and the associated resource consumption graph, assuming that all activities are scheduled as early as possible. What is the largest resource consumption at any point in time?

Solution: (a) See Fig. 8.17.
(b) The critical path includes the activities A, C, D, and E. Its length is 15.
(c) See Fig. 8.18.
The largest resource consumption at any point in time is 80.

Problem 3 (*PERT* network): A project has been subdivided into five activities. Their immediate predecessors and the estimated activity durations (optimistic, most likely, and pessimistic) are shown in Table 8.9.

(a) Draw the project network. Based on the mean activity durations, calculate all earliest and latest starting and finishing times, and the total floats of all activities. What is the critical path? What is its duration?
(b) Calculate the variance and standard deviation on the critical path.
(c) Calculate the probability that the project will be finished between 11 and 15 days.
(d) What is the probability that the project will be finished within 14 days? What is the probability that the project will be finished in exactly 14 days?

8.4 The PERT Method

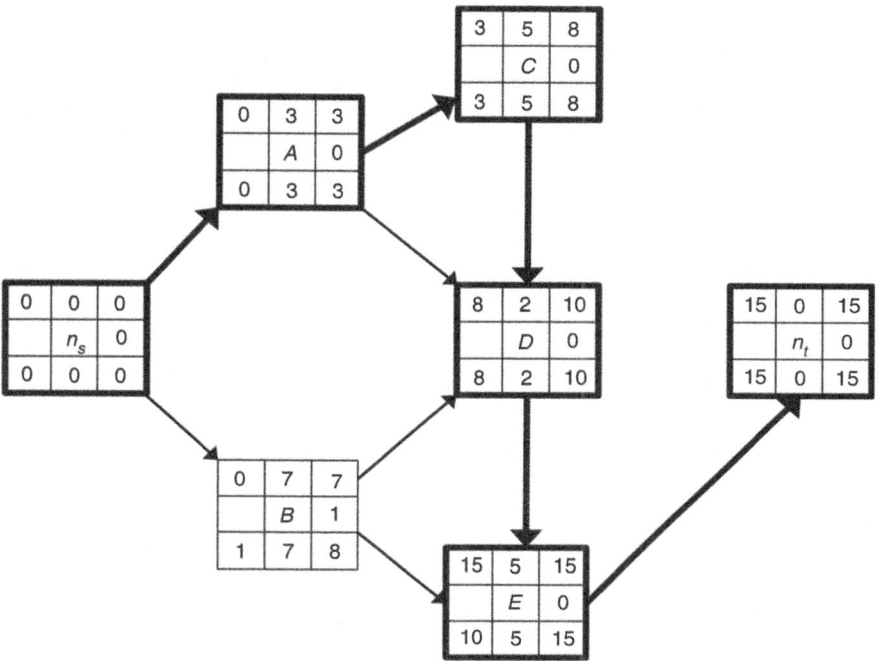

Fig. 8.17 Project network for Problem 2

Solution: (a) See Fig. 8.19.
The critical path is n_s–B–C–D–E–n_t. Its duration is 14 days.
(b) $\sigma^2 = \frac{1}{36}[16 + 4 + 0 + 400] = \frac{420}{36}$, $\sigma = 3.4157$.
(c) $P(11 \leq X \leq 15) = P(11 \leq X \leq 14) + P(14 \leq X \leq 15)$. Finding z-scores z_1 and z_2 for the two ranges results in $z_1 = \frac{14-11}{3.4157} = .8783$, leading to 31.01%, and $z_2 = \frac{15-14}{3.4157} = .2928$, leading to 11.52%, for a total of 42.53%.
(d) $P(X \leq 14) = 0.5$. $P(X = 14) = 0$.

Problem 4 (Gantt chart, resource requirement graph, and PERT network): A project has been subdivided into five activities. Their immediate predecessors and the estimated activity durations (optimistic, most likely, and pessimistic) are shown in Table 8.10.

(a) Draw the project network. On the basis of the expected durations, calculate all earliest and latest starting and finishing times, and the total floats of all activities. What is the critical path? What is its duration?
(b) Calculate the variance and standard deviation on the critical path.
(c) Calculate the probability that the project will be finished between 14 and 18 days.

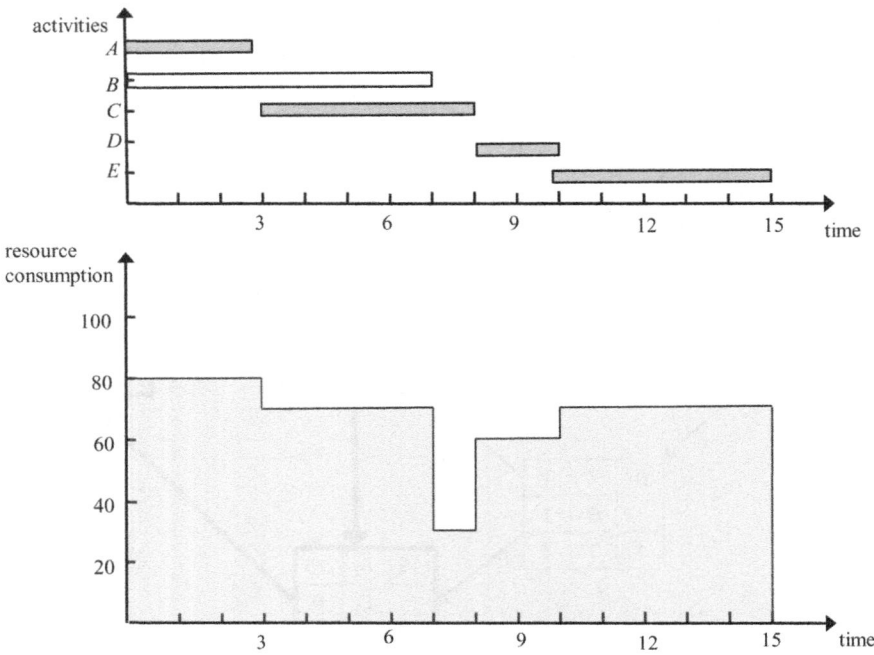

Fig. 8.18 Gantt chart and resource requirement graph for Problem 2

Table 8.9 Details about the network in Problem 3

Activity	Immediate Predecessor	Estimated duration (in days)
A	–	1, 2, 3
B	–	1, 3, 5
C	A, B	3, 4, 5
D	A, C	2, 2, 2
E	C, D	1, 2, 21

(d) What would happen to the result under (c), if the time estimates of activity D were to be revised to 3, 6, and 9? Explain in one short sentence.

(e) Consider the starting and finishing times calculated in (a) as well as resource requirements of 20, 50, 30, 40, and 60, respectively. On that basis, draw a Gantt diagram given that all activities are scheduled as early as possible. What is the highest resource requirement at any one time during the project?

(f) Repeat question (e) for the "Latest possible" scheduling rule.

Solution: (a) See Fig. 8.20.
The unique critical path is n_s–A–E–n_t, and its length is 16.

8.4 The PERT Method

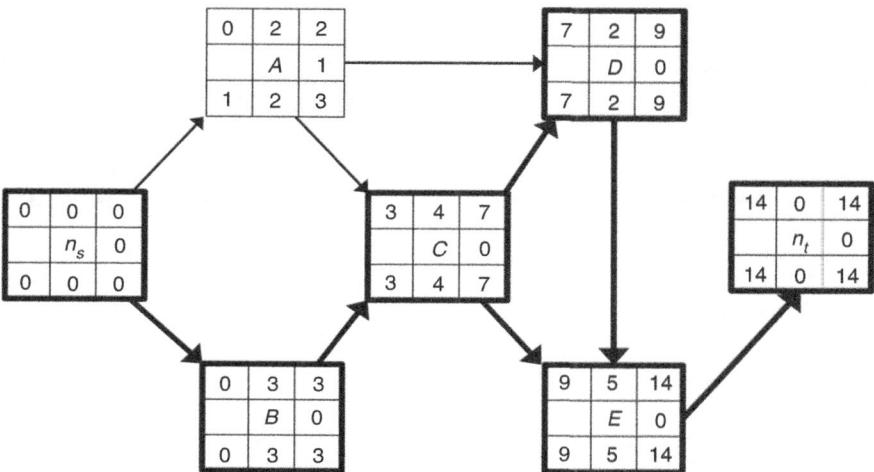

Fig. 8.19 Project network for Problem 3

Table 8.10 Details about the network in Problem 4

Activity	Immediate Predecessor	Estimated duration (in days)
A	–	2, 6, 16
B	–	2, 2, 8
C	A	6, 8, 10
D	B	5, 6, 7
E	A, B	9, 9, 9

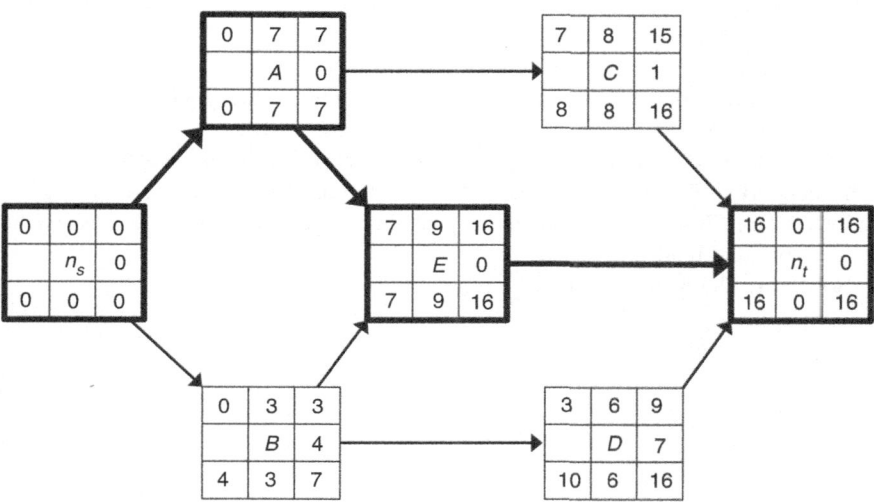

Fig. 8.20 Project network for Problem 4

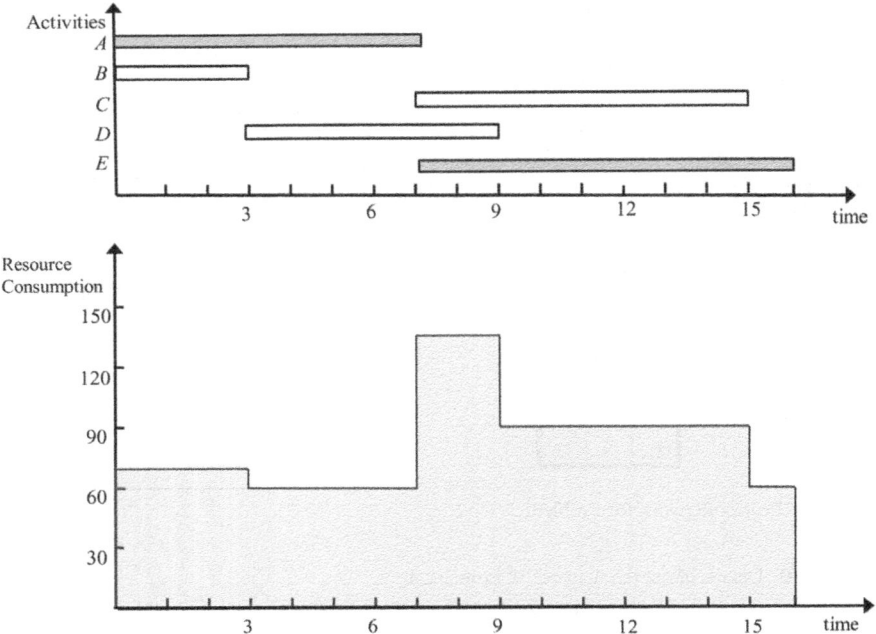

Fig. 8.21 Gantt chart and resource requirement graph for Problem 4 (e)

(b) and (c) $\sigma^2 = \frac{1}{36}(196 + 0) \approx 5.4444$ and $\sigma \approx 2.3333$. Then $z = \frac{18-16}{2.3333} = .8571$, and $P(14 \leq X \leq 18) = 2(.3043) = 60.86\%$.

(d) The change does not affect the mean of activity D, only its variation, and since activity D was not on the critical path before and will not be on it after the modification, the results will not change.

(e) See Fig. 8.21.

The highest resource requirement at any point in time during the project is 130.

(f) See Fig. 8.22.

Problem 5 (Alternative linear programming formulation): Consider again the CPM network in Fig. 8.3. Define variables x_j as the earliest possible finishing time of activity j (i.e., $EF(n_j)$) and formulate the problem with constraints for all arcs (i, j) in the network that read $x_j \geq x_i + d_j$.

Solution: The formulation of the problem is

8.4 The PERT Method

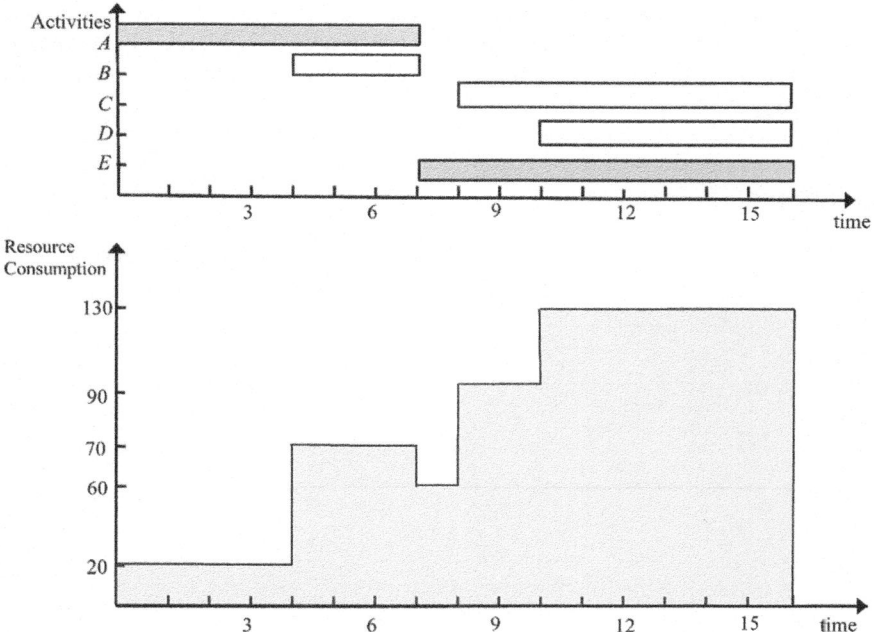

Fig. 8.22 Gantt chart and resource requirement graph for Problem 4 (f)

$P : \text{Min } x_t$

s.t. $x_s = 0$

$x_A \geq x_s + 5$

$x_B \geq x_A + 3$

$x_C \geq x_A + 7$

$x_C \geq x_B + 7$

$x_D \geq x_B + 4$

$x_E \geq x_B + 6$

$x_E \geq x_C + 6$

$x_F \geq x_C + 4$

$x_F \geq x_D + 4$

$x_F \geq x_E + 4$

$x_G \geq x_D + 2$

$x_H \geq x_F + 9$

$x_H \geq x_G + 9$

$x_I \geq x_F + 6$

$x_I \geq x_G + 6$

$x_J \geq x_I + 2$

$x_t \geq x_H$

$x_t \geq x_J$

$x_j \geq 0$ for all j.

Solving the problem, we identify the excess variables with a value of zero and positive shadow prices. Here, they belong to constraints 1, 2, 3, 5, 8, 11, 13, and 18. These constraints belong to arcs in the network, which, if put together, result in the critical path $s - A - B - C - E - F - H - t$.

Reference

Eiselt HA, Sandblom C-L (2004) Decision analysis, location models, and scheduling problems. Springer, Berlin

Machine Scheduling

9

The subject of this chapter is the allocation of *jobs* (or *tasks*) to *processors* (or *machines*). These terms should be understood in the widest possible sense: in the case of a doctor treating patients, the doctor is the processor and the patients represent the tasks; in the case of a tax audit, the auditor is the processor (or the "machine"), while the individual cases are the tasks to be processed. This allocation is to be made so as to optimize some objective. We will discuss a number of these objectives below.

This problem is somewhat reminiscent of the production scheduling problem introduced in Chap. 2, and the project planning problems discussed in Chap. 8. The main feature of the machine scheduling problems in this chapter is that they include a *sequencing* component, which determines the *order* in which tasks are processed on a machine, and a *scheduling* component that determines at what *time* the processing of a task begins and ends on a machine. Often, the term scheduling is meant to include both the sequencing and scheduling parts of the allocation problem.

This chapter deals exclusively with deterministic scheduling problems, i.e., situations in which all parameters are assumed to be known with certainty. This is not to mean that these are the only relevant scheduling problems: on the contrary, in many important real-world applications, some of the parameters involved in the problem are uncertain. However, the structure of probabilistic problems is no different from that of deterministic problems (other than the fact that probabilistic problems are typically much more difficult than their deterministic counterparts), and what we attempt to convey in this chapter is a general idea of scheduling models, their applications, and degree of difficulty.

The first section of this chapter introduces the basic concepts of scheduling models. The remaining three sections of this chapter deal with different scheduling scenarios with increasing degree of difficulty.

9.1 Basic Concepts of Machine Scheduling

As mentioned above, each machine scheduling problem features n tasks (jobs) T_1, $T_2, \ldots, T_n$ that are to be performed on m machines (processors) $P_1, P_2, \ldots, P_m$. Processing a task on a machine will take a certain (and known) amount of time, which is referred to as the *processing time*. In particular, we define p_{ij} as the processing time of task T_j on machine P_i for all combinations of tasks and machines. In case there is only a single machine, or all machines have the same processing time for a job, we simplify the notation to p_j.

We can then distinguish between three broad categories of models. The first category assumes that there is only a single machine, we refer to this as *single-machine scheduling*. The second category has multiple machines working in parallel, all able to perform the same function. This type of model is called a *parallel machine scheduling problem*. There are two subcategories of parallel machine scheduling. In the first, parallel machines are *identical* (meaning that all machines take the same amount of time to process any given task), or they are *unrelated*, which indicates that it takes different amounts of time to process any given task on different machines.

The third broad category of models includes *dedicated machine scheduling models*. The main characteristic of these models is that not all machines have the capability to process all jobs, and not all jobs need to be processed on all machines. In particular, we distinguish between three subcategories. The first category is an *open shop*. In an open shop, each task must be processed by all machines, but there is no specific order in which the processing will take place. The second category comprises *flow shops*, in which each job is to be processed on all machines, and each task is processed by the machines in the same specified order. Finally, there are *job shops*, in which each job needs to be processed on a specific set of machines, and the processing order is also job specific.

In addition to the processing time p_{ij} introduced above, we may also have a number of additional parameters. The first such parameter is the *ready time* (also referred to as *arrival time* or *release time*) r_j, which indicates the time at which task T_j is ready for processing on the machines. In the simplest case, $r_j = 0$ for all tasks T_j, meaning that all jobs are ready when the scheduling process starts. The second additional parameter is the *due time* d_j, which is the time at which task T_j *should* have finished all processing. This could be a time specified in a contract or some other delivery time that was agreed upon. If the task is not completed by that time, then there may be a penalty for the delay.

Once all jobs have been scheduled on the machines, we can use a number of different properties inherent in any given schedule. The first is the *completion time* c_j of job T_j, which is the time at which task T_j is actually completely finished. The second property is the *flow time* f_j of task T_j. The flow time is formally defined as $f_j = c_j - r_j$, and it can be thought of as the time that a job is in the system, either waiting to be processed or being processed. (This time is reminiscent as the "time in the system" W_s in the analysis of waiting lines; see Chap. 15). Finally, there is the *lateness of a task* T_j, which is defined as $\ell_j = c_j - d_j$, and *tardiness*, which is

9.1 Basic Concepts of Machine Scheduling

expressed as $t_j = \max\{\ell_j, 0\}$. The lateness of a task is the time that elapses after the due date and the actual completion date of a task. For late jobs, lateness and tardiness are the same. For jobs that are completed before their due date, the lateness becomes negative, while their tardiness equals zero.

There are many different criteria that may be used to optimize scheduling systems. Three of the most popular such criteria are introduced here. First, there is the *makespan* (or *schedule length*) $C_{\max}$. It is formally defined as $C_{\max} = \max_j \{c_j\}$, and it expresses the time at which the last of the tasks has been completed. Such a measure is meaningful if a project can only be considered completed, if all of its individual tasks have been completed (similar to project networks; see Chap. 3). A typical example is the processing of machine parts that have been ordered by a customer. All of them will be shipped in a box, and the box can only be released for transportation, once all of the individual machine parts are included.

The second criterion is the *mean flow time F*. The mean flow time is formally defined as the unweighted average $F = \frac{1}{n}(f_1 + f_2 + \ldots + f_n)$. (As a matter of fact, by virtue of the definition of flow time, it is easy to demonstrate that the mean flow time differs from the mean completion time $\frac{1}{n}(c_1 + c_2 + \ldots + c_n)$ only by the constant $\frac{1}{n}(r_1 + r_2 + \ldots + r_n)$.) The mean flow time refers to the average time that a job is in the system either waiting to be processed or being processed. The mean flow time is a meaningful measure in instances such as a maintenance or repair system, in which a machine is not available if it is waiting for repair or being repaired.

The third and last criterion in this context is the *maximal lateness* $L_{\max}$. The maximal lateness is defined as $L_{\max} = \max_j \{\ell_j\}$, and it expresses the longest lateness among any of the jobs. This criterion is applicable in case some lateness is unavoidable or deemed acceptable, but the decision maker attempts to ensure that very long delays beyond the due date are avoided.

Before discussing any details of specific scheduling models, we should point out some general features inherent in scheduling problems. In this type of model, there is a fairly fine line that separates models that are rather easy to solve (some of which are represented in this chapter), while others, seemingly straightforward extensions, are very difficult. In those cases, exact methods that find optimal solutions will take a very long time, which may—depending on the individual circumstances—render them impractical. In such cases, decision makers will resort to heuristic methods that have the advantage of providing (hopefully good) solutions quickly, but the obvious disadvantage of not necessarily resulting in an optimal solution. If a difficult problem has to be solved in real time, the use of a heuristic is imperative; if some time is available, an exact algorithm may be employed.

9.2 Single-Machine Scheduling Models

The models in this section deal with the simplest of scheduling problems: there is only a single machine on which tasks are to be processed. Before investigating the solutions that result from the use of the three criteria presented in the introduction above, we may introduce another wrinkle in this seemingly primitive scenario. In particular, we may consider modes that allow the *preemption* of a task, while others do not. In case preemption is permitted, this means that each task has associated with it a priority, and if a task with a higher priority becomes available at a time when a task with a lower priority is being processed, then processing on the lower-level task stops, and the higher-level task is processed first. Examples of preemption abound: consider the case of a surgeon who is dealing with a broken leg as another patient with a heart attack arrives. Rather than referring to the usual "first come, first served" rule (inviting juicy lawsuits), most surgeons would probably stabilize the broken leg and deal with the heart patient first. Similar preemptions are found for police officers, who would interrupt a routine investigation to attend to a robbery in progress, or a plumber, who will interrupt the installation of a water pump in a residence to attend to a broken main. In this section, we restrict ourselves to cases, in which preemption is not permitted.

Minimizing the makespan in case of a single machine is not meaningful, as each sequence of tasks will result in the same value of $C_{\max}$. More specifically, $C_{\max}$ equals the sum of processing times of all tasks.

The objective that minimizes the mean flow time F is not as straightforward. However, it is not difficult either, as it has been shown that the simple *Shortest Processing Time (SPT) algorithm* solves the problem optimally. The algorithm can be summarized by a simple rule. Note that we assume that all release times are zero.

SPT Algorithm: Schedule the task with the shortest processing time first. Delete the task and repeat the procedure until all tasks have been scheduled.

Rather than illustrate the *SPT* algorithm by an example, we will first introduce a minor extension of the rule. In particular, suppose that the decision maker has not only processing times p_j to consider, but there are also weights w_j associated with the tasks. Typically, a weight could symbolize the cost of a task as it is processed on a machine, given that different tasks are associated with different costs. The objective is then to minimize the average weighted flow time, defined for task T_j as $w_j f_j$. The weighted generalization of the *SPT* algorithm can then be stated as

WSPT Algorithm (Smith's Ratio Rule): Schedule the task with the shortest weighted processing time p_j/w_j first. Delete the task and repeat the procedure until all tasks have been scheduled.

As a numerical illustration, consider:

9.2 Single-Machine Scheduling Models

Table 9.1 Processing times and downtime cost for Example 1

Job #	T_1	T_2	T_3	T_4	T_5	T_6	T_7
Service time (minutes)	30	25	40	50	45	60	35
Downtime cost (\$ per minute)	2	3	6	9	4	8	3
Weighted processing time p_j/w_j	15	$8\frac{1}{3}$	$6\frac{2}{3}$	$5\frac{5}{9}$	$11\frac{1}{4}$	$7\frac{1}{2}$	$11\frac{2}{3}$

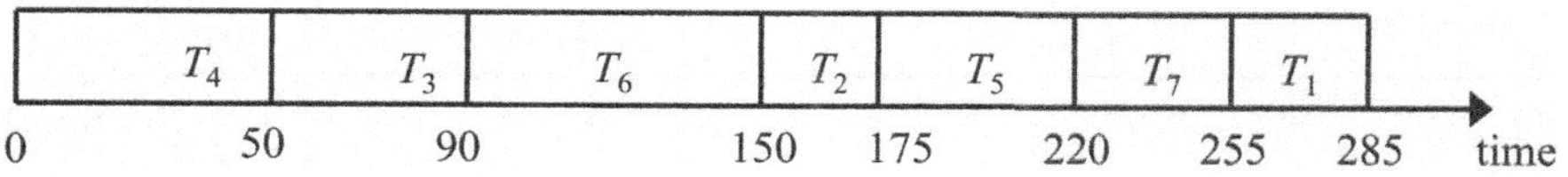

```
| T_4    | T_3   | T_6    | T_2   | T_5  | T_7  | T_1 |
0        50      90       150    175    220    255   285   time
```

Fig. 9.1 Gantt chart for Example 1

Example 1 There are seven machines in a manufacturing unit. Scheduled maintenance has to be performed on these machines once in a while. The repairman has identified the processing times required for the maintenance. Costs are incurred for downtime, regardless of whether the machine waits for service or is being served. These costs differ between the machines. The estimated processing times of the machines, the downtime costs, and the weighted processing times are summarized in Table 9.1.

Applying the *WSPT* algorithm, we first schedule task T_4 (which has the lowest weighted processing time of $5\frac{5}{9}$), followed by T_3 with the next lowest weighted processing time of $6\frac{2}{3}$, followed by T_6, T_2, T_5, T_7, and T_1. The schedule is shown in the *Gantt chart* (named after the American engineer Henry L. Gantt (1861–1919; see also Sect. 8.3), who developed these charts in 1917) in Fig. 9.1.

The tasks T_4, T_3, T_6, ..., T_1 now have idle times of 0, 50, 90, 150, 175, 220, and 255, respectively. Adding the processing times to the idle times results in 50, 90, 150, 175, 220, 255, and 285, respectively. Multiplying these by the individual per-minute costs and adding them up results in a total of \$4930.

Consider now the objective that minimizes maximal lateness L_{max}. Again, this problem turns out to be easy from a computational point of view. A simple method was developed in the mid-1950s by Jackson, which is now commonly referred to as the *earliest due date algorithm* (or *EDD algorithm* or *Jackson's rule*). It finds an optimal solution and can be stated as follows. Again, the release times of all tasks are zero.

EDD Algorithm: Schedule the task with the earliest due date first. Delete the task and repeat the procedure until all tasks have been scheduled.

We will explain this rule by means of

Example 2 The accounting department of a large firm processes bookkeeping jobs for various divisions of the firm. At present, they have identified eleven tasks, which

Table 9.2 Data for Example 2

Job #	T_1	T_2	T_3	T_4	T_5	T_6	T_7	T_8	T_9	T_{10}	T_{11}
Processing time (hrs)	6	9	4	11	7	5	5	3	14	8	4
Due dates	25	15	32	70	55	10	45	30	30	80	58

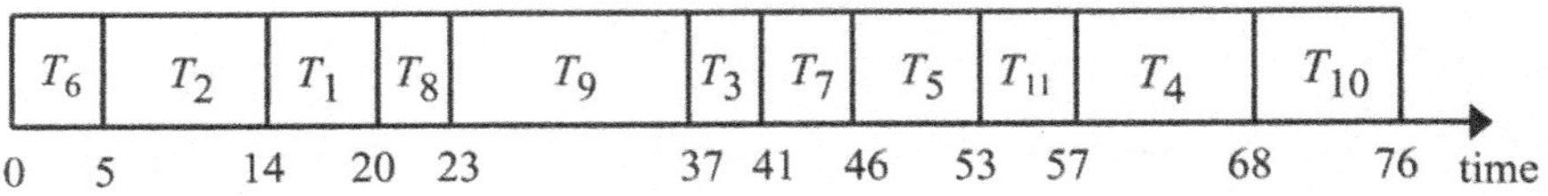

T_6	T_2	T_1	T_8	T_9	T_3	T_7	T_5	T_{11}	T_4	T_{10}

0   5   14 20 23   37 41 46   53 57   68   76   time

Fig. 9.2 Gantt chart for Example 2

are to be completed by a single team, one after another. The processing times and the due dates for the individual jobs are shown in Table 9.2.

The *EDD* rule starts by scheduling task T_6 first (its due date is 10, the earliest of all due dates), followed by T_2, T_1, and so forth. The Gantt chart for the schedule obtained by the *EDD* rule is shown in Fig. 9.2 (where the tie between tasks T_8 and T_9 that have the same due dates is broken arbitrarily).

Simple inspection reveals that the tasks T_6, T_2, T_1, and T_8 are finished before the due date. Task T_9 is late by 7 hrs, T_3 is late by 9 hrs, T_7 is late by 1 hr, and T_5, T_{11}, T_4, and T_{10} are again finished before their due dates. This means that the maximal lateness occurs for job T_3, so that $L_{max} = 9$. Had we broken the tie in favor of T_9 rather than T_8, task T_9 would have been completed at time 34 (4 hrs late), and task T_8 would have been finished at time 37 (7 hrs late). Otherwise, the schedule would have been identical to that shown in Fig. 9.2, with $L_{max} = 9$ being still defined by job T_3.

The discussion in this chapter may leave readers with the impression that single-machine scheduling problems are easy. This is, however, not the case. Consider again Example 2, but as an objective, we now use the total tardiness rather than the maximal tardiness. In other words, our objective is now $L_{sum} = t_1 + t_2 + \ldots + t_{11}$ rather than $L_{max} = \min\{\ell_1, \ell_2, \ldots, \ell_{11}\}$. We can find an optimal solution to the problem by formulating an integer programming problem. In order to do so, we first use variables $t_1, t_2, \ldots, t_{11}$ for the tardiness of the individual tasks. We then define variables $x_1, x_2, \ldots, x_{11}$ for the actual starting times of the eleven tasks. Furthermore, we need to define zero-one variables y_{ij} as

$$y_{ij} = \begin{cases} 1, & \text{if job } T_i \text{ directly precedes job } T_j \\ 0, & \text{otherwise} \end{cases},$$

where the subscripts i and j each can assume any value between 1 and 11, except that $i \neq j$, resulting in $11^2 - 11 = 110$ zero-one variables y_{ij} in addition to the eleven continuous variables $t_1, \ldots, t_{11}$ and the eleven continuous variables $x_1, \ldots, x_{11}$ for a total of 132 variables.

9.2 Single-Machine Scheduling Models

The objective simply minimizes the sum of tardiness $t_1 + t_2 + \ldots + t_{11}$. As far as constraints are concerned, we first must define tardiness, which is done by stating that tardiness of a job can be expressed as its starting time plus its processing time (resulting in its finishing time) minus the due date, assuming that the resulting figure is positive. As an example, consider job T_1, whose tardiness can be written as $t_1 = x_1 + p_1 - d_1$, provided that $t_1 \geq 0$; otherwise, t_1 is set to zero. This can be achieved by writing $x_1 + p_1 - d_1 \leq t_1$, as whatever value the left-hand side of this inequality assumes, the objective function will ensure that the value of t_1 is chosen as small as possible. This constraint is written for each of the eleven tasks separately.

We then have to ensure that a job must be completely finished before the next job in line can begin being processed. Suppose that we have scheduled job T_1 directly before job T_2. The pertinent constraint will then be $x_1 + p_1 \leq x_2 + M(1 - y_{12})$, where $M >> 0$ is a suitably chosen large number. This constraint can be explained as follows. The left-hand side expresses the finishing time of job T_1, while the right-hand side shows the starting time of job T_2 plus M, if T_1 is not scheduled before T_2 and 0 if it is. In other words, if T_1 is scheduled before T_2, then this constraint requires that the starting time of T_2 is at least as large as the finishing time of T_1 (i.e., T_2 cannot start before T_1 is finished), while in case that T_1 is not scheduled before T_2, then the right-hand side of the inequality is very large, ensuring that it is satisfied regardless of the values of the variables. The problem can then be written as follows:

$$P : \text{Min } z = t_1 + t_2 + \ldots + t_{11}$$
$$\text{s.t. } x_1 + p_1 - d_1 \leq t_1$$
$$\vdots$$
$$x_{11} + p_{11} - d_{11} \leq t_{11}$$

$$x_1 + p_1 - x_2 \leq M(1 - y_{12})$$
$$\vdots$$
$$x_1 + p_1 - x_{11} \leq M(1 - y_{1,11})$$
$$x_2 + p_2 - x_1 \leq M(1 - y_{21})$$
$$\vdots$$
$$x_{11} + p_{11} - x_{10} \leq M(1 - y_{11,10})$$

$$x_1, x_2, \ldots, x_1 \geq 0$$
$$t_1, t_2, \ldots, t_{11} \geq 0$$
$$y_{11}, y_{12}, \ldots, y_{1,11}, y_{21}, y_{22}, \ldots y_{11,11} = 0 \text{ or } 1.$$

Note that the size of the integer programming problem for a scheduling model with n tasks includes $n^2 - n$ zero-one variables, $2n$ continuous variables, and n^2 structural constraints plus nonnegativity and zero-one conditions. It is apparent that the formulation becomes unwieldy for large numbers of tasks, thus possibly requiring the use of heuristic algorithms.

9.3 Parallel Machine Scheduling Models

All scheduling models in this section have in common that the tasks can now be processed on more than one machine. In general, we assume that a given number m of machines are available. We further assume that these machines are identical in the sense that not only can all machines process each of the tasks, but it takes the same amount of time to process a task, regardless of the machine it is processed on.

First consider the objective of minimizing makespan $C_{\max}$. It can be demonstrated that this problem is very difficult from a computational point of view, even for just two machines. This means that we typically have to resort to heuristics to solve the problem (except in cases, in which there is ample time to find exact solutions). The most popular heuristic method for this type of problem is the *longest processing time first* (or *LPT*) algorithm. This heuristic method belongs to the class of *list scheduling methods*. All list scheduling methods first produce a priority list of tasks, starting with the job that is assigned the highest priority. Using this list and starting with the task that has the highest priority, jobs are then assigned one at a time to the first available machine. In particular, the longest processing time first algorithm can be described by the following rule:

> *LPT Algorithm:* Put the tasks in order of nonincreasing processing times. Starting at the top, assign the first available task to the first available machine. Repeat the procedure until all tasks have been scheduled.

Example 3 In order to demonstrate the way the algorithm works, consider again Example 1 above, but assume that now we have three equally qualified servicemen available to perform the maintenance tasks. Putting the seven tasks in order of their processing time, starting with the longest, we obtain the sequence T_6, T_4, T_5, T_3, T_7, T_1, and T_2 with processing times of 60, 50, 45, 40, 35, 30, and 25 min. In the beginning, all three machines are available, so we first assign the longest task T_6 to machine P_1. (Note that this machine will become available again at time 60, when T_6 is completely processed.)The next task in line is T_4, which is assigned to machine P_2, which is available now. (This machine will become available again at time 50, when T_4 is completely processed.) The next task in line is T_5 and it is assigned to machine P_3. This machine will become available again at time 45. The next task to be scheduled is now T_3. The three machines become available again at 60, 50, and 45, so that assigning T_3 to the next available machine means that it is scheduled on P_3. This process continues until all jobs are scheduled. The actual schedule is shown in the Gantt chart in Fig. 9.3, where the shaded areas indicate the idle time on the machines.

Note that the schedule length is $C_{\max} = 110$. It is worth mentioning that this schedule is not optimal, which is not surprising, since the *LPT* algorithm, which it was determined with, is not an exact algorithm but a heuristic. Incidentally, the optimal solution schedules T_6 and T_7 on machine P_1, jobs T_4 and T_5 are processed on

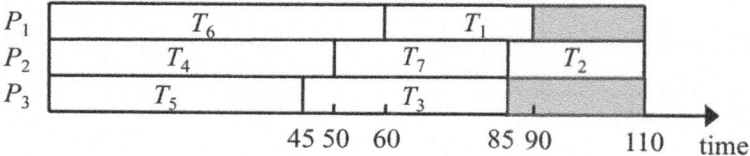

Fig. 9.3 Gantt chart for Example 3

machine P_2, and tasks T_3, T_1, and T_2 are processed on P_3. This schedule has no idle time at all and all machines finish at time 95. Note that the optimality of a schedule does *not* necessarily require that there is no idle time. On the other hand, if there is no idle time on any of the machines, the schedule must obviously be optimal. However, it is clear that if for instance, there are three machines and one of the jobs has a processing time longer than one-third of the total processing times of all the jobs, then every possible schedule must include some idle time on some machine.

Recall that the *LPT* algorithm is a heuristic algorithm and as such it is not guaranteed to find an optimal solution, i.e., a solution that truly minimizes the schedule length, or makespan, C_{max}. Instead, we may try to find an estimate of how far from an optimal solution the heuristic solution found by the *LPT* algorithm might be. Such an estimate is conveniently defined in terms of a *performance ratio* $R = C_{max}(\text{heuristic})/C_{max}(\text{optimum})$, where $C_{max}(\text{heuristic})$ is the schedule length obtained by the heuristic method and $C_{max}(\text{optimum})$ is the true minimal schedule length of the particular problem at hand. Since C_{max} is a minimization objective, $C_{max}(\text{optimum}) \leq C_{max}(\text{heuristic})$, so that $R \geq 1$, and the smaller the value of R, the closer the obtained schedule will be to the true minimum. It has been proved that the performance ratio R_{LPT} for the *LPT* algorithm applied to an n-job, m-machine problem satisfies the inequality

$$R_{LPT} \leq \frac{4}{3} - \frac{1}{3m}.$$

In Example 3 above, the 3-machine problem has a boundary value of $R_{LPT} = 1.2222$, while the actual solution found deviates by $110/95 = 1.1579$, i.e., by 15.79% from the optimal solution.

For a two-machine ($m = 2$) problem, the bound translates into a worst-case performance bound of $R_{LPT} = \frac{4}{3} - \frac{1}{3(2)} = 7/6 \approx 1.167$, meaning that in the worst case, the *LPT* algorithm will find a schedule that is 16.7% longer than that of the optimal solution. The fact that this bound is actually tight (i.e., can be reached) is shown in the following:

Example 4 Let a two-machine, five-job scheduling problem have the processing times 3, 3, 2, 2, and 2, respectively. Applying the *LPT* heuristic to this problem results in the schedule shown in Fig. 9.4a, whereas an optimal schedule for the same problem is shown in Fig. 9.4b. With $C_{max} = 7$ for the *LPT* schedule shown in (a), and

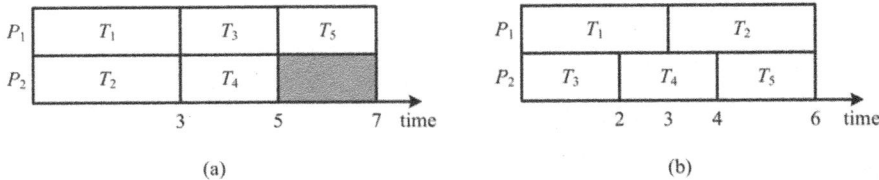

Fig. 9.4 Nonoptimal (**a**) and optimal (**b**) schedule for Example 4

$C_{\max} = 6$ for the optimal schedule shown in (b), we see that the worst-case bound of $R_{LPT} = 7/6$ is actually achieved.

Having shown that the worst-case scenario can occur for the performance ratio bound of $R_{LPT} = \frac{4}{3} - \frac{1}{3m}$, we can look at the positive side and conclude that for a two-machine problem an *LPT* schedule can never be poorer than 16.7% above the optimal value $C_{\max}$. For a three-machine problem, this becomes slightly worse with $R_{LPT} = \frac{4}{3} - \frac{1}{3(3)} = 11/9 \approx 1.222$, i.e., $C_{\max}$ will never be more than 22% longer than its optimal value (see Example 3 above). For four machines, we obtain 25%, and for five machines 26.7%; for a large number of machines, the value approaches $33\frac{1}{3}\%$.

When scheduling tasks on several parallel processors, it may sometimes be possible and advantageous to allow *preemption*. In such a case, each task assigned to a processor may be preempted, i.e., stopped, at any time and restarted later, at no additional cost, on any (possibly another) processor. It turns out that preemption makes the problem of minimizing the schedule length $C_{\max}$ on parallel machines easy. The so-called *Wrap-Around Rule* of McNaughton finds an optimal schedule when preemption is permitted. More specifically, assume that there are m parallel processors on which n tasks are to be performed with processing times p_j, $j = 1, 2, \ldots, n$. Clearly, no schedule exists with a makespan $C_{\max}$ shorter than the longest of the processing times $\max_{1 \leq j \leq n} \{p_j\}$. It is also apparent that $C_{\max}$ cannot be shorter than the mean processing time of all jobs, i.e., $\frac{1}{m} \sum_{j=1}^{n} p_j$. Therefore, we must have

$$C_{\max} \geq \max \left\{ \max_{1 \leq j \leq n} \{p_j\}, \frac{1}{m} \sum_{j=1}^{n} p_j \right\}.$$

The Wrap-Around Rule of McNaughton will actually achieve this lower bound. The algorithm can be described as follows. Recall that there are m parallel machines, n jobs with processing times $p_1, p_2, \ldots, p_n$, and the objective is to determine a schedule that minimizes the schedule length $C_{\max}$, while allowing the preemption of jobs.

McNaughton's Wrap-Around Rule: First sequence the tasks in some arbitrary order, thus obtaining a sequence of length $p_1 + p_2 + \ldots + p_n$ time units. Then compute

9.3 Parallel Machine Scheduling Models

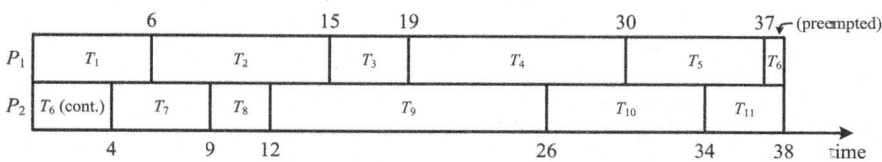

Fig. 9.5 Optimal schedule for Example 5

$$C^*_{max} := \max \left\{ \max_{1 \le j \le n} \{p_j\}, \frac{1}{m} \sum_{j=1}^{n} p_j \right\}$$

and break the time sequence at the points $iC^*_{max}, i = 1, 2, \ldots, m-1$. Now schedule all tasks in the interval $[(i-1)C^*_{max}; iC^*_{max}]$ on processor P_i, $i = 1, 2, \ldots, m$, noting that preempted tasks may be at the beginning and/or the end of each processor schedule. Finally, any task that is preempted and accordingly processed on two different processors P_i and P_{i+1} will have to be processed first on P_{i+1}, then preempted, and finished on P_i.

The Wrap-Around Rule can be shown to find a schedule whose length equals C^*_{max}. It is clear that there will be no idle time on any of the processors, thanks to the option of preempting tasks, unless there is any job with a processing time longer than $\frac{1}{m} \sum_{j=1}^{n} p_j$.

Example 5 Consider the processing times for the eleven tasks in Example 2, but ignore the due dates. If the tasks are to be processed on two processors (teams in our example) with a minimal schedule length, we first compute

$$C^*_{max} = \max \left\{ \max_{1 \le j \le 11} \{p_j\}, \tfrac{1}{2} \sum_{j=1}^{11} p_j \right\} = \max \{p_9, \tfrac{1}{2}(76)\} = \max \{14, 38\} = 38 .$$

The Wrap-Around Rule will then produce the optimal schedule displayed in Fig. 9.5.

In a practical application of this optimal schedule, job T_6 would start on processor P_2 at time $t = 0$ and preempted at $t = 4$; it would then continue being processed on P_1 at $t = 37$ and finished at $t = 38$.

Assume now that three processors (or teams) were available, and that the processing time of job T_9 increases from $p_9 = 14$ to $p_9 = 34$. We then obtain $C^*_{max} = \max \{34, \tfrac{1}{3}(96)\} = \max \{34, 32\} = 34$. Some idle time is now inevitable due to the long processing time p_9. Using the Wrap-Around Rule, we obtain the schedule displayed in Fig. 9.6.

For the preempted jobs in this optimal schedule, job T_5 will commence being processed at $t = 0$ on P_2, preempted at $t = 3$, and then continued on P_1 at $t = 30$, until it is finished at $t = 34$. Job T_9 will start being processed on P_3 at $t = 0$ and being processed completely without being preempted, until it is finished at $t = 34$ on P_3.

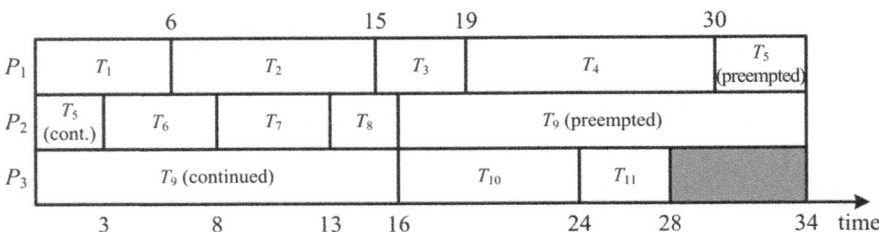

Fig. 9.6 Optimal schedule with three processors

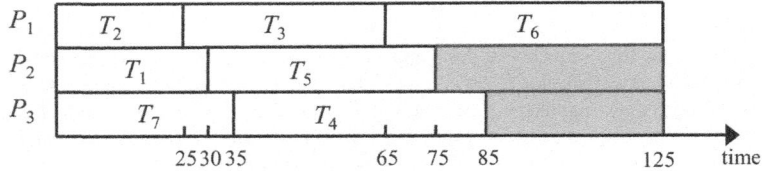

Fig. 9.7 Optimal schedule for minimizing mean flow time

Job T_{10} is being processed on P_2 from $t = 16$ to 24, immediately followed by T_{11} from $t = 24$ to 28, after which P_2 is idle until the end of the schedule at $t = 34$.

Consider now the second of our objectives, which is to minimize mean flow time F. It has been shown for the case of identical parallel processors, preemption is not profitable. Therefore, we will consider only nonpreemptive schedules. The problem can then be solved to optimality by means of a fairly simple technique. It assumes that all tasks are ready at the beginning of the process, meaning that the mean flow time reduces to the mean completion time.

Recall that there are n jobs and m machines. We can then formulate the following method:

An algorithm for minimizing mean flow time: First sort the jobs in order of nondecreasing processing time (ties are broken arbitrarily). Renumber them as $T'_1, T'_2, \ldots, T'_n$. Then assign tasks $T'_1, T'_{1+m}, T'_{1+2m}, \ldots$ to machine P_1, tasks $T'_2, T'_{2+m}, T'_{2+2m}, \ldots$ to machine P_2, tasks $T'_3, T'_{3+m}, T'_{3+2m}, \ldots$ to machine P_3, and so forth. The tasks are processed in the order that they are assigned in.

In order to illustrate this algorithm, consider again Example 1 above. Recall that the reordered sequence of tasks is $(T'_1, T'_2, T'_3, T'_4, T'_5, T'_6, T'_7) = (T_2, T_1, T_7, T_3, T_5, T_4, T_6)$ with the processing times (25, 30, 35, 40, 45, 50, 60). Given that we have three machines, we assign to machine P_1 the tasks T'_1, T'_4, and T'_7 (or, renumbering them again, T_2, T_3, and T_6), machine P_2 is assigned the jobs T'_2 and T'_5 (i.e., T_1 and T_5), and machine P_3 will process jobs T'_3 and T'_6, i.e., T_7 and T_4. The resulting optimal schedule is shown in Fig. 9.7.

The mean completion time of the solution in Fig. 9.7 is $F = \frac{1}{7} \times [25 + 65 + 125 + 30 + 75 + 35 + 85] = 440/7 \cong 62.8571$. The seemingly straightforward extension to the model with nonidentical ready times renders the problem very difficult from a computational point of view. This is yet another indication how fine a line there is between (sometimes very) easy and (sometimes very) difficult problems.

Finally, the minimization of the maximal lateness in case of parallel machines turns out to be difficult, and we leave its discussion to specialized books, such as Eiselt and Sandblom (2004).

9.4 Dedicated Machine Scheduling Models

This section deals with different types of dedicated machine scheduling models. The first such model includes an open shop. Recall that in an open shop, each task must be processed on each of a number of different machines, performing different operations. The sequence of machines, in which the jobs are processed, is immaterial. Here, we will deal only with the case of two machines, which happens to be easy, while problems with three or more machines are difficult. Minimizing the schedule length (makespan) C_{max} is easy. Optimal schedules can be found by means of the *Longest Alternate Processing Time (LAPT) algorithm*. It can be described as follows:

> *LAPT Algorithm:* Whenever a machine becomes idle, schedule the task on it that has the longest processing time *on the other machine*, provided the task has not yet been processed on that machine and is available at that time. If a task is not available, the task with the next longest processing time on the other machine is scheduled. Ties are broken arbitrarily.

In order to illustrate the method, consider:

Example 6 In an automotive assembly shop, each semi-finished product goes through two phases, assembly of individual components, and checking of the components. The sequence in which these tasks are performed is immaterial. The times (in minutes) that it takes to assemble and check the six products are shown in Table 9.3.

Using the *LAPT* algorithm, we begin by scheduling a task on machine P_1. The task with the longest processing time on P_2 is T_3, so this job is scheduled first on P_1. The next step is to schedule a task on P_2 which is now (we are still at time zero) idle. The task with the longest processing time on P_1 is again T_3, which is not available at this time, so that we schedule the task with the next longest processing time on P_1 next. This is either T_1 or T_4. Arbitrarily choose T_1. With tasks T_3 and T_1 being scheduled on the two machines, T_1 is the first task whose processing is finished at

Table 9.3 Data for Example 6

Job #	T_1	T_2	T_3	T_4	T_5	T_6
Processing time on P_1	30	15	40	30	10	25
Processing time on P_2	35	20	40	20	5	30

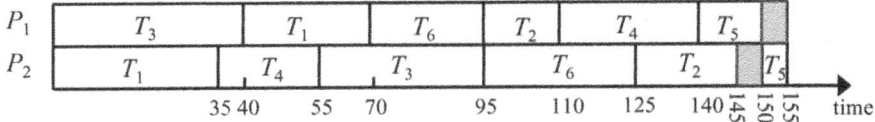

Fig. 9.8 Optimal schedule for Example 6

time 35, and machine P_2 becomes idle again. At this time, the available task with the next longest processing time on P_1 is T_4, which is then scheduled next on P_2. The process continues in this fashion, and the resulting optimal schedule is shown in Fig. 9.8, where the shaded areas toward the end again indicate idle times on the machines. The total schedule length is $C_{\max} = 155$ min.

There are no simple extensions of this model that are computationally easy. The other two objectives (i.e., those that minimize mean completion time and maximal lateness) are both very difficult from a computational point of view, even for two machines. For their discussion, we refer to the advanced literature on the subject.

The second dedicated machine scheduling model is a flow shop model, i.e., a model in which each task has to be processed by all machines, but in the same, prespecified order. The schedule in Fig. 9.8 does not satisfy this condition; note that for instance job T_3 is processed on P_1 first and later on P_2, while task T_1 is processed on P_2 first and then on P_1. Similar to the case of open shops, there are very few cases that are easy to solve. Among them is the case of two machines, for which the makespan is to be minimized. The solution algorithm is the famed *Johnson's rule* that was first described in the early 1950s. Assuming that all jobs have to be processed on P_1 first and then on P_2, and the processing time of task T_j is p_{1j} on machine P_1 and p_{2j} on machine P_2, the algorithm can be summarized as follows:

> *Johnson's Algorithm:* For all jobs whose processing time on P_1 is the same or less than their processing time on P_2 (i.e., $p_{1j} \leq p_{2j}$), determine the subschedule S_1 with the tasks in nondecreasing order of their p_{1j} values. For all other jobs (i.e., tasks for which $p_{1j} > p_{2j}$), determine the subschedule S_2 with all tasks in nonincreasing order of their p_{2j} values. The sequence of jobs is then (S_1, S_2).

Example 7 As a numerical example for Johnson's rule, consider again the scenario in Example 6, but with the proviso that each job has to be assembled first and then it is checked, i.e., it is processed on machine P_1 first before it can be processed on P_2. The tasks for which the condition $p_{1j} \leq p_{2j}$ holds are T_1, T_2, T_3, and T_6, while the jobs with $p_{1j} > p_{2j}$ are T_4 and T_5. Putting the former four tasks in nondecreasing order of the processing times on P_1 results in the subsequence $S_1 = (T_2, T_6, T_1, T_3)$, while the latter two jobs in nonincreasing order of their processing time on P_2 are put in the sequence $S_2 = (T_4, T_5)$. The resulting sequence is $(T_2, T_6, T_1, T_3, T_4, T_5)$, and the corresponding schedule is shown in the Gantt chart in Fig. 9.9.

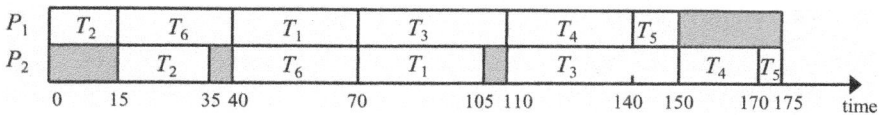

Fig. 9.9 Optimal schedule for Example 7

It is apparent that the schedule length is $C_{max} = 175$. Note the increase in the schedule length in comparison to the same example, in which the sequence of processing in the machines is not fixed (i.e., the open shop) shown in Fig. 9.8. Given that a flow shop is more restrictive than an open shop (it has the additional constraint that all tasks have to be performed in the same order on the machines), the increase of the schedule length from 155 to 175 min is not surprising.

Simple extensions of the problem to more than two machines as well as the application of the mean flow time and the minimization of tardiness are computationally much more difficult. Again, we refer readers to the specialized literature.

The last model in this chapter deals with a job shop. Recall that by definition of a job shop, not all tasks need to be performed on all machines, and the sequence in which a job is processed on the machines is job specific. Again, due to the inherent complexity of the problem at hand, we will restrict ourselves to the problem with two machines and the objective that minimizes the makespan. For this type of problem, Jackson described an exact algorithm in 1955. Note that Jackson's method uses Johnson's (flow shop) algorithm as a subroutine. The method can be described as follows:

Jackson's Job Shop Algorithm: Subdivide the set of jobs into four subcategories:
J_1 includes all jobs that require processing only on machine P_1,
J_2 includes all jobs that require processing only on machine P_2,
J_{12} is the set of jobs that require processing on machine P_1
first and then on P_2, and
J_{21} is the set of jobs that need to be processed on P_2 first
and then on P_1.
Apply Johnson's rule to the jobs in the set J_{12}, resulting in the sequence S^{12}. Then apply Johnson's rule to the jobs in the set J_{21}, but with the processing times p_{1j} and p_{2j} exchanged. The result is the subsequence S^{21}. All jobs in the two sets J_1 and J_2 are sequenced in arbitrary order (e.g., with those jobs that have smaller subscripts scheduled first). We denote their sequences by S^1 and S^2, respectively. The jobs are then sequenced as follows: the job order on machine P_1 is (S^{12}, S^1, S^{21}), while the job order on machine P_2 is (S^{21}, S^2, S^{12}).

Example 8 As a numerical illustration, we will use a modification of Example 6. The pertinent information is found in Table 9.4.

First defining the sets, we have $J_1 = \{T_4\}$, $J_2 = \{T_3\}$, $J_{12} = \{T_2, T_5\}$, and $J_{21} = \{T_1, T_6\}$. Since J_1 and J_2 include only a single task each, their respective subsequences are $S^1 = (T_4)$ and $S^2 = (T_3)$. Applying Johnson's algorithm to the tasks

Table 9.4 Data for Example 8

Job #	T_1	T_2	T_3	T_4	T_5	T_6
Processing time p_{1j}	30	15	–	30	10	25
Processing time p_{2j}	35	20	40	–	5	30
Processing sequence	P_2, P_1	P_1, P_2	P_2	P_1	P_1, P_2	P_2, P_1

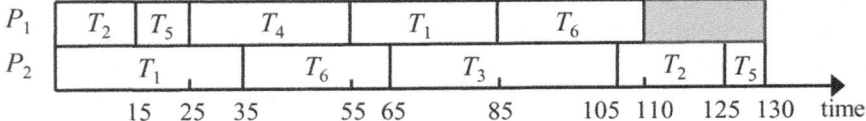

Fig. 9.10 Optimal schedule for Example 8

in the set J_{12}, we obtain the sequence $S^{12} = (T_2, T_5)$. Consider now the set J_{21}. Applying Johnson's rule to the two tasks in the set, viz., T_1 and T_6, with their processing times p_{1j} and p_{2j} switched, we obtain the subsequence $S^{21} = (T_1, T_6)$. Following Jackson's job shop algorithm, the overall sequence on machine P_1 is then $(S^{12}, S^1, S^{21}) = (T_2, T_5, T_4, T_1, T_6)$, while the overall sequence on machine P_2 is $(S^{21}, S^2, S^{12}) = (T_1, T_6, T_3, T_2, T_5)$. The appropriate Gantt chart is shown in Fig. 9.10. It turns out that the overall schedule length $C_{max} = 130$ min. We also note that processor P_1 is idle for 20 min at the end of the schedule.

Exercises

Problem 1 (single- and parallel machine scheduling): A machine scheduling problem has ten jobs with given processing times of 3, 1, 4, 1, 5, 9, 2, 6, 5, and 3 hrs, respectively.

(a) Assume that the jobs are to be performed on a single machine and that the objective is to minimize mean flow time, find an optimal schedule, and draw the corresponding Gantt chart.
(b) Assume that due dates for the ten jobs are given as 14, 4, 10, 19, 40, 32, 13, 31, 7, and 26 hrs, respectively. Use the earliest due date algorithm to find a schedule that minimizes the maximal lateness L_{max}. Show the corresponding Gantt chart. What is the minimal value of L_{max}, and for which job does it occur? Which jobs are late?
(c) Assume that there are two parallel machines to process the jobs. Trying to minimize the schedule length, schedule the jobs using the *LPT* algorithm. Display the corresponding Gantt chart. What is the performance ratio?
(d) Given two parallel machines as under (c), what is the schedule that minimizes the mean flow time? Display the corresponding Gantt chart.
(e) Reconsider questions (c) and (d) given that there are now three rather than two parallel machines.

9.4 Dedicated Machine Scheduling Models

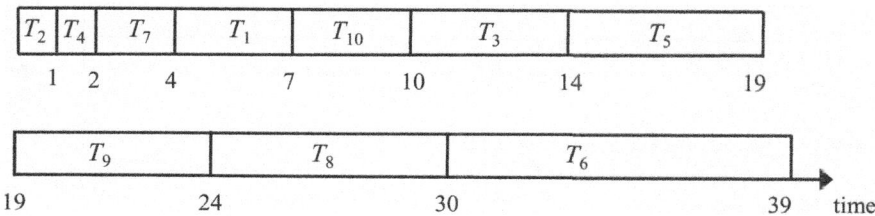

Fig. 9.11 Optimal schedule for Problem 1 (a)

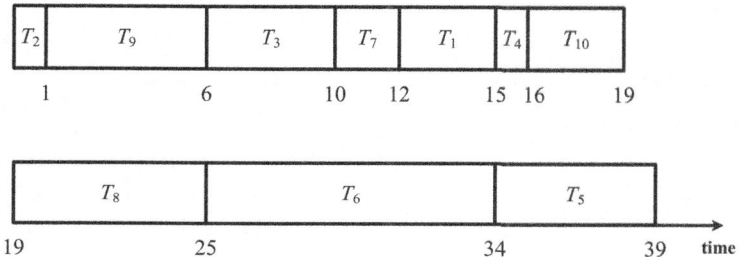

Fig. 9.12 Optimal schedule for Problem 1 (b)

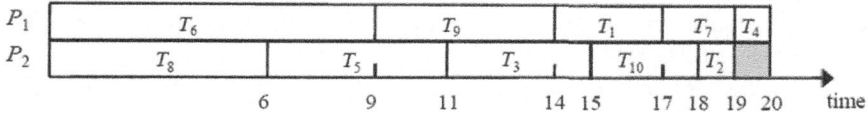

Fig. 9.13 Optimal schedule for Problem 1 (c)

Solution:

(a) Assuming ready times $r_j = 0$, the mean flow time F is minimized by the shortest processing time (*SPT*) algorithm. The sequence of jobs is $T_2, T_4, T_7, T_1, T_{10}, T_3, T_5, T_9, T_8$, and T_6. The corresponding Gantt chart is shown in Fig. 9.11.
The minimal mean flow time for this schedule is then $F = \frac{1}{10} \times (1 + 2 + 4 + 7 + \ldots + 30 + 39) = 15$. Since T_1 and T_{10} both have processing times of 3 hrs each, they may be swapped in the optimal schedule, thus creating alternative optimal solutions. A similar argument applies to the pairs T_2 and T_4, and T_5 and T_9.

(b) Using the *EDD* algorithm, we obtain the optimal schedule shown in Fig. 9.12. It turns out that task T_1 is late by 1 hr and task T_6 is late by 2 hrs. All other tasks are finished on time (but T_3 only just). The lateness L_{max} therefore equals 2 hrs, obtained for T_6.

(c) With two machines P_1 and P_2, the longest processing time *LPT* algorithm minimizes the makespan C_{max}.

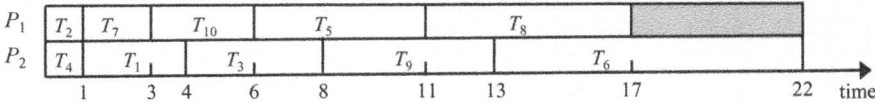

Fig. 9.14 Optimal schedule for Problem 1 (d)

Fig. 9.15 Optimal schedule for Problem 1 (e)

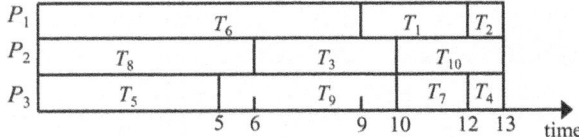

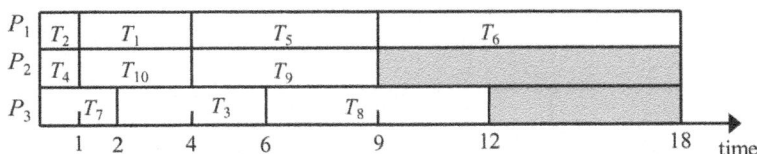

Fig. 9.16 Minimizing mean flow time for problem 1 (c)

The makespan is 20 hrs. In the schedule shown in Fig. 9.13, one of the machines is idle for 1 hr.

(d) With two machines, find a schedule that minimizes the mean flow time. The Gantt chart is shown in Fig. 9.14.

The schedule length is 22 hrs, and the mean flow time is $F = \frac{1}{10} \times (1 + 1 + 3 + 4 + \ldots + 17 + 22) = 8.6$, and one processor is idle for 5 hrs. Since the schedule is optimal, the performance ratio equals 1.

(e) With three machines, we use the *LPT* rule as a heuristic to minimize $C_{\max}$. The Gantt chart is shown in Fig. 9.15.

The schedule has a makespan of 13 hrs. Since there is no idle time, the schedule must be optimal. The schedule is shown in Fig. 9.16.

The schedule length is 18 hrs, and the mean flow time is $F = 6.6$. Note that there is significant idle time.

Problem 2: (Parallel machine scheduling with preemption): Consider Example 5 in Sect. 9.3.

(a) Using the original data, apply the Wrap-Around Rule to find a schedule that minimizes makespan, assuming that preemption is allowed and that there are three, rather than two, processors.

(b) Consider again the example as under (a). Using the increased processing time of job T_9, solve the problem with two, rather than three, processors.

9.4 Dedicated Machine Scheduling Models

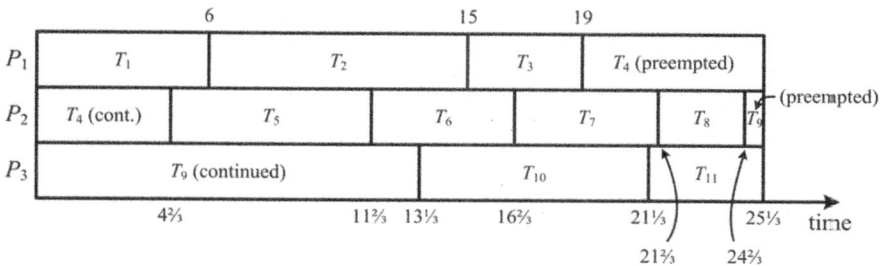

Fig. 9.17 Schedule for Problem 2 (a)

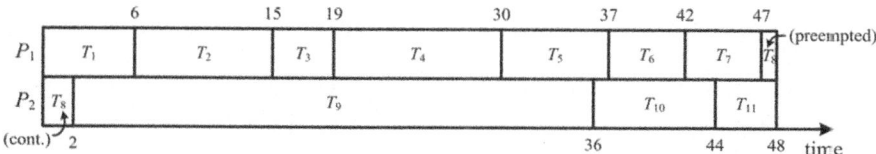

Fig. 9.18 Schedule for Problem 2 (b)

Table 9.5 Data for Problem 3

Blood sample	T_1	T_2	T_3	T_4	T_5
Processing time on P_1	8	9	7	9	3
Processing time on P_2	2	3	8	4	6

Solution:

(a) First, we compute $C^*_{\max} = \max\left\{\max_{1 \leq j \leq 11}\{p_j\}, 1/3 \sum_{j=1}^{11} p_j\right\} = \max\{p_9, 1/3(76)\} = \max\{14, 25 1/3\} = 25 1/3$. Using the Wrap-Around Rule we then obtain the schedule displayed in the Gantt chart in Fig. 9.17.
An implementation of this schedule would start T_4 on P_2, preempt it at $t = 4 2/3$, and then resume T_4 on P_1 at time $t = 19$. Task T_9 then starts at P_3 at $t = 0$ and is then preempted at $t = 13 1/3$, after which it resumes at $t = 24 2/3$ on P_2.

(b) We now find $C^*_{\max} = \max\{34, 1/2(96)\} = 48$, and the Wrap-Around Rule generates the schedule shown in Fig. 9.18.
For a practical implementation of this schedule, task T_8 will be started on processor P_2 at time $t = 0$, will be preempted at $t = 2$, and then resumed on P_1 at time $t = 47$.

Problem 3 (open shop and flow shop scheduling): In a hospital laboratory, there are two machines testing patient blood samples. Table 9.5 shows the number of minutes required on each machine to process the samples.

(a) Assume that the order of processing the blood samples on the two machines is arbitrary. Schedule the testing on the two machines so as to minimize the

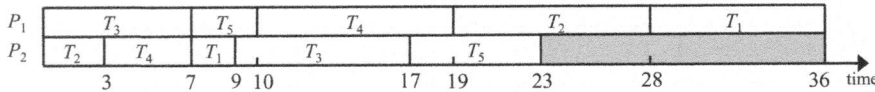

Fig. 9.19 Schedule for Problem 3 (a)

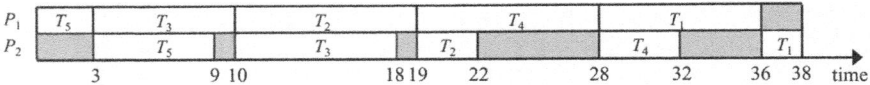

Fig. 9.20 Schedule for Problem 3 (b)

schedule length. Display the optimal schedule in a Gantt chart. Indicate the schedule length as well as the idle time.
(b) Assume now that all blood samples must be processed on P_1 before they can be processed on P_2. Redo part (a) with these new assumptions.
(c) Are the optimal schedules in (a) and (b) unique? Discuss. There is no need to display Gantt charts.

Solution:

(a) The *LAPT* algorithm is used to obtain the optimal schedule shown in Fig. 9.19. The minimal schedule length is $C_{\max} = 36$ min. There is no idle time on P_1, while P_2 is idle at the end for 13 min. This is an open shop model.
(b) Johnson's rule is used to obtain the optimal schedule shown in Fig. 9.20. The minimal schedule length is now $C_{\max} = 38$ min. Processor P_1 has an idle time of 2 min at the very end of the schedule, whereas P_2 has five separate idle time periods, totaling 15 min. This is a flow shop model.
(c) In (a), tasks T_2 and T_4 could swap positions in the schedule of P_1 (and on P_2 for that matter) without consequences regarding the schedule length. There are several other changes that would not destroy optimality. In (b), tasks T_2 and T_4 could also swap positions on P_1, necessitating modifications on processor P_2.

Reference

Eiselt HA, Sandblom C-L (2004) Decision analysis, location models, and scheduling problems. Springer, Berlin

Decision Analysis 10

Everywhere in the world, at each moment, millions of people make their own decentralized decisions: when to get up in the morning, what tie to wear, what to eat for lunch or dinner, what to do in the evening (go to the theater or watch television), where to vacation, and many more. Similarly, firms will decide which mode of transportation to use when routing their products to customers, where to locate regional distribution centers, what new product lines to develop, etc. This chapter first introduces the main elements of decision analysis, and then offers some visualizations of decision analysis problem. This is followed by a discussion of some simple decision rules, sensitivity analyses, and a discussion of the value of information. The chapter wraps up the discussion by some thoughts on utility theory.

10.1 Introduction to Decision Analysis

In order to put decision making into a general framework, we must first distinguish between the three major elements of decision analysis: the finite set of *decisions* available to the decision maker (whom we will think of as "us"), the *states of nature*, which are the decisions available to our opponent (more about that below), and the *outcome* that results from the combination of our decision and that of our opponent. Typically, the outcome is given in monetary terms, and it is usually referred to as the *payoff*.

The type of problem under consideration in this chapter is not a philosophical investigation into the decision-making process; instead, it refers to a very specific scenario that is prevalent in decision-making circumstances. In particular, we assume that there are a finite number of decisions at our disposal. Among these choices, the decision maker's task is to choose exactly one. This type of situation is often referred to as a *selection problem*. Returning to the capital budgeting decision in the introduction to integer programming in Chap. 5, we can define a binary variable y_j for each decision, such that the variable assumes a value of one if we

Table 10.1 Payoffs

	s_1	s_2	s_3	s_4
d_1	3	−2	4	6
d_2	2	0	−4	1
d_3	5	2	0	−3

make decision j, and 0 otherwise. A selection problem with n possible decisions will then feature the constraint $y_1 + y_2 + \ldots + y_n = 1$.

However, in contrast to problems that can simply be formulated as integer programming problems, there are two possible extensions that typically arise. The first extension involves the evaluation of a decision on more than a single criterion. As an example, consider a department store that considers changing the layout of its store. There are, of course, the costs of such a decision. But there is more: there is the changed customer flow that may result in higher exposure of some goods to customers and resulting potential higher sales of these products (the main reason for a change in layout), the (temporary) confusion of customers who may refrain from purchasing at the store, the potential retraining of employees if the layout change was in conjunction with new products added to the goods available at the store, and so forth. Selection problems of this type, in which multiple attributes of decisions are considered, are called *multiattribute decision-making* problems (or *MADM*). Problems of this type are discussed in Chap. 11. On the other hand, suppose that we only consider a single criterion, but the outcome of our decision is no longer certain. For instance, consider a major public project, such as the construction of a new airport or a new sewage treatment plant for a major city. There are the uncertainties of future inflation rates, the potential of strikes, possible future changes of interest rates, major weather disruptions, and a myriad of other unforeseen events that can and most likely will change the finishing time of the project and its cost. This is the standard scenario in *decision analysis* or, as it is frequently called, *games against nature*. The name stems from game theory and can be explained as follows. Consider a standard two-person game with two players, one decision maker (us), and the other being our opponent (nature). Each of the two players has a number of possible actions at his disposal: the decision maker has the decisions, while nature controls the "states of nature," nature's equivalent of the decision maker's decision choices. This is the scenario examined in this chapter.

There is, however, a fundamental asymmetry in games against nature. First of all, the combination of the decision maker's choice and nature's state of nature will determine the outcome *for the decision maker* (nature will face no outcome). Secondly, the decision maker will examine the possible outcomes of his decisions before choosing one, while nature does not consider the outcomes, but chooses her strategies according to some probability distribution. This is the reason why the decision maker is usually called "intelligent" (we prefer to think of it as "rational"), while nature is referred as a "random player."

As an illustration, consider the following numerical example with three decisions d_1, d_2, and d_3, and four states of nature s_1, s_2, s_3, and s_4. The payoffs are shown in Table 10.1.

The decision maker could argue that d_1 is best, as it provides a reasonable payoff for three states of nature, while its largest possible loss is -2 and as such not as bad as the losses that can occur with the other two decisions. If the decision maker were to choose d_1, while nature would randomly choose s_3, then the decision maker would obtain a payoff of 4.

It is important to realize that nature's decision is either made simultaneously with that of the decision maker without cooperation, or, equivalently, the decision maker chooses first, followed by nature's choice. As in all game-theoretic situations, it is crucial to specify which player knows what and when. This also marks the distinction concerning the level of certainty the decision maker has about nature's choice. As usual, consider the extremes first. If the decision maker knows with certainty what nature is going to do, we face a decision problem under certainty. Given the scenario of selection problems, this means that the decision maker knows which column of the payoff matrix will apply. In the example of Table 10.1, suppose that the decision maker knows that nature will choose s_1. This means that the consequences of the decisions are known with certainty: choosing d_1 will result in a payoff of 3, choosing d_2 will result in a payoff of 2, and choosing decision d_3 will yield a payoff of 5. Clearly, the decision maker's payoff is maximized by the payoff of 5, which we arrive at by choosing d_3. This is the optimal solution, and the problem is solved. (Before we continue with different levels of knowledge, note that the decision maker has jurisdiction only over his choices, e.g., d_3, but he cannot choose the payoff directly.)

On the other extreme is *uncertainty*. In decision making under uncertainty, the decision maker has absolutely no idea about nature's choice. Uncertainty is quite rare; it may occur in the performance of new and untried products, behavior of customers in new and untested markets, and other situations, in which no information is provided. As in all decision-making situations, if the level of input into the problem is low, the output will have to make do with simplistic rules. This is precisely what happens in this situation as will be seen below.

Clearly, there is much territory between certainty and uncertainty. One milestone in between is decision making under *risk*. In decision making under risk, the decision maker is assumed to know the probability distribution used by nature. Examples would include past weather observations for farmers, predictions concerning customer behavior based on similar situations, and so forth. Rules for decision making under risk are described below.

10.2 Visualizations of Decision Problems

Before getting into specific rules for different situations, we would like to describe some ways to visualize decision-making scenarios. Different visualizations on different levels are available. For the macro view, there are *influence diagrams*. The idea is to show the basic interdependencies between decision and outcome, while ignoring details. Their biggest strength is clarity, which is achieved by their concentration of fundamental relations and their resulting small size. On the other

Table 10.2 Decision, events, and consequences for the influence diagram

Decision	Random event	Consequence
Add electronics department	General economic conditions	Profit
Relocate department into a separate building	Local acceptance of services	

hand, *decision trees* are used for the micro view. They include specific decision rules at each step of the way. Consequently, they tend to be large and cumbersome to deal with.

In order to demonstrate influence diagrams, we first make a list with three columns. The first column includes the decision maker's possible decisions, the second column represents the random events that somehow influence the decisions and the outcomes, and the last column includes the consequences that are a result of the decisions and the random events. As an illustration, consider a department store that contemplates adding an electronics department to its services. In case the introduction is accepted by the public, management considers relocating that department into a separate building. The aforementioned listing is shown in Table 10.2.

In order to visualize the problem, we can create an influence diagram, in which our decisions are shown by rectangles, random events are shown as circles, and consequences are shown by triangles. We then add directed arcs, so that an arc from some node i to another node j exists, if it is believed that i influences j. The influence diagram for our problem could look as shown in Fig. 10.1.

The broken arcs in the figure are somewhat tenuous: they indicate the belief that local acceptance of an electronics department or store is influenced by the existence of an electronics department in our department store and our competitors' reaction to our introduction of the department.

On the other hand, we could zoom in and outline our decisions, our competitors' decisions, and economic conditions in a decision tree. Decision trees have decisions and events listed next to the arcs. A square node indicates that we will make a decision, a round node means that a decision is made not within our control (i.e., a random event), and a triangular node denotes the end of this branch of the tree. Normally, there will be an indication next to a triangular node what the payoff is to us at that point. For now, we will concentrate on the structure of the tree and return to the numerical aspects of decision trees in a detailed discussion in Sect. 10.5. A possible decision tree for our sample problem is shown in Fig. 10.2.

The figure of the decision process indicates that after we have decided to expand, our competitor(s) could either not react, also expand their stores accordingly, or build a superstore. While these are not random events, they are shown here as such as these decisions are not within our control. In case our competitors do not react, our profit will depend on the state of the economy, which is shown in the Figure simply as "up" or "down." (This exceedingly simplistic notion has been used so as to save space—it should have become clear by now that decision trees tend to get very large even if the players do not have very many options.) In case our competitor(s) expand, we can either do not react ourselves or build the planned new building. Each of our decisions will be followed by a random event, at which nature decides which turn the economy takes. Finally, if the competitors have decided to build a superstore, we

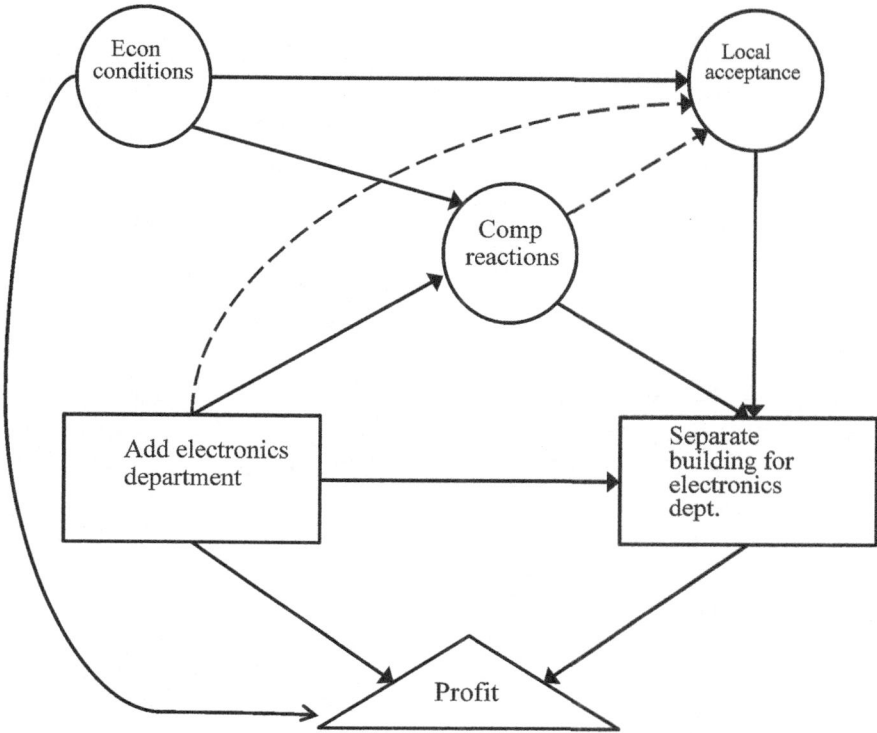

Fig. 10.1 Influence diagram

may either withdraw due to limited options to raise capital or build the planned addition. Again, each decision is followed by the economy going up or down.

10.3 Decision Rules Under Uncertainty and Risk

Our discussion below will be based on a numerical example with the payoff matrix

	s_1	s_2	s_3
d_1	2	-2	5
d_2	0	-1	7
d_3	2	1	1
d_4	2	-3	4

Before performing any calculations, it is useful to first examine the payoff matrix regarding *dominances*. A decision i (indicating a row) dominates a decision k (a different row) if all elements in row i are at least as large as the corresponding payoffs in row k. In other words, for each state of nature, decision i is at least as good as decision k. If this is the case, then decision k can be deleted from consideration.

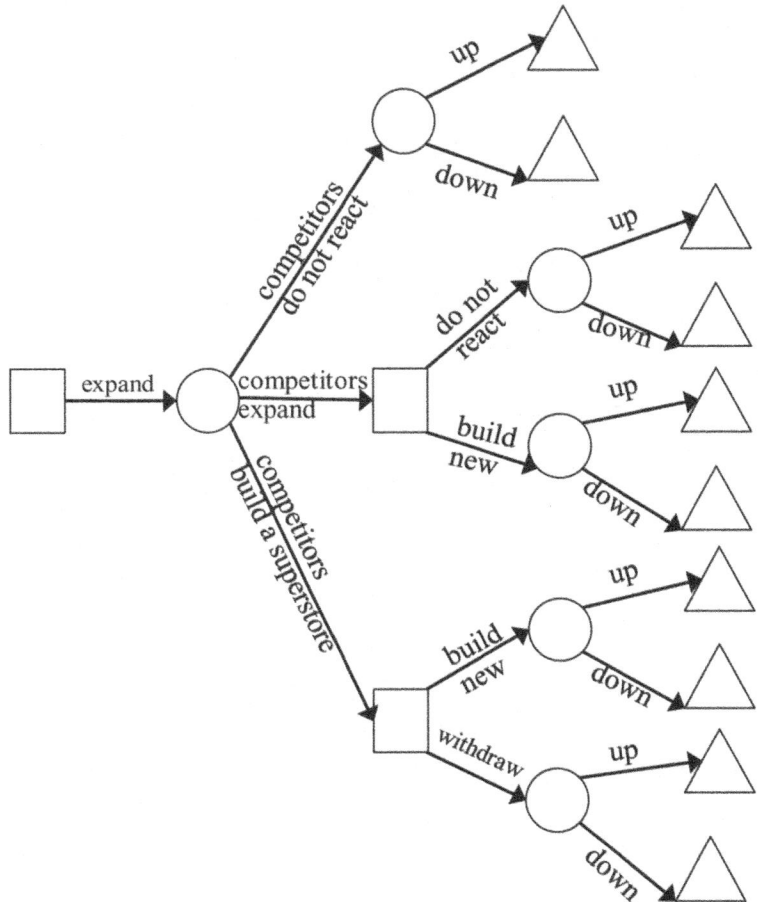

Fig. 10.2 Decision tree for sample problem

The determination whether or not dominances exist in a problem requires pairwise comparisons. In our example, let us first compare decisions d_1 and d_2. Given the first state of nature s_1, decision d_1 results in a payoff of 2, while d_2 nets us only 0, so d_1 is preferred. However, given s_2, decision d_1 results in a loss of 2, while d_2 results in a loss of only 1, so that d_2 is preferred. This means that neither decision is consistently better than the other. Comparing d_2 and d_3 also results in no dominance (d_3 is preferred in case s_1 or s_2 occurs, while d_2 is preferred to d_3 in case s_3 comes up). However, the picture changes when comparing d_1 and d_4. Here, it is apparent that d_1 and d_4 are equally good given s_1, while d_1 is better than d_4 in case of s_2 and s_3, so that d_1 dominates d_4; thus, d_4 can be deleted. The examination would have to continue comparing the remaining pairs of decisions.

Given m decisions, $\frac{1}{2}m(m-1)$ pairs of decisions will have to be compared. If any dominances are missed by accident, no harm is done: the model will be a bit bigger

10.3 Decision Rules Under Uncertainty and Risk

than it needs to be, but no "reasonable" rule will choose a dominated decision. Note that we cannot apply the concept of dominance to the states of nature (i.e., the columns of the payoff matrix). The reason is that the concept of dominances is based on a comparison of the payoffs, i.e., a rational decision maker, and nature is no such player.

Consider now a decision-making problem under uncertainty. One simple decision rule will attempt to guard against the worst case. It has a variety of names, *Wald's rule* named after its inventor, the *pessimistic rule* based on the mindset of the decision maker, and the *maximin rule* based on the way the rule works. First of all, we will determine the *anticipated payoffs* associated with our decisions. In our example, a pessimist choosing d_1 would assume that the worst case applies and the payoff will be -2. Similarly, decision d_2 would result in a payoff of -1, and so forth. The vector of anticipated payoffs would then be $\mathbf{a} = [-2, -1, 1, -3]^T$. Formally, these are the row minima of the payoff matrix. Choosing the best among these decisions will then be d_3 as it leads to the maximum payoff among the anticipated payoffs. This is the reason for calling the rule a "maximin" rule.

While the rule surely protects against the worst case, it has a number of shortcomings. The most predominant problem with it is its exclusive focus on the worst case. For instance, if someone had to pick up his multimillion-dollar winnings from the lottery office, the worst-case rule would suggest that he not do that: while walking or driving to the office, he might get hit by a truck and die, and this worst case has to be avoided. Since it is unknown how likely (or, in this case, how unlikely) such an incident would be, a decision made on the basis of Wald's rule would try to prevent it.

Another, very similar, rule is the *optimist's rule*. An optimist's anticipated payoff would include the best possible payoff for each decision, i.e., the row maxima. In our example they are $\mathbf{a} = [5, 7, 2, 4]$. The optimist would then choose the decision with the highest anticipated payoff, leading to a *maximax rule*. In our illustration, d_2 would be the optimist's choice. This Pollyanna-inspired rule suffers from the same limitations as Wald's rule, except that it has replaced guarding against the worst case by anticipation of the best case.

A third rule for making decisions under uncertainty was independently developed by Savage and Niehans. It is called the *minimax regret criterion*, and it has become the basis of what is often referred to as *robust optimization*. The idea is to compare for each state of nature the payoff the decision maker gains with his decision and the best possible payoff that could have been obtained given the same state of nature. In our numerical example, we compute the regret of decision d_2 given the second state of nature s_2. The payoff to the decision maker is -1. However, if the decision maker had just known in advance which state of nature would occur (the second), he could have his best response d_3, which would have led to a payoff of 1, the highest payoff given s_2. The difference between the actual payoff and the best possible payoff under that state of nature gives a regret of $1 - (-1) = 2$. These regrets are calculated for all pairs of decisions and states of nature, and they form the *regret matrix*. In our example, the regret matrix is

$$R = \begin{bmatrix} 0 & 3 & 2 \\ 2 & 2 & 0 \\ 0 & 0 & 6 \\ 0 & 4 & 3 \end{bmatrix}.$$

Note that regrets are never negative. Furthermore, there is always at least one zero in each column of the regret matrix (belonging to the element that determines the column maximum). The decision maker can then apply any rule on the regret matrix rather than the original payoff matrix. If we were to use the pessimist's rule onto the regret matrix R, we would anticipate regrets of $r = [3, 2, 6, 4]$. Note that these are the row maxima, not the row minima as used above. The reason is that a payoff matrix includes payoffs that the decision maker would like to maximize, whereas the regret matrix features regrets that, similar to costs, the decision maker would like to minimize. Among the anticipated regrets, the decision maker will then choose the lowest, which in our example is d_2, a decision leading to an anticipated regret of 2. Applied to the regret matrix, Wald's rule is a minimax rule in contrast to the maximin version that is used in case a payoff matrix is given.

Consider now decision making under risk. In this scenario, we can associate with each state of nature a probability p_j, which has been determined by past observations. *Bayes's rule* is then used to compute the expected values and choose the decision that leads to the maximum expected payoff, making Bayes's criterion a *weighted maximum rule*. In our example, suppose that the probabilities of the three states of nature have been determined as $p = [0.5, 0.3, 0.2]$. The *expected payoffs* (or *expected monetary values EMV*) are then

$$\text{EMV}(d_1) = 2(.5) - 2(.3) + 5(.2) = 1.4,$$
$$\text{EMV}(d_2) = 0(.5) - 1(.3) + 7(.2) = 1.1,$$
$$\text{EMV}(d_3) = 2(.5) + 1(.3) + 1(.2) = 1.5, \quad \text{and}$$
$$\text{EMV}(d_4) = 2(.5) - 3(.3) + 4(.2) = 0.9.$$

Thus, the anticipated payoffs are 1.4, 1.1, 1.5, and 0.9, so that the decision maker will choose d_3, a decision that has the highest expected payoff of 1.5. In case of a tie, it is possible to use secondary criteria.

Bayes's rule has been popularized by what is known as the *Newspaper Boy Problem*. The decision of the newspaper boy concerns the number of newspapers he will purchase in the presence of demand uncertainty. The probabilities of the demand are based on past experience. If the boy buys too many papers on a slow day, he will have papers left over, which will have to be disposed of for some low salvage value. On the other hand, if he purchases too few and the paper turns out to have some interesting stories, he will not have enough papers to sell, so that not only does he lose business today, but may irritate customers who may purchase their papers elsewhere in the future. These opportunity costs also have to be included in the model.

10.3 Decision Rules Under Uncertainty and Risk

As an illustration, consider the following numerical:

Example: A newspaper boy knows that he can sell either 10, 20, 30, or 40 newspapers on any given day (barring days with major headlines, such as assassinations, wars, or the latest replacements of body parts of some actress). The boy will purchase a newspaper for 20¢ and then sell for 90¢. The salvage value of an unsold newspaper is 5¢, while the opportunity cost for newspapers has been estimated to be 15¢ for each newspaper that could have been sold but was not, due to the lack of supply. The payoff matrix for the newsboy problem is then:

$$\mathbf{A} = \begin{array}{c} \\ d_1 \\ d_2 \\ d_3 \\ d_4 \end{array} \begin{array}{cccc} s_1 & s_2 & s_3 & s_4 \\ \left[\begin{array}{cccc} \$7.00 & 5.50 & 4.00 & 2.50 \\ 5.50 & 14.00 & 12.50 & 11.00 \\ 4.00 & 12.50 & 21.00 & 19.50 \\ 2.50 & 11.00 & 19.50 & 28.00 \end{array}\right] \end{array}$$

where the decisions d_1, d_2, d_3, and d_4 refer to the newspaper boy buying 10, 20, 30, and 40 newspapers, while the states of nature s_1, s_2, s_3, and s_4 refer to a demand of 10, 20, 30, and 40, respectively. (Note that it is generally not necessary that the set of decisions equals the set of possible states of nature.) All entries on the main diagonal refer to cases, in which the sale equals the number of newspapers that were purchased, while all entries above the main diagonal involve some unsatisfied demand with its opportunity costs, while the entries below the main diagonal involve a surplus of newspapers, so that salvage values have to be applied. Given probabilities of $\mathbf{p} = [0.6, 0.2, 0.1, 0.1]$ for the four states of nature, the four decisions have expected payoffs of \$5.95, \$8.45, \$8.95, and \$8.45, so that the newspaper boy should buy 30 newspapers and expect an average daily payoff of \$8.95.

Applications of this type occur in many circumstances, in which we are dealing with perishable goods. An excellent example is the airline industry. Seats in an airplane are a perishable commodity, as, once the airplane door closes, an empty seat is worthless. The airline's decision problem is then to choose among their aircraft the type that best fits the expected demand on each of the routes.

Let us now return to our discussion of different approaches to decision making. In decision making under risk, there is also the possibility to use *target values*. The idea with this approach is to choose the decision that provides the highest probability that the payoff does not fall short of a predetermined target value. To illustrate, use again the example introduced earlier in this section. Suppose that a target value $T = 1$ has been chosen. Decision d_1 will then achieve this target value only if nature chooses either s_1 or s_3 as her strategy, which will happen with a probability of 0.5 and 0.2, respectively. This means that when using d_1, there is a probability of 0.7 that the target value is reached or exceeded.

When using decision d_2, the target value $T = 1$ will be achieved only if s_3 comes up, which happens with a probability of 0.2. Similarly, decision d_3 reaches the target value in case nature plays s_2 or s_3, so that the probability of a payoff of at least

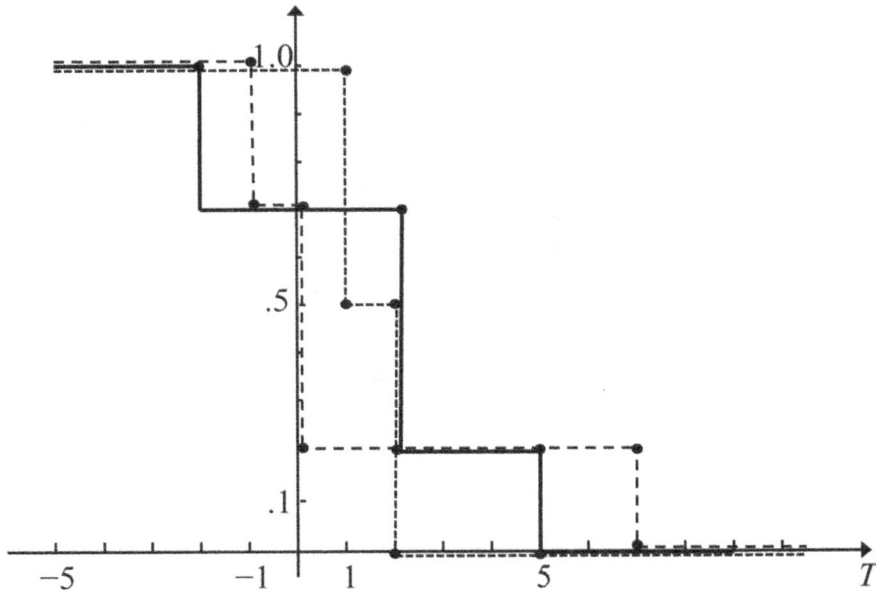

Fig. 10.3 Probability of achieving target value T

T equals 0.5, and for d_4 the probability is 0.7. The decision maker will then choose the decision that maximizes the probability of getting at least $T = 1$, which is done by choosing either d_1 or d_4. Note that one of those optimal choices is the dominated decision d_4. This is possible; however, a dominated decision can never be the *unique* optimum for any "reasonable" decision rule.

In order to derive a simple decision tool, we plot the probability that a decision achieves at least a prespecified target value T against the full range of target values T. For our example, Fig. 10.3 provides the graph in question. In particular, the solid line shows the achievements of decision d_1, the broken line is for d_2, and the dotted line shows the results for d_3. For clarity, we ignore the dominated decision d_4.

Given the graph in Fig. 10.3, we can very simply determine which decision has the highest probability of reaching a target value. For instance, if the target value were $T = -1.5$, then decisions d_2 and d_3 both have a probability of "1" to achieve this value, while decision d_1 has only a probability of 0.7 of achieving this payoff. In other words, we are interested in the "upper envelope," i.e., the highest of all of the functions. Given that, we can determine which function is highest and summarized it in the following decision rule:

If $T < -2$, any decision will achieve the target.
If $T \in [-2, -1]$, d_2 and d_3 are best. Both decisions will reach the target with a probability of 1.
If $T \in [-1, 1]$, d_3 is best. It reaches the target with a probability of 1.

If $T \in [1, 2]$, d_1 is best. It reaches the target with a probability of 0.7.
If $T \in [2, 5]$, d_1 and d_2 are best. Both achieve the target with a probability of 0.2.
If $T \in [5, 7]$, d_2 is best. It achieves the target with a probability of 0.2.
If $T > 7$, none of the decisions will be able to reach the target.

A loose summary will indicate that d_3 is best for low target values, d_1 is best for intermediate target values, and d_2 is best for high target values. This is, of course, not surprising: decision d_3 has no extremes on the low end, while d_2 does have an extreme possible payoff on the high end.

10.4 Sensitivity Analyses

This section examines two types of sensitivity analyses. The first (simpler) case assumes that a payoff, i.e., one element of the payoff matrix, is no longer known with certainty. The idea is to develop simple decision rules that provide guidance to the decision maker in case the payoff changes. Again, we will base our arguments on the example introduced at the beginning of this chapter, which is shown again here for convenience.

	s_1	s_2	s_3
d_1	2	−2	5
d_2	0	−1	7
d_3	2	1	1
d_4	2	−3	4
p	0.5	0.3	0.2

Suppose now that there is some uncertainty concerning the payoff of the second decision in case of the third state of nature. Given that we expect the actual payoff to be somewhere between 5 and 10, we can then write the payoff as $a_{23} = 7 + \varepsilon$ with an unknown $\varepsilon \in [-2, 3]$. The expected payoffs are then

$$EMV = \begin{bmatrix} 1.4 \\ 1.1 + .2\varepsilon \\ 1.5 \\ .9 \end{bmatrix}.$$

Here, we can ignore decisions d_1 and d_4 as, regardless of the value of ε, decision d_3 is better than those. This leaves us with the comparison between d_2 and d_3. Figure 10.4 plots the expected monetary values of the two decisions as a function of the change ε.

The two payoff curves $EMV(d_2)$ and $EMV(d_3)$ intersect at $\varepsilon = 2$. To the left of $\varepsilon = 2$, the payoff is higher for decision d_3, while to the right of $\varepsilon = 2$, the payoff with d_2 is higher. This leads us to the following decision rule:

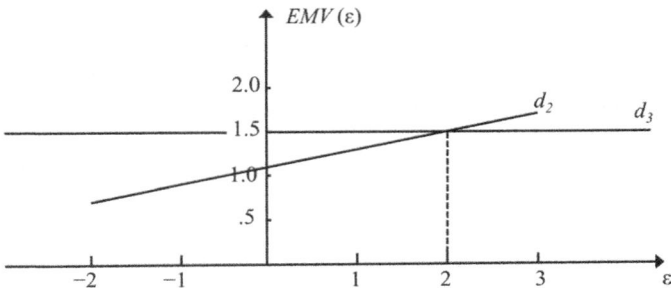

Fig. 10.4 *EMV* as a function of a payoff change

If $\varepsilon \geq 2$ (or, alternatively, $a_{23} \geq 9$), then decision d_2 is best.
If $\varepsilon \leq 2$ (or, alternatively, $a_{23} \leq 9$), decision d_3 is best.

Next, consider the possibility that there is some uncertainty surrounding a probability estimate. This case is somewhat more difficult conceptually, as the increase or decrease of a single probability will necessarily imply that other probabilities change as well, based on the simple fact that the sum of probabilities equals one. Back in our numerical example with the original payoffs, we are now no longer certain about the probability of s_1.

As an example, we may assume that an increase of p_1 by some unknown value ε may reduce the probabilities of all other states of nature by the same amount, and similar for a decrease of p_1. If this assumption were reasonable, we then have probabilities $[p_1 + \varepsilon, p_2 - \frac{1}{2}\varepsilon, p_3 - \frac{1}{2}\varepsilon]$, or, with the values of p_1, p_2, and p_3, we have $[0.5 + \varepsilon, 0.3 - \frac{1}{2}\varepsilon, 0.2 - \frac{1}{2}\varepsilon]$. Given these probabilities, we can then again compute the expected monetary values, which are

$$EMV(\varepsilon) = \begin{bmatrix} 1.4 + .5\varepsilon \\ 1.1 - 3\varepsilon \\ 1.5 + 1\varepsilon \\ .9 + 1.5\varepsilon \end{bmatrix}.$$

Suppose now that it has estimated that p_1 will assume a value somewhere between 0.3 and 0.6. In other words, starting with its present value of $p_1 = 0.5$, the change $\varepsilon \in [-0.2, +0.1]$. We can now again plot the expected monetary values of the decisions as functions of ε. This is shown in Fig. 10.5.

For any value of ε we are interested in the highest expected payoff, i.e., in the point on the highest curve (something called the upper envelope, shown here by the broken line). We observe in Fig. 10.5 that the decisions d_1 and d_4 are dominated by d_3, leaving d_2 and d_3 as the only decisions of interest. The two functions intersect where their payoffs are equal, i.e., at the point at which $1.1 - 3\varepsilon = 1.5 + 1\varepsilon$. Solving for ε, we obtain $\varepsilon = -0.1$ This leads to the following decision rule:

10.4 Sensitivity Analyses

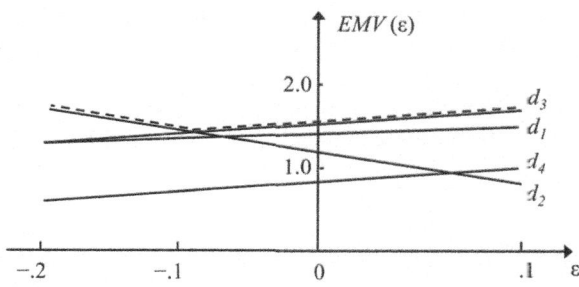

Fig. 10.5 *EMV* as a function of a probability change

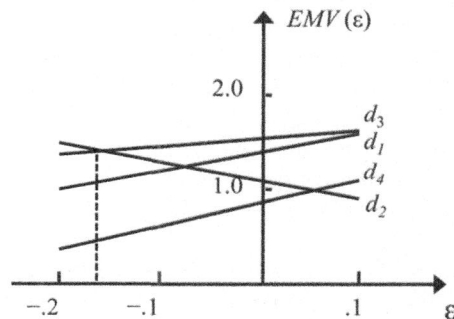

Fig. 10.6 *EMV* as a function of alternative probability change

If $\varepsilon \leq -0.1$ (or, alternatively, $p_1 \leq 0.4$), then decision d_2 is best.
If $\varepsilon \geq -0.1$ (or, alternatively, $p_1 \geq 0.4$), then decision d_3 is best.

Note that it is not generally true that decision rules such as this consist of only two parts. It is possible that any number of the existing decisions may be best for some range of changes.

It is, of course, possible that a change of p_1 does not affect the remaining probabilities equally. For instance, it could be estimated that an increase of p_1 by some unknown value ε will decrease p_2 by $2/3 \varepsilon$, while p_3 will decrease by $1/3 \varepsilon$. The expected payoffs are then

$$\begin{bmatrix} 1.4 + \frac{5}{3}\varepsilon \\ 1.1 - \frac{5}{3}\varepsilon \\ 1.5 + \varepsilon \\ .9 + \frac{8}{3}\varepsilon \end{bmatrix}.$$

The payoff functions (as functions of ε) are shown in Fig. 10.6.

As above, we assume that $\varepsilon \in [-0.2, +0.1]$. A similar analysis to that provided above reveals that decision d_3 dominates d_1 and d_4, leaving the decision maker with d_2 and d_3. Again, we are interested in the upper envelope, shown here by the broken line. The expected monetary values of the two decisions are equal if $1.1 - 5/3\varepsilon = 1.5 + \varepsilon$, i.e., for $\varepsilon = -0.15$. This leads to the following decision rule:

If $\varepsilon \leq -0.15$ (or, equivalently $p_1 \leq 0.35$), then decision d_2 is optimal.
If $\varepsilon \geq -0.15$ (or, equivalently, $p_1 \geq 0.35$), then decision d_3 is optimal.

10.5 Decision Trees and the Value of Information

In this section, we will again consider decision-making problems under risk. In addition to the decision rules discussed in the previous section, we will determine the value of information that goes beyond the probabilities for the states of nature that we continue to assume to be known. We will commence our discussion with an extreme case known as the *expected value of perfect information (EVPI)*. Clearly, in reality no information is perfect, but this value provides an upper bound for the value of any information, as no information can be worth more than perfect information. Since it is easy to compute, it provides the decision maker with a ballpark figure. In simple words, the *EVPI* is the difference of the payoff with perfect information and the best we can do without any information beyond what is included in the standard setting. As an illustration, consider again our example. Recall that with the probabilities of 0.5, 0.3, and 0.2 of the three states of nature, the highest expected monetary value was $EMV^* = 1.5$, which was achieved by choosing decision d_3, where we use an asterisk to indicate optimality. This is the best the decision maker can do without additional information. Consider now perfect information. It means that the decision maker will know in advance which state of nature will occur. It is important to realize that this does *not* mean that the decision maker can change the probabilities of the states of nature—all we assume is that the decision maker knows which state of nature occurs before he makes his own decision.

In our numerical example, the best response to the first state of nature s_1 is to use d_1, d_3, or d_4; each of these responses will result in a payoff of 2 to the decision maker. Similarly, if the decision maker knows that s_2 occurs, his best response is to choose d_3, which results in a payoff of 1. Finally, if nature chooses s_3 and the decision maker knows about it beforehand, the best response is d_2, netting 7. The payoff matrix $\mathbf{A}$ is shown again below with the starred element indicating those payoffs that result from the decision maker's best response to nature's action.

$$\mathbf{A} = \begin{bmatrix} 2^* & -2 & 5 \\ 0 & -1 & 7^* \\ 2^* & 1^* & 1 \\ 2^* & -3 & 4 \end{bmatrix}$$

Given the known probability distribution $\mathbf{p} = [0.5, 0.3, 0.2]$, we can state that 50% of the time, nature chooses s_1 and the decision maker obtains a payoff of 2 (the first column), 30% of the time, nature chooses s_2 and the decision maker's best

10.5 Decision Trees and the Value of Information

Table 10.3 Conditional probabilities $P(I|s)$

	s_1	s_2	s_3
I_1	0.6	0.9	0.2
I_2	0.4	0.1	0.8

reaction results in a payoff of 1, and finally, 10% of the time nature chooses s_3 and the decision maker's response is d_2, resulting in a payoff of 7. Hence, the *expected payoff with perfect information* is $EPPI = 2(0.5) + 1(0.3) + 7(0.2) = 2.7$. The expected value of perfect information is then $EVPI = EPPI - EMV^* = 2.7 - 1.5 = 1.2$. As indicated above, this is an upper bound on the amount of money that the decision maker should be prepared to pay for any type of additional information.

Consider now a situation, in which the information provided to the decision maker is imperfect. Imperfect information is usually provided by indicators. As an example, consider the price of an individual stock. It is usually not possible to get any direct information, so that we must rely on a proxy, such as demand for products of firms in that industry or manufacturers' receipts. Clearly, such proxies are only of value if there is a link between them and the state of nature we want to forecast. For instance, it would be meaningless to forecast the probability of sales of a new camera by using the demand for potatoes. The stronger the links between a proxy and the state of nature it is used to forecast, the closer we will be to perfect information. The strength of the link between indicator and the state of nature is typically provided by a table of conditional probabilities. For our example, assume there are two indicators I_1 and I_2, whose links to the three states of nature are shown in Table 10.3.

In other words, given that the first state of nature s_1 will actually occur, indicator I_1 has a 60% chance of coming up, and similar for the other values. It is apparent that the probabilities in each column add to one. (In the extreme case, there would be three indicators and the probabilities $P(I_1|s_1) = P(I_2|s_2) = P(I_3|s_3) = 1$ and all other conditional probabilities equal to zero. Then the three indicators are sure predictors of the states of nature, and we have again perfect information. This is the limiting case.)

Before performing any computation, we will first depict the structure of the decision-making process in the form of a *decision tree*. The general structure of such a tree is shown in Fig. 10.7.

In Fig. 10.7, we start on the left with the square node (the root of the tree), indicating that we have to make a decision first. Our decision is whether or not to solicit additional information. The lower branch shows that no additional indication is sought, and it will end up with what we have already done earlier when applying Bayes's rule. Often, this part of the tree is deleted and its outcome at the end of the first branch is shown as EMV^* ($= 1.5$ in our example).

Consider now the upper branch that indicates that we ask for additional information. This will be followed by a random event according to which some indicator comes up. This is shown by the circular node, followed by branches for all indicators $I_1, \ldots, I_p$. Once we have received an indicator (meaning we now have additional information), we must make one of our decisions $d_1, \ldots, d_m$. This is shown again as a decision node, followed by arcs, one for each of our decisions. Finally, once we have

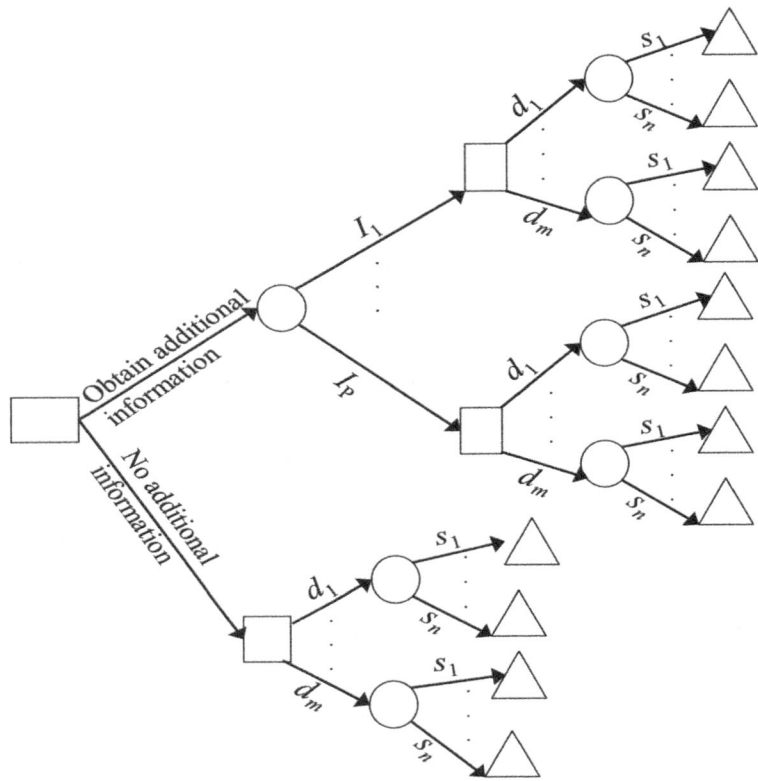

Fig. 10.7 Decision tree involving additional information

made a decision, a random event, i.e., one of the states of nature, will occur. The end of the sequence of decisions and events is marked by a triangle, next to which we will place the outcome for that particular scenario. It is important to realize that each endpoint marked with a triangle actually symbolizes a scenario that includes the entire sequence of decisions and events from the root of the tree to that triangle. For instance, the topmost triangle on the right of the tree indicates that we asked for additional information, received the indicator I_1, made decision d_1, and then state of nature s_1 occurred.

Once the structure of the tree has been determined, we need to put some numbers into the tree. As already mentioned above, the payoffs are taken directly from the payoff matrix and put next to the triangles on the right side of the decision tree. What are now needed are probabilities. More specifically, we need two types of probabilities. The first types are associated with the random events that determine which of the indicators come up. They are called *indicator probabilities* $P(I_k)$. So far, we do not have these probabilities. The second type of probabilities is associated with the states of nature occurring at the very end of the decision tree, just before the

10.5 Decision Trees and the Value of Information

Table 10.4 Computation of $P(I_1)$ and $P(s|I_1)$

s	$P(s)$	$P(I_1|s)$	$P(I_1|s)P(s)$	$P(s|I_1)$
s_1	0.5	0.6	0.30	0.4918
s_2	0.3	0.9	0.27	0.4426
s_3	0.2	0.2	0.04	0.0656
			$P(I_1) = 0.61$	

payoffs are due. At first glance, it would appear that the probabilities that we used in Bayes's rule (so-called *prior probabilities*) $P(s)$ should be used. This is, however, not the case. The reason is that a state of nature occurs after an indicator has come up. And the whole point of indicators is that they are not independent from the states of nature.

So what we need are the so-called *posterior probabilities* $P(s|I)$. In other words, these are conditional probabilities that specify the likelihood that a state of nature occurs, given that an indicator has come up earlier. And while it may appear that we are given the posterior probabilities in a table such as Table 10.3, this is not the case: while both are conditional probabilities, Table 10.3 includes probabilities of the type $P(I|s)$; posterior probabilities are $P(s|I)$. In other words, the probabilities will have to be inverted, which is done by what is known as *Bayes's theorem*. This theorem (or rather conversion rule) is explained in Appendix C of this book. We will use it here in a very convenient computational scheme shown below. And a by-product of the conversion is the set of indicator probabilities that we also need to put numbers on our decision tree.

The computational scheme that determines indicator probabilities and posterior probabilities deals with each of the indicators separately. Consider again our numerical example and use only the first indicator I_1. Table 10.4 shows the computational scheme we use as it applies to the indicator I_1. The first column lists the states of nature, and the second column includes their prior probabilities. The third column includes the conditional variables that relate to the indicator under consideration (here I_1) and all states of nature. In other words, the third column is nothing but the first row in Table 10.3. Following Bayes's rule, we then multiply the elements in the second and third columns and put them in the fourth column. Their sum, shown at the bottom of column four, is then the indicator probability for the indicator under consideration, here $P(I_1)$. Finally, the posterior probabilities in the last column are obtained by dividing each element of column four by the indicator probability, i.e., $P(s_i|I_1) = P(I_1|s_i)P(s_i)/P(I_1)$.

This procedure provides us with both the indicator probabilities and the posterior probabilities needed to complete the numerical information required in the decision tree. Similar computations are then performed for the second indicator. The results are shown in Table 10.5. Note that the sum of indicator probabilities must equal one. The same applies to the posterior probabilities in the rightmost columns of Tables 10.4 and 10.5.

The decision tree with all numerical information is then shown in Fig. 10.8.

At this point, all numerical information is available, and we need to discuss a technique that determines the optimal course of action.

Table 10.5 Computation of $P(I_2)$ and $P(s|I_2)$

s	$P(s)$	$P(I_2|s)$	$P(I_2|s)P(s)$	$P(s|I_2)$
s_1	0.5	0.4	0.20	0.5128
s_2	0.3	0.1	0.03	0.0769
s_3	0.2	0.8	0.16	0.4103
			$P(I_2) = 0.39$	

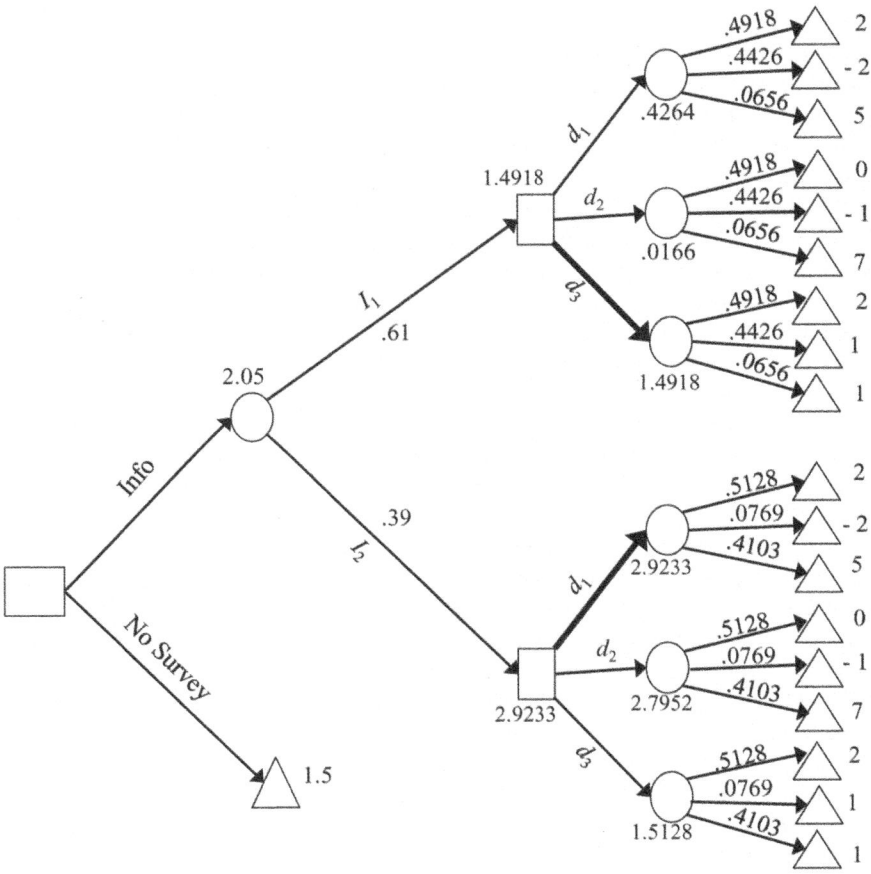

Fig. 10.8 Decision tree with numerical information included

The general idea to deal with decision trees is to use a recursive procedure. This procedure starts from the leaves of the tree and works back to the root. There are two rules in doing so:

Rule 1: Moving back into an event node, we take expected values.
Rule 2: Moving back into a decision node, we choose the decision with the highest expected payoff.

10.5 Decision Trees and the Value of Information

Applying the two rules to our example, we start in the northeast corner of Fig. 10.8. Following the top three branches with the payoffs of 2, −2, and 5 at their respective ends, backward, we reach a decision node. This means that we use the posterior probabilities of the branches to calculate expected payoffs at the event node at the beginning of the branches. In this case, we compute $2(0.4918) - 2(0.4426) + 5(0.0656) = 0.4264$. A similar process is used for the other branches.

The next step starts from the (total of six) event nodes whose expected payoffs have just been computed. Starting again from the top with the three event nodes labeled with 0.4264, 0.0166, and 1.4918, we will follow the arcs that lead into these nodes one step backward. Since this step leads into a decision node, we will choose the decision that results in the highest expected payoff. In this case, the highest payoff is 1.4918 and the decision that leads to this payoff is d_3. Similarly, comparing the expected payoff in the lower part of the diagram (2.9233, 2.7952, and 1.5128), the highest expected payoff is 2.9233 and the decision leading to this payoff is d_1. The arcs of the two decisions chosen in this step are in bold.

We now have two decision nodes labeled with 1.4918 and 2.9233, respectively. Moving back another step means going into an event node, a task performed by computing the expected payoff using the indicator probabilities 0.61 and 0.39. This results in the expected payoff with imperfect information $EPII = 2.05$. Similar to the definition of the expected value of perfect information (the difference between the expected payoff with and without this information), we can now define the *expected value of imperfect information*, most frequently called *expected value of sample information EVSI*. The *EVSI* is defined as the difference between the expected payoff to the decision maker with and without sample information. Formally, we obtain $EVSI = EPII - EMV^* = 2.05 - 1.5 = 0.55$. This is the highest amount that we should be prepared to pay for this information. In other words, if the sample information costs no more than 0.55, we should purchase it and proceed with the upper part of the tree. Otherwise, we should not buy it and follow the "no survey" part of the tree.

The final step consists of the computation of the *efficiency E*. The efficiency measures how close the sample information is in comparison with perfect information, i.e., $E = EVSI/EVPI$. In our example, we obtain $E = 0.55/1.2 = 0.4583$. Loosely speaking, this means that the sample information is about 45% perfect. Note that the value of sample information depends on the strength of the link between the indicators and the state of nature. If the indicators are very strong, the efficiency will be close to one; if they are very weak, it will be close (or equal to) zero. Since *EVSI* is never negative and it can never exceed the value of *EVPI*, the efficiency is always a number between 0 and 1. Loosely speaking, it indicates the value of sample information as a proportion of perfection.

Consider now the same example with two different indicators, whose strengths are shown in Table 10.6.

The indicator probabilities and posterior probabilities for this example are calculated and shown in Tables 10.7 and 10.8 for the indicators I_1 and I_2, respectively.

Table 10.6 Conditional probabilities $P(I|s)$

	s_1	s_2	s_3
I_1	0.9	0.6	0.2
I_2	0.1	0.4	0.8

Table 10.7 Computation of $P(I_1)$ and $P(s|I_1)$

s	$P(s)$	$P(I_1|s)$	$P(I_1|s)\,P(s)$	$P(s|I_1)$
s_1	0.5	0.9	0.45	0.6716
s_2	0.3	0.6	0.18	0.2687
s_3	0.2	0.2	0.04	0.0597
		$P(I_1) = 0.67$		

Table 10.8 Computation of $P(I_2)$ and $P(s|I_2)$

s	$P(s)$	$P(I_2|s)$	$P(I_2|s)\,P(s)$	$P(s|I_2)$
s_1	0.5	0.1	0.05	0.1515
s_2	0.3	0.4	0.12	0.3636
s_3	0.2	0.8	0.16	0.4848
		$P(I_2) = 0.33$		

Constructing a decision tree similar to that in Fig. 10.8 and performing the backward recursion results in $EPII = 2.12$, so that $EVSI = 2.12 - 1.5 = 0.62$ and $E = 0.62/1.2 = 0.5167$, slightly more efficient than in the original example.

In case the indicators are totally random, we would expect the value of this information to be zero. It can be readily seen that this is indeed the case. To illustrate this, consider again the above example and suppose now that there are three indicators. The conditional probabilities $P(I_k|s)$ all equal $1/3$. This results in all indicator probabilities equaling $1/3$ as well, while the posterior probabilities equal the prior probabilities. Inserting them into the decision tree, we find that in the first step of the backward recursion, we obtain the same expected payoffs we would as if Bayes's rule were used and, in the next step when making the decision, we will choose the best Bayesian decision with a payoff of EMV^* in each of the three cases. The next step will multiply this value by the indicator variables by $1/3$, resulting again in EMV^*, so that $EPII = EMV^*$ and $EVSI = 0$.

An interesting case occurs in the other extreme. It is apparent that if we were to use a forecasting institute whose forecast is always correct the value of this information equals the value of perfect information. On the other hand, imagine an institute whose advice is always wrong. At first glance, it would appear that the value of such advice equals zero. This is, however, not correct. As a matter of fact, such information is also perfect, as we can rely on it: whenever they say one thing, we know that the opposite applies. It is the consistency between what is predicted and what actually happens that really counts.

10.6 Utility Theory

Utilities have been used for a long time by economists. Particularly noteworthy are the analyses by the psychologists Kahneman and Tversky in the 1970s. The main idea is to express the usefulness of a product or a service to the individual decision maker. In order to illustrate the concept, consider the following argument. If the expected value were to apply, then a decision maker would be indifferent to the choice of a certain $50,000 gift and a lottery that pays $100,000 with a 50% chance and a zero payoff with a 50% chance. The expected value in both cases is the same: $50,000. However, most decision makers would prefer the certain $50,000.

Let us then take this argument a step further. Suppose we were to offer the aforementioned lottery—a $100,000 payoff with a 50% chance and a zero payoff with a 50% chance—to a decision maker and inquire what amount of money received with certainty he were to consider equivalent to playing the lottery. This value is called the *certainty equivalent*. The certainty equivalent is typically determined by a string of questions that narrow down the value. For instance, we would describe the lottery to the decision maker and offer, say, $45,000 for certain. Would he take the $45,000? If so, we renege on our offer and offer only $40,000 instead. This process continues until the certainty equivalent is found. For many people, the certainty equivalent is quite low; some go as low as $20,000. This shows a behavioral trait referred to as *risk aversion*. As a rule, if a decision maker's certainty equivalent is less than the expected value of the lottery, the decision maker is *risk averse*. If the certainty equivalent is higher than the expected value of the lottery, the decision maker is *risk seeking* (gamblers are a typical example), while if a decision maker's certainty equivalent equals the expected value of a lottery, he is called *risk neutral*. The graph in Fig. 10.9 plots a dollar value against the decision maker's certainty equivalent of the lottery.

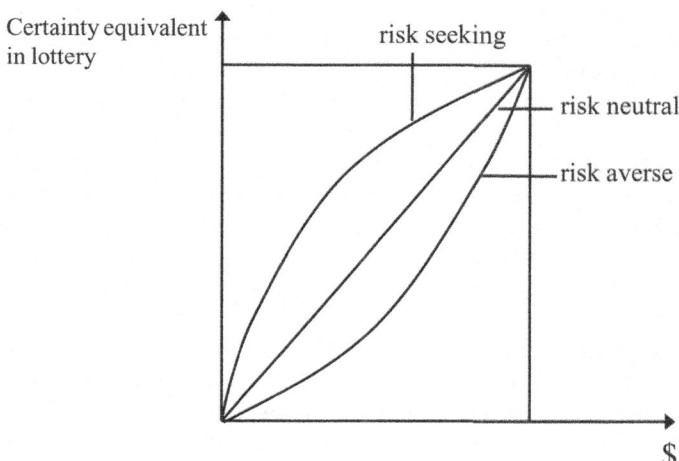

Fig. 10.9 Certainty equivalent as a function of probabilities in the lottery

Fig. 10.10 Concave utility function

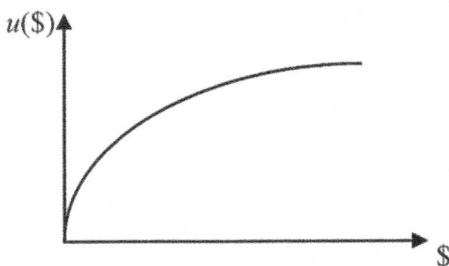

Risk aversion is the raison d'être for insurance companies that, if taken over all those insured by them, take in more in premiums (certain cost) than they pay out (uncertain benefits to their members). Real-life utility functions tend to be convex for smaller dollar values and concave for larger values. The reason is that having or not having a small amount makes very little difference to most people, while owning a very large amount will change their life, e.g., enable them to retire early. This is, of course, how lotteries make a living.

Note that it is important to realize when asking these questions that many times people who claim they would rather take a risk and gamble than accept a fairly small certainty equivalent would change their mind once the actual amount of cash is put in front of them, taking the decision out of a purely "theoretical" realm into the hands-on practical world. Also note that the certainty equivalents are specific to a decision maker and are not transferable, as individuals differ in their acceptance of risk.

Another way to plot a utility function is this. Suppose that an individual owns nothing at all, so that the first (few) dollar(s) enable him to survive, so they have a very high value, i.e., a high utility u. Subsequent dollars, while very welcome, have less and less utility attached to them; i.e., they are said to have decreasing marginal values. In the end—to go to an extreme—the difference between winning ten or 11 million dollars in the lottery is negligible. Hence, utility functions tend to be concave as shown in Fig. 10.10.

Once the certainty equivalents have been determined, they can be used to replace the actual payoffs, so that the problem is then to maximize the expected utility. This concept is also able to deal with cases, in which information concerning the likelihood of states of nature is known but may be ignored by the decision maker. As an example, consider the case of a patient, whose physician has the choice of a number of drugs. Assume that one drug may be able to reduce the pain somewhat without known side effects, while another may not only eliminate the pain, but also the cause—with the possibility of major side effects that include death. If the latter event has only a tiny probability of occurring, the expected "payoff" to the patient may be such that the second and more effective may be chosen. However, the physician may choose to either ignore the probabilities and use Wald's rule so as to minimize the worst-case damage and choose the former less effective drug, or, similarly, may assign a very high negative "payoff" to the possibility of major side

10.6 Utility Theory

effects, resulting also in the former drug being chosen. Assigning very high costs (typically shown as $M \gg 0$) to options is a technique that is also used in optimization under the name penalty costs, whenever options or situations are to be avoided, while still using objectives that maximize the sum of benefits.

Exercises

Problem 1 (influence diagram): Consider the following situation. Jill lives presently in Missoula, Montana, where she has a fairly boring job. She has heard that there are many opportunities in Denver, Colorado, and she plans to go there, possibly resettle there, and buy a small house for herself. She plans to use her annual vacation to look things over in Denver.

Develop an influence diagram and a decision tree for the problem.

Solution: The list of potential events and the two graphs below indicate just some possible set of interdependencies of the decisions and events.

Decision	Random event	Consequence
Travel to Denver and look for a job	Economic situation	Salary (or net worth)
Resettle in Denver	Get job offer	
Purchase a house in Denver		

The influence diagram for this list is shown in Fig. 10.11.

A possible decision tree that describes the sequence of events is shown in Fig. 10.12.

Problem 2 (single-stage decision making under uncertainty and risk): Consider the payoff matrix of a game against nature.

	s_1	s_2	s_3	s_4	s_5	s_6
d_1	5	-2	7	1	0	-6
d_2	6	0	3	-5	8	1
d_3	1	4	0	0	-1	0
d_4	4	-2	3	-5	6	0

(a) Explain in one short sentence the concept of a dominated decision. Are there any dominated decisions in this example?
(b) How would an optimist decide?
(c) How would a pessimist decide?
(d) Find an optimal strategy using the minimax regret criterion.
(e) Suppose that the decision maker has been informed that the states of nature occur with probabilities of 0.3, 0.2, 0.1, 0.1, 0.05, 0.25. How would a risk-neutral decision maker decide?
(f) Construct the graph that plots the probabilities that a decision achieves a target value T for all target values between -10 and 10 for all decisions. What is the optimal decision for $T = 3\frac{1}{2}$? For $T = 5\frac{1}{2}$?

Fig. 10.11 Influence diagram for Problem 1

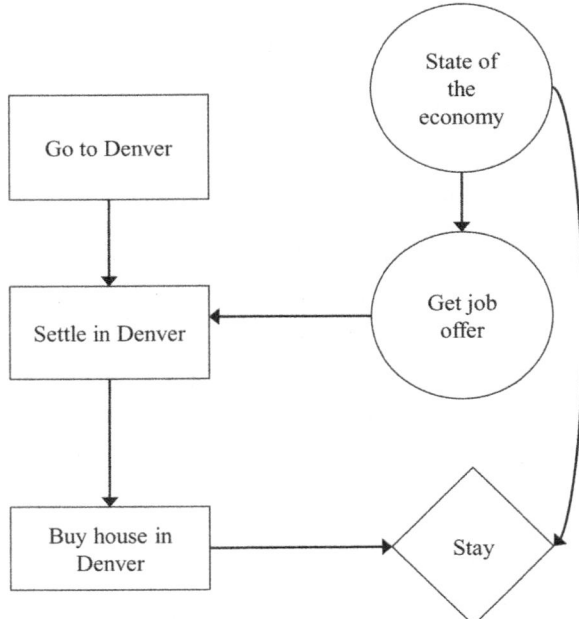

Solution:

(a) A decision dominates another if it is better or the same for all states of nature. In this example, decision d_2 dominates d_4.
(b) An optimist anticipates payoffs of 7, 8, 4, and 6, so that d_2 would be chosen.
(c) A pessimist anticipates payoffs of $-6, -5, -1, -5$, so that d_3 would be chosen.
(d) The regret matrix is $\mathbf{R} = \begin{bmatrix} 1 & 6 & 0 & 0 & 8 & 7 \\ 0 & 4 & 4 & 6 & 0 & 0 \\ 5 & 0 & 7 & 1 & 9 & 1 \\ 2 & 6 & 4 & 6 & 2 & 1 \end{bmatrix}$ with anticipated regrets of 8, 6, 9, 6, so that the (pessimistic) choice is d_2 or d_4 with regret of 6.
(e) The anticipated payoffs in Bayes's model are 0.4, 2.25, 1.05, and 0.9, so that the decision maker will choose d_2.
(f) In Fig. 10.13, we show d_1 as a solid line, d_2 as a broken line, and d_3 as a dotted line; the dominated decision d_4 is deleted for clarity.

For $T = 3\tfrac{1}{2}$, decision d_1 is best, followed by d_2 and d_3. For $T = 5\tfrac{1}{2}$, d_2 is best and then d_1 and d_2.

Problem 3 (sensitivity analysis): Consider a decision problem with three decisions and four states of nature. The payoff matrix is as follows:

10.6 Utility Theory

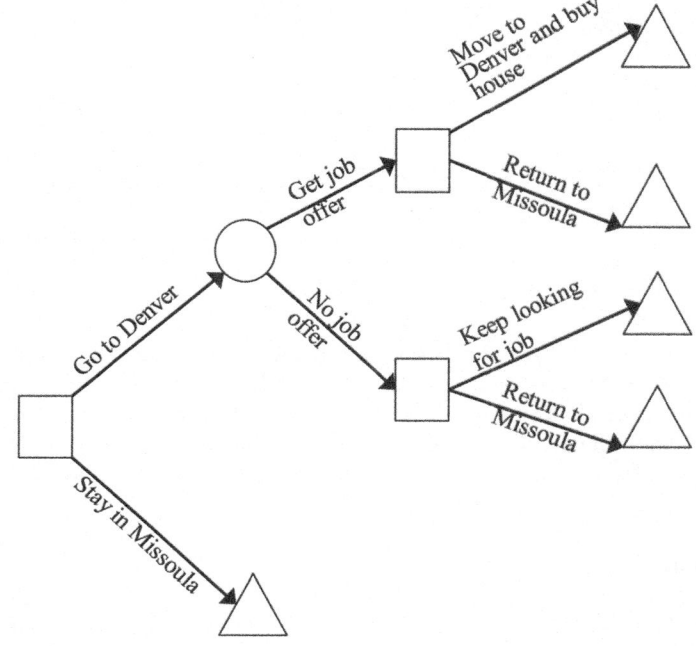

Fig. 10.12 Decision tree for Problem 1

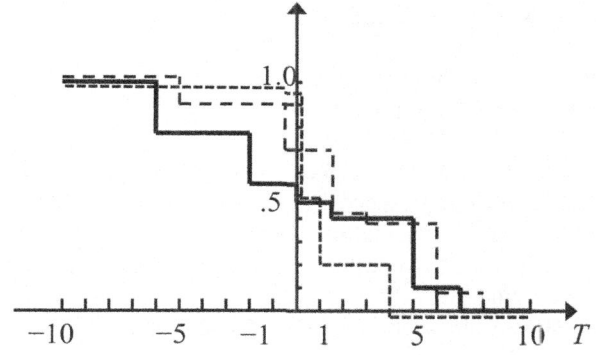

Fig. 10.13 Probability of achieving a target value for Problem 2

$$\begin{bmatrix} 2 & 0 & -2 & 4 \\ 1 & 1 & 2 & -3 \\ 2 & 3 & -2 & 1 \end{bmatrix}.$$

Furthermore, suppose that the probabilities of the four states of nature are 0.4, 0.1, 0.3, and 0.2.

Fig. 10.14 *EMV* as a function of ε, for Problem 3 (a)

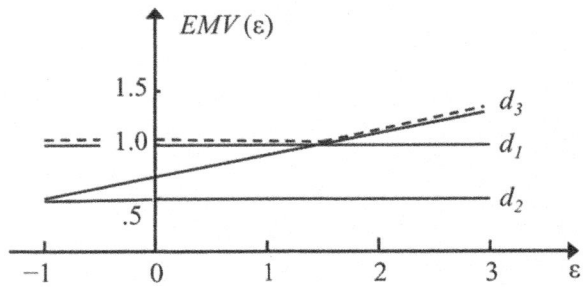

(a) Perform a sensitivity analysis on a_{34}, the payoff that results from decision d_3 coupled with the fourth state of nature. It has been estimated at $a_{34} \in [0, 4]$.
(b) Back to the original situation, perform a sensitivity analysis on p_2. It is assumed that (*i*) for each unit of increase of p_2, the probability p_1 decreases twice as much as each of p_3 and p_4, and (*ii*) that p_2 may decrease by as much as 0.05, while it may increase by at most 0.2.

Solution:

(a) Given the payoff $a_{34} = 1 + \varepsilon$, we obtain expected monetary values of 1.0, 0.5, and $0.7 + 2\varepsilon$ for the three decisions. Given that $a_{34} \in [0, 4]$, or, equivalently, $\varepsilon \in [-1, 3]$, Fig. 10.14 shows the expected monetary values of the three decisions as functions of ε and the upper envelope in the form of the broken line.
This leads to the following decision rule:

If $a_{34} \leq 2.5$, (or, equivalently, $\varepsilon \leq 1.5$), choose d_1.
If $a_{34} \geq 2.5$ (or, equivalently, $\varepsilon \geq 1.5$), choose d_3.

(b) The updated probabilities are $\mathbf{p} = [0.4 - \frac{1}{2}\varepsilon, 0.1 + \varepsilon, 0.3 - \frac{1}{4}\varepsilon, 0.2 - \frac{1}{4}\varepsilon]$, and the resulting expected monetary values are $EMV(\varepsilon) = 1.0 - 1.5\varepsilon$, $0.5 + 0.75\varepsilon$, and $.7 + 2.25\varepsilon$ for the three decisions. Figure 10.15 shows the expected monetary values of the three decisions as functions of ε and the upper envelope is shown by the broken line.
This results in the following decision rule:

If $p_2 \leq 0.18$ (or, equivalently, $\varepsilon \leq 0.08$), then choose d_1.
If $p_2 \geq 0.18$ (or, equivalently $\varepsilon \geq 0.08$), then choose d_3.

Problem 4 (expected value of perfect and sample information): The Canadian McMoose chain of fast-food outlets is deciding how to keep up with the changing tastes of its customer base. They have narrowed down their choices to the following three decisions: $d_1 =$ completely redecorate the existing franchises, $d_2 =$ rebuild the

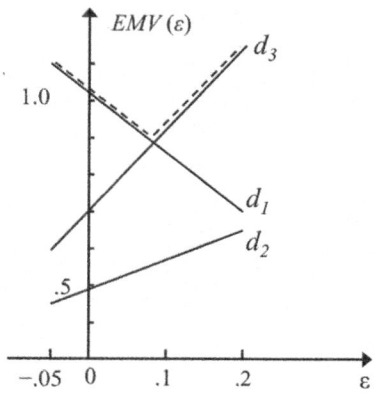

Fig. 10.15 EMV as a function of ε, for Problem 3 (b)

outlets, and d_3 = modify the existing decor slightly to emphasize the "mooseyness" of the outlet. The chain faces different states of the economy s_1, s_2, and s_3, indicating a level state, a slight upturn, and a significant upturn. The payoffs for all combinations of decisions and states of the economy are shown in the following matrix:

	s_1	s_2	s_3
d_1	5	4	2
d_2	−4	2	9
d_3	3	8	1

The probabilities for the three states of the economy have been determined as 0.3, 0.2, and 0.5.

(a) Determine the expected payoffs for the three decisions and choose the most preferred decision on that basis.
(b) What is the expected value of perfect information?
(c) The McMoose management considers hiring a research institute to obtain more detailed information about the state of the economy. They use two indicators I_1 and I_2 for their forecast. These two indicators are linked to the state of the economy as shown in the following table of conditional probabilities $P(I|s)$:

	s_1	s_2	s_3
I_1	0.5	0.1	0.8
I_2	0.5	0.9	0.2

Construct the decision tree for this problem and determine the expected value of sample (imperfect) information. If the research institute charges 1.4 for their services, should they be hired? Explain in one very short sentence.

(d) What is the efficiency of the sample information?

Solution:
(a) The expected payoffs are 3.3, 3.7, and 3.0, so that they would choose d_2 and get 3.7.
(b) $EVPI = 7.6 - 3.7 = 3.9$.
(c)
For I_1:

s	$P(s)$	$P(I_1\|s)$	$P(I_1\|s)P(s)$	$P(s\|I_1)$
s_1	0.3	0.5	0.15	0.2632
s_2	0.2	0.1	0.02	0.0351
s_3	0.5	0.8	0.40	0.7018
			$P(I_1) = 0.57$	

For I_2:

s	$P(s)$	$P(I_2\|s)$	$P(I_2\|s)P(s)$	$P(s\|I_2)$
s_1	0.3	0.5	0.15	0.3488
s_2	0.2	0.9	0.18	0.4186
s_3	0.5	0.2	0.10	0.2326
			$P(I_2) = 0.43$	

The decision tree for the problem is then shown in Fig. 10.16.
$EVSI = 5.0301 - 3.7 = 1.3301$, which is less than the amount requested by the institute, so do not hire them.
(d) The efficiency is $E = 1.3301/3.9 = 0.3411$.

Problem 5 (expected value of perfect and sample information): The Australian automobile manufacturer Australomobil must decide whether or not to manufacture the transmissions of their "Pithecus" model in-house (d_1) or to subcontract them out (d_2). The company faces different levels of demand for the Pithecus that are defined as s_1, s_2, s_3, and s_4. The payoffs for all combinations of decisions and levels of demand are shown in the following table:

	s_1	s_2	s_3	s_4
d_1	4	-1	-2	4
d_2	-4	2	1	3

(a) Are there any dominances in the payoff matrix? Explain in one short sentence.
(b) How would a pessimist decide? What is the optimal decision under the regret criterion? In both cases, what are the anticipated payoffs?
(c) Given prior probabilities of 0.2, 0.3, 0.4, and 0.1 for the states of nature, what is the optimal decision with Bayes's criterion? What is the expected payoff?
(d) What is the expected value of perfect information?

10.6 Utility Theory

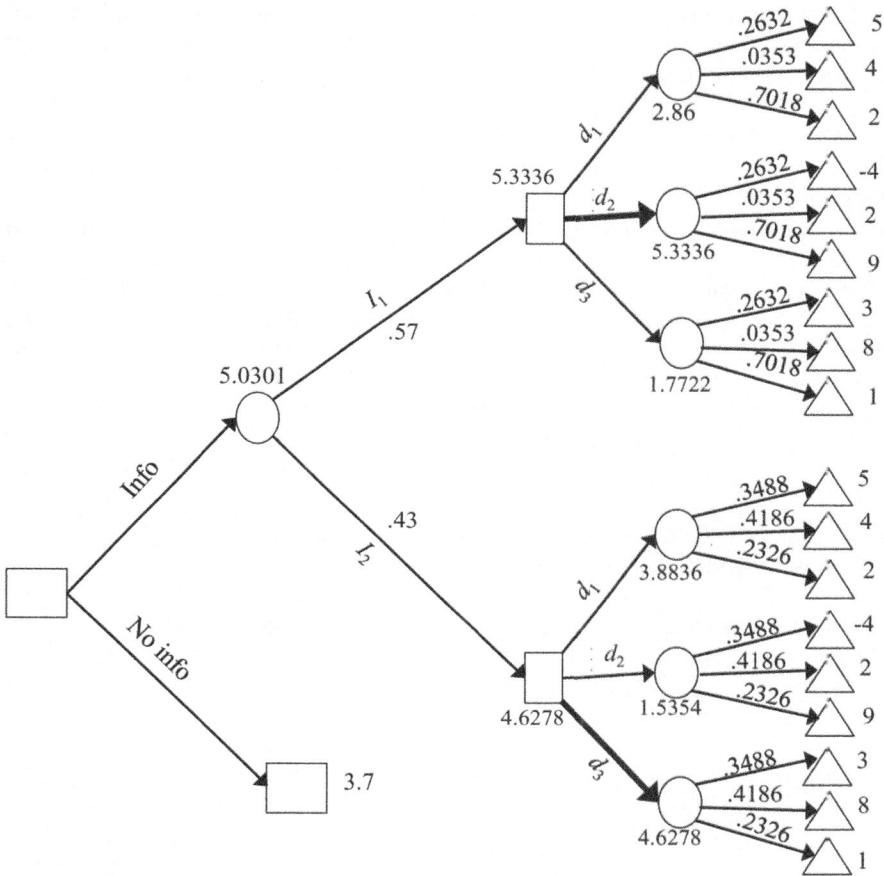

Fig. 10.16 Decision tree for Problem 4

(e) The Australomobil management considers hiring a research institute to obtain more detailed information about the future level of demand. They use two indicators I_1 and I_2 for their forecast. These two indicators are linked to the level of demand as shown in the following table of conditional probabilities $P(I|s)$:

	s_1	s_2	s_3	s_4
I_1	0.4	0.8	0.9	0.5
I_2	0.6	0.2	0.1	0.5

Construct the decision tree for this problem and determine the expected value of sample (imperfect) information. If the research institute charges 0.7 for their services, should they be hired? Explain in one very short sentence.

(f) What is the efficiency of the sample information?

Solution:

(a) There are no dominances. For s_1, d_1 is better than d_2, but for s_2, d_2 is preferred over d_1.

(b) A pessimist will anticipate payoffs of -2 and -4, respectively. He will choose d_1 and anticipate a payoff of -2. For the regret criterion, we set up the regret matrix

$$\mathbf{R} = \begin{bmatrix} 0 & 3 & 3 & 0 \\ 8 & 0 & 0 & 1 \end{bmatrix},$$

so that the (anticipated) maximal regrets are 3 and 8. The decision maker will then choose d_1 and anticipate a regret of 3.

(c) The expected payoffs are 0.1 and 0.5, so that they would choose d_2 and get 0.5.

(d) $EVPI = 2.2 - 0.5 = 1.7$.

(e)

s	$P(s)$	$P(I_1|s)$	$P(I_1|s)P(s)$	$P(s|I_1)$
s_1	0.2	0.4	0.08	0.1096
s_2	0.3	0.8	0.24	0.3288
s_3	0.4	0.9	0.36	0.4932
s_4	0.1	0.5	0.05	0.0685
			$P(I_1) = 0.73$	

s	$P(s)$	$P(I_2|s)$	$P(I_2|s)P(s)$	$P(s|I_2)$
s_1	0.2	0.6	0.12	0.4444
s_2	0.3	0.2	0.06	0.2222
s_3	0.4	0.1	0.04	0.1481
s_4	0.1	0.5	0.05	0.1852
			$P(I_2) = 0.27$	

The decision tree associated with this problem is shown in Fig. 10.17.

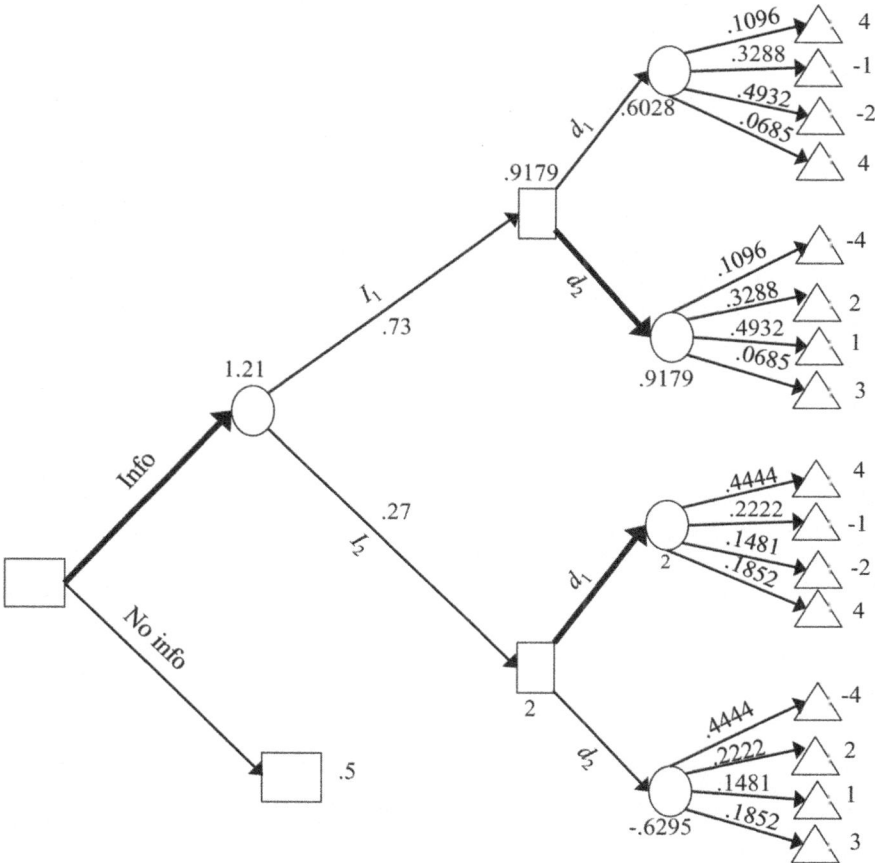

Fig. 10.17 Decision tree for Problem 5

$EVSI = 1.21 - 0.5 = 0.71$, which is slightly more than the amount requested by the institute, so they should hire them.

(f) The efficiency is $E = 0.71/1.7 = 0.4176$.

Multiattribute Decision Making   11

The models and methods in this chapter all consider problems, in which the consequences of a decision are no longer one-dimensional, while, in contrast to the scenario in decision analysis, the outcomes are deterministic. In other words, if we were to, say, change the composition of a soft drink we manufacture or change the features on a cell phone, we do not just deal with profit as a result of this decision, but face changing customer acceptance (and demand) for the product, different costs, changing market share, customer satisfaction, and other factors, all of which will influence short- and long-term viability of the firm. The models discussed in this chapter are similar to those in the chapter on multiobjective optimization, except that the models in this chapter are discrete: we face only a finite (and usually fairly small) number of choices.

11.1 The General Model and a Generic Solution Method

Throughout this section, we assume that the decision maker has a finite number of possible decisions $d_1, d_2, \ldots, d_m$ that are to be evaluated on a number of criteria $c_1, c_2, \ldots, c_n$. All decisions are then evaluated on all criteria in a matrix (not unlike the payoff matrices in decision analysis). Without loss of generality, we assume that all criteria are of the utility type, i.e., more is always better. In case some of the criteria originally denote costs or distances, i.e., criteria for which less is better, we may convert them to utilities by multiplying them by some negative value or subtracting them from some large number.

In order to demonstrate some of the basic concepts, consider the problem of choosing a house. The decision maker has identified some of the criteria most important to him, which include the size of the house, the size of the lot, the taxes charged most recently, the price of the house, and the distance between the house and his place of work. Furthermore, the decision maker has chosen seven possible choices from a multiple listing service that provides some basic information

Table 11.1 Data for the sample problem

Decision	Size (sq ft)	Lot size (sq ft)	Taxes ($)	Asking price ($)	Distance to work (mi)
d_1	2154	8820	3829	308,500	3.8
d_2	2600	14,445	7264	669,000	3.7
d_3	2500	15,791	3480	284,900	6.0
d_4	2300	10,496	5601	419,000	3.5
d_5	3250	24,412	3275	595,000	10.6
d_6	3600	7.4 acres	4719	699,900	7.5
d_7	1982	10,678	6253	624,500	4.8

Table 11.2 Revised evaluations for sample problems

Decision	Size (sq ft)	Lot size (sq ft)	Monthly cost ($)	Distance to work (mi)
d_1	2154	8820	840.33	3.8
d_2	2600	14,445	2.027.83	3.7
d_3	2500	15,791	752.25	6.0
d_4	2300	10,496	1264.25	3.5
d_5	3250	24,412	1510.42	10.6
d_6	3600	7.4 acres	1893.00	7.5
d_7	1982	10,678	1832.33	4.8

(which is how these actual data have been determined). The information is summarized in Table 11.1.

Often, decision makers will also consider additional criteria, such as the number of bedrooms, bathrooms, and garages. All of these properties have two garages, four or five bedrooms, and 2½ to 3 baths, so that we have eliminated those criteria from further consideration. In general, decision makers should restrict themselves to, say, no more than six criteria, so as to avoid clutter and correlation. This limitation has been suggested by psychological research.

Clearly, there is usually a correspondence between the size of the house and the number of rooms in it. When considering correlated criteria, it is typically a good idea to include only one of those criteria and thus avoid considering and counting them multiple times.

Along similar lines, both asking price for the property and taxes contribute to the financial burden to be borne by the buyer. We can combine the two by computing the annual or monthly cost of each of the properties. In order to do so, we first assess the amount of money the decision maker can provide in cash. Assume that this amount is $100,000. The remaining part of the cost of the house will then have to be financed, which we assume can be done at the (somewhat favorable) rate of 3% p.a. Formally, we can write the monthly cost of the projects as (cost) = 0.0025 (price − 100,000) + 0.08333 (taxes), where the coefficient of the taxes simply converts the annual tax to a monthly cost. The revised evaluations of the projects are then shown in Table 11.2.

11.1 The General Model and a Generic Solution Method

Table 11.3 Reduced evaluation matrix for sample problem

Decision	Size (sq ft)	Lot size (sq ft)	Monthly cost ($)	Distance to work (mi)
d_1	2154	8820	840.33	3.8
d_3	2500	15,791	752.25	6.0
d_4	2300	10,496	1264.25	3.5
d_5	3250	24,412	1510.42	10.6

At this point, the decision maker will have to decide which of the projects are actually financially feasible, considering his income. Assume that the decision maker has decided that a monthly financial load beyond $1600 is not feasible. (This, by the way, will include potential increases of the interest rate in future years, when our decision maker has not yet paid back a significant amount on the principal.) This will eliminate options 2, 6, and 7. The reduced evaluation matrix is then shown in Table 11.3.

The next step is to convert the evaluations in Table 11.3 to utilities. This is an important and necessary step, as some of the criteria are considered better, if the numerical evaluations are larger (within certain limits, this applies to the first two criteria), while other criteria are considered superior, if their evaluations are smaller (the latter two criteria in our example). Furthermore, most people would agree that for the size of the house, the simple "bigger is better" rule does not necessarily apply: while a house with 1000 sq. ft. is preferred to one with 600 sq. ft. by most people, a house of the size 50,000 sq. ft. is not necessarily considered preferable (forget about affordable) to one with 3000 sq. ft.

This decision maker does not wish to purchase a house of less than 1200 sq. ft., his utility increases linearly between 1200 and 2400 sq. ft. from a utility of 0 to a utility of 0.8, and from there it increases to a utility of 1, which it reaches at 3000 sq. ft. Beyond that, the utility stays at that level. Denoting by x the size of the house and by $u(x)$ the utility of the size, we can write

$$u(x) = \begin{cases} 0, \text{ if } x \leq 1,200 \\ -0.8 + 0.0006667x, \text{ if } x \in]\,1,200;\,2,400] \\ \dfrac{1}{3,000}x, \text{ if } x \in]\,2,400;\,3,000] \\ 1, \text{ if } x > 3,000. \end{cases}$$

A similar argument applies to the size of the lot. Here, our decision maker will not consider very small lots of 5000 sq. ft. or less, and his utility $u(y)$ for a lot of size y is summarized in the step function

Table 11.4 Utility matrix for sample problem

Decision	Size	Lot size	Financial load	Distance to work
d_1	0.6361	0.4	0.6843	0.9927
d_3	0.8333	0.7	0.7392	0.9742
d_4	0.7334	0.7	0.3493	0.9941
d_5	1.0000	0.9	0.1007	0.8686

$$u(z) = \begin{cases} 0, \text{if } x \in [0; 5,000] \\ 0.4, \text{if } x \in]\, 5,000; 10,000] \\ 0.7, \text{if } x \in]\, 10,000; 20,000] \\ 0.9, \text{if } x \in]\, 20,000; 43,560] \\ 1.0, \text{if } x \in]\, 1 \text{ acre}; 30 \text{ acres}] \\ 0.9, \text{if } x > 30 \text{ acres}. \end{cases}$$

Similarly, the utility of the cost of the property is captured by the utility function $u(z) = 1 - 0.0001z - 0.000000328z^2$, given a cost of z, and for a distance w between a property and the decision maker's place of work, the utility is $u(w) = 1 - .0001w - .0001w^2 - .0001w^3$. The utility matrix for the four remaining properties is then shown in Table 11.4.

This is the table we will work on with the generic procedure. The procedure itself is easily described: Assign a weight to each criterion, then aggregate the utility by using these weights (typically the weighted sum), and then choose the highest of the resulting weighted average utilities. This "simple weighted average" process is just about identical to Bayes's rule in decision analysis that computes expected values, except that the weights in this process are no probabilities.

In this example, assume that the decision maker has decided that the weights of the four criteria are 0.25, 0.1, 0.5, and 0.15, respectively, with the obvious emphasis on the cost of the property. The average utilities of the four decisions are then 0.6901, 0.7941, 0.5771, and 0.5206, respectively, making the second choice (the decision d_3) the clear winner.

As usual, since decision makers cannot be expected to specify exact values of the weights, we perform sensitivity analyses by changing the weights somewhat and determine if and how the decisions fare in this modified scenario. Suppose now that the decision maker were to put less weight on the cost and more on the size of the house and the lot, e.g., by using weights 0.35, 0.15, 0.35, 0.15. The weighted average utilities are then 0.6710, 0.8015, 0.6331, and 0.6505, so that the second decision (d_3) is still the best by a long shot, indicating the stability of the decision.

One possibility to visualize different utilities in multiattribute problems are *spider plots*. To prepare for the graph, the criteria have to be either all of the maximization or all of the minimization variety. In a spider plot, each criterion is assigned one dimension. Given m criteria, the axes are graphed with $360/m$ degrees between them. In our example, this is $360/4 = 90$ degrees. For each criterion, we then plot points on

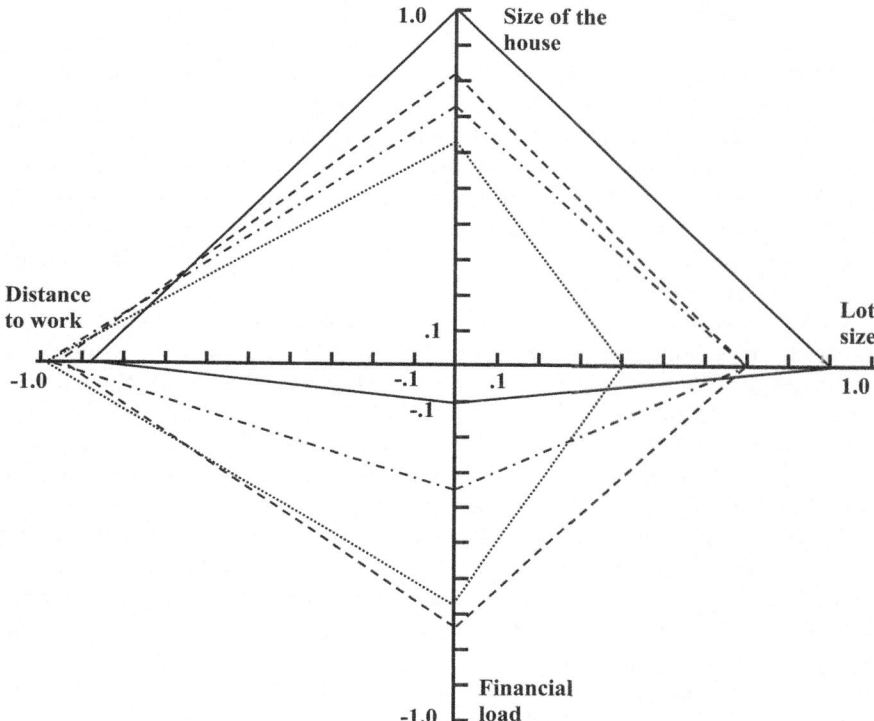

Fig. 11.1 Spider plot

the axes depending on the attributes. Finally, the points of one criterion on neighboring axes are connected. For the utility matrix of our example in Table 11.4, the spider plot is shown in Fig. 11.1, where the decisions d_1, d_3, d_4, and d_5 are shown as dotted, dashed, dashed-dotted, and solid lines, respectively.

Spider plots allow an easy assessment of different decisions. For instance, if the area of one decision completely covers the area of another decision, the latter decision is dominated. In our example d_4 is almost dominated by d_3. It also visualizes the strengths of individual decisions. For instance, decision d_5 in our example has stronger mass in the top part of the plot, i.e., it performs very well when it comes to the size of the house, whereas it is weak at the bottom, i.e., when it comes to the financial load.

11.2 TOPSIS

The *TOPSIS* (*T*echnique for *O*rder of *P*reference by *S*imilarity to *I*deal *S*olution) was first described by Hwang and Yoon (1981). It evaluates decisions on the basis of the distances between their attributes and some ideal points: the closer a decision's attributes are to the chosen ideal point, the better a decision is. In other words, we consider the distance as a disutility that we attempt to minimize. Even at this level, it is apparent that at least two user-defined measures will influence the solution: the choice of an ideal point and the yardstick used to measure the distance between the attributes of a decision and the ideal point.

We may explain the concept by means of a simple example. A student is in the process to choose one among four institutions of higher learning for his studies. In five topics, the choices are the "Business Undergraduate Necessary Knowledge" (BUNK), the "Online University of Indeterminate Junior Academics" (OUIJA), the Commonwealth Undergraduate Teaching Education (CUTE), and the "Business Undergraduate Liberal Learning" (BULL). Given these choices (for simplicity, we will refer to them as d_1, d_2, d_3, and d_4), the student wants to choose one of these institutions to learn about technical issues related to business. The attributes of the four institutions regarding the five programs are shown in Table 11.5, where all scores are expressed on a scale between 0 and 100, where 100 means that the course perfectly suits our student's interests, while 0 means that our student is not interested in that academy's offering at all.

Suppose now that the student wants to learn all about optimization, decision analysis, and logistics, some background statistics, and some basic finance. Given that, he has determined his ideal point to be [60, 100, 90, 80, 20], and that deviations of an attribute are measured by Euclidean distances. The negative deviations (we consider only these, as a positive deviation from a desirable goal carries no penalty) are squared, summed up, and then the square root is taken. For decision d_1, we obtain $\sqrt{\left((0)^2 + (100-60)^2 + (90-40)^2 + (80-30)^2 + 0^2\right)} = 81.24$. Similarly, for the other three decisions, we obtain disutilities/penalties of 45.83, 88.88, and 45.83, respectively, so that decisions d_2 and d_4 are considered equally good and best among the available decisions. At this point, the decision maker could either use secondary criteria or apply different distances: weighted Euclidean distances (square root of the weighted sum of deviations), (weighted) Manhattan distances, or others. For details on distance functions, readers are referred to Chap. 7. Taking unweighted

Table 11.5 Strengths of the four institutions regarding the topics

Decision	Statistics	Optimization	Decision analysis	Logistics	Finance
d_1	70	60	40	30	90
d_2	50	60	70	80	20
d_3	70	80	20	30	10
d_4	40	70	70	60	20

Manhattan distances (the sum of positive deviations from the goal) results in deviations of 140, 70, 150, and 90, respectively, so that the second decision performs best, and our protagonist will attend OUIJA.

11.3 The Analytic Hierarchy Process

Similar to the generic procedure shown in the previous section, the analytic hierarchy process is a tool that also allows decision makers to choose among decision alternatives that are to be evaluated on multiple criteria. The technique was developed by Thomas Saaty in the late 1970s (Saaty 1980), based on the premise that decision makers are not really able to make choices among more than two alternatives at a time (which is what we had to do in the generic method, when assigning utilities to all decisions). This principle of *pairwise comparisons* has been advocated for a very long time. While convenient for the user, it may (and usually does) create inconsistencies that will have to be dealt with.

The analytic hierarchy process (or *AHP*, as it is sometimes called) allows a variety of levels of objects that are arranged in a hierarchy. The exact way a problem is structured will, of course, depend on the specific situation. In the simplest case, we have the overall objective on the highest level, followed by the individual criteria that are used to evaluate the individual decision on the second level, which, in turn, are located on the third level.

Formally, for each criterion k, we will set up a matrix $\mathbf{C}^k$, whose entry in row i and column j will indicate how much the decision maker prefers decision d_i over decision d_j by just considering criterion c_k. Given m possible decisions, each of the matrices $\mathbf{C}^k$ has m rows and m columns.

In case quantitative evaluations of the decisions on the criterion are available, their ratios will be used. For instance, if one decision produces sales of 120 and another decision results in expected profits of 80, then the first decision is preferred over the second decision $120/80 = 1.5$ times. This value will then be used in the matrix of pairwise comparisons. For qualitative data, the author of the original study suggested to use the values shown in Table 11.6.

Note that the numerical values associated with the verbal statements are rather arbitrary; in our opinion they appear extreme. For instance, if a decision maker were

Table 11.6 Quantitative values for preferences

Numerical value of preference of d_i over d_j	Interpretation
1	The decisions d_i and d_j are considered *equal*
3	Decision d_i is *preferred* over decision d_j
5	Decision d_i is *strongly preferred* over decision d_j
7	Decision d_i is *very strongly preferred* over decision d_j
9	Decision d_i is *extremely preferred* over decision d_j

to "strongly" prefer decision d_i over decision d_j, it would appear that considering d_i being twice as important as d_j expresses this fact rather well and there is no need of using a factor of 5 as suggested by Saaty.

Pairwise comparisons between decisions are performed for all of the q given criteria. This results in matrices $\mathbf{C}^1, \mathbf{C}^2, \ldots, \mathbf{C}^q$. In addition, a matrix $\mathbf{C}$ must be set up, which includes the pairwise comparisons between the individual criteria. For instance, if the decision maker considers the criterion "customer satisfaction" twice as important as the criterion "sales," then the value of "2" will be included in this matrix. We assume that the matrix is pairwise consistent, in that if a decision d_i is considered to be x times as good as decision d_j, then we imply that decision d_j is considered to be $1/x$ times as good as decision d_i.

The task at hand is now to describe a technique that will help the decision maker to make a decision on the basis of these evaluations. The first step in one version of the analytic hierarchy process, called the *normalization technique*, is to create new matrices $\widetilde{\mathbf{C}}^1, \widetilde{\mathbf{C}}^2, \ldots, \widetilde{\mathbf{C}}^q$ and $\widetilde{\mathbf{C}}$ from each of the existing matrices $\mathbf{C}^1, \mathbf{C}^2, \ldots, \mathbf{C}^q$ and $\mathbf{C}$ by dividing each of its elements by its respective column sum. This is nothing but a normalization procedure, which results in matrices, all of whose column sums equal one. We then construct a utility matrix $\mathbf{U}$ that has as many rows as we have decisions (i.e., m rows) and as many columns as we have criteria (i.e., q columns). The first column of matrix $\mathbf{U}$ contains the averages of the rows of matrix $\widetilde{\mathbf{C}}^1$, the second column of $\mathbf{U}$ contains the averages of the rows of the matrix $\widetilde{\mathbf{C}}^2$, and so forth. Similarly, we obtain the vector of weights $\mathbf{w}$, whose elements are the averages of the rows of the matrix $\widetilde{\mathbf{C}}$. We are now ready to compute the vector of values of the decisions $\mathbf{v}(d) = \mathbf{Uw}$. In other words, the values of decision d_1 are obtained by multiplying the first row of the utility matrix with the vector of weights $\mathbf{w}$, the expected value of the second decision is obtained by multiplying the second row of $\mathbf{U}$ with $\mathbf{w}$, and so forth.

This procedure is best explained by an

Example: The manager of a resort faces the task of defining its target group. He will either cater to families with children (with the appropriate kids' facilities), or to couples who want to experience nature (with guided walks and "soft" adventure trips), to young individuals (with rock climbing and white-water rafting), or to seniors (with appropriate leisure facilities for entertainment such as Bingo). These four decisions d_1, d_2, d_3, and d_4 are mutually exclusive and collectively exhaustive. The decisions are to be evaluated on two criteria c_1 and c_2. The two criteria are market share (and with it the potential for long-term viability) and profit (short-term benefit); they are deemed about equally important. The decision maker has been asked to compare the individual strategies by pairwise comparisons and has come up with the following results.

11.3 The Analytic Hierarchy Process

$$\mathbf{C}^1 = \begin{bmatrix} 1.0000 & 1.2000 & 2.0000 & .9000 \\ .8333 & 1.0000 & 2.0000 & .8000 \\ .5000 & .5000 & 1.0000 & .5000 \\ 1.1111 & 1.2500 & 2.0000 & 1.0000 \end{bmatrix} \text{ and}$$

$$\mathbf{C}^2 = \begin{bmatrix} 1.0000 & .3000 & .5000 & 1.0000 \\ 3.3333 & 1.0000 & 1.6000 & 3.0000 \\ 2.0000 & .6250 & 1.0000 & 2.0000 \\ 1.0000 & .3333 & .5000 & 1.0000 \end{bmatrix}, \text{ as well as}$$

$$\mathbf{C} = \begin{bmatrix} 1 & 1 \\ 1 & 1 \end{bmatrix}.$$

For instance, as far as criterion c_1 is concerned, the decision maker considers decision d_1 twice as valuable as decision d_3 (as shown by the value of "2" in the first row and the third column of matrix $\mathbf{C}^1$). On the other hand, decision d_1 is considered only half as good as decision d_3 when looking at the second criterion. This is shown in row 1, column 3, of matrix $\mathbf{C}^2$, which shows a value of "0.5."

The column sums of the matrix $\mathbf{C}^1$ are 3.4444, 3.95, 7, and 3.2, respectively, the column sums of the matrix $\mathbf{C}^2$ are 7.3333, 2.2583, 3.6, and 7, respectively, and the column sums of the matrix $\mathbf{C}$ are both 2. Dividing the elements of the matrices $\mathbf{C}^1$, $\mathbf{C}^2$, and $\mathbf{C}$ by their respective column sums, we obtain the normalized matrices $\widetilde{\mathbf{C}}^1, \widetilde{\mathbf{C}}^2$, and $\widetilde{\mathbf{C}}$ shown below.

$$\widetilde{\mathbf{C}}^1 = \begin{bmatrix} .2903 & .3038 & .2857 & .2813 \\ .2419 & .2532 & .2857 & .2500 \\ .1452 & .1266 & .1429 & .1563 \\ .3226 & .3165 & .2857 & .3125 \end{bmatrix} \text{ and}$$

$$\widetilde{\mathbf{C}}^2 = \begin{bmatrix} .1364 & .1328 & .1389 & .1429 \\ .4545 & 4428 & .4444 & .4286 \\ .2727 & .2768 & .2778 & .2857 \\ .1364 & .1476 & .1389 & .1429 \end{bmatrix}, \text{ as well as}$$

$$\widetilde{\mathbf{C}} = \begin{bmatrix} \frac{1}{2} & \frac{1}{2} \\ \frac{1}{2} & \frac{1}{2} \end{bmatrix}.$$

The row averages of the matrices $\widetilde{\mathbf{C}}^1$ and $\widetilde{\mathbf{C}}^2$ are then collected in the columns of the utility matrix $\mathbf{U} = \begin{bmatrix} .2903 & .1378 \\ .2577 & .4426 \\ .1428 & .2783 \\ .3093 & .1415 \end{bmatrix}$, while the row averages of the matrix $\widetilde{\mathbf{C}}$ are collected in the weight vector $\mathbf{w} = \begin{bmatrix} .5 \\ .5 \end{bmatrix}$. Multiplying the rows of the matrix $\mathbf{U}$ by the weight vector $\mathbf{w}$ results in the values that are associated with the four decisions

$$\mathbf{v}(d) = \begin{bmatrix} .2141 \\ .3502 \\ .2106 \\ .2254 \end{bmatrix}.$$

It is apparent that the decision maker is very much in favor of decision d_2; i.e., the firm will be best off if they cater to couples with a love for nature. The three remaining decisions have much lower scores.

The next question then concerns the validity of the scores provided by the decision maker. One measure to evaluate them is the consistency of the decision maker's rating. While pairwise consistency is, by assumption, always satisfied, more complex consistency is easily violated. For instance, suppose that the decision maker considers a decision d_2 three times as good as decision d_5, which, in turn, he considers to be twice as good as decision d_7. This implies that the decision maker believes that decision d_2 is six times as good as decision d_7. However, many decision makers will not use this value but, say, argue that d_2 is five times as good as d_7, thus introducing inconsistency. Small degrees of inconsistency are tolerable and will, in most cases, not influence the decision. However, it is important to flag such instances. The only remedy, though, will be for the decision maker to rethink the problem and attempt to be more consistent in his evaluations.

There is a variety of measures that may be used to evaluate the degree of consistency. While the originator of the analytic hierarchy process devised a so-called *consistency index*, we will use a simple measure from elementary statistics. We first observe that if the decision maker's evaluations are completely consistent, then all matrices $\widetilde{\mathbf{C}}^1, \widetilde{\mathbf{C}}^2, \ldots, \widetilde{\mathbf{C}}^q$ and $\widetilde{\mathbf{C}}$ will have identical elements in each of the rows. Typically, this is not the case. Smaller variations will be tolerable, but if large deviations show, the problem should be returned to the decision maker for reevaluation. The magnitude of the variation can be evaluated by the *coefficient of variation*. In order to do so, we first calculate the mean and the variance (and standard deviation) of the elements in each row of the matrices $\widetilde{\mathbf{C}}^1, \widetilde{\mathbf{C}}^2, \ldots, \widetilde{\mathbf{C}}^q$ and $\widetilde{\mathbf{C}}$, and then divide the standard deviation by the mean. The resulting coefficients of variation are then examined. If their magnitude is excessive, e.g., larger than, say, ten percent, the evaluations in that matrix should be returned to the decision maker for inspection.

In order to explain the concept, we return to the above

Example: First consider the matrix $\widetilde{\mathbf{C}}^1$. The rows of the matrix have means of 0.2903, 0.2577, 0.14275, and 0.3093, while its standard deviations are 0.008432, 0.016682, 0.010612, 0.014106, so that the coefficients of variation are 0.029049, 0.064734, 0.074340, and 0.045606, respectively. In other words, none of the

11.3 The Analytic Hierarchy Process

Table 11.7 Evaluations for Problem 1

	Price ($)	Gas mileage (mpg)	Comfort
Chrysler Prowler	50,000	12	Medium
Ford Ranger	20,000	25	Poor
Cadillac DeVille	60,000	18	Excellent

coefficients of variation are larger than 7.5%, making the decision maker's evaluations with respect to the first criterion acceptable.

Considering now the second criterion and its associated normalized matrix $\widetilde{\mathbf{C}}^2$, we find its coefficients of variation as 0.02671, 0.02086, 0.01692, and 0.02997, respectively, with a maximum variation of 3% from the mean. This is rather consistent, and no corrective action is recommended.

Finally, the matrix $\widetilde{\mathbf{C}}$ is completely consistent, as its elements are identical within each row.

Exercises

Problem 1 (Generic method, simple weighted average): Consider the problem of purchasing a vehicle. The decision maker has narrowed down the problem to three choices and three criteria. The evaluations of the vehicles are summarized in Table 11.7.

It can reasonably be assumed that the decision maker has utility functions

$$u(x) = 1 - \frac{2,778}{10^{10}} x^2$$

for a purchase price of x, and

$$u(y) = 1 - 50 y^{-2}$$

for a gas mileage of y. For the qualitative scores regarding the comfort, the decision maker uses the so-called 5-point Likert scale: terrible—poor—medium—good—excellent in increments of 0.25.

(a) Set up the utility matrix.
(b) Given that gas mileage and comfort combined are deemed equally important and that combined, they are assessed as important as the purchase price, determine the decision maker's weights of the criteria.
(c) What are the values (utilities) of the individual decisions and how would you advise the decision maker?
(d) Assume now that the decision maker decided to sell the car in 3 years' time, after having driven for 20,000 miles per year, which would cost him $4 per gallon. For simplicity, we will ignore interest and inflation. The resale values of the three vehicles have been estimated at 40%, 60%, and 70% of their present purchase price. The decision maker will then only consider two criteria, viz., the cost of operating the vehicle for 3 years and the comfort the vehicle provides. The utility of the costs is expressed as $u(z) = 1 - 2e^{-\frac{40}{z}}$, where z denotes the cost

(in $1000) of the vehicle during the 3 years. The decision maker considers costs three times as important as comfort. Set up the modified utility matrix and use the generic method to make a recommendation to the decision maker.

Solution:
(a) Given the utility functions, the utility matrix

$$\mathbf{U} = \begin{bmatrix} .3055 & .6528 & .5 \\ .8889 & .92 & .25 \\ 0 & .8457 & 1 \end{bmatrix}.$$

(b) $w_1 = .5$, $w_2 = .25$, $w_3 = .25$.
(c) $\mathbf{v}(d) = (.44095, .73695, .461425)$ and the analyst would recommend that the decision maker purchase the Ford Ranger.
(d) The loss of value of the three vehicles is $50{,}000 - 20{,}000 = \$30{,}000$, $20{,}000 - 12{,}000 = \$8000$, and $60{,}000 - 42{,}000 = \$18{,}000$, respectively. The gas consumptions of the three vehicles during the 3 years are $\frac{1}{12}60{,}000 = 5{,}000$, $\frac{1}{25}60{,}000 = 2{,}400$, and $\frac{1}{18}60{,}000 = 3{,}333.33$ gallons, respectively, for costs of \$20,000, \$9600, and \$13,333.33, respectively. The combined costs for the three vehicles are then \$50,000, \$17,600, and \$31,333.33, so that the utilities with the function provided above are .1649, .7061, and .4767. The evaluation matrix with columns for cost and comfort is then

$$\begin{bmatrix} .1013 & .5 \\ .7939 & .25 \\ .4420 & .1 \end{bmatrix}.$$

The weights specified by the decision maker are $w_1 = .75$ and $w_2 = .25$, so that the weighted average utilities of the three automobiles are .2010, .6579, and .3565, which gives a large advantage to the Ford Ranger.

Problem 2 (TOPSIS): A family needs to decide where to go for breakfast. There are four acceptable restaurants, the "Grill & Chill," the "Prime Roster," the "Simmering Heights," and the "Head Cookie" (denoted by $d_1, \ldots, d_4$). The important components of the breakfast are the potatoes, the eggs, the bacon, the toast, and the value for money. Based on past experience, the family has put together evaluations of the five individual components, which are shown as normalized utilities in Table 11.8.

The family has decided that the best possible point is $[1, 1, \ldots, 1]$, and they believe that deviations from the ideal point are measured well by weighted Euclidean distances, given that they have decided that while the deviations of the quality of

11.3 The Analytic Hierarchy Process

Table 11.8 Utilities for the breakfast example

Decision	Potatoes	Eggs	Bacon	Toast	Value for money
d_1	.9	.8	.6	.5	.6
d_2	.6	.5	.4	.3	1.0
d_3	.8	.9	.5	.4	.5
d_4	.6	.7	.7	.8	.5

potatoes, eggs, and bacon from the ideal are about equally important, they are three times as important as the deviation of toast from the ideal, and half as important as the deviation of the value for money from the ideal point. Where should the family eat?

Solution: We determine that the weights are 1 for toast, 3 for potatoes, eggs, and bacon each, and 6 for the value. The weighted deviations from the ideal point for the four restaurants are

d_1 : Deviation

$$= \sqrt{\left(3(1-.9)^2 + 3(1-.8)^2 + 3(1-.6)^2 + 1(1-.5)^2 + 6(1-.6)^2\right)}$$

$$= 1,3565,$$

d_2 : Deviation

$$= \sqrt{\left(3(1-.6)^2 + 3(1-.5)^2 + 3(1-.4)^2 + 1(1-.3)^2 + 6(1-1)^2\right)}$$

$$= 1.6733,$$

d_3 : Deviation

$$= \sqrt{\left(3(1-.8)^2 + 3(1-.9)^2 + 3(1-.5)^2 + 1(1-.4)^2 + 6(1-.5)^2\right)}$$

$$= 1.6613,$$

and

d_4 : Deviation

$$= \sqrt{\left(3(1-.6)^2 + 3(1-.7)^2 + 3(1-.7)^2 + 1(1-.8)^2 + 6(1-.5)^2\right)} = 1.6.$$

Other than the "Grill & Chill" (the first restaurant), the other eateries are fairly close in their evaluations. Formally d_4, i.e., the "Head Cookie," would be chosen, but as the assessments are so close, we would consider secondary objectives, which include important considerations that have not been included in this model.

Problem 3 (AHP, normalization technique): An investor may invest his available funds in blue-chip stocks, real estate, and bonds. These investments are to be evaluated on two criteria, viz., short-term profits and long-term viability, with the former being evaluated to be 1.5 times as important as the latter. Pairwise preference statements have been recorded as follows:

$$\text{Criterion 1: } \mathbf{C}^1 = \begin{bmatrix} 1 & 3 & 3 \\ 1/3 & 1 & 2 \\ 1/3 & 1/2 & 1 \end{bmatrix}, \text{ criterion 2: } \mathbf{C}^2 = \begin{bmatrix} 1 & 1/2 & 1/5 \\ 2 & 1 & 1/2 \\ 5 & 2 & 1 \end{bmatrix}, \text{ and the}$$

relation between the two criteria is expressed in the matrix $\mathbf{C} = \begin{bmatrix} 1 & \frac{3}{2} \\ 2/3 & 1 \end{bmatrix}$.

(a) Use the analytic hierarchy method to determine a ranking of the three decisions.
(b) Examine the consistency in the matrices $\mathbf{C}^1$ and $\mathbf{C}^2$, using a quantitative criterion. What are your conclusions?

Solution:

(a) The normalized matrices are

$$\tilde{\mathbf{C}}^1 = \begin{bmatrix} .6000 & .6667 & .5000 \\ .2000 & .2222 & .3333 \\ .2000 & .1111 & .1667 \end{bmatrix}, \tilde{\mathbf{C}}^2 = \begin{bmatrix} .1250 & .1429 & .1176 \\ .2500 & .2857 & .2941 \\ .6250 & .5714 & .5882 \end{bmatrix} \text{ and}$$

$$\tilde{\mathbf{C}} = \begin{bmatrix} .6 & .6 \\ .4 & .4 \end{bmatrix}.$$

The row averages of these normalized matrices are the columns of the utility matrix $\mathbf{U}$ and the weight vector $\mathbf{w}$, respectively, so that

$$\mathbf{U} = \begin{bmatrix} .5889 & .1285 \\ .2518 & .2766 \\ .1593 & .5949 \end{bmatrix} \text{ and } \mathbf{w} = \begin{bmatrix} .6 \\ .4 \end{bmatrix}.$$

We then obtain $\mathbf{v}(d) = [.4047, .2617, .3335]$, so that we recommend the first decision (investment in blue-chip stocks) to the decision maker.

(b) As far as consistency is concerned, we compute coefficients of variation of .20156, .2316, and .2302 for matrix $\tilde{\mathbf{C}}^1$, i.e., an average variation of more than 20%. This would give us sufficient reason to return the evaluation matrix $\mathbf{C}^1$ to the decision maker for reevaluation.

Consider now the matrix $\widetilde{\mathbf{C}}^2$. The three rows have coefficients of variation of .1432, .1197, and .0652 with an average of 10.94%, which appears fairly high, but more or less acceptable. The matrix $\mathbf{C}$ has zero variability, and it is consistent.

References

Hwang CL, Yoon K (1981) Multiple attribute decision making: methods and applications. Springer, New York

Saaty TL (1980) The analytic hierarchy process: planning, priority setting, resources allocation. McGraw-Hill, New York

Inventory Models 12

Worldwide, companies hold billions of dollars in inventories. The main reason is to create a buffer that balances the differences between the inflow and outflow of goods. Inventories can be thought of as water tanks: there may be a constant inflow of water that is pumped into the tank by a pump, while the outflow is low at night and high in the morning (when people get up, take a shower, etc.); it then decreases significantly until the demand again increases in the evening (when people come home, do laundry, etc.), just to fall off again for the night. Other, popular, examples include grocery stores whose inventories consist of various foodstuffs awaiting sale to its customers. Here, the delivery of the goods is in bulk whenever a delivery truck arrives, while the demand is unknown and erratic. In the case of hospitals, they have in stock medical supplies, bed linen, and blood plasma. Again, the demand for these items is uncertain and may differ widely from 1 day to the next.

All these instances have a few basic features in common. They have a supply, a demand, and some costs to obtain, keep, and dispose of inventories. The next section introduces a number of parameters and variables that are typically found in inventory models. Section 12.2 describes a basic inventory model, and the subsequent sections deal with a variety of extensions of the basic model.

12.1 Basic Concepts in Inventory Planning

For many organizations, inventories represent a major capital cost, in some cases the dominant cost, so that the management of this capital becomes of the utmost importance. When considering the inventories, we need to distinguish different classes of items that are kept in stock. In practice, it turns out that about 10% of the items that are kept in stock usually account for something in the order of 60% of the value of all inventories. Such items are therefore of prime concern to the company, and the stock of these items will need close attention. These most important items are usually referred to as "*A items*" in the *ABC* classification system developed by the General Electric Company in the 1950s. The items next in line are the B items, which are of intermediate importance. They typically represent 30% of the items, corresponding to about 30% of the total inventory value. Clearly, B items do require some attention, but obviously less than A items. Finally, the bottom 60% of the items are the C items. They usually represent maybe 10% of the monetary value of the total inventory. The control of C items in inventory planning is less crucial than that of the A and B items. The models in this chapter are mostly aimed at A items.

The importance can be seen that as of 2021, the total business inventories in the USA were evaluated at 2 trillion dollars. Given the economic importance of the management of inventories, a considerable body of knowledge has developed as a specialty of operations research. We may mention *just-in-time* (*JIT*) systems that attempt to keep inventory levels in a production system at an absolute minimum and put to work in Toyota's so-called *kanban* system. There is also *material requirements planning* (*MRP*) aimed at using the estimated demand for a final product in order to determine the need for materials and components that are part of a final product. *Multi-echelon* and *supply-chain management* systems also consider similar aspects of production-inventory control systems. Such topics are beyond the scope of this text, in which we can only cover some basic inventory models.

Throughout this chapter, we will deal with inventory models that concern just a single item. Consider an item for which the demand per period (typically a year, but other time frames can easily be accommodated) is known or estimated to be D units. Unless otherwise specified, the parameter D is assumed to be constant over time.

The number of items in stock is depleted over time by the demand. On the other hand, the stock is also increased from time to time by additions caused by deliveries, referred to as *orders*. Typically, replenishments are assumed to be instantaneous (such as the arrival of goods by the truckload), resulting in sudden jumps in the inventory level, whereas deliveries to satisfy demand are typically assumed to be gradual. The order quantity is denoted by Q. In the models presented in this book, Q turns out to be constant over time, given that the parameters of the model do not change.

Related to the order quantity is also the *lead time* t_L. The lead time is defined as the time that elapses between the placement of an order and the moment that the shipment actually arrives and is available on the shelf.

Figure 12.1 may illustrate the general inventory and ordering process. The inventory level drops gradually from some point in time until it reaches a time t_0. At that point, a new shipment is received. The order and the resulting shipment are of magnitude Q_1. The inventory either shoots up immediately as shown here or increases gradually. The moment the new shipment has arrived, customers continue to take units out of the warehouse, so that the inventory level drops down again. At the point t_1 another shipment, this time of magnitude Q_2, arrives. Again, the inventory level drops down until it reaches zero at time t_2, and after that, we encounter shortages. At point t_3, a new shipment of magnitude Q_3 units comes in. The first units are used to satisfy the backorders, and beyond that, the inventory is built up again, and the next cycle begins.

12.1 Basic Concepts in Inventory Planning

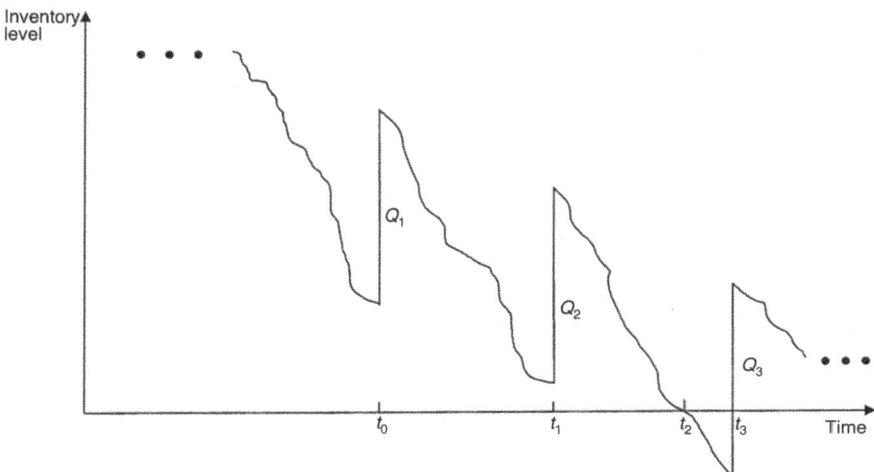

Fig. 12.1 The general inventory and ordering process

The cost for carrying one unit of the good in inventory for one period is called the unit *carrying* or *holding cost*. We will denote this cost by the parameter c_h. Occasionally, the carrying cost c_h will be specified as a proportion of the inventory value rather than per unit of inventory. This will be clearly indicated whenever it applies. The total carrying cost is therefore related to the inventory actually on hand and will be zero for those time periods when the inventory level equals zero. Carrying costs not only include the rental of warehouses, heating, cooling, lighting, security, insurance, obsolescence, etc., but, as probably the most important component, the cost of tied-up capital. Take, for instance, a car dealership with, say, 50 new vehicles on the lot, each priced a modest $30,000. With an interest rate of 5%, the annual holding costs are $75,000, equivalent to no less than 2½ vehicles. Similarly, imagine that the inventory comprises, say, cell phones. Given that new models are introduced once a year, the value of the inventory will drop dramatically a year later, as the market for old cell phones is severely limited.

In contrast, the *unit ordering costs* c_o are defined as the cost of placing an order and having it delivered. Ordering costs include administrative costs as well as transportation costs. Ordering costs are usually considered to be independent of the size of the order, but since transportation costs are involved, one may question this assumption. However, we may argue that transportation costs may not be directly related to the size of the order: if the order is delivered by container or truck, the rate charged is usually not much dependent on whether the container is one-quarter or three-quarters full; a similar argument applied to truckloads. Additionally, if the transportation costs were to relate to each item, then—assuming that the demand must be satisfied—the total ordering cost for a period with a given demand D would be the same, regardless of whether there are many small or few large orders. Being a constant, this factor would then not affect the optimization.

Other cost types exist as well. For instance, in case of backorders, we allow shortages, assuming that the sales are not lost, but will be fulfilled, once a new order comes in. We need to note that these shortages are planned and not due to an unexpected discrepancy between expected and real sales. The *unit shortage costs* c_s are defined as the cost of being short one unit for one time period (typically, but not necessarily, a year). As such the unit shortage costs are an exact counterpart of holding costs: while unit holding costs express the costs of *having* one unit in stock for 1 year, shortage costs measure the cost of *not having* one unit in stock for a year.

12.2 The Economic Order Quantity (*EOQ*) Model

The most basic inventory model, the *economic order quantity* or *EOQ* model, has an interesting history. Although the *EOQ* formula was first published in 1913 by Ford Whitman Harris, it has been known under other names such as Camp's formula and Wilson's lot size formula in the 1920s and 1930s. A good historical account is provided in an article by Erlenkotter (1990), indicating deliberate attempts in the past to deny Ford Whitman Harris the credit of originating the *EOQ*.

The assumptions of the basic economic order quantity are:

- The inventory consists of one single unperishable good held in one location.
- The demand rate for the item is constant over time and demand must be satisfied exactly.
- Units are ordered from a single supplier in the same amount each time. For now, we assume that replenishment is instantaneous, even though that is not really necessary as we will show later.
- There are no quantity discounts.
- Stockouts are not allowed and the demand must be satisfied completely.
- The planning horizon is infinite, and all model parameters are stationary, i.e., they do not change over time.

It is apparent that the problem has two components: *how much* to order and *when*. The question of when to order is simply resolved in case of instantaneous replenishments: since the inventory level is not allowed to become negative, and since it does not make sense for a replenishment to occur while the inventory level is positive, it must be optimal to place an order (and immediately receive the shipment), when the inventory level reaches zero. Instead of looking at the time when to reorder, we use the inventory level as a proxy for the time clock.

Since the order size was assumed to be the same each time, the inventory levels over time will have the characteristic sawtooth pattern shown in Fig. 12.2, where the time between two consecutive replenishment times is referred to as the *inventory cycle length* t_c, while the *order quantity* is Q.

We now wish to minimize the total inventory-related costs TC per period (year), which is the sum of ordering costs and holding costs for the period, as well as the *purchasing costs*.

12.2 The Economic Order Quantity (EOQ) Model

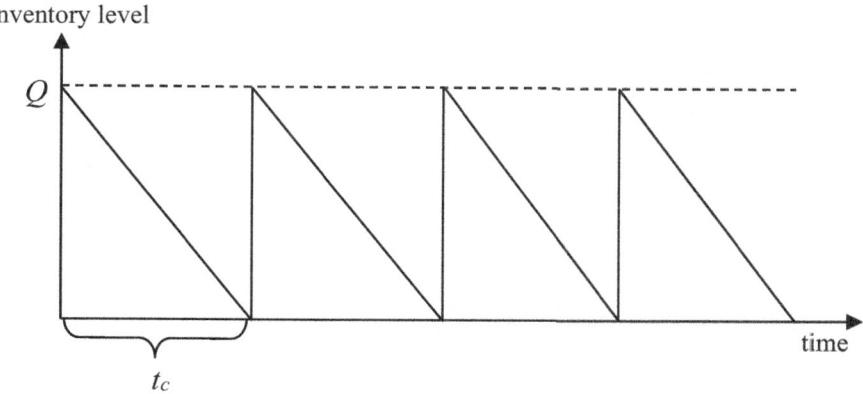

Fig. 12.2 Sawtooth inventory pattern

First, consider the purchasing costs. Assuming that the annual demand is D and the per-unit cost of an item are p, the total purchasing costs are pD. Assuming that the demand must be completely satisfied and there are no price discounts, the purchasing costs are actually a constant, so that they do not influence the optimization and can therefore be left out of the analysis.

Next, there are the ordering costs. If we were to place N orders, each of size Q, within the planning period, then the total amount ordered is NQ, which must equal the demand D: since stockouts are not allowed, NQ cannot be less than D, and with $NQ > D$ we would be carrying more inventory than needed, incurring unnecessary costs. With $NQ = D$, we find that $N = D/Q$ equals the number of orders per period. Incidentally, since t_c is the length of one inventory cycle, of which there are N in one period, $t_c N = 1$ period, so that $t_c = 1/N = Q/D$. Recalling that the cost for one order is c_o, we conclude that total annual ordering costs are $c_o N = c_o D/Q$.

Since holding costs are charged per unit in inventory (the vertical axis in the diagram in Fig. 12.2), we will need to compute the total area under the sawtooth curve, which can be seen to be ½Q. In other words, the average inventory level is ½Q, implying that the total annual holding costs are c_h(½Q). Hence, the total inventory costs per period are

$$TC = c_o D/Q + 1/2 c_h Q.$$

This expression makes intuitive sense: the larger the order quantity Q, the lower the total ordering costs, since larger, but fewer, orders are being placed. However, the holding costs will be higher, since more inventory is being kept in stock (on average). This is illustrated in Fig. 12.3, where the abscissa measures the order quantity and the ordinate measures the costs. The broken line (a rational function of type "constant divided by the order quantity Q") represents the ordering costs, while the linear function (the broken line) represents the inventory holding costs. The solid line is then the sum of the ordering and holding costs. If we were to include the

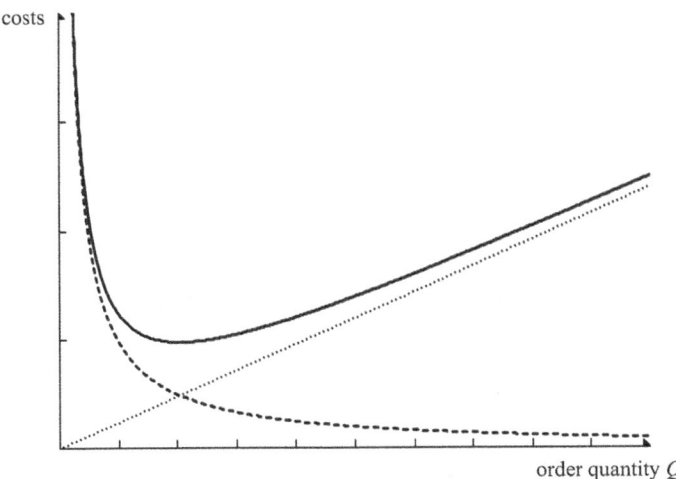

Fig. 12.3 Ordering, holding, and total inventory costs

constant purchasing costs as well, the resulting total cost curve would simply move up, but do not change in shape.

To find the value Q^* that minimizes total inventory cost TC, we find the derivative with respect to Q, which results in

$$TC' = -c_o D/Q^2 + \tfrac{1}{2} c_h.$$

Setting this expression equal to zero and solving for the variable Q results in the optimal order quantity

$$Q^* = \sqrt{\frac{2Dc_o}{c_h}}.$$

(Technically, we also have to check the second derivative in order to ensure that the optimal order quantity results in minimal, rather than maximal, inventory costs. It does.) The expression for Q^* is referred to as the *economic order quantity EOQ*. Inserting this order quantity into the cost function, we obtain the total inventory costs at optimum, which are

$$TC(Q^*) = \sqrt{\tfrac{1}{2} D c_o c_h} + \sqrt{\tfrac{1}{2} D c_o c_h} = \sqrt{2 D c_o c_h}.$$

Note that it just so happens that holding and ordering costs are always equal at optimum regardless of the value of the parameters D, c_o, and c_h; however, the solution Q^* is *not optimal simply because* holding and ordering costs are equal.

Given Q^*, we can then determine the related variables $N^* = D/Q^*$ and the optimal inventory cycle length $t_c^* = 1/N^* = Q^*/D$.

Example: A retail store faces a demand of $D = 800$ car battery chargers per year. The cost of placing an order for the chargers is $c_o = \$100$ and the holding cost is $c_h = \$4$ per charger per year. The economic order quantity is therefore $Q^* = \sqrt{\frac{2(800)(100)}{4}} = 200$ chargers, and the total cost is $TC^* = \sqrt{2(800)(100)(4)} = \$800$. The optimal number of orders to be placed throughout the year is $N^* = D/Q^* = 800/200 = 4$ orders, and the optimal inventory cycle length is $t_c^* = Q^*/D = 200/800 = 1/4\,\text{year} = 3$ months. The total ordering costs are then $4(100) = \$400$, which is half of the total cost; the other half is made up by the total holding costs.

A useful feature of the economic order quantity is its insensitivity to errors in the input data (i.e., the parameters). In our example, assume that the annual demand was erroneously estimated to be 920 chargers, instead of the correct amount of 800 chargers, i.e., a 15% overstatement. Using the *EOQ* formula, we obtain the nonoptimal and erroneous value Q^* from the expression $Q^* = \sqrt{\frac{2(920)(100)}{4}} \cong 214.5$, which is an overstatement by slightly more than 7%, only about half of the original relative error. One can show that due to the square root of the formula, relative input data errors result in relative *EOQ* errors of only about half the size, for reasonable errors (say, 30% or less). Checking the total inventory costs that result from the wrong data, we find (using the true value of $D = 800$ and the erroneous order quantity $Q^* = 214.5$) that $TC(Q^* = 214.5) = \frac{100(800)}{214.5} + 1/2(4)(214.5) = \$801.96$, which deviates only very marginally from the true value of $\$800$.

12.3 The Economic Order Quantity with Positive Lead Time

We will now carry our discussion further and extend the model to the situation where a positive lead time t_L elapses from the moment an order is placed until the quantity ordered has arrived and been added to the inventory.

It is apparent that the decision *when* to order will in no way affect the decision *how much* to order. In other words, the optimum order quantity Q^* still applies. Instead of looking at the clock time for the point at which to place an order, we will observe the inventory level and determine the order time in terms of the *reorder point R*: when the on-hand inventory level decreases to the level R, an order is placed, which will arrive after a delay of t_L time units; at this point, the inventory level, for optimal performance, should have reached zero. Putting a reorder decision on the basis of the quantity still at hand rather than on the basis of a specific date is much more flexible, particularly in practical situations, in which demand is not continuous.

In our analysis, we will consider two different cases, depending on the length of the lead time t_L, which differ on the basis of the relation between lead time and cycle time.

Case 1: $t_L \leq t_c^*$, i.e., the lead time is less than or equal to the optimal inventory cycle length. The demand during the lead time is then $t_L D$, and it follows that if an

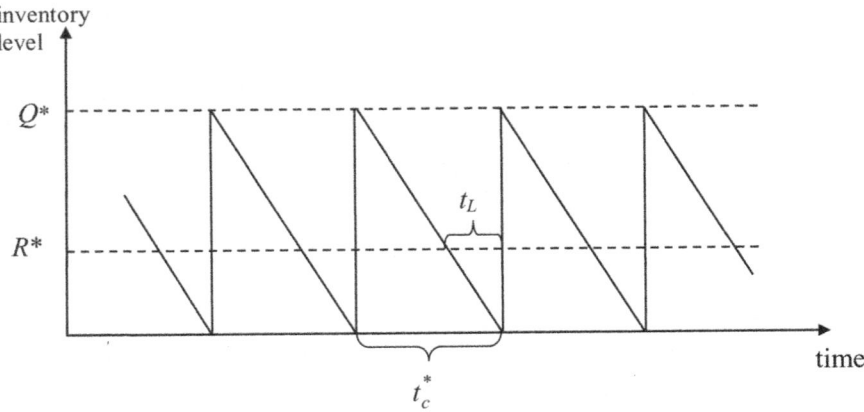

Fig. 12.4 Lead time less than or equal to optimal inventory cycle length

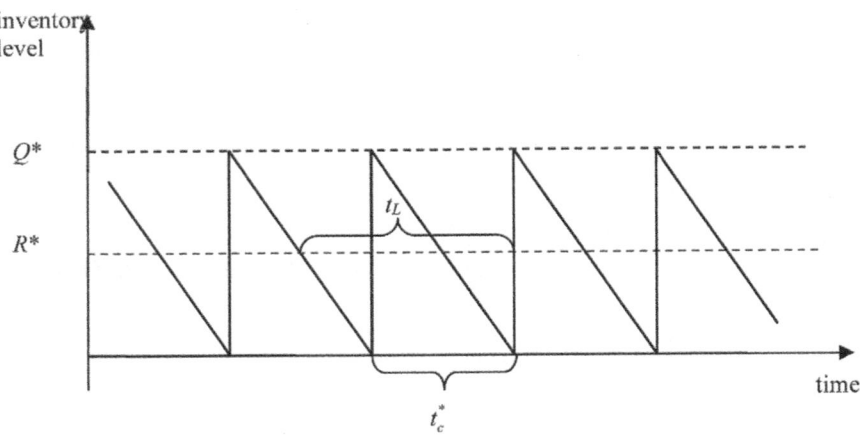

Fig. 12.5 Lead time greater than optimal inventory cycle length

order of size Q^* is placed when the inventory level reaches $R^* = t_L D$, then the replenishment will arrive exactly at the time when the inventory on hand has been depleted, which is neither too soon, nor too late. This situation is depicted in Fig. 12.4.

Case 2: $t_L > t_c^*$, i.e., the lead time is greater than the optimal inventory cycle length. The demand during the lead time is still $t_L D$, but since $t_L > t_c^*$, it follows that $t_L D > t_c^* D = Q^*$, which is the highest level of inventory on hand that we will ever reach. The arrival of an order will therefore occur during a subsequent inventory cycle and not during the cycle in which it was ordered. This situation is illustrated in Fig. 12.5, where $t_c^* < t_L < 2t_c^*$.

12.3 The Economic Order Quantity with Positive Lead Time

In general, there could be several replenishments occurring during the lead time. This number is actually $\lfloor \frac{t_L}{t_c^*} \rfloor$ (the "floor of the number"), i.e., the ratio $\lfloor \frac{t_L}{t_c^*} \rfloor$ rounded down to the nearest integer. In Fig. 12.5, the floor equals 1, so that the replenishment arrives at the end of the inventory cycle following after the one during which it was ordered. In general, we obtain the relation

$$R^* = t_L D - \left\lfloor \frac{t_L}{t_c^*} \right\rfloor Q^*.$$

It can easily be demonstrated that this expression will cover both cases above. Since for $t_L < t_c^*$ (as in Case 1), $\lfloor \frac{t_L}{t_c^*} \rfloor = 0$, so that $R^* = t_L D$, which is the expression derived for that case.

Example: Consider the battery charger example above with $D = 800$, $c_o = \$100$, and $c_h = \$4$, for which we have obtained an optimal order quantity of $Q^* = 200$ at an annual cost of $TC^* = 800$. Given now a lead time of 2 months, i.e., 1/6 of a year, we have $t_L = 1/6 < 1/4 = t_c^*$, so that Case 1 applies. Here, we find the optimal reorder point as

$$R^* = t_L D = 1/6(800) = 133\frac{1}{3} \text{ units.}$$

On the other hand, with a lead time of $t_L = 4$ months, i.e., 1/3 of a year, we have $t_L = 1/3 > 1/4 = t_c^*$, so that Case 2 applies, and we find that

$$R^* = 1/3(800) - \left\lfloor \frac{1/3}{1/4} \right\rfloor (200) = 66\frac{2}{3} \text{units.}$$

As mentioned earlier, these shipments will not arrive in the inventory cycle they are ordered in, but in the next cycle.

A practical way to implement the reordering in Case 1 above where $t_L \leq t_c^*$ is the *two-bin system*. Upon replenishment, the order quantity Q^* is physically separated and put into two storage bins. The first bin has a capacity of $Q^* - R^*$ units, while the second bin holds R^* units. The demand for items in stock is then satisfied exclusively from the first bin until it is empty. At that time, only the R^* items in the second bin remain and the inventory manager will reorder the item. From this point onward, all subsequent demand is now satisfied from the second bin. Given the assumptions made in this section, the second bin will be depleted exactly at the time when the next shipment arrives and the process is repeated.

A similar argument allows the two-bin system to be implemented in Case 2 with $t_L > t_c^*$. Supermarkets use a "virtual" version of the two-bin system by electronically monitoring inventory levels by counting items that pass through the checkout counters. Orders are then automatically triggered when the appropriate reorder point has been reached.

12.4 The Economic Order Quantity with Backorders

Assume now that we allow backorders, so that the inventory level may become negative, in the sense that unsatisfied demand is recorded or "backordered," to be satisfied immediately upon replenishment of the inventory. That way, there will be no lost sales. Such planned shortages are considered to incur a shortage cost per unit and per period. These unit shortage costs will be denoted by c_s. These costs consist of the inconvenience to the customer of the unsatisfied demand and they will be difficult to estimate in practice. The costs could also include special handling costs that are incurred due to the preferential delivery to the customer of the backordered units when they become available. As stated before, unit shortage costs are very similar to unit holding costs: while unit holding costs express the costs to *have* one unit of the good in stock for a year, unit shortage costs express the costs of *not having* one unit of the good in stock for a year.

Assuming a repetitive situation, the net inventory level will be as in Fig. 12.6, where S denotes the amount of the maximal planned shortage. Note in the figure that the maximal inventory directly after replenishment is no longer Q as in the standard model, but $Q - S$, as the stockouts are satisfied first before new inventory is built up.

In Fig. 12.6, the parameter t_1 denotes the length of time during an inventory cycle during which the inventory level is nonnegative, i.e., when there is no stockout. On the other hand, the parameter t_2 denotes the length of time during which there is no stock at hand. Clearly, $t_1 + t_2 = t_c$. Using the geometric relationship of the two similar (shaded) triangles in Fig. 12.6, we find that $t_1/t_2 = (Q - S)/S$. We will use this expression below.

We will now determine optimal levels of the order size Q and the optimal largest shortage level S simultaneously by minimizing total inventory costs. These costs now include not only ordering and holding costs as before, but also shortage costs. As in the standard economic order quantity model, the annual ordering costs

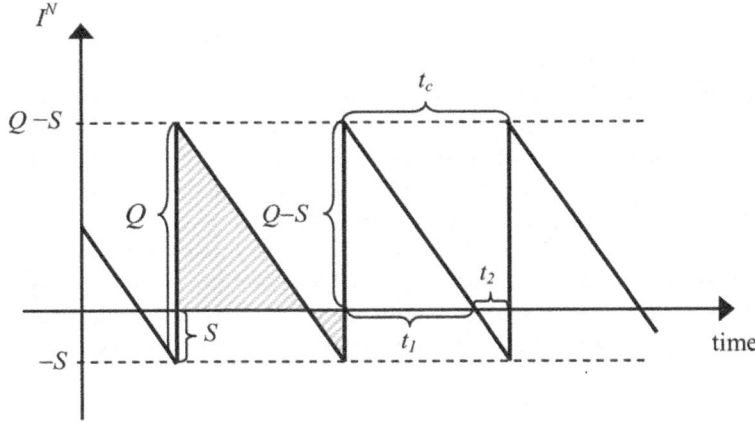

Fig. 12.6 Inventory with planned shortages

are c_oD/Q. As far as the carrying costs are concerned, we find that the average inventory level is obtained by averaging the inventory level during the time that no stockouts occur. This weighted average is $\frac{1}{2}(Q - S)$ during the time t_1, while the inventory level during the time stockouts occur is zero for the duration t_2. After some calculations, this leads to inventory holding costs of $c_h = \frac{(Q-S)^2}{2Q}$.

We can now deal with the average shortage in a similar fashion. The average annual shortage is $\frac{1}{2}S$ during the time t_2, which is when we have shortages. This leads to total shortage costs of $c_s \frac{S^2}{2Q}$. The total inventory costs are therefore

$$TC(Q, S) = c_o \frac{D}{Q} + c_h \frac{(Q - S)^2}{2Q} + c_s \frac{S^2}{2Q}.$$

Using partial derivatives, we can show that the total inventory costs are minimized for

$$Q^* = \sqrt{\frac{2Dc_o}{c_h} \frac{c_h + c_s}{c_s}} \text{ and } S^* = \sqrt{\frac{2Dc_o}{c_s} \frac{c_h}{c_h + c_s}} = \frac{c_h}{c_h + c_s} Q^*.$$

Example: Using the basic *EOQ* example above with its parameters $D = 800$, $c_o = \$100$, and $c_h = \$4$ to which we add unit shortage costs of $c_s = \$6$ per unit and year, we can then determine the optimal order quantity as $Q^* = \sqrt{\frac{2(800)(100)}{4} \frac{4+6}{6}} \approx 258.20$ units.. The optimal shortage can then be determined as $S^* = \frac{c_h}{c_h + c_s} Q^* \approx 103.28$. The total costs are then $TC(Q^*, S^*) = 309.84 + 185.90 + 123.94 = \$619.68$.

Note that while this model includes more cost items than the basic *EOQ* model, the same demand and cost parameters cost less in this model. The reason is that this model allows the decision maker an added possibility, viz., to run planned shortages. Nobody says that we have to have shortages (this model certainly allows not having any), but here it is cost-effective to plan some shortages. As a matter of fact, the relation between the unit shortage cost and the unit holding cost will determine the magnitude of the planned shortage: if c_s/c_h is large, then shortages are expensive and the solution will include only minor shortages. If, on the other hand, c_s/c_h is small, then shortages are comparably cheap, and the model will prescribe large shortages.

To push that argument even further, suppose that shortage costs would increase beyond all reasonable limits, i.e., $c_s \rightarrow \infty$. Then the "correction factor" $\frac{c_h+c_s}{c_s}$ in the order quantity root will approach 1, so that the order quantity Q^* will assume the same value as in the basic *EOQ* model. At the same time, the magnitude of the planned shortage will tend to zero, as the shortage costs appear only in the denominator of the formula. Thus, it becomes clear that the basic model is just a special case of the model with shortages, given that shortage costs are infinitely high. This result is nothing but an application of the usual principle of penalty costs: if there is something that we do not want, assign a very high penalty to it, and as a result, the optimizer will not include the very expensive option in the solution. Another extreme case exists if c_s approaches zero. In this case, $S^* = Q^*$, and Q^* will assume the largest possible value, which is the demand for all future periods. The important part is that in this case, there will be no inventory at all.

12.5 The Economic Order Quantity with Quantity Discounts

So far we have assumed that the unit purchasing cost p is constant and independent of the order size Q. Recall that in the original economic order quantity model, the costs actually were the sum of ordering, holding, and purchasing costs, viz.,

$$TC(Q,p) = c_o D/Q + \tfrac{1}{2} c_h Q + pD.$$

However, we argued, for a fixed price p, the purchasing costs during the period under consideration are pD, which is a constant, which does not influence the solution, so that we could (and did) ignore the purchasing costs. In practice, however, many suppliers will offer incentives for purchases of larger quantities in the form of lower unit costs. The basic economic order quantity model described in Sect. 12.2 can easily be modified to take such quantity discounts into consideration. For simplicity, we will restrict ourselves to the standard model with no shortages allowed and three price levels: the original non-discounted price and two discount levels. It is straightforward to extend the model to any number of discount levels.

Before we proceed, we have to make a minor modification. Recall that the main component of the holding costs is the cost for tied-up capital. Given that we paid a fixed price for the good so far, this cost could simply be expressed as a dollar amount for each unit in stock. Since we are now paying a price that is no longer fixed but does depend on the discount level that we choose, the unit holding costs have to be redefined. This is most easily done by defining c_h as a proportion of the unit purchasing price p. Given that, the economic order quantity is redefined as

$$Q^* = \sqrt{\frac{2Dc_o}{c_h p}}.$$

We will denote the given (non-discounted) price level as p_0, the price with the small discount as p_1, and the price given the large discount as p_2. Clearly, $p_0 > p_1 > p_2$. (In general, we can have as many price levels as desired.) The rationale behind this scenario is simple. For the regular price of p_0, we can obtain any quantity we desire. In order to convince our supplier to sell the goods for us at the lower price of p_1, we have to purchase at least a certain quantity of goods. This quantity will be called Q_1. Going one step further, we can ask our supplier to let us have the goods even cheaper at a price of p_2, which he may agree to, but only if we order at least Q_2 units with the obvious condition that $Q_2 > Q_1$.

At this point, we have a cost function for each of the price levels. This situation is shown in Fig. 12.7.

The three cost functions are shown as $TC(p_0)$, $TC(p_1)$, and $TC(p_2)$, respectively. The dots on the cost functions denote their respective optimal points, at which the

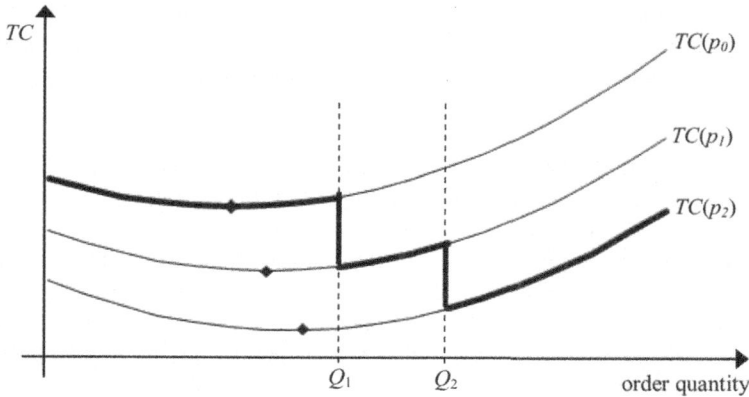

Fig. 12.7 Inventory cost functions for different price levels

costs are minimized. The minimal quantity levels that allow for the discounts are also shown.

Consider now the actual costs that we incur as the order quantity gradually increases. For very low order quantities, we must be on the highest cost curve, as we do not qualify for a discount. As we increase the value of Q, the costs decrease and reach their minimum at the dot on the $TC(p_0)$ curve. As Q increases further, the costs increase as well until we reach Q_1. At this point, we do qualify for the small discount, so that our actual costs jump down onto the cost curve $TC(p_1)$. As this point happens to be to the right of the minimum on this function, the costs increase as Q increases. This process continues until we reach Q_2, the value that allows us to obtain the second (larger) discount. Again, the costs drop at this point onto the third and lowest cost function $TC(p_2)$. Increasing the value of Q further increases the total costs.

The piecewise nonlinear cost function is shown as a bold line in Fig. 12.7. In order to determine the order quantity with the overall minimal costs, we have to examine each cost curve separately. More specifically, we determine the optimal order quantity at each price level, and then compare them and choose the option with the lowest costs.

First consider the highest cost curve without a price discount. We simply determine the point of lowest cost with the EOQ and record the associated cost. This is the optimal solution given the option of paying the regular price p_0.

We then continue to examine the costs incurred when paying the price p_1. Again, we determine the optimal quantity at this price by solving the economic order quantity with this price. (Note that with decreasing prices, the optimal order quantity increases slightly, as the expression in the denominator $c_h p$ decreases.) We then have to determine whether or not this quantity permits us to obtain the discount. If so, we have found our optimal order quantity at this level. If this is not the case, we have to move out of the optimum, but, as the function is increasing the farther we move out of the optimum, just as much as required to qualify for the discount.

In our illustration in Fig. 12.7, the optimal order quantity is less than Q_1, the lowest quantity that qualifies for the price p_1. We thus increase the order quantity to Q_1 and determine the costs at that point. This is the optimal order quantity *given the price* p_1.

This process is repeated for all discount levels. Once this has been accomplished, we simply compare the best-known costs at each price level and choose the overall minimum. This is our optimal solution; in Fig. 12.7 it is Q_2.

This process can be illustrated by the following numerical:

Example: A company faces an annual demand for 10,000 footballs. The purchasing costs are $2 per football, the holding costs are 5% of the purchasing price per football and year, and the costs of placing one order are $80. The supplier now offers a ½% discount in case the company orders at least 6000 units. As an alternative, the supplier also offers a 1% discount, if the company orders at least 15,000 units. Consider all alternatives, compute the total costs in each case and make a recommendation.

The parameters of the problem include $D = 10,000$, $c_h = 5\%$ of p, and $c_o = \$80$.

Case 1: No discount, so that $p_0 = \$2$. Then $c_h = \$0.10$, and we use the *EOQ* to compute the order quantity as $Q^* = 4000$ with costs of $TC^* = 200 + 200 + 20,000 = \$20,400$.

Case 2: Small discount, so that $p_1 = \$1.99$. Then $c_h = \$0.0995$, and the solution of the *EOQ* is $Q^* = 4010.038$. This quantity does not qualify for the discount, so that we have to move out of the optimum just as much as necessary to qualify for the discount. Hence, we set $Q := 6000$, for which we then obtain costs of $TC(6000) = 133.33 + 298.50 + 19,900 = \$20,331.83$.

Case 3: Bigger discount, so that $p_2 = \$1.98$. Then $c_h = \$0.099$, and the solution of the *EOQ* is $Q^* = 4020.15$. This quantity does not qualify for the discount, so that we have to move out of the optimum just as much as necessary to qualify for the discount. Hence, we set $Q: = 15,000$, for which we then obtain costs of $TC(15,000) = 53.33 + 742.50 + 19,800 = \$20,595.83$.

Comparing the three options, Case 2 offers the lowest total costs, so that we should order 6000 footballs, obtain a ½% discount, and incur total costs of $20,331.83.

12.6 The Production Lot Size Model

As an alternative to ordering models of the type discussed so far in this chapter, we may produce the desired items ourselves. In such a case, the items will not arrive in one bulk as they do in case of orders, but they arrive one piece at a time from our machines. Assume for the time being that we cannot regulate the speed with which our machines produce the units: we either turn the machine on, in which they churn out r units per day (this is our production rate), or we turn off the machine, in which case we make nothing. Recall that our annual demand was assumed to be of magnitude D, from which we can easily compute the daily demand d, which is $D/365$, $D/360$, or $D/250$ (working days), depending on the decision maker's

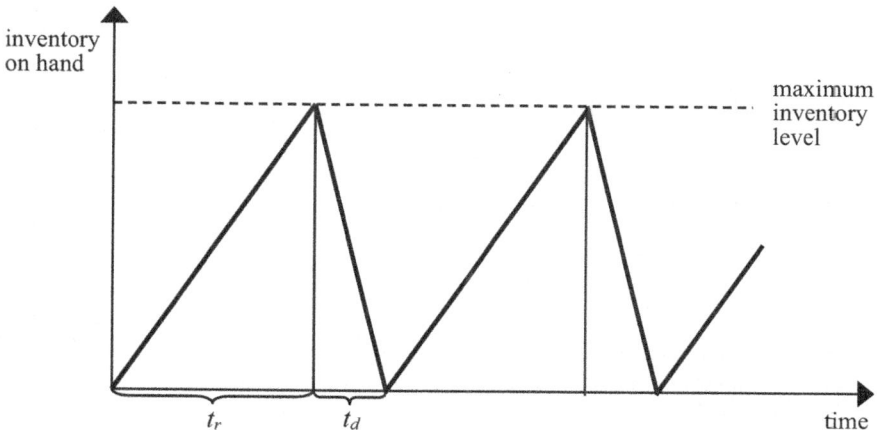

Fig. 12.8 Inventory levels in the production lot size model

specifications. Before engaging in any computations, it is necessary to determine whether or not the system has any feasible solutions. The simple *regularity condition* is that $r \geq d$. If this condition is not satisfied, then it will not be possible to satisfy the total demand, and we have to find ways that allow us to do so. The following arguments assume that the regularity condition is satisfied.

Batch or *intermittent production* as described above occurs in many vertically integrated companies, where the ordered items are produced internally. A production run can then be considered an order, with the production run size corresponding to the order size Q, and the production setup costs corresponding to the ordering costs c_o.

Using an argument similar to that in Sect. 12.2, we note that D/Q is the number of setups or production runs per period, so that the total setup costs are $c_o(D/Q)$. As far as the carrying costs are concerned, we will consider the *production phase* t_r (the phase during with production and demand occur) and the *demand phase* t_d (the phase during which production does not occur, while demand occurs as usual) separately. In the production phase, inventory accumulates at the rate of $(r - d)$. We notice that the duration of the production phase is $t_r = Q/r$, so that the maximal level of inventory at the end of each production run will be $(r - d)Q/r$. During the demand phase, the inventory, which starts with a level of $(r - d)Q/r$, decreases to zero in a linear fashion at a rate of d, so that the slope of the function in Fig. 12.8 during that phase is $-d$. Note that the main difference between this model and the basic *EOQ* model is the gradual increase of the inventory level in this case. The average inventory level during the entire cycle of duration $t_c = t_r + t_d$ is then ½ $(r - d)Q/r$. Therefore, the total carrying costs per period are ½$c_h(r - d)Q/r$. As a result, the total production- and inventory-related costs are

$$TC = c_o D/Q + \tfrac{1}{2} c_h (r-d) Q/r.$$

Following the same procedure applied to the standard economic order quantity in Sect. 12.2, we find the derivative with respect to the single variable Q, which results in $TC' = -c_o D/Q^2 + \tfrac{1}{2} c_h (r-d)/r$. Setting the derivative equal to zero results in the unique optimal lot size of

$$Q^* = \sqrt{\frac{2 D c_o}{c_h} \frac{r}{r-d}}.$$

(Again, we have to check the second derivative so as to ensure that the lot size actually minimizes the costs. It does.)

We can now illustrate the economic lot size by means of a numerical.

Example: A bottling plant faces an annual demand of 200,000 bottles of a certain type. It can produce these bottles at a rate of 1000 bottles per day during each of the 300 working days in a year. Setup costs for a production run are $1000, and each bottle has a carrying cost of 10¢ per bottle and year.

We first check whether or not the regularity conditions holds. Here, we have $r = 1000 > 666.67 = 200,000/300 = d$, so that the condition is indeed satisfied. Thus, we can compute the optimal quantity made during one production run as $Q^* = \sqrt{\frac{2(200,000)(1,000)}{0.10} \frac{1,000}{1,000-666.67}} \cong 109,545$ bottles. The corresponding costs are $TC(Q^*) = \$3651.50$.

An interesting observation concerns the relation between the optimal lot size developed above and the basic economic order quantity. Assuming that we can equate the production setup cost and the unit order cost in the two respective models, we find that the optimal value of Q in the lot size model is never smaller than the order quantity in the economic order quantity, and that the costs in the lot sizing model are never larger than those of the *EOQ*. As a matter of fact, the economic order quantity can be seen as a special case of the lot sizing model with an infinite production capacity (as exemplified by the fact that an order in the *EOQ* arrives with infinite speed). Increasing r to arbitrarily high values has the expression $r/(r-d)$ approach one, so that the lot size formula reduces to the standard economic order quantity. Applying this argument to our numerical example, we find that the economic order quantity with the same parameters as those used in the example equals $Q^* = 63,245.55$, a policy that costs $TC(Q^*) = \$6324.56$.

Along similar lines, it is also interesting to note that more capable machines, i.e., those with higher production rates, incur higher costs. As a matter of fact, the machine with the "optimal" production rate has $r = d$, so that inventories are unnecessary, as customers satisfy their demand at the same rate the machine produces the goods. This is yet another example for optimal solution that fitting the production to the demand results in solutions with the lowest cost (for another example, see the "technology choice" example in the linear programming formulations).

In conclusion, just like the economic order quantity, the production lot size model has the attractive property of being robust, i.e., quite insensitive to changes of the parameters (input data).

12.7 The Economic Order Quantity with Stochastic Lead Time Demand

So far in this chapter we have assumed a deterministic environment, in which all relevant data are known with certainty, and in which the consequences of our actions are completely predictable. We will now extend our analysis to situations involving uncertainty and begin with the simple but important case, in which the demand during the lead time is a random variable. However, we assume that the demand during the lead time follows a discrete probability distribution that is known to us. The random behavior of the demand may cause undesired and unplanned stockouts and surpluses. While the demand is irregular throughout, we are only concerned about the irregularity that occurs between the time that we have placed an order (i.e., after the reorder point has been reached), and the time that the next shipment arrives.

Whenever an unplanned surplus occurs due to the irregularity of the demand, we have in fact carried more inventory than was actually needed, thus incurring unnecessary carrying costs, while in case of an unplanned stockout, there will be a *penalty cost* c_p charged for each unit we are out of stock. Note that c_p is assumed to be independent of the length of time that we are out of stock. This is in contrast to the shortage costs c_s for the backorder model of Sect. 12.4, where the shortage cost was defined per quantity unit *and* per time unit. The difference between c_p and c_s in the backorder model and this model is that the shortages in Sect. 12.4 were planned deliberately, while the shortages in this section occur because there is a higher-than-expected demand during the lead time t_L.

Formally, we define c_p as the penalty cost per unit and stockout. Furthermore, we have $\overline{D} = E(D)$ as the expected value of demand per year, the lead time demand d_L (a random variable), and the expected value of d_L is $\overline{d}_L = E(d_L)$. The (discrete) probability distribution of the lead time demand d_L is $p(d_L)$, while $F(d_L)$ is the cumulative probability distribution of d_L. We will restrict our discussion to the case, in which d_L is a discrete random variable. Furthermore, in this simple model, the length of the lead time t_L is still deterministic, i.e., fixed and known to the decision maker. We also assume that $R - \overline{d}_L \geq 0$; i.e., on average, there is still a positive inventory level when replenishment occurs. If this condition were not to be required, we would, on average, run out of stock at the end of each cycle. Therefore, we may regard the quantity $R - \overline{d}_L$ as the amount of stock that is kept at all times. For this reason, this quantity is usually referred to as the expected *safety stock* or *buffer stock*.

12.7.1 A Model that Optimizes the Reorder Point

The objective in this section is to minimize the sum of the carrying costs for the expected safety stock plus the expected penalty costs for stockouts. This sum will be denoted by $TC_1(R, Q)$, since it depends on the reorder point R as well as on the order quantity Q. To start, we will simply use the order quantity Q_{EOQ}, which was obtained independently of the reorder point by way of the economic order quantity. This can be justified because of the robustness of the economic order quantity formula. We then obtain the partial cost function

$$TC_1(R, Q_{EOQ}) = c_h(R - \bar{d}_L) + c_p \left(\frac{\bar{D}}{Q_{EOQ}}\right) \sum_{d_L > R} (d_L - R) p(d_L),$$

where $Q_{EOQ} = \sqrt{\frac{2\bar{D}c_o}{c_h}}$, and where the first part of the relation is the cost for carrying the safety stock. The summation in the second part of relation is taken over all instances, in which shortages occur, so that we compute the expected shortage level. Differentiating $TC_1(R, Q_{EOQ})$ with respect to R and setting the resulting expression to zero yields the condition for the optimal reorder point R^*, which is

$$P[d_L > R^*] = \frac{c_h Q_{EOQ}}{c_p \bar{D}}.$$

As $P[d_L > R^*] = 1 - P[d_L \leq R^*] = 1 - F(R^*)$, we obtain

$$F(R^*) = 1 - \frac{c_h Q_{EOQ}}{c_p \bar{D}}.$$

Since we have assumed that d_L is a discrete random variable, its cumulative distribution function F will be a step function that assumes only discrete values in the interval $[0, 1]$. Therefore, it is unlikely that the right-hand side of the above equation will equal one of these discrete values. As a way out of this dilemma, we let R^* denote the smallest value that satisfies the inequality

$$F(R^*) \geq 1 - \frac{c_h Q_{EOQ}}{c_p \bar{D}}.$$

Note that we only have to consider the possible values of d_L for R^*.

In order to illustrate the above discussion, consider the following numerical

Example: Consider again the battery charger example of Sect. 12.2 with a demand of $D = 800$, ordering costs of $c_o = \$100$, and holding costs of $c_h = \$4$ per charger per year. Furthermore, assume that the penalty costs are $c_p = \$5$ per

12.7 The Economic Order Quantity with Stochastic Lead Time Demand

charger and stockout. Suppose that the expected annual demand is $\overline{D} = 800$. Suppose that the demand during lead time has the following probability distribution.

d_L (units)	$p(d_L)$	$F(d_L)$
70	0.1	0.1
75	0.2	0.3
80	0.2	0.5
85	0.3	0.8
90	0.2	1.0

The economic order quantity in this example equals $Q_{EOQ} = 200$ units, so that $1 - \frac{c_h Q_{EOQ}}{c_p \overline{D}} = 1 - \frac{4(200)}{5(800)} = 0.8$, and since the smallest value of d_L with $F(d_L) \geq 0.8$ equals 85, we have $R^* = 85$. As the expected demand $\overline{d}_L = E(d_L) = \sum_x x p_L(x) = 81.5$, the expected safety stock will equal $R^* - \overline{d}_L = 85 - 81.5 = 3.5$ units. The carrying cost for the expected safety stock is then $c_h (R^* - \overline{d}_L) = 4(85 - 81.5) = \$14$, and the expected penalty cost is $c_p \left(\frac{\overline{D}}{Q_{EOQ}}\right) \sum_{d_L > R^*} (d_L - R^*) p(d_L) = 5 \frac{800}{200} \times (90 - 85)(0.2) = \$20$. Note that stockouts occur only if $d_L > R^* = 85$, which happens only in case $d_L = 90$, an occurrence that has a probability of 0.2.

12.7.2 A Stochastic Model with Simultaneous Computation of Order Quantity and Reorder Point

We can now refine the above model and determine the order quantity Q and the reorder point R simultaneously. For that purpose, we consider the expected total cost of ordering, carrying, and penalty, i.e.,

$$TC_2(Q, R) = c_o \frac{\overline{D}}{Q} + c_h(\tfrac{1}{2}Q + R - \overline{d}_L) + c_p \left(\frac{\overline{D}}{Q}\right) \sum_{d_L > R} (d_L - R) p(d_L).$$

Using partial differentiation with respect to Q and to R and setting the result to zero yields

$$Q^* = \sqrt{\frac{2\overline{D}}{c_h} \left(c_o + c_p \sum_{d_L > R^*} (d_L - R^*) p(d_L)\right)} \text{ and}$$

$$F(R^*) \geq 1 - \frac{c_h \overline{Q}}{c_p \overline{D}}.$$

Again, it is understood that R^* is taken to be the smallest value that satisfies the inequality. The above two relations should be solved simultaneously, which is difficult, since Q^* and R^* appear in both. Instead, we will use an iterative procedure that shuttles between these two relations. It commences with an order quantity Q^*, uses the second of the two relations to determine a reorder point R^*, uses this reorder point in the first relation to compute a revised value of Q^*, and so forth until the process converges and the numbers do not change anymore. As an aside, note again that if the penalty costs get very large, the order quantity reduces again to the standard economic order quantity of the basic model.

This process will be illustrated in the following numerical:

Example: Consider again the situation of the example in the previous section with a demand of $D = 800$, ordering costs of $c_o = \$100$, holding costs of $c_h = \$4$ per charger per year, penalty costs $c_p = \$5$ per charger and stockout, and the above probability distribution of the demand.

Again, we obtain $Q^* = 200$, so that $R^* = 85$, just as in the previous procedure. Using the modified economic order quantity, we then find a revised value of Q^* as

$$Q^* = \sqrt{\frac{2(800)}{4}[100 + 5(5)(0.2)]} \approx 204.94 \text{ units.}$$

Using this revised order quantity in the latter of the two relations, we find that $F(R^*) \geq 1 - \frac{(4)(204.94)}{5(800)} \approx .795$, so that $R^* = 85$ again, and thus the procedure terminates.

Comparing the results for Q^* and R^* of the simple model in the previous subsection and the refined approach in this subsection, we notice that in both cases the reorder point is $R^* = 85$ units, whereas the order quantity is $Q^* = 200$ units in the simple model (obtained by using the standard economic order quantity), while it is not very different at $204.95 \approx 205$ units in the refined model. Again, this demonstrates the robustness of the economic order quantity formula.

12.8 Extensions of the Basic Inventory Models

This section offers an outlook on some inventory policies. Following the standard terminology (which is in some conflict of the symbols that we have used so far), we define s as the reorder level (what we have referred to so far as the reorder point, i.e., the inventory level at which an order is placed), R as the intervals at which the inventory level is checked, and S as the inventory level we have directly after a replenishment.

We now distinguish between periodic and continuous review systems. In a *periodic review system*, we check the inventory levels at regular intervals R (e.g., hourly, daily, or weekly), while in a *continuous review system*, we continuously watch the inventory level.

An *order-point, order-quantity*, or (s, Q) *policy* involves continuous review (i.e., $R = 0$) at which time an order of a given magnitude Q is placed whenever the inventory reaches a prespecified reorder level s. An example of an (s, Q) policy is the two-bin system described in Sect. 12.3.

12.8 Extensions of the Basic Inventory Models

An *order-point, order-up-to-level*, or *(s, S) policy* is another continuous review policy. Its key is an inventory level S that is specified by the inventory manager. This is an inventory level to be attained directly after a shipment is received. So, once the reorder point s is reached, an order of size $S - s$ is placed, which then increases the inventory level to S. This may be a reasonable policy in case the demand is irregular, so that at the time that an order is placed the inventory level may suddenly dip below s, at which time the regular order quantity may then not be sufficiently large.

A *periodic review, order-up-to-level, replenishment cycle*, or *(R, S) policy* is a periodic review policy. At each review instant (which occurs at intervals of length R time units) an order is placed of a size that raises the inventory level to S.

In addition, there are hybrid policies such as the (R, s, S) policy, where at review time nothing is done, if the inventory level is above s, whereas if it is at or below the level s, an order is placed to increase the inventory level to the magnitude S.

Each of the above policies has its own advantages and drawbacks, and it depends on the practical situation at hand which one is the most appropriate choice. Typically, continuous review policies such as the (s, Q) and (s, S) policies are suitable for the A items in the ABC classification introduced at the beginning of this chapter. For B and C items, the cost of continuous review of the inventory level may not be justified, so that periodic review policies may make more sense. For C items, the review interval length R may be set, so that the review is done less frequently for these items of minor value. Modern computerized inventory control systems are of great help with any inventory system.

Exercises

Problem 1 (*EOQ* with positive lead time, shortages, production lot size): A retailer faces an annual demand for 2,400,000 shirts with the "Mumbo Jumbo Man-Savior of the Universe" logo. It costs $450 to place a single order and the costs for keeping a single shirt in stock for an entire year are 60¢.

(a) How many units should the retailer order each time an order is placed and what are the associated costs?

(b) Given the result under (a), how many orders should be placed and what is the time between two consecutive orders (given a 360-day year)? What is the reorder point if the lead time were 20 days?

(c) Assume now that it is possible to allow shortages, given that the portion of the demand that cannot immediately be satisfied will be satisfied immediately after the next shipment arrives. It has been estimated that the associated loss of goodwill equals costs of 80¢ per shirt and year. Compute the order quantity, the maximum shortage, and the associated costs.

(d) What would happen to the results in (c) if the unit shortage costs were to increase by, say, 10¢? Explain in one short sentence, indicating the reason why. Calculations are not required.

(e) Suppose now that the retailer were to purchase the equipment to make the shirts in-house. The machine is capable of making 10,000 shirts per day (again based on a 360-day year). What is the number of shirts made in each production run? What are the total costs?

(f) How would the results under (e) change if the capacity of the machine under (e) were not 10,000 units per day but only 6000? Explain in one short sentence.

Solution:

(a) $D = 2,400,000$, $C_o = 450$, $C_h = 0.6$, so that $Q^* = 60,000$ and $TC^* = \$36,000$.

(b) Then $N^* = 40$ and $t_c^* = 1/40$ [years] $= 9$ [days]. Reorder point $R = 20(2,400,000/360) - \lfloor \frac{20}{9} \rfloor 60,000 = 13,333.33$.

(c) $Q^* = 79,372.54$ and $S^* = 34,016.80$. Costs $TC =$ (holding costs) + (ordering costs) + (shortage costs) $= 7775.27 + 13,606.72 + 5831.45 = \$27,213.44$.

(d) If c_s increases, shortages become more expensive, so that the order quantity Q^* and the maximum shortage S^* both decrease, while the total costs will increase.

(e) The regularity condition is satisfied. $Q^* = 103,923.05$ and $TC^* = 10,392.30 + 10,392.30 = \$20,784.60$.

(f) The regularity condition is violated, i.e., the machine capacity is insufficient to satisfy the demand.

Problem 2 (*EOQ*, positive lead time, shortages, production lot sizing): A retailer faces an annual demand for 4000,000 pairs of sneakers with the "King Bong" logo. It costs \$100 to place a single order and the costs for keeping a single pair of sneakers in stock for an entire year are 50¢.

(a) How many pairs of sneakers should the retailer order each time an order is placed and what are the associated costs?

(b) Given the result under (a), how many orders should be placed and what is the time between two consecutive orders (given a 250-day year)? What is the reorder point if the lead time were 8 days?

(c) Assume now that it is possible to allow shortages, given that the portion of the demand that cannot immediately be satisfied will be satisfied immediately after the next shipment arrives. It has been estimated that the associated loss of goodwill equals costs of 20¢ per pair of sneakers and year. Compute the order quantity, the maximum shortage, and the associated costs.

(d) What would happen to the results in (c) if the unit shortage costs were to decrease by, say, 5¢? Explain in one short sentence, indicating the reason why. Calculations are not required.

(e) Suppose now that the retailer were to purchase the equipment to make the sneakers in-house. The machine is capable of making 20,000 pairs per day (again based on a 250-day year). How many pairs of sneakers are made in each production run? What are the total costs?

12.8 Extensions of the Basic Inventory Models

(f) How would the results under (e) change if the capacity of the machine under (e) were not 20,000 units per day but 25,000? Explain in one short sentence. No calculations are necessary.

Solution:

(a) $D = 4,000,000$, $C_o = 100$, $C_h = 0.5$, so that $Q^* = 40,000$ and $TC^* = \$20,000$.
(b) Then $N^* = 100$ and $t_c^* = 1/100$ [years] $= 2.5$ [days]. Reorder point $R^* = 8\,(4,000,000/250) - \lfloor \frac{8}{2.5} \rfloor 40,000 = 8,000$
(c) $Q^* = 74,833.15$ and $S^* = 53,452.25$. Costs $TC = $ (holding costs) + (ordering costs) + (shortage costs) $= 1527.21 + 5345.22 + 3818.02 = \$10,690.45$.
(d) If c_s decreases, shortages become even cheaper, so that the order quantity Q^* and the maximum shortage S^* both increase, while the total costs will decrease.
(e) The regularity condition is satisfied. $Q^* = 89,442.72$ and $TC^* = 4472.14 + 4472.14 = \$8944.28$.
(f) The order quantity will decrease and the inventory-related costs will increase.

Problem 3 (quantity discounts): The annual demand for a product is 9000, and the unit price of the product is \$5. The holding costs are estimated to be 10% of the price of the product, and the costs of placing a single order are \$1000.

(a) Calculate the optimal order quantity and the associated costs.
(b) The supplier now offers a discounted price of \$4.80 if we order at least 8000 units at a time. Should we take the offer? What are the associated costs?
(c) Our supplier adds another offer: a price of \$4.75, if we purchase at least 18,000 units (so that we have to order only every other year). Should we take this offer?

Solution:

(a) With a unit price of $p_0 = \$5$, we obtain $Q^* = \sqrt{\frac{2(9,000)(1,000)}{(.1)(5)}} = 6,000$ with $TC^* = 1500 + 1500 + 45,000 = \$48,000$.
(b) The discounted price of $p_1 = \$4.80$ leads to an optimal quantity of $Q^* = 6123.72$. This quantity does not qualify for a discount, so that we set Q: $= Q_1 = 8000$. The associated costs are $TC(Q_1) = 1125 + 1920 + 43,200 = \$46,245$. As the total costs are lower at this price level, we should take the discount and order 8000 units at a time, which will cost us \$46,245.
(c) The deeply discounted price of $p_2 = \$4.75$ leads to an optimal order quantity of $Q^* = 6155.87$, still not enough to qualify for the discounted price level. As a result, we must increase the order quantity to the smallest level that allows the discount, i.e., $Q = Q_2 = 18,000$. At this level, the total costs are $TC(Q_2) = 500 + 4275 + 42,750 = \$47,525$, less than paying the full price, but not as good as the smaller discount determined under (b). Hence, the overall best option is to order 8000 units each time we place an order, a policy that will cost us \$46,245 in each cycle.

Problem 4 (*EOQ* with stochastic lead time demand): Let the annual demand for a certain item be 1000 units in the planning period. The holding costs are $5 per unit, and the cost of placing an order is $36 per order. The penalty cost for stockouts is $3 per unit and stockout. There is a demand for 55 units during the lead time with a probability of 20%. The demand is 60 with a probability of 0.4, it is 65 with a probability of 0.30, and it is 70 with a probability of 0.10.

(a) Calculate the expected demand during lead time.
(b) Determine the economic order quantity and the resulting reorder point.
(c) Find the order quantity and the reorder point by simultaneous computation.
(d) What is the buffer stock?
(e) What are the minimal expected ordering, holding, and penalty costs?

Solution:

(a) $E(d_L) = (55)(0.2) + (60)(0.4) + (65)(0.3) + (70)(0.1) = 61.5$.

(b) $Q_{EOQ} = \sqrt{\frac{2(1,000)36}{5}} = 120$ units, and with $F(R^*) \geq 1 - \frac{5(120)}{3(1,000)} = 1 - 0.2 = 0.8$, so that $R^* = 65$.

(c) Revising $Q^* = \sqrt{\frac{2(1,000)}{5}[36 + 3(5)0.1]} \approx 122.47$ units, $F(R^*) \geq 1 - \frac{5(122.47)}{3(1,000)} \approx 0.7959$, so that $R^* = 65$ units.

(d) The buffer stock is $R^* - \bar{d}_L = 65 - 61.5 = 3.5$ units.

(e) Costs TC_2 = (ordering costs) + (holding costs) + (penalty costs) = 293.95 + 323.68 + 12.25 = $629.87.

Reference

Erlenkotter D (1990) Ford Whitman Harris and the economic order quantity model. Oper Res 38: 937–946

Stochastic Processes and Markov Chains 13

Some of the previous chapters have dealt with random events in an ad hoc fashion. This chapter will deal with such events in a systematic way. In general, in stochastic processes, events occur over time. Time can be dealt with either in continuous fashion, or in discrete fashion. In the continuous case, we may look at the speed of an automobile at any given point in time or at the inventory level of a product in a supermarket at any time. In the discrete case, speed or inventory level are observed only during specific points in time, e.g., each minute, once a week or at similar intervals. In this chapter, we only deal with discrete-time models. The following three sections will introduce some of the basic ideas of stochastic processes and Markov chains.

13.1 Basic Ideas and Concepts

Consider first the random events that take place. Using the same examples as above, there is an infinite number of different speeds that a vehicle could be moving at, while the demand for a product may be very large but is hardly infinite. Some types of events are much more restrictive: as an example, consider a light bulb. It will always be in exactly one of two "states of nature," in that it either works, or it does not. This is referred to as the *state space*, i.e., the number of different states the "system" can possibly be in. As already hinted at, the individual states are similar to the states of nature in decision analysis, see Chap. 10 in this volume. This chapter deals only with processes that have a finite state space.

Each event in this *discrete-time, finite state space process* is then a random variable X_t that depends on the time t at which it is observed. As an illustrative example, consider a used car. The vehicle is in one of four states: it either runs well (state s_1), it runs with minor problems (state s_2), it runs with major problems (state s_3), or it fails altogether (state s_4). At any point in time, the vehicle is in exactly one of these four states. It stands to reason that the state that the vehicle is in 1 year does depend on the state the car was in the year before. More specifically, we can define

© The Author(s), under exclusive license to Springer Nature Switzerland AG 2022
H. A. Eiselt, C.-L. Sandblom, *Operations Research*, Springer Texts in Business and Economics, https://doi.org/10.1007/978-3-030-97162-5_13

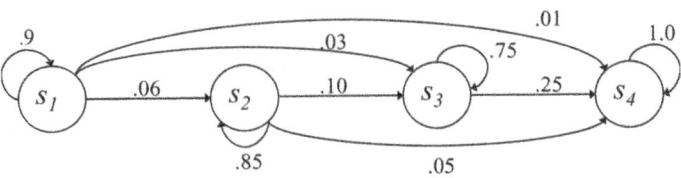

Fig. 13.1 Transition diagram for automobile example

transition probabilities p_{ij}, which indicate the probability that the vehicle is in state j, given that it was in state i in the previous year. (We assume that no repairs are performed.) It is apparent that the transition probabilities are conditional probabilities of the type $p_{ij} = P(X_{t+1} = j | X_t = i)$, or in simple words, the probability that the random variable is in state s_j in year $t + 1$, given that it was in state s_i in year t. As a numerical illustration, consider the matrix $\mathbf{P} = (p_{ij})$ of transition probabilities

$$\mathbf{P} = \begin{bmatrix} 0.90 & 0.06 & 0.03 & 0.01 \\ 0 & 0.85 & 0.10 & 0.05 \\ 0 & 0 & 0.75 & 0.25 \\ 0 & 0 & 0 & 1.00 \end{bmatrix}.$$

For example, if the vehicle has been running well in the previous year, then there is a 90% chance that it will be running smoothly in this year as well, as shown by the element p_{11}. Or, if the vehicle experiences minor problems this year, then there is a 5% chance that it will fail altogether next year, as shown by the element p_{24}. We should point out that a similar example is known in the literature as the "bad debt" example, in which the states of a debt are "paid in full," "in good standing," "in arrears," and "gone to collection."

A few features of this particular transition matrix are noteworthy. First of all, note that all elements below the main diagonal (i.e., in the lower left corner) are zero. This simply means that, given that no repairs are performed, the vehicle will never improve. Secondly, we can observe that the sum of probabilities in each row of the transition matrix equals one. This is always the case, as the transition probabilities are conditional probabilities, and as such, given that we are in a row at time t, we *must* choose a successor state for time $t + 1$. Thirdly, notice that once the car is in state s_4 (i.e., the vehicle fails), then it will never get out of it again. Thus, the state s_4 is referred to as an *absorbing state*. The example presented here is a stochastic process with the *Markovian property* that holds, if the present state of the process depends only on the state of the system immediately prior to this and the transition probabilities, meaning that the system does not have a (long) memory. As a historical aside, Andrey Andreyevitch Markov, 1856–1922, was a Russian mathematician who made important contributions to the field of probability and statistics.

A nice visual representation is the *transition diagram*. In this diagram, the nodes represent the states of the process, and the arcs represent transitions with a positive

13.1 Basic Ideas and Concepts

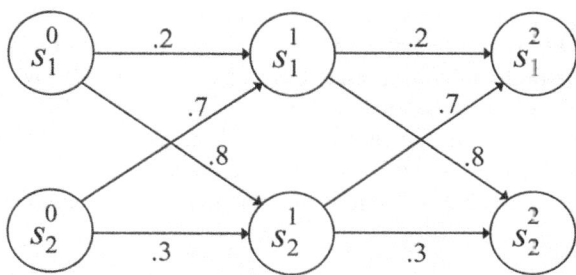

Fig. 13.2 Time—state graph for stock price example

probability. The transition diagram for our automobile example is shown in Fig. 13.1.

Note that absorbing states have only arcs leading into them, but not out, at least not to other states. Here, we also assume that the process is *stationary*, meaning that the transition probabilities do not change over time.

An obvious question to be asked is what happens, if we go through more than one transition. In other words, the transition probabilities tell us what the likelihood is to be in one state one period after the process starts. But what about two or more periods? This is the question we discuss in the next paragraphs. In order to facilitate the discussion, consider a simple example involving stock prices. In particular, we only consider an upward movement of the price and a downward movement. The transition probabilities are shown in the matrix

$$\mathbf{P} = \begin{bmatrix} 0.2 & 0.8 \\ 0.7 & 0.3 \end{bmatrix}.$$

In other words, if the stock went up today, then there is a 20% chance that it will go up again tomorrow, while there is an 80% chance that the stock's price will decline. Similarly, if the price of the stock decreased today, then there is a 70% chance that it will increase tomorrow and a 30% chance that it will decrease tomorrow. When analyzing changes after multiple periods, we could use a *time-state graph* as shown in Fig. 13.2, where the nodes s_i^t indicate the state of nature i at the end of period t, and the arcs denote the possible transitions, and their values are the transition probabilities. As the transition probabilities are stationary, an arc from, say, s_i^t to s_j^{t+1} will have the same value as, say an arc from s_i^{t+2} to s_j^{t+3}.

Considering Fig. 13.2, assume now that we are presently in an upswing of the stock price, i.e., in state s_1^0. To find the probability that there will be an increase in stock prices 2 days later, we will have to examine all paths from the present state s_1^0 to the state s_1^2. Here, there are exactly two paths: the first path leads from s_1^0 to s_1^1 and on to s_1^2, while the second path leads from s_1^0 first to s_2^1 and then to s_1^2. Note that the first path considers a price increase on the first day, while the second has a decrease on the first day. The probability is then calculated as the sums of probabilities of all paths from s_1^0 to s_1^2, which, in turn, are computed as the product of all transition probabilities along the path. In this example, the probability on the former path s_1^0, s_1^1,

s_1^2 is $(.2)(.2) = .04$ (meaning that there is only a 4% chance of two price increases in a row), while the latter path s_1^0, s_2^1, s_1^2 has a probability of $(.8)(.7) = .56$, meaning that the probability of a price decrease on day one, followed by a price increase on day two has a probability of 56%. This means that we obtain the transition probability that that state changes from a price increase on day t to a price increase on day $t + 2$ as $0.04 + 0.56 = 0.60$. Similarly, we compute the remaining probabilities, resulting in the matrix $\mathbf{P}^2$, which indicates the transition probabilities from day t to day $t + 2$. It is

$$\mathbf{P}^2 = \begin{bmatrix} 0.6 & 0.4 \\ 0.35 & 0.65 \end{bmatrix}.$$

This procedure can be repeated for any number of days. While this is possible, the procedure is extremely awkward. We can achieve the same results by simple matrix multiplication. In particular, we can obtain $\mathbf{P}^2 = \mathbf{PP}$, $\mathbf{P}^3 = \mathbf{P}^2\mathbf{P} = \mathbf{PPP}$, and so forth, so that we obtain $\mathbf{P}^r = \mathbf{PPP}\ldots\mathbf{P}$, i.e., the matrix $\mathbf{P}$ multiplied itself r times. For instance, in the stock example, we obtain

$$\mathbf{P}^3 = \mathbf{P}^2\mathbf{P} = \begin{bmatrix} 0.6 & 0.4 \\ 0.35 & 0.65 \end{bmatrix} \begin{bmatrix} 0.2 & 0.8 \\ 0.7 & 0.3 \end{bmatrix} = \begin{bmatrix} 0.4 & 0.6 \\ 0.525 & 0.475 \end{bmatrix}.$$

This means that given that there has been an increase in prices on day one, then there is a 40% chance that there will be a rise in prices again 2 days later, and similar for the other elements of the matrix.

We may also be interested in the *mean first passage time* r_{ij}, which is the expected number of transitions needed to reach state j from state i for the first time. For $i = j$, i.e., elements r_{ii}, we talk of the *mean recurrence time*, which is then the expected number of transitions occurring after leaving state i, until we return to this state for the first time. For the car example above the element r_{23} would denote the expected number of transitions required to reach state 3 ("major problems") for the first time after being in state 2 ("minor problems"). For the stock price example, the element r_{11} denotes the mean number of days between two consecutive days with upward price moves.

To compute the mean first passage times r_{ij}, assume that the current state is "i." From this state, we will reach state j in a single step with probability p_{ij}. Moving from state i to state k with $k \neq j$ in a single step, and from there, an average of r_{kj} steps from state k to state j. Therefore

$$r_{ij} = p_{ij}(1) + p_{i1}(1 + r_{1j}) + p_{i2}(1 + r_{2j}) + \ldots + p_{i,j-1}(1 + r_{j-1,j})$$
$$+ p_{i,j+1}(1 + r_{j+1,j}) + + \ldots + p_{im}(1 + r_{mj}),$$

assuming that there are a total of m states. However, since $p_{i1} + p_{i2} + \ldots + p_{im} = 1$, we can simplify the above expression to

$$r_{ij} = 1 + p_{i1}r_{1j} + \ldots + p_{i,j-1}r_{j-1,j} + p_{i,j+1}r_{j+1,j} + \ldots + p_{im}r_{mj}.$$

As a numerical example, consider the stock price example, for which we obtain

13.1 Basic Ideas and Concepts

$$r_{11} = 1 + p_{12}r_{21},$$
$$r_{12} = 1 + p_{11}r_{12} \text{ (or, equivalently } r_{12} = 1/(1-p_{11})),$$
$$r_{21} = 1 + p_{22}r_{21} \text{ (or, equivalently, } r_{21} = 1/(1-p_{22})), \text{ and}$$
$$r_{22} = 1 + p_{21}r_{12}.$$

It follows that $r_{11} = 1 + p_{12}/(1-p_{22})$ and $r_{22} = 1 + p_{21}/(1-p_{11})$. For example, the mean number of days between two consecutive days with price increases will be $r_{11} = 1 + 0.8/(1-0.3) = 15/7 \approx 2.14$ days. Furthermore, after a day with price increases, the expected number of days until a price decrease occurs for the first time is $r_{12} = 1/(1-p_{11}) = 1 - (1-0.2) = 5/4 = 1.25$ days.

For the car example, we obtain 16 equations for the mean first passage times r_{ij}, making them a bit harder to compute. Actually, with the particular feature of deteriorating conditions leading to the absorbing state of "fail," $r_{44} = 1$, and $r_{41} = r_{42} = r_{43} = \infty$.

So far, we have only dealt with conditional probabilities, i.e., *given* that a certain state of nature prevails in the beginning, we computed the probability that some (other) state of nature occurs after a given number of periods. Below, we start the process with an initial probability distribution that assigns an unconditional probability u_i^0 to each state of nature s_i. These probabilities are then collected in the initial probability row vector $\mathbf{u}^0 = [u_1^0, u_2^0, \ldots, u_m^0]$, assuming that there is a total of m states of nature. The unconditional probabilities that describe the likelihood that the system is in one of the states of nature after 1 week is $\mathbf{u}^1$, which can be computed by simple vector-matrix multiplication as $\mathbf{u}^1 = \mathbf{u}^0 \mathbf{P}$. In general, we have

$$\mathbf{u}^r = \mathbf{u}^{r-1}\mathbf{P} = \mathbf{u}^0\mathbf{P}^r.$$

As an illustration, consider again the car example. Recall that its single-stage transition matrix was

$$\mathbf{P} = \begin{bmatrix} 0.90 & 0.06 & 0.03 & 0.01 \\ 0 & 0.85 & 0.10 & 0.05 \\ 0 & 0 & 0.75 & 0.25 \\ 0 & 0 & 0 & 1.00 \end{bmatrix}$$

with the states of nature being s_1: running well, s_2: running with minor problems, s_3: running with major problems, and s_4: not running. Suppose now that the car is initially running with minor problems, i.e., $\mathbf{u}^0 = [0, 1, 0, 0]$. After 1 year, we have the probabilities

$$\mathbf{u}^1 = \mathbf{u}^0\mathbf{P} = [0,1,0,0] \begin{bmatrix} 0.90 & 0.06 & 0.03 & 0.01 \\ 0 & 0.85 & 0.10 & 0.05 \\ 0 & 0 & 0.75 & 0.25 \\ 0 & 0 & 0 & 1.00 \end{bmatrix} = [0, 0.85, 0.10, 0.05],$$

i.e., there is no possibility that the vehicle with perform perfectly, there is an 85% chance that it will continue to run with minor problems, there is a 10% chance that the problems are now major, and there is a 5% chance that the vehicle will fail altogether. For the second year, we obtain

$$\mathbf{u}^2 = \mathbf{u}^1\mathbf{P} = [0, 0.85, 0.10, 0.05] \begin{bmatrix} 0.90 & 0.06 & 0.03 & 0.01 \\ 0 & 0.85 & 0.10 & 0.05 \\ 0 & 0 & 0.75 & 0.25 \\ 0 & 0 & 0 & 1.00 \end{bmatrix}$$

$= [0, 0.7225, 0.16, 0.1175]$, indicating that there is now a 72.25% chance that the minor problems will persist, a 16% chance that the problems are now major, and an 11.75% chance that the vehicle will fail. Repeating the calculations, we can compute probabilities in this way for any future year.

13.2 Steady-State Solutions

Assuming that the process converges, we will call the resulting solution a *steady-state solution*. Clearly, a steady-state solution is an ideal concept that is not very likely going to be realized in practice: take, for instance, the used car example. While we may keep the vehicle for a long time, the time will be finite, while a steady state, a state that no longer depends on the initial conditions, is typically reached only after an infinite number of transitions. Still, the steady state is an important concept that will tell us to what a process converges, provided that it converges at all. A sufficient condition for the existence of a steady state is given if each state can be reached from each other state on a path that has positive probability.

We showed in the previous section that $\mathbf{u}^n = \mathbf{u}^0\mathbf{P} = \mathbf{u}^{n-1}\mathbf{P}$, and if n tends to infinity, we obtain $\mathbf{u}^\infty = \mathbf{u}^\infty\mathbf{P}$. To distinguish the steady-state solutions from all others, it is customary to replace $\mathbf{u}^\infty$ by the row vector π, so that the steady-state solutions will satisfy $\pi = \pi\mathbf{P}$. In addition, we have to ensure that the sum of all elements in π equals 1, as all components of π are probabilities that are mutually exclusive and collectively exhaustive.

In the used car example, the system of simultaneous linear equations is

$$\pi_1 = 0.9\pi_1$$
$$\pi_2 = 0.06\pi_1 + 0.85\pi_2$$
$$\pi_3 = 0.03\pi_1 + 0.10\pi_2 + 0.75\pi_3$$
$$\pi_4 = 0.01\pi_1 + 0.05\pi_2 + 0.25\pi_3 + \pi_4$$
$$\pi_1 + \pi_2 + \pi_3 + \pi_4 = 1.$$

The first equation requires that $\pi_1 = 0$, inserting this result in the second equation leads to $\pi_2 = 0$, and using this result in the third equation leads to $\pi_3 = 0$. The fourth equation then reduces to the tautological identity $\pi_4 = \pi_4$, but the last equation then helps to solve the system with $\pi_4 = 1$. This is the obvious result mentioned earlier in this section: whatever state the vehicle is in now, eventually it will be in a failed state.

The stock example provides another illustration of the concept. It is easily seen that a steady state is certain to exist. In order to determine the (unconditional) steady-state probabilities, we solve the system

$$[\pi_1, \pi_2] = [\pi_1, \pi_2] \begin{bmatrix} 0.2 & 0.8 \\ 0.7 & 0.3 \end{bmatrix},$$

which can be written as

$$\pi_1 = 0.2\pi_1 + 0.7\pi_2$$
$$\pi_2 = 0.8\pi_1 + 0.3\pi_2, \text{ coupled with}$$
$$\pi_1 + \pi_2 = 1.$$

The system has steady-state probabilities $\pi_1 = 7/15$ and $\pi_2 = 8/15$, or $\pi = [0.4667, 0.5333]$. In other words, in the long run, we can expect stock prices to rise about 47% of the time, while we can expect them to drop about 53% of the time.

One can show that for the mean recurrence time r_{ii}, the relationship $r_{ii} = 1/\pi_i$ holds. For our stock price example, $\pi_1 = 7/15$ means that $r_{11} = 15/7$ days, in agreement with what we calculated in Sect. 13.1. Furthermore, $r_{22} = 1/\pi_2 = 15/8 = 1.875$ days will be the mean time between two consecutive days of price decreases.

13.3 Decision Making with Markov Chains

This section demonstrates how we can use the results obtained in the previous two sections in the context of decision making. To explain, consider again the used car example of the previous sections. Suppose now that there are three automobiles for sale. They are of the same make and model and the transition probabilities in the matrix **P** above apply to all of them. One vehicle is in perfect running order, and it sells for $5000. The second vehicle runs with minor problems, and it sells for $4000, while the third vehicle has major problems and it costs $2000. After 2 years we want to sell the vehicle again. Its price then (as it does now) will depend on its state. In particular, it has been estimated that a vehicle in perfect condition will sell for $3500, one with minor problems sells for $2500, and a car with major problems sells for $500. We assume that there is no difference between the cars in maintenance costs and that no repairs are made (which is not really realistic). Which car should we purchase?

We can address the problem by considering each option, one at a time. If we purchase the vehicle that is in perfect condition, then we decide that $\mathbf{u}^0 = [1, 0, 0, 0]$. After 1 year, we obtain $\mathbf{u}^1 = [0.9, 0.06, 0.03, 0.01]$ and after 2 years we have $\mathbf{u}^2 = [0.81, 0.105, 0.0555, 0.0295]$. In other words, when we attempt to sell the vehicle, it will still be in perfect shape with a probability of 81%, it will have minor problems with a probability of 10.5%, and so forth. The expected price we can sell the vehicle for is then $3,500(0.81) + 2,500(0.105) + 500(0.0555) + 0(0.0295) = \$3,125.25$, resulting in a loss of $5000 - 3125.25 = \$1874.75$.

The other two vehicles are dealt with similarly. If we purchase the car with minor flaws, we decide to choose $\mathbf{u}^0 = [0, 1, 0, 0]$ and obtain the probability vectors $\mathbf{u}^1 = [0, 0.85, 0.10, 0.05]$ after 1 year and $\mathbf{u}^2 = [0, 0.7225, 0.16, 0.1175]$ after 2 years of ownership, so that the expected price of the car at the time of sale is $1886.25, which means a loss of $2113.75.

Finally, consider the car with major flaws. Deciding to purchase it in the beginning means to set $\mathbf{u}^0 = [0, 0, 1, 0]$, so that by the end of year 1, we have $\mathbf{u}^1 = [0, 0, 0.75, 0.25]$, and by the end of the second year (or, equivalently, at the beginning of year 3), we have probabilities $\mathbf{u}^2 = [0, 0, 0.5625, 0.4375]$, so that the expected price of sale is $281.25 for an expected loss of $1718.75. Given our assumptions, all options result in losses, but our best bet would be to purchase the car with major flaws.

Consider now the possibility of either purchasing a warranty repair policy for $100 per month or pay the $1800 repair bill, whenever state s_4 occurs. In either case, whenever the system enters state s_4, (the car fails), it stays in this state for one period during which it is repaired and, at the end of this period, returns to s_1 (runs without problems). The transition matrix then changes to

$$\mathbf{P} = \begin{bmatrix} 0.90 & 0.06 & 0.03 & 0.01 \\ 0 & 0.85 & 0.10 & 0.05 \\ 0 & 0 & 0.75 & 0.25 \\ 1 & 0 & 0 & 0 \end{bmatrix}.$$

As opposed to the previous case without repairs, this transition matrix does not have an absorbing state. Finding the steady-state probabilities for this case, we obtain $\pi = [\pi_1, \pi_2, \pi_3, \pi_4] = [\frac{50}{89}, \frac{20}{89}, \frac{14}{89}, \frac{5}{89}] \approx [.5618, .2247, .1573, .0562]$. As above, we can purchase a vehicle in state s_1, s_2, and s_3 for $5000, $4000, and $2000, respectively, while we now add that a failed vehicle can be purchased for $1000. (Earlier, it made no sense to purchase a vehicle in state s_4, as there were no

13.3 Decision Making with Markov Chains

repairs.) This means that in the long run, the annual expected cost of repairing the vehicle is $0.0562(1800) = \$101.16$ if we pay ourselves, or $100 for the policy, which makes the purchase of the policy slightly superior.

Consider now again the case, in which four vehicles of the same make and model are offered, one in each of the four states. The purchase prices and the repair costs are the same as above. The idea is to purchase the vehicle at the end of year 0 (or, equivalently, the beginning of year 1) and sell it again at the end of year 3 (the beginning of year 4). The price for which the vehicle can be sold at that time has been estimated to be $2500 if it is running perfectly (state s_1), $1500 if it runs with minor problems (state s_2), $800 in case it has major problems (state s_3), and $100 if it fails (state s_4). Analyzing this case necessitates the use of solution trees rather than the computation of the states $\mathbf{u}^0$, $\mathbf{u}^1$, $\mathbf{u}^2$, and $\mathbf{u}^3$. The reason is that we can, for instance, purchase a vehicle with major flaws today and sell it at the end of year 3 with major flaws. However, it may happen that this vehicle simply stayed in this state, or it failed, we repaired it and it was perfect for 1 year, and then it degenerated again to the state with major problems. The two cases have the same initial state and the same state at the end of the planning period, but they incur very different costs.

In particular, we will need four probability trees, one for each vehicle we can purchase. For reasons of limitations of space, we display only the solution tree for the vehicle that is initially running with major problems. The different states are abbreviated as P (perfect running condition, state s_1) Mi (state s_2, minor problems), Ma (state s_3, major problems) and F (state s_4, fail). All possible sequences of states that may occur are shown in Fig. 13.3. The two numbers next to the last set of nodes when we sell the vehicle denote the probability that this string of events occurs, and the costs occurred throughout the time that we own the vehicle. The latter includes the purchase price plus repairs, if any, minus the price we obtain when we sell the car. As an example, consider the string of events that can be described as $P - F - P - Mi$. In other words, the car runs perfectly when we purchase it, it then fails in the next year, after which it runs perfectly again, and it exhibits minor problems after that. The probability for such a sequence of events is $(0.01)(1)(0.06) = 0.0006$, and the costs include the purchase price of $5000, the repair bill of $1800, and the sales price of $1500, resulting in overall costs of $5300. The expected costs for purchasing a vehicle in perfect condition and keeping it for 3 years are then $2888.42.

We can then perform a similar analysis with a vehicle that is initially in state s_2 (running with minor problems). The tree is a bit smaller, and the expected costs are $3650.21. Purchasing a car that has major problems will cost $1375.92, while buying a vehicle that fails and must be repaired right away will have expected costs of $564.15, making this the preferred strategy of the buyer.

Exercises

Problem 1 (brand switching, computation of steady-state probabilities): Three major motel chains compete for the lucrative economy market: Sunny 6, Cloudy 7, and Rainy 8. It has been observed that customers who stay in one of the motels usually patronize the same chain again, except if they perceive the service to be poor.

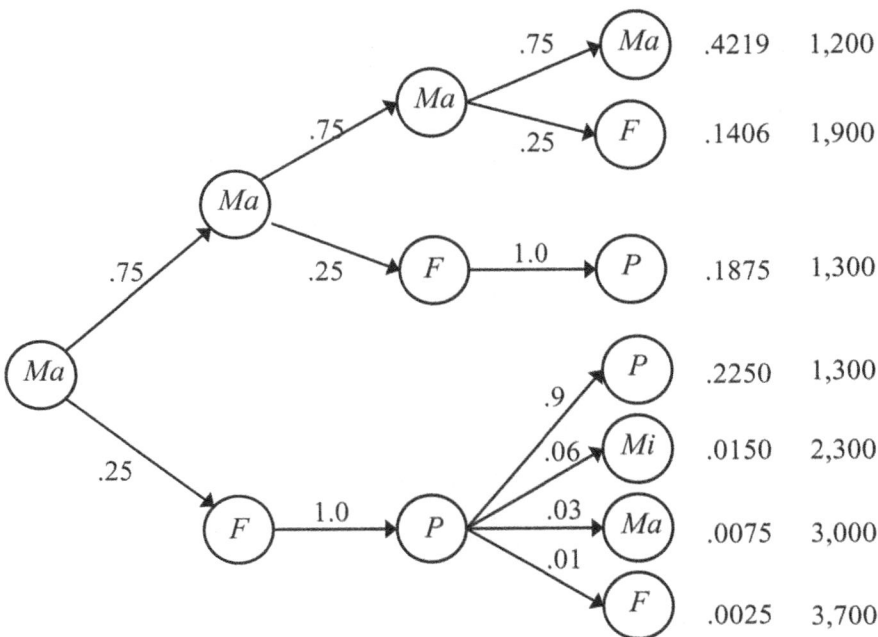

Fig. 13.3 Probability tree for vehicle with major problems

In particular, in case of Sunny 6, 10 percent of the time, the service is perceived to be poor, in which case customers change to one of the other two chains with equal probability. In case of Cloudy 7, service is perceived to be satisfactory 80 percent of the time; if it is not, customers switch to one of the other two chains with equal probability. Finally, 90 percent of the time service at Rainy 8 is deemed to be ok; if it is not, customers always switch to Sunny 6.

(a) Set up the transition matrix **P**.
(b) In the long run, what percentage of customers patronizes the three motels? What are the mean recurrence times for the three motels?
(c) Management of the Cloudy 7 chain perceives that its customer loyalty is not as good as one could wish. By using of better management techniques, they may be able to reduce the proportion of poor service to ten percent (again, in case of poor service, customers switch to the other chains with equal likelihood). What are the new steady-state proportions?
(d) Given that one percent of business is worth $500,000, what is the maximum amount that the management of Cloudy 7 should be prepared to pay to implement the new management techniques that result in the improved service?

13.3 Decision Making with Markov Chains

Solution:

(a) The transition matrix is

$$\mathbf{P} = \begin{bmatrix} .9 & .05 & .05 \\ .1 & .8 & .1 \\ .1 & 0 & .9 \end{bmatrix}.$$

(b) The steady-state probabilities are $\pi = [4/8, 1/8, 3/8] = [0.5000, 0.1250, 0.3750]$. With $r_{ii} = 1/\pi_i$ we conclude that $r_{11} = 2$, $r_{22} = 8$, and $r_{33} = 8/3 \approx 2.67$ transitions. Therefore, a customer staying at a Sunny 6 motel will on average first return to a Sunny 6 motel two transitions later.

(c) The revised transition matrix is

$$\mathbf{P} = \begin{bmatrix} .9 & .05 & .05 \\ .05 & .9 & .05 \\ .1 & 0 & .9 \end{bmatrix}.$$

The steady-state probabilities are $\pi = [4/9, 2/9, 3/9] = [0.4444, 0.2222, 0.3333]$.

(d) The market share of the Cloudy 7 chain has increased by 9.7222%, which is worth $4,861,100, which is the maximal amount the management of Cloudy 7 should pay for the change.

Problem 2 (criminal recidivism, value of a policy change): Consider a city of 500,000 people, whose inhabitants have been classified by the authorities as either "not criminal," "misdemeanor," or "felon." Long-term studies indicate that the transition of an individual from one state to another from 1 year to the next is shown in the transition matrix

$$\mathbf{P} = \begin{bmatrix} 0.95 & 0.04 & 0.01 \\ 0.50 & 0.40 & 0.10 \\ 0.10 & 0.30 & 0.60 \end{bmatrix}.$$

City council contemplates a new program that costs $70,000,000 and changes the transition probabilities to

$$\mathbf{P}' = \begin{bmatrix} 0.95 & 0.04 & 0.01 \\ 0.70 & 0.25 & 0.05 \\ 0.20 & 0.40 & 0.40 \end{bmatrix}.$$

Assume that an individual in the "misdemeanor" category costs about $1000 annually, while the costs of a felon are $10,000 each year.

(a) What is the probability that a felon does not commit any crime in the next 2 years with and without the crime prevention initiative?
(b) What is the probability that an individual in the "misdemeanor" category does not commit any crime in the next 3 years with and without the crime prevention initiative?
(c) What are the steady-state probabilities with and without the initiative? Compare the long-term costs with and without the initiative.
(d) What are the mean recurrence times for the three categories with and without the incentive? How many years, on average, does it take for an individual to move from the "felon" to the "misdemeanor" category, without the crime prevention initiative?

Solution:

(a) The probability of the string "felon"—"no crime"—"no crime" is $(0.1)(0.95) = 0.095$ without the initiative and $(0.2)(0.95) = 0.19$ with it.
(b) (b) The probability of the string "misdemeanor"—"no crime"—"no crime"—"no crime" is $(0.5)(0.95)(0.95) = 0.45125$ without and $(0.7)(0.95)(0.95) = 0.63175$ with it.
(c) The steady-state probabilities are $\pi = [\pi_1, \pi_2, \pi_3] = [210/239, 19/239, 10/239] \approx [0.87866, 0.07950, 0.04184]$ without the initiative and $[\pi_1, \pi_2, \pi_3] = [860/935, 56/935, 19/935] \approx [0.91979, 0.05989, 0.02032]$ with it. Without the initiative, the expected costs per individual are $497.908, while they are $263.102 with it. For the city of 500,000, this means costs of $248,954,000 without the initiative and $131,551,000 with it, a savings of $117,403,000. Since the savings exceed the costs of $70,000,000, the initiative should be introduced.
(d) Using the values for the steady-state probabilities computed in (c) and the fact that $r_{ii} = 1/\pi_i$, we obtain mean recurrence times for the three categories without the incentive as $r_{11} = 239/210 \approx 1.138$, $r_{22} = 239/19 \approx 12.58$, and $r_{33} = 239/10 = 23.9$ years. Considering the lifespan of individuals, mean recurrence times in excess of twenty years may be less meaningful. If the crime prevention program goes ahead, the mean recurrence times will change to $r_{11} = 935/860 \approx 1.087$, $r_{22} = 935/56 \approx 16.70$, and $r_{33} = 935/19 \approx 49.21$ years. Again, caution must be advised when interpreting the practical aspects of long mean recurrence times.

To compute the expected number of years to move from the "felon" category to the "misdemeanor" category without the initiative, we need to compute r_{32}. This can be accomplished by solving the following system of two equations in two unknowns:

$$r_{12} = 1 + p_{11}r_{12} + p_{13}r_{32}, \text{ and}$$
$$r_{32} = 1 + p_{21}r_{12} + p_{33}r_{32}.$$

13.3 Decision Making with Markov Chains

Solving this system with the probabilities shown in the matrix $\mathbf{P}$ (i.e., $p_{11} = 0.95$, $p_{13} = 0.01$, $p_{31} = 0.10$, and $p_{33} = 0.6$), we obtain $r_{12} = 410/19 \approx 21.58$ years and $r_{32} = 150/19 \approx 7.89$ years, which will be the expected time for an individual to move from the "felon" to the "misdemeanor" category.

Reliability Models   14

In the previous chapter on stochastic processes, we considered a system, which could be in different states in any one time period, and in which the transition from one state to another can occur randomly. Reliability theory in its simplest form includes a number of independent components, each of which is in exactly one of two states: it functions properly, or it does not. Our task is to calculate the probability that the entire system works. The system under consideration may describe a life-changing or life-saving scenario. Consider, for instance, a passenger elevator in a building. Its proper functioning may determine whether or not the passenger live to tell of their experience. On the other hand, if a hardware or software component of a personal computer crashes, it could probably result in some inconvenience to the user but would not be fatal. In reality, deadly accidents related to elevator malfunctions are quite rare: worldwide, there are about 27 elevator deaths per year. On the other hand, computer crashes are an almost everyday occurrence.

Given the tremendous differences between the consequences in these two, admittedly extreme, examples, the setup of the two systems will differ: one will include extensively tested components with high reliability and/or backup systems in case of a failure, while computer systems, and sometimes emergency use medications, are put on the market as beta versions that have not gone through rigorous testing.

Reliability theory will determine the probability that a system functions well, given the probabilities of all of its components. The system may be small with only a few components, or highly complex with thousands or even millions of components. Each of its components may actually be a system in itself with many subcomponents, which, in turn, may consist of sub-subcomponents and so forth. This is the reason why we refer to a hierarchy of systems, and it is up to the modeler to choose the level, at which a reliability analysis of a system and its components is to be undertaken.

The first section of this chapter studies how the reliability of a system depends on the configuration and reliability of its components. The second section introduces time aspects and investigates how the breakdown of components and the entire systems occur over time from a probabilistic view. Section 14.3 extends the probabilistic analysis by defining a so-called hazard function, which is subsequently used

© The Author(s), under exclusive license to Springer Nature Switzerland AG 2022
H. A. Eiselt, C.-L. Sandblom, *Operations Research*, Springer Texts in Business and Economics, https://doi.org/10.1007/978-3-030-97162-5_14

to compute the reliability over time for different probability distributions of component lifetime. The remaining two sections consider redundancy and standby systems, as well as how reliability can be estimated.

Modern reliability theory has its roots in statistical quality control, developed in the 1930s by W. Shewhart, H.F. Dodge and H.G. Romig, all at Bell Labs at the time. Another line of thought, dating to the 1950s, is due to W. Weibull, whose distribution bearing his name is used in connection with the hazard function. For a full account of reliability, see, e.g., the text by E.A. Elsayed (2020).

14.1 Fundamentals of Reliability Systems

The proper functioning of a system will depend on the functioning and arrangement of its components. The reliability R of a system refers to the probability that the system will function properly over some given period of time. Assuming that the system consists of n components, each with an individual reliability $p_j, j = 1, \ldots, n$, the system reliability depends on the p_j values as well as the system structure (or configuration). A common way to display the structure is by way of a network representation called a *block diagram*. We are located on the left side of the diagram and have a message that is to be delivered through the network to the right side. For each component, we know the probability that the message is processed correctly. The task is now to send the message through the network, so that it arrives on the right side of the system. For the transmission of the message, there are some rules. Whenever there is a fork, the message will be duplicated and sent along all forks. Whenever a message enters a component, the probability that it leaves the component in readable form equals the probability of the component to function. The probability that at least one message arrives at the end of the process on the right side of the diagram is then the reliability of the system.

One of the simplest such structures is the *series* structure, in which case the system functions if and only if all of its components function. A typical example concerns a vehicle. Suppose that we consider the motor (1), the brakes (2), and the steering system (3), with individual reliabilities of $p_1 = 0.8, p_2 = 0.9, p_3 = 0.9$. A block diagram of the series structure is shown in Fig. 14.1.

Statistically, the requirement that (component 1 must function) *and* (component 2 must function) *and* ... is expressed as the product of the individual probabilities, here $R = p_1 p_2 p_3 = (0.8)(0.9)(0.9) = 0.6480$, a fairly dismal performance characteristic. Note that the series in the block diagram expresses solely the fact that all components must function for the entire system to work properly. The interpretation is similar to that of a project network, in which tasks shown in series must all be

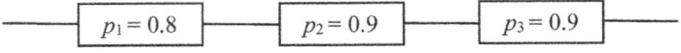

Fig. 14.1 The series structure of the example

14.1 Fundamentals of Reliability Systems

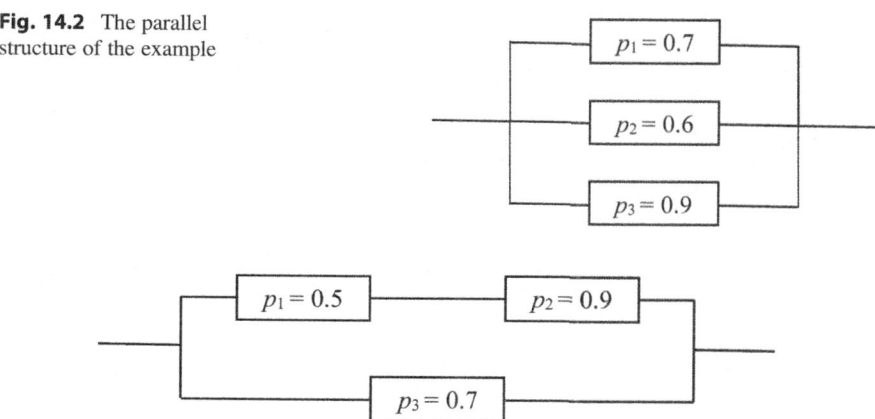

Fig. 14.2 The parallel structure of the example

Fig. 14.3 A mixed system

performed (where in the reliability block diagrams the sequence does not play a role).

On the other hand, individual components may be backups or alternatives of each other. For instance, we may use three cables 1, 2, and 3 in a cable car that is pulled by a wire rope, such as the one in San Francisco. Suppose now that the reliability of each cable is $p_1 = 0.7$, $p_2 = 0.6$, and $p_2 = 0.9$, respectively. The overall system functions if at least one of the system functions. This is ensured in all cases except if *all* systems fail. This means that the failure probability, the complement of reliability, is $1 - R$, which is defined as the probability that all components fail, i.e., $(1 - p_1)(1 - p_2)(1 - p_3) = (0.3)(0.4)(0.1) = 0.012$, so that the reliability of the system is $R = 0.988$ or 98.8%. In a block diagram, this situation is displayed in a *parallel* structure as shown in Fig. 14.2.

For the general serial structure, we can easily show that $R \leq \min_{1 \leq j \leq n} p_j$ ("a chain is no stronger than its weakest link"), so that if any $p_j = 0$, then $R = 0$ as well. On the other hand, for the general parallel structure, we find that $R \geq \max_{1 \leq j \leq n} p_j$, so that if any $p_j = 1$, then $R = 1$.

Clearly, few systems in engineering are pure series or pure parallel systems. As an example, consider a mixed system with three components 1, 2, and 3, which functions if either components 1 and 2 both function, and/or component 3 functions. Assuming that the reliabilities of the three components are 0.5, 0.9, and 0.7, this situation is shown in Fig. 14.3.

The probability that components 1 and 2 are both working is $(0.5)(0.9) = 0.45$. Now, the probability that the system functions equals [(Prob that 1 works) *and* (Prob that 2 works)] *or* [Prob that 3 works]. Thus $1 - R = (1 - 0.45)(1 - 0.7) = 0.165$, or simply $R = 83.5\%$.

A general *parallel-series system* contains m parallel paths, each consisting of n components arranged in series, as shown in Fig. 14.4.

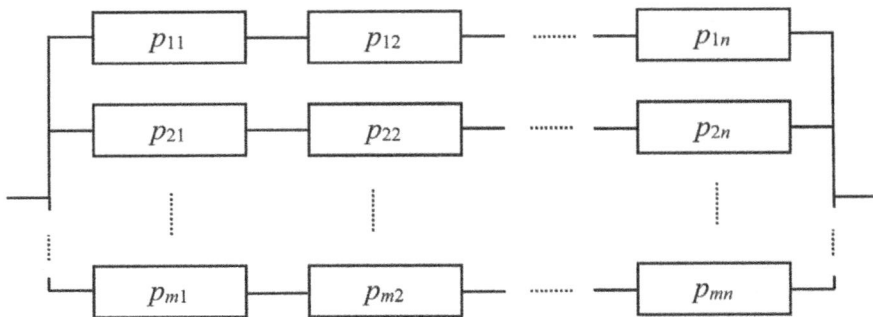

Fig. 14.4 A general parallel-series system

The system shown in Fig. 14.4 is actually quite general. In case the parallel paths have different lengths, we simply introduce fictitious "dummy" components, with reliability of 1, so that each of the parallel paths has the same length.

The reliability of a parallel-series system is calculated by first determining the reliability of each series subsystem (the product of the individual reliabilities) and then the unreliability $1 - R$ as the product of the unreliabilities of each series subsystem.

As an example of parallel-series systems, consider an oceanographic research vessel operating in a remote location at sea without any satellite, VHF, or cell phone coverage. The results of its scientific measurements need to be transmitted to its onshore base location by radio, a so-called single-sided band (SSB) shortwave. Its messages consist of seven consecutive data components, each of which has only a 90% probability of being transmitted correctly; additionally, all seven pieces of data need to be successfully radioed for the message to be received by the base station. At the base station, any incoming defective message will be automatically filtered, ensuring that only correctly received information is accepted. Our task is now to determine how many times a message needs to be sent so as to reach a 95% probability that a correct message is received at the base station. Alternatively, we are interested in our course of action in case a 99% reliability is required.

We can model this situation as a parallel-series reliability system with m parallel paths, each indicating a full transmission of the entire message. Each such path consists of seven data components. With $n = 7$ for each path and m parallel paths, the reliability of any single path (i.e., the reliability that one complete message is transmitted properly) equals $(0.9)^7 = 0.4782969$. Then the unreliability of the entire system given that the entire message is transmitted four times equals $1 - R = (0.4782969)^4 = 0.0523$, so that with four full messages transmitted, the probability that at least one complete message is received properly is $R = 92.59\%$. For $m = 5$, we obtain a reliability of 0.9614, so that we need to transmit the full message at least five times in order to obtain a probability of more than 95% that the message is received properly. Increasing m further, we determine that it takes 8 transmissions of the message to achieve a reliability of $R = 99.45\%$, the smallest number of transmissions that achieves at least 99% reliability.

14.1 Fundamentals of Reliability Systems

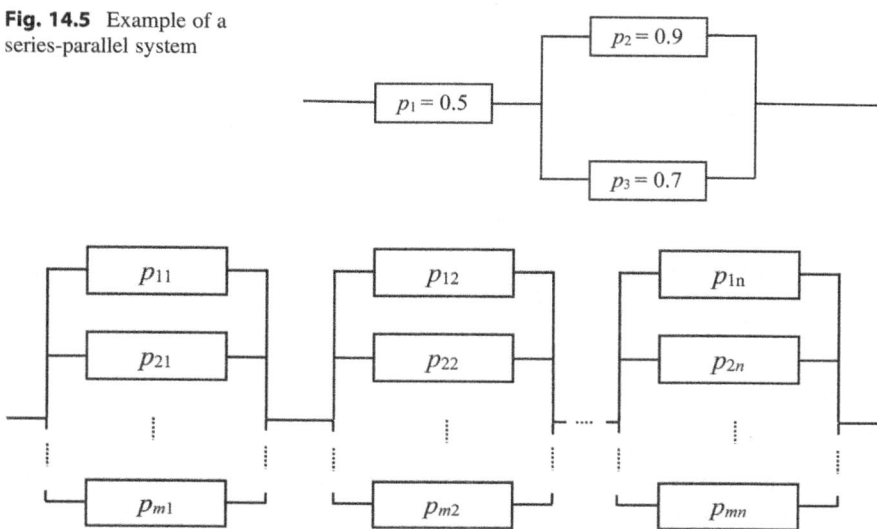

Fig. 14.5 Example of a series-parallel system

Fig. 14.6 A general series-parallel system

Another type of organizational system is a *series-parallel* system. These types of arrangements are essentially parallel subsystems arranged in series. A simple example is shown in Fig. 14.5, for which we calculate the reliability of the system as follows. The probability that both parallel components 2 and 3 do not work is $(1 - 0.9)(1 - 0.7) = 0.03$, so that the reliability of the parallel segment is 0.97. The reliability is then $(0.5)(0.97) = 0.485$.

A general series-parallel system contains n subsystems, arranged in series fashion, each of which is a parallel system of m components as shown in Fig. 14.6.

Again, the system in Fig. 14.6 is quite general. It can easily accommodate different numbers of parallel components. In such a case, we simply add fictitious "dummy" components with zero reliability, so that all parallel subsystems have the same number of components.

Yet another type of reliability systems are *k-out-of-n* systems, which function if and only if at least k out of the n system components function. Special cases are n-out-of-n systems, which are nothing more but series systems, while 1-out-of-n systems are simply parallel systems. As an example, for a 2-out-of-3 system with component reliabilities $p_1 = 0.7$, $p_2 = 0.8$, and $p_3 = 0.9$, respectively, we can compute the system reliability R by observing that no more than one component can fail, and we therefore consider the four cases: (1) none of the components fails, (2) only the first component fails, (3) only the second component fails, and (4) only the third component fails. The reliability of the system is then

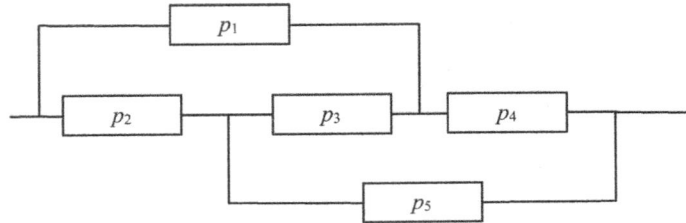

Fig. 14.7 A 5-component system

$$R = (0.7)(0.8)(0.9) + (0.3)(0.8)(0.9) + (0.7)(0.2)(0.9) + (0.7)(0.8)(0.1)$$
$$= 0.902.$$

Given the complexity of this and similar systems, we may resort to a brute-force enumeration method, often referred to as a *Boolean truth method*. It simply enumerates all possibilities the n components of the system can be in, which, given that each component either works properly or fails, is 2^n, which already reveals the weakness of the technique: even for a small system with only, say, $n = 10$ components, there are already $2^{10} = 1024$ different scenarios.

As an example, consider a large aircraft with eight engines, four under each wing (such as the B-52 Stratofortress). Suppose that on some long flight, each of the independently functioning engines has a 98% chance of working properly. i.e., $p = 0.98$. For the flight mission to be successful, at least three engines on each of the two wings need to be working during the entire flight. The question is now what the probability for a successful mission is.

In order to solve the problem, we first enumerate all cases, in which the entire system works properly. First, there is the case in which all eight components work and that is just a single instance. The probability of this happening is $(0.98)^8 = 0.85076302$. Secondly, there are those cases in which exactly one engine fails. There are eight such cases, each of which has a probability of $(0.02)(0.98)^7 = 0.01736251$ for a total of 0.13890009. Finally, there are all those cases, in which there are two failures altogether, one on each side. There are four possible cases of a single failure on each side coupled with four cases with a single failure on the other side of the plane for a total of 16 cases. Each such case has a probability of $(0.98)^6(0.02)^2 = 0.00035434$ for a total of 0.00566939. The 25 cases under consideration have thus a total probability of 99.533%. Even though this probability appears quite high as it is, the probability that each engine functions properly is much higher than 98% in reality, thus having a dramatically higher probability of a mission being successful.

Finally, there are systems that do not fall into any categories we have covered so far. One such example is shown in Fig. 14.7 which shows a five-component system.

Again, while applying the Boolean truth method may be tedious, it may be possible to economize the calculations by some ad-hoc measures. In the example of Fig. 14.7, we may exclude scenarios in which components 1 and 2 are both not functioning properly, as they yield a zero contribution to the overall reliability as do states, in which either components 1, 3, and 5, or components 4 and 5 both do not work.

There are methods that can handle complex systems more efficiently. They involve paths and cuts in the block diagrams akin to the techniques covered in Chap. 6 on network models, but they are beyond the scope of this introductory text.

14.2 Time Aspects of Reliability

In the previous section, we dealt with the possibility of failure of components over a given time span of operation. Clearly, a component is more likely to break down over a longer time period than during a shorter one, and in this section, we will discuss how time affects reliability. Specifically, observe a component j and observe its performance over time. Building on our discussion of stochastic processes in the previous chapter, we can model the behavior of the component as a life-and-death process, i.e., a stochastic process with only two states: life, i.e., working, and death, i.e., not working, and with the only transitions being those from life to life, life to death, and death to death. Associated with this life-and-death process is the random variable that measures the time that elapses between some point t in time (typically $t = 0$) at which the component is operating (alive) and the time at which it fails (dies). We will refer to this random variable as the *time-to-failure* (*TTF*), *lifetime*, or simply *life*, and denote it by X_j; similarly for the *system life* X. We will also involve the cumulative life distributions

$$F_j(t) = F_{X_j}(t) := P(X_j \leq t) \text{ and}$$
$$F(t) = F_X(t) := P(X \leq t).$$

These denote the respective probabilities that the component and system survive until time t, so that $F_j(0) = F(0) = 0$. We will also make use of their respective probability density functions $f_j(t) = \frac{dF_j(t)}{dt}$ (so that $F_j(t) = \int_0^t f_j(x)dx$), and $f(t) = \frac{dF(t)}{dt}$ (so that $F(t) = \int_0^t f(x)dx$). It follows that $p_j(t) = P(\text{component } j \text{ still functions at time } t) = P(X_j > t) = 1 - P(X_j \leq t) = 1 - F_j(t)$, and $R(t) = P$ (the system still functions at time t) $= P(X > t) = 1 - P(X \leq t) = 1 - F(t)$. This is equivalent to stating that the reliability $p_j(t)$ of component j (or the reliability $R(t)$ of the system), expresses the probability that it will function properly over the time period $[0, t]$. Clearly, the reliability will therefore depend on the length of the time period $[0, t]$, as well as the life distribution of X_j (or X). Specifically, for a series system, we now find the reliability $R(t)$ over the time span $[0, t]$ as $1 - F(t) = R(t) = p_1(t)p_2(t)\ldots p_n(t) = (1 - F_1(t))(1 - F_2(t))\ldots(1 - F_n(t))$, and therefore $F(t) = 1 - (1 - F_1(t))(1 - F_2(t))\ldots(1 - F_n(t))$.

Similarly, for a parallel system we obtain $F(t) = 1 - R(t) = (1 - p_1(t))(1 - p_2(t))\ldots(1 - p_n(t)) = F_1(t)F_2(t)\ldots F_n(t)$. As an example of a lifetime distribution, we will use the exponential distribution X with the parameter λ, which has the density function $f(t) = \lambda e^{-\lambda t}$ and cumulative distribution function $F(t) = 1 - e^{-\lambda t}$ (see Appendix C). We write $X \sim Exp(\lambda)$. This distribution is applicable to electronic devices, LED lights, and similar objects, which do not suffer from mechanical or any other kind of wear as opposed to those that deteriorate and work less and less well. As a numerical example, consider a series system that consists of two components with lifetime distributions $X_1 \sim Exp(\lambda_1)$ and $X_2 \sim Exp(\lambda_2)$, respectively. For the system lifetime X we then obtain the cumulative distribution $F(t) = 1 - (1 - F_1(t))(1 - F_2(t)) = 1 - e^{-\lambda_1 t}e^{-\lambda_2 t} = 1 - e^{-(\lambda_1+\lambda_2)t}$, so that $X \sim Exp(\lambda_1 + \lambda_2)$. In general, for an n-component series system with components that are $X_j \sim Exp(\lambda_j)$, $j = 1, \ldots, n$, respectively, we will similarly obtain the system life $X \sim Exp\left(\sum_{j=1}^{n} \lambda_j\right)$.

Since the expected value of an exponential distribution $Exp(\lambda)$ is $E(X) = 1/\lambda$ (see Appendix C), we find that the mean lifetime of a system with $Exp(\lambda_j)$ components is $E(X) = \left(\sum_{j=1}^{n} \lambda_j\right)^{-1}$. As an example, if a two-component series system has exponential lifetime components of 2 and 3 months, respectively (i.e., $\frac{1}{\lambda_1} = 2$ and $\frac{1}{\lambda_2} = 3$), then the system mean lifetime will be $\frac{1}{\lambda_1+\lambda_2} = \frac{1}{\frac{1}{2}+\frac{1}{3}} = 1.2$ months.

Turning to parallel systems, assume that a two-component system has components with lifetimes $X_1 \sim Exp(\lambda_1)$ and $X_2 \sim Exp(\lambda_2)$, respectively. Then we find $F(t) = F_1(t)F_2(t) = \left(1 - e^{-\lambda_1 t}\right)\left(1 - e^{-\lambda_2 t}\right) = 1 - e^{-\lambda_1 t} - e^{-\lambda_2 t} + e^{-(\lambda_1+\lambda_2)t}$, so that for the density function $f(t)$, we obtain $f(t) = F'(t) = \lambda_1 e^{-\lambda_1 t} + \lambda_2 e^{-\lambda_2 t} - (\lambda_1 + \lambda_2)e^{-(\lambda_1+\lambda_2)t}$, from which we conclude that $E(X) = \int_0^\infty xf(x)dx = \int_0^\infty x\lambda_1 e^{-\lambda_1 x}dx + \int_0^\infty x\lambda_2 e^{-\lambda_2 x}dx - \int_0^\infty (\lambda_1 + \lambda_2)e^{-(\lambda_1+\lambda_2)x}dx = \frac{1}{\lambda_1} + \frac{1}{\lambda_2} - \frac{1}{\lambda_1+\lambda_2}$. In general, for an n-component parallel system with component lifetimes $X_j \sim Exp(\lambda_j)$, $j = 1, \ldots, n$, we similarly find

$$F(t) = F_1(t)F_2(t)\ldots F_n(t) = \left(1 - e^{-\lambda_1 t}\right)\left(1 - e^{-\lambda_3 t}\right)\ldots\left(1 - e^{-\lambda_n t}\right),$$

from which the density function $f(t)$ and then the system mean lifetime $E(X)$ can be determined. As an illustration, consider again the previous example with two components and their 2- and 3-month lifespan, but assume now that the structure is parallel rather than in series. The system mean life is then calculated as $E(X) = \frac{1}{\lambda_1} + \frac{1}{\lambda_2} - \frac{1}{\lambda_1+\lambda_2} = 3.8$ months. This is considerably longer than the 1.2 months in case of the series system.

As another example, we consider the mixed structure displayed in Fig. 14.3 of the previous section, but with exponentially distributed exponent lives, whose means are 5, 6, and 7 months, respectively, as shown in Fig. 14.8.

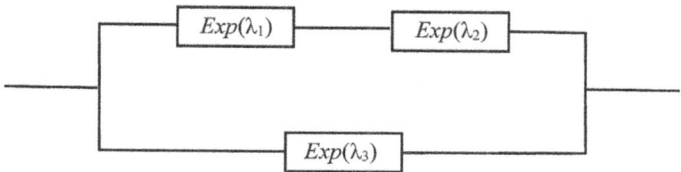

Fig. 14.8 A mixed structure with exponential component lives

For this system, we will determine the distribution of the system lifetime and its expected value. We will also calculate the probability that the system life is no longer than 5 months, at least 4 months, and between 6 and 8 months, respectively. From the example in Fig. 14.3, we know that $1 - R(t) = [1 - p_1(t)p_2(t)][1 - p_3(t)]$, so that $F(t) = [1 - (1 - F_1(t))(1 - F_2(t))]F_3(t) = [1 - e^{-\lambda_1 t}e^{-\lambda_2 t}][1 - e^{-\lambda_3 t}] = 1 - e^{-\lambda_3 t} - e^{-(\lambda_1+\lambda_2)t} + e^{-(\lambda_1+\lambda_2+\lambda_3)t}$ and therefore $f'(t) = \lambda_3 e^{-\lambda_3 t} + (\lambda_1 + \lambda_2)e^{-(\lambda_1+\lambda_2)t} - (\lambda_1 + \lambda_2 + \lambda_3)e^{-(\lambda_1+\lambda_2+\lambda_3)t}$. This allows us to conclude that the mean system life is $E(X) = \frac{1}{\lambda_3} + \frac{1}{\lambda_1+\lambda_2} - \frac{1}{\lambda_1+\lambda_2+\lambda_3} = \frac{1}{7} + \frac{1}{\frac{1}{5}+\frac{1}{6}} - \frac{1}{\frac{1}{5}+\frac{1}{6}+\frac{1}{7}} \approx 7.76$ months. Furthermore, $P(X \leq 5) = F(5) = 1 - e^{-5/7} - e^{-(1/5+1/6)5} + e^{-(1/5+1/6+1/7)5} \approx 0.42885$, i.e., 42.9%. Also, $P(X \geq 4) = 1 - F(4) = e^{-4/7} + e^{-22/15} - e^{-217/105} \approx 66.5\%$. Finally, $P(6 \leq X \leq 8) = F(8) - F(6) \approx 13.30\%$.

We end this section with some notation and some useful mathematical results. The expected system life $E(X)$ is sometimes also called *mean time-to-failure* (MTTF), i.e., $MTTF = E(X) = \int_0^\infty x f(x) dx$ and it is possible to demonstrate that $MTTF = \int_0^\infty R(x) dx$. Furthermore, if systems or components are replaced as soon as they fail, we also speak of system or component *mean time between failures* MTBF. In the above example, $MTTF \approx 7.765$ months. Another reliability measure is the *mean residual life* $MRL(t)$, which is the expected remaining life of a system or component that has survived until time $t \geq 0$, i.e., $MRL(t) = E(X - t \mid X \geq t)$. It is possible to show that $MRL(t) = \frac{1}{R(t)} \int_0^\infty x f(x) dx - t$. Using the concept of reliability $R(t)$ rather than the lifetime X, it is straightforward to obtain alternative expressions for MTTF.

14.3 Failure Rates and the Hazard Function

While the previous section investigated the issue of how long a reliability system might be working, we here examine the risk of a breakdown in a functioning system. More specifically, assume that a system has survived until a given time t, the question is then what its probability is to fail within Δt time units, i.e., within the time interval $[t, t + \Delta t]$, where Δt is sufficiently small. Given a random variable

Fig. 14.9 Bathtub shape failure rate

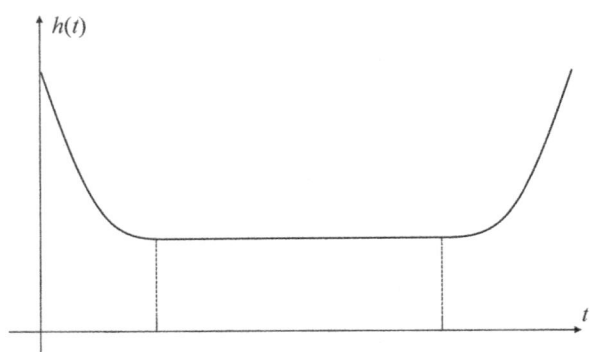

X that denotes system life, the conditional probability that the system life is somewhere in the interval $[t, t + \Delta t]$ is then $P(t < X \leq t + \Delta t | X > t) = \frac{P(t < X \leq t + \Delta t)}{P(X > t)}$. Since $P(X > t) = R(t)$ and $P(t < X \leq t + \Delta t) \approx f(t)\Delta t$ for small values of Δt, the conditional probability in question is approximated by $\frac{f(t)}{R(t)} \Delta t$. We define $h(t) := \frac{f(t)}{R(t)}$ as the *hazard rate, hazard function*, or (*instantaneous*) *failure rate* of the system. In other words, $h(t)\Delta t$ is the conditional probability that a system that has survived until time t, will fail during the interval $[t, t + \Delta t]$. The hazard rate is of major importance when dealing with the breakdown risk of systems. We can, of course, also consider the hazard rate $h_j(t)$ of individual components with probability density function $f_j(t)$ and reliability $R_j(t)$, as $h_j(t) := \frac{f_j(t)}{R_j(t)}$. Since $h(t) = \frac{f(t)}{R(t)} = \frac{F'(t)}{1-F(t)}$, we see that the hazard function $h(t)$ is completely determined by the lifetime distribution $F(t)$. Conversely, one can show that $F(t)$ is completely determined by $h(t)$, so that either $F(t)$ or $h(t)$ can be used to describe any system.

Example: For a system life X that is exponentially distributed with the parameter λ, the hazard rate becomes $h(t) = \frac{f(t)}{1-F(t)} = \frac{\lambda e^{-\lambda t}}{1-(1-e^{-\lambda t})}$, so that the hazard rate is constant over time and equals the parameter λ. Conversely, given a constant hazard rate λ, we can infer that the random variable X is exponentially distributed with the parameter λ. Therefore, an exponentially distributed system life has a constant failure rate and is the only distribution with this property.

In general, reliability systems will have a time-dependent failure rate $h(t)$, which typically has a "bathtub" shape similar to that shown in Fig. 14.9. As the Figure shows, the failure rate is initially declining. This phase is often referred to as "early failure period," "break-in period," "infant mortality," "debugging," "shake-down' or "burn-in period," just to name a few. What then follows is a period with more or less constant failure rate for a mature system. In some mechanical systems of poor quality, this period may be missing altogether with wear-out weaknesses occurring right after the break-in period. Finally, the failure rate increases in the wear-out period due to ageing. In some electronic systems, the wear-out period is missing altogether: as an example, LED lightbulbs are typically rated 25,000–40,000 hrs (or 12.5–20 years, given an 8 hrs burn time per day 5 days of the

14.3 Failure Rates and the Hazard Function

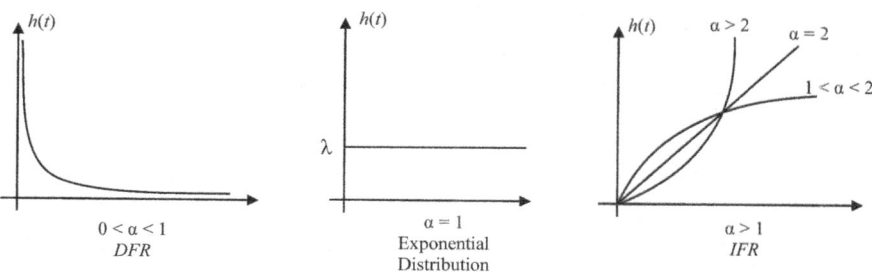

Fig. 14.10 Hazard rates for the Weibull distribution

week), before a catastrophic event, i.e., a failure occurs. Note though, that over time, dimming and color shifts do occur.

As discussed above, the exponential distribution is the only distribution with a constant failure rate. Furthermore, it is the only distribution that is "memoryless," i.e., the remaining lifetime of an item of any age is the same as a remaining lifetime of a new item. Using terminology from the previous section, we can state that $MRL(t) = MTTF$ for all $t \geq 0$, which can also be expressed as "old items are as good as new ones" or "items don't age."

Lifetime distributions with a nonconstant failure rate are sometimes classified as having a *decreasing failure rate* (DFR) or *increasing failure rate* (IFR), depending on the case. In this context, the *Weibull distribution*, a generalization of the exponential distribution, is quite useful. Its cumulative distribution is $F(t) = 1 - e^{-(\lambda t)^\alpha}$, where α and λ are given positive parameters (see Definition C7), so that its hazard rate is $h(t) = \alpha t (\lambda t)^{\alpha-1}$. One can show that for the Weibull distribution, we have $MTTF = \frac{1}{\lambda}\Gamma(1 + 1/\alpha)$, where Γ denotes the gamma function (see Definition C8). Figure 14.10 displays the hazard rate for the Weibull distribution, given a number of different values of the parameter α. Note that for $\alpha = 1$, we obtain the exponential distribution.

For the special case of $\alpha = 2$, we obtain $F(t) = 1 - e^{-(\lambda t)^2}$, which is the so-called *Rayleigh distribution* with its linearly increasing hazard rate $h(t) = 2\lambda^2 t$. This distribution has applications in fields as diverse as ballistics and oceanography.

As an example, a component with a failure rate of $h(t) = t$ has a Rayleigh distribution with $\alpha = 2$ and $2\lambda^2 = 1$, so that $\lambda = 1/\sqrt{2}$. Its lifetime distribution is then $F(t) = 1 - e^{-\frac{1}{2}t^2}$ and its $MTTF = \sqrt{2}\,\Gamma(1 + \frac{1}{2}) = \sqrt{\pi/2} \approx 1.2533$.

The *gamma distribution* $\Gamma(\alpha, \lambda)$ is another common lifetime distribution with shape and scale parameters α and λ. Its probability density functions is $f(t) = \frac{\lambda}{\Gamma(\alpha)}(\lambda t)^{\alpha-1}e^{-\lambda t}$ (see Definition C9). For $n = \alpha$ with an integer value for n, we obtain the *Erlang distribution* and for $\alpha = 1$, the exponential distribution $Exp(\lambda) = \Gamma(1, \lambda)$. One can show that the sum of n independent $Exp(\lambda)$ variables is $\Gamma(n, \lambda)$-distributed, a very useful result in replacement theory.

Example: A component of a reliability system has an exponential lifetime with a mean of 8 days. As soon as the component fails, it is replaced by a new component that is identical to the failed piece, and we are now interested in the distribution, the expected value and standard deviation of the time until the nth component fails. We can answer this question, realizing that with $Exp(\lambda)$ individual component lives, the system life will be $\Gamma(n, \lambda) = \Gamma(n, 1/8)$. Therefore, $E(X) = n/\lambda = 8n$ days and $\sigma(X) = 8\sqrt{n}$. With this information, we can furthermore compute the probability such that the probability that the second component fails within 10 days. With $X \sim \Gamma(2, 1/8)$, $f(t) = \frac{\lambda}{\Gamma(2)} \lambda t e^{-\lambda t} = \lambda^2 t e^{-\lambda t} = \frac{t}{64} e^{-\frac{1}{8}t}$ and by integration, we obtain $P(X < 10) = \int_0^{10} f(t)dt = 1 - \frac{9}{4}e^{-\frac{5}{4}} \approx 0.3554$, i.e., there is a 35.5% risk that the second component fails within 10 days.

14.4 Redundancy and Standby Systems

When designing a system, one important consideration regarding the system's reliability is the course of action to be taken in case the system fails. One way to deal with failure is the introduction of redundancy. Loosely speaking a component is referred to as redundant, if its failure does not cause the entire system to fail. Redundancy works in parallel systems and in "k out of n" systems with $k < n$. On the other hand, a series system has no redundancy since any component failure will inevitably cause the entire system to fail.

In general, there are two types of redundancy:

- *Active redundancy*, where all redundant components are in service, and
- *Inactive redundancy* or *nonactive standby*, which occurs in three different versions:
  1. *Hot standby*, where the redundant component is in working mode and has the same failure rate as if it had been in service
  2. *Warm standby*, where the redundant component is working at a reduced level and with a failure rate less than if it had been in full service, and
  3. *Cold standby*, where the redundant component is idle with a zero failure rate.

Figure 14.11 illustrates the difference between active and inactive redundancy in reliability systems.

As an example of a standby arrangement, consider a factory that uses electricity in a chemical production process, where the redundant component is an electric power generator. In case the process is sensitive, the power generator may very well be

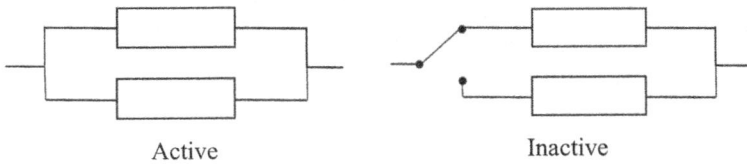

Active   Inactive

Fig. 14.11 Active and inactive redundancy systems

running continuously in hot standby mode. If the process can tolerate a brief interruption of power, a warm standby with the generator at idling speed may be sufficient. In a cold standby mode, the generator would be started only when in need of service.

Another classification of redundant system components distinguishes between repairable and non-repairable systems. For instance, *repairable components* include power generators, vehicle brake systems, or airplane jet engines. On the other hand, non-repairable components include telecommunications satellites in orbit, printed electronic circuits, and lightbulbs.

14.5 Estimating Reliability

Given a system of components that are arranged in some given configuration, we may wish to estimate the reliability, the failure rate, or the lifetime distribution of the system or of any of its individual components. Such estimations are done by so-called *reliability testing*. This may, for instance, be required in the process of acceptance sampling of lots of components that are supposed to meet specifications expressed in terms of reliability, *MTTF*, or similar measures. For lifetime testing of components, we would take a random sample of n components from a given lot, test them, and record their life (failure) times $t_1, t_2, \ldots, t_n$. The life testing can be done by *operational life testing (OLT)* by using typical operating conditions, *accelerated life testing (ALT)* by employing operating conditions, in which failures occur sooner than in the actual use of the component, or *highly accelerated life testing (HALT)* at very high stress levels. Accelerated testing may be needed, if results need to be obtained quickly, or if the components that are being tested have a very long lifespan, such as the aforementioned *LED* light bulbs. Using the collected data, such as the reliability, failure rate, or lifetime distribution, parameters of the tested components can then be statistically estimated.

For practical purposes, the life testing process may have to be stopped before all of the n sampled units have failed. This is referred to as *truncation*, which occurs either at a prespecified time T or after a prespecified number r of components have failed. With $t_1, t_2, \ldots, t_r$ denoting the respective failure times of the units, we have $t_1 \leq t_2 \leq \ldots \leq t_r \leq T$. If the units have an exponentially distributed lifetime $Exp(\lambda)$, one can show that the mean lifetime can be estimated by the expression $\frac{1}{r} \times (t_1 + t_2 + \ldots t_r + T(n-r))$.

Example: Nine components with identically distributed exponential lives are tested. Failures occur at 27.2, 29.5, 32.6, 36.8, 36.8, 37.9, and 38.3 hrs, respectively, at which time the test is terminated. The component mean life is then estimated by $\frac{1}{7}(27.2 + 29.5 + 32.6 + 36.8 + 36.8 + 37.9 + 38.3 + 38.3(9-7)) = 45.1$ hrs—a surprisingly high figure. The reliability of an individual component at time t is therefore $R(t) = e^{-t/45.1}$, so that at time $t = 40$, a component has the reliability $R(40) = e^{-40/45.1} = 41.2\%$.

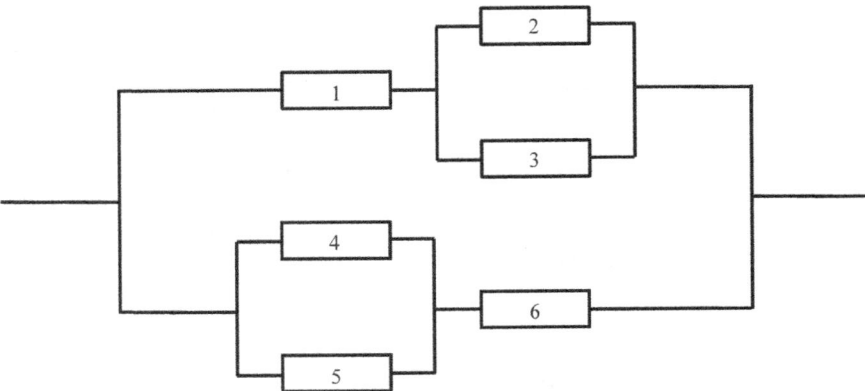

Fig. 14.12 A reliability structure for Problem 1

Expressions can be obtained for estimating the parameters of the Weibull and Gamma distributions as well, but they are considerably more complicated and outside the scope of this book.

Exercises

Problem 1 (System reliability) A certain system has the reliability structure shown in Fig. 14.12.

Here, we have $p_1 = p_4 = p_5 = 0.9$ and $p_2 = p_3 = p_6 = 0.8$.

(a) Calculate the system reliability.
(b) Suppose that one of the six components is to be removed (i.e., its reliability is downgraded to 0). Evaluate the system reliability for each of the six possible, resulting configurations and choose the best.
(c) Suppose that one of the six components is to have its reliability upgraded to 1.0. Evaluate the system reliability for each of the possible six resulting configurations and choose the best.

Solutions

(a) We find $R = 1 - \{1 - p_1(1 - (1 - p_2)(1 - p_3))\}\{1 - (1 - (1 - p_4)(1 - p_5))p_6\} = 0.971712$ or 97.2%.
(b) Performing the required computations, the results are summarized in Table 14.1.

Therefore, the best options are to let $p_4 = 0$ or $p_5 = 0$, resulting in a maximal reliability of $R = 0.96192$, i.e., 96.2%, which is only a slight deterioration from the 97.2% found under (a).

14.5 Estimating Reliability

Table 14.1 System reliability with one component removed

$j, p_j = 0$	R
1	0.792
2	0.94176
3	0.94176
4	0.96192
5	0.96192
6	0.864

Table 14.2 System reliability with one component that cannot fail

$j, p_j = 1$	R
1	0.99168
2	0.9792
3	0.9792
4	0.9728
5	0.9728
6	0.99864

(c) Performing the required computations, the results are summarized in Table 14.2.

The best option is $p_6 = 1$ and it yields the maximum reliability $R = 0.99864$, i.e., 99.9%, a significant improvement over the reliability of 97.2% found in (a).

Problem 2 (System lifetime)

A radio communications satellite is powered by seven solar power cells. The satellite will function if and only if at least four of the seven cells function, i.e., have not yet broken down. (In reliability terms, such systems are referred to as a "four out of seven system.") The cells break down independently of each other and over time. Once a cell has broken down, it will remain inoperative. The probability that a cell breaks down in any given year is 4%. Denote by X_n the number of cells still working after n years and by Y_n the number of cells that have broken down after n years.

(a) What are the distributions of X_1 and Y_1?
(b) What are the expected values and standard deviations of X_1 and Y_1?
(c) What is the probability that the satellite still functions after one year?
(d) What are the distributions of X_2 and Y_2?
(e) Calculate the probability that the satellite still functions after 2 years.

Solutions

(a) We find that X_1 is binomially distributed with $Bin(7, 0.96)$ and that Y_1 is $Bin(7, 0.04)$, noting that $Y_1 = 7 - X_1$.
(b) $E(X_1) = 6.72$, $\sigma(X_1) = \sqrt{0.2688} \approx 0.518$, $E(Y_1) = 0.28$, $\sigma(Y_1) = \sqrt{0.2688} \approx 0.518$.
(c) The satellite reliability over a one-year time span is $R = P(X_1 \geq 4) = P(Y_1 \leq 3) \approx 0.999919$ or 99.99%.

(d) X_2 is $Bin(7, 0.96^2) = Bin(7, 0.9216)$, and Y_2 is $Bin(7, 1 - 0.96^2) = Bin(7, 0.0784)$ where we note that $Y_2 = 7 - X_2$.

(e) The satellite reliability over a 2-year life span is $R = P(X_2 \geq 4) = P(Y_2 \leq 3) \approx 0.99891$ or 99.89%.

Problem 3 (System lifetime) A reliability system consists of two identical components arranged in series. The components have exponentially distributed lifetimes with distribution parameter λ.

(a) Determine the lifetime density function as well as the average system life.
(b) Assume now that the two components are arranged in parallel instead of in series. Determine the system lifetime density functions as well as the average system life.
(c) What is the ratio of the average system lives for the configurations in (a) and (b) above?
(d) For the case with three identical components, determine the system lifetime density function using the series as well as the parallel arrangement.

Solution:

(a) Since the component lives X_1 and X_2 are both $Exp(\lambda)$, we find $F_1(t) = F_2(t) = 1 - e^{-\lambda t}$. Therefore, for the series system lifetime X_{ser} we obtain $F_{ser}(t) = 1 - (1 - F_1(t))(1 - F_2(t)) = 1 - e^{-2\lambda t}$, so that $f_{ser}(t) = 2\lambda e^{-2\lambda t}$. Consequently, $E(X_{ser}) = 1/2\lambda$.

(b) For the parallel system lifetime X_{par}, we find $F_{par}(t) = F_1(t)F_2(t) = (1 - e^{-\lambda t})^2$, so that $f_{par}(t) = 2(1 - e^{-\lambda t})\lambda e^{-\lambda t} = 2\lambda e^{-\lambda t} - 2\lambda e^{-2\lambda t}$, and therefore $E(X_{par}) = 2/\lambda - 1/2\lambda = 3/2\lambda$.

(c) The ratio becomes $E(X_{par})/E(X_{ser}) = \frac{3/2\lambda}{1/2\lambda} = 3$, which shows that the parallel system has an average life three times as long as the series system.

(d) For a 3-component series system, we calculate $F_{ser}(t) = 1 - (1 - F_1(t))(1 - F_2(t))(1 - F_3(t)) = 1 - e^{-3\lambda t}$ and consequently, $f_{ser}(t) = 3\lambda e^{-3\lambda t}$ and $E(X_{ser}) = 1/3\lambda$. For the 3-component parallel systems, we obtain $F_{par}(t) = F_1(t)F_2(t)F_3(t) = (1 - e^{-\lambda t})^3$ and therefore $f_{par}(t) = 3(1 - e^{-\lambda t})^2 \lambda e^{-\lambda t}$. Although it was not asked for in the question, one can show that $E(X_{par}) = 11/6\lambda$, so that the ratio is 11/2 for the 3-component case. Actually, for the n-component case, the ratio is $n\left(1 + \frac{1}{2} + \frac{1}{3} + \ldots + \frac{1}{n}\right)$.

Problem 4 (System reliability, lifetime, and hazard function) A system that consists of four identical components will function if and only if at least three of the four components function.

(a) Express the system reliability R as a function of the component reliability p.
(b) Assume that the components have exponentially distributed lifetimes with some distribution parameter λ. Determine the system lifetime density function as well

as the system average life and compare it to the individual components' average life.
(c) Based on the results found under (b), what is the system hazard rate function? Determine if the system is *DFR*, *IFR*, or neither.

Solutions

(a) The system reliability is given by $R = 4p^3(1-p) + p^4 = 4p^3 - 3p^4$.
(b) With $F_i(t) = 1 - e^{-\lambda t}$, we obtain $p_i(t) = 1 - F_i(t) = e^{-\lambda t}$, and therefore $R(t) = 1 - F(t) = 4p_i(t)^3 - 3p_i(t)^4 = 4e^{-3\lambda t} - 3e^{-4\lambda t}$, and $f(t) = F'(t) = 12\lambda(e^{-3\lambda t} - e^{-4\lambda t})$. For the system average life $E(X)$ we find that $E(X) = (4)\left(\frac{1}{3\lambda}\right) - (3)\left(\frac{1}{4\lambda}\right) = \frac{7}{12\lambda}$, i.e., slightly more than half the individual component average life $1/\lambda$.
(c) The hazard rate is $h(t) = \frac{f(t)}{R(t)} = \frac{12\lambda(e^{-3\lambda t} - e^{-4\lambda t})}{4e^{-3\lambda t} - 3e^{-4\lambda t}} = \frac{12\lambda(1 - e^{-\lambda t})}{4 - 3e^{-\lambda t}}$. Differentiating, we determine that $h'(t) = \frac{12\lambda^2 e^{-\lambda t}}{(4 - 3e^{-\lambda t})^2} > 0$. Hence, the system has an increasing hazard rate, i.e., it is *IFR*.

Reference

Elsayed EA (2020) Reliability engineering, 3rd edn. Wiley, Hoboken, NJ

Waiting Line Models 15

Waiting line or queuing systems are pervasive. Many of us remember the long lineups in front of stores in the Soviet Union and Vietnam, and we have all experienced lineups in banks and supermarkets, but there are many more instances with waiting lines: think, for instance, about traffic lights, where drivers line up and wait, files that wait for processing in the inbox at a clerk's workstation, or telephone calls that are put in a queue. Queuing systems were first examined by Agner Krarup Erlang (1878–1929). Erlang was a Danish mathematician, who worked for the Copenhagen Telephone Company. One of the questions he faced during this time was to determine the number of telephone circuits that are required to provide an acceptable level of service.

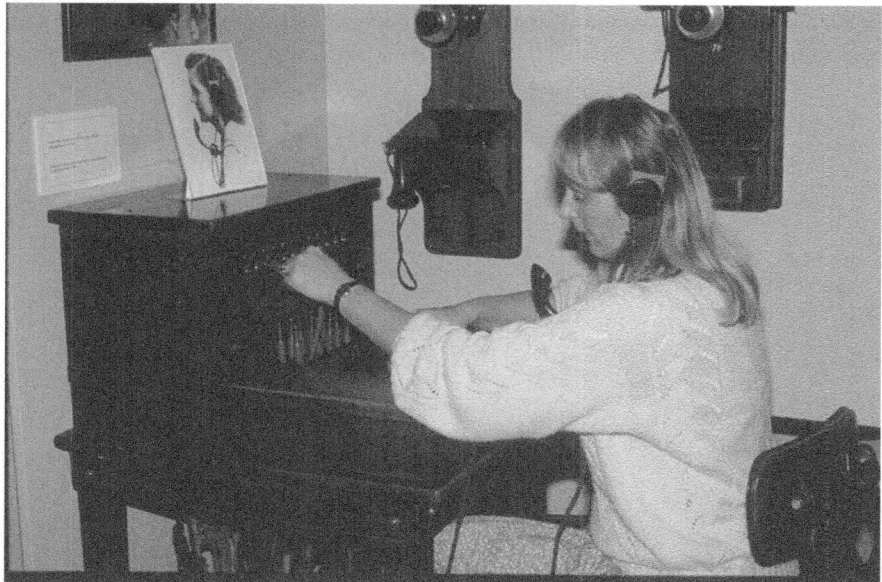

Phone switchboard operator

© The Author(s), under exclusive license to Springer Nature Switzerland AG 2022
H. A. Eiselt, C.-L. Sandblom, *Operations Research*, Springer Texts in Business and Economics, https://doi.org/10.1007/978-3-030-97162-5_15

15.1 Basic Queuing Models

In order to avoid discussing only special cases, we will formalize queuing systems as follows. All entities that are in need of service of some kind will be referred to as *customers*, while the service is performed at *service stations*. The process can then be thought of as follows. Customers are in a *calling population*. Once they are in need of service (here, for simplicity, we will consider only a single type of service customers are interested in), they will approach the service system, where they will line up. When it is the customer's turn, he will be served, after which the customer will leave the service system and rejoin the calling population. The structure of this process can be visualized in Fig. 15.1.

The service system will require some further specifications. First of all, each service system has only a single waiting line. A service system typically consists of a number of c parallel service stations, each assumed to perform the same type of service. Parallel service stations are usually referred to as *channels*. In some instances, one channel consists of a series of service stations: imagine entering a building, where a potential customer first has to be cleared by security, then being directed to a general secretary, from where the service continues to the department director's secretary, and finally on to the department director. At each station, the customer may be asked to leave the system, e.g., for not clearing security, the unavailability of a service station, and for other reasons. Multi-phase systems can be very complex and will not be discussed here.

In order to categorize queuing systems, Kendall (1918–2007) devised a taxonomy in 1953 that, in a variety of versions, some of which were added by Lee and Taha in 1968, is the standard to this day. The original system consists of three descriptors, and it has been extended to at most six components. In order to be as

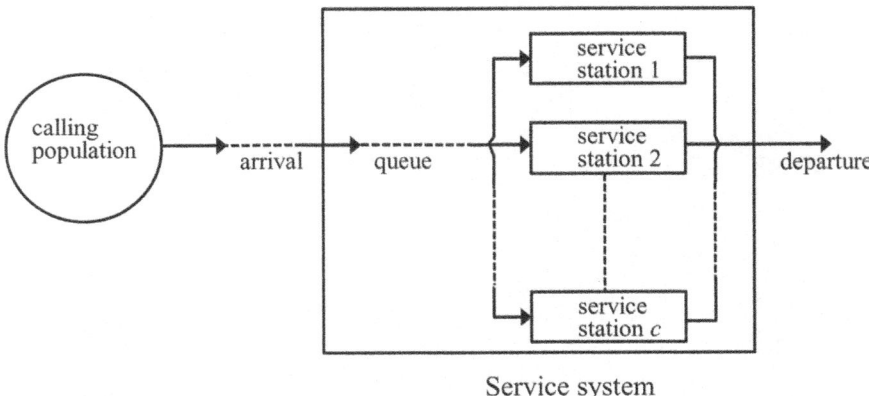

Fig. 15.1 Queuing system

15.1 Basic Queuing Models

general as possible, we introduce the complete six-component system first, but then use only the more compact 3-component system later on. The notation is

A/S/C K/N/D,

where each letter is a placeholder for one component of waiting lines. These components are now described in some detail. (We should note, though, that some publications arrange the last three components in a different order.)

The symbol *A* describes the arrival process. We will use the symbol "*M*," if arrivals are random and follow a so-called Poisson process (where the "*M*" stands for "Markovian" or "memoryless"). The Poisson distribution (see Appendix C in this book) describes such a process. In the ubiquitous Poisson process, events such as customer arrivals, occur randomly over time. Specifically, during a time interval of length t, the number of events will follow a Poisson distribution $Po(\lambda t)$, where λ is the parameter of the process. Furthermore, the time between two consecutive events then follows an exponential distribution with the same parameter λ, i.e., $Exp(\lambda)$. Other popular processes include "*D*" or deterministic arrivals (actually: constant arrivals) in which the time between arrivals is constant and known with certainty, as is the case on an assembly line), "*G*," where arrivals follow some general distribution, for which only some key parameters are known, e.g., mean and variance, or "*GI*," which symbolizes a general independent distribution.

The second component "*S*" symbolizes the service process. Again, the letter "*M*" symbolizes Markovian or memoryless, i.e., exponential service time, while "*D*" "*G*," and "*GI*" symbolize deterministic/constant, general, and general independent service rates, respectively.

The letter *C* indicates the number of parallel service stations that are alternatives of each other.

The letter *K* in the extended version of the taxonomy describes the number of customers that can be accommodated in the entire system. This includes the number of customers who wait as well as those who can be served. In case this descriptor is not used, it is assumed that *K* is infinite. While no real-life system has infinite capacity, assuming an infinite *K* simplifies the analysis tremendously and is usually very close to the true results as soon as *K* exceeds 30 or 40.

The symbol *N* denotes the size of the calling population. As in the case of the capacity of the system, *N* will be assumed to approach infinity if not otherwise specified. Again, for reasonably large values of *N*, we may assume its value to be sufficiently close to infinity so as to simplify the computations.

Finally, the symbol *D* denotes the queuing discipline. Typical disciplines are *FCFS* or *FIFO* (first come, first served or first in, first out), *LCFS* or *LIFO* (last come, first served or last in, first out), or *SIRO* (service in random order). An important category are priority queues, in which some customers receive preferential treatment. The most prominent example of priority queues occurs in health care, where more serious cases, as classified by the triage nurse, will be treated first. In case no queuing discipline is specified, it is assumed that the *FCFS* discipline applies.

In general, it will be useful to think about waiting lines as buffers between arrival and service. A good image would be one of a water tank. While a waiting line (similar to an inventory) grows with the inflow, it shrinks whenever service is provided and customers leave the system.

In queuing theory, we distinguish between transient states and steady states. While a *steady state* occurs if the system has been running for a very long time, *transient states* are still, at least to some degree, dependent on the initial state of the system. A simple example is the opening of a store in the morning. Initially, no one is in the system and waiting times will be short. However, as the store operates for some time, the system becomes more congested and is no longer dependent on the opening conditions.

The key task of queuing theory is to compute measures of interest from some key input factors or queue characteristics. The main characteristics of a queue are those described in the taxonomy. Measures of interest include average waiting times, the probability that a newly arriving customer will have to wait, the average length of a queue, and others. In order to formalize our discussion, we use the following conventions about notation (which, incidentally, are almost all standard in the pertinent literature):

λ denotes the mean *arrival rate*, measured in [customers/hour]. It is the average number of customers who *actually arrive* at the system in the specified amount of time.

μ is the mean *service rate*, measured in [customers/hour]. It is the average number of customers who *can be served* by a single service station.

It is worth noting that while λ expresses an actual observable fact, μ indicates a capability. The actual number of customers served does not only depend on the service station's capabilities, but also on the number of customers who desire service.

The inverse values of λ and μ also have important interpretations. Suppose that a service station can deal with $\mu = 12$ customers per hour, then the inverse value is $1/\mu = 1/12$ [hours/customer] $= 5$ [minutes/customer]. This is the average *service time*. The inverse value of the arrival rate can be interpreted similarly. If the average arrival rate is $\lambda = 10$ [customers/hour], then $1/\lambda = 1/10$ [hours/customer] $= 6$ [minutes/customer], meaning that on average, 6 min elapse between two successive arrivals. This is referred to as the (average) *interarrival time*.

It is apparent that, in case of a single service station, the arrival rate cannot exceed the service rate. If it would, then there are more arrivals than can be handled by the service station, so that—given infinite patience of the customers without balking (leaving a queue immediately upon arrival because of the expected long waiting time) or reneging (leaving the queue after a while as the anticipated future waiting time appears too long)—the waiting line will grow towards infinity. This gives rise to a regularity condition. In order to express it in a compact form, we define the *traffic intensity* (sometimes also referred to as *utilization rate*) $\rho = \lambda/\mu$. A feasible system (i.e., a system that has a steady state) must then have $\rho < c$, where c denotes

15.1 Basic Queuing Models

Table 15.1 Steady-state formulas for the $M/M/1$ queuing model

$P_0 = 1 - \rho$	$L_s = \frac{\lambda}{\mu - \lambda}$	$W_s = \frac{1}{\mu - \lambda}$
$P_n = P_0 \rho^n$	$L_q = \frac{\rho \lambda}{\mu - \lambda} = \frac{\rho^2}{1 - \rho}$	$W_q = \frac{\rho}{\mu - \lambda}$

the number of parallel service stations. As an example, consider a system with a single service station that faces an arrival rate of $\lambda = 24$ customers per hour. For it to be feasible, the service rate must be $\mu > 24$, or, equivalently, the average service time cannot exceed 150 sec.

On the output side are the measures that we are interested in and that we can compute. They include:

P_n: the probability that there are n customers in the system (with the important special case of $n = 0$ and P_0, i.e., the probability that the system is idle)
W_s: the average waiting time per customer
W_q: the average time a customer spends in the queue
L_s: the average number of customers in the system, and
L_q: the average number of customers in the queue.

Before stating formulas for these measures, there are some general relations that hold in queuing, regardless of the specific system under consideration. The first such relation is

$$W_s = W_q + 1/\mu. \tag{15.1}$$

Simply stated, the relation expresses that the total time a customer spends in the system equals the waiting time plus the service time. Another relation is known as *Little's formula* (based on the work by J.D.C. Little who published the formula in 1961), which states that

$$L_\bullet = \lambda W_\bullet, \tag{15.2}$$

where the "$\bullet$" stand for either the subscript "s" or the subscript "q". This formula provides a convenient way to compute the number of customers in the queue and in the system from the average time customers spends in the queue or in the system, respectively (or vice versa).

The simplest queuing model is the $M/M/1$ model, in which the number of customer arrivals are random and follow a Poisson process (making the interarrival times exponentially distributed), and the service time is exponential (so that the service rate again follows a Poisson distribution). The key queuing formulas for this model are shown in Table 15.1.

As an illustration, consider the following

Example: Customers arrive at the counter of a bank at a rate of 30 per hour. Arrivals are random and service time is exponential, so that we are dealing with an $M/M/1$ model. The clerk's average service time is 80 sec. Putting the parameters in their required form, we glean $\lambda = 30$ and $\mu = 45$ from this information. As $\rho =$

$\lambda/\mu = 30/45 = \tfrac{2}{3} < 1$, the system does have a steady state. The probability that the bank teller is idle is $P_0 = 1 - \rho = 1/3$. The probability that at least two customers are waiting equals the probability that there are at least three customers in the bank or, formally, $P_3 + P_4 + P_5 + \cdots = 1 - P_0 - P_1 - P_2 = 1 - \tfrac{1}{3} - \tfrac{1}{3}(\tfrac{2}{3})^1 - \tfrac{1}{3}(\tfrac{2}{3})^2 \cong \tfrac{8}{27} \approx .2963$ or slightly less than one-third. On average, there are $L_q = 1.3333$ customers waiting in line and the average time a customer spends in the system is $W_s = 1/15$ [hr] $= 4$ min.

An interesting case arises when the decision maker specifies the service level and determines bounds for the capabilities of the servers. Suppose that in an *M/M/1* system with $\lambda = 20$, the decision maker specifies that the probability that there are three or more customers in the system should not exceed 95%. The probability of three or more customers in the system is again $1 - P_0 - P_1 - P_2 = 1 - (1 - \rho) - \rho(1 - \rho) - \rho^2(1 - \rho) = 1 - 1 + \rho - \rho + \rho^2 - \rho^2 + \rho^3 = \rho^3$. As this probability should not exceed 95%, we obtain the condition $\rho^3 = \tfrac{\lambda^3}{\mu^3} \leq .95$, or, as $\lambda = 20$, $\mu^3 > 8000/0.95$ or $\mu > 20.3449$. The reason that it is sufficient that the service rate barely exceeds the arrival rate is that the probability of three or more customers in the system is very small.

Suppose now that the service rate is no longer random but that it follows some general distribution. All we know about this distribution is that the mean service time is $1/\mu$ and the variance of the distribution equals σ^2. This means that we are dealing with an *M/G/1* model, for which some rather elegant formulas are available. Again, as in all single channel systems in a steady state, $P_0 = 1 - \rho$. Furthermore, the *Pollaczek-Khintchine formula* developed in 1930 is

$$L_q = \frac{\lambda^2 \sigma^2 + \rho^2}{2(1 - \rho)}. \qquad (15.3)$$

The values of L_s, W_q, and W_s can then be computed based on the general relations (15.1) and (15.2). Before demonstrating this model on a numerical example, note that the *M/M/1* model is a special case of the *M/G/1* model with $\sigma^2 = 1/\mu^2$. Replacing σ^2 in the above expression results in $L_q = \tfrac{\rho^2}{1-\rho}$, which is the standard formula of the *M/M/1* model. Similarly, we observe that in the case of the *M/D/1* model, i.e., the queuing model with deterministic service time, the variance $\sigma^2 = 0$, so that the Pollaczek–Khintchine formula reduces in this case to $L_q = \tfrac{\rho^2}{2(1-\rho)}$. Observe that the number of customers waiting in the case of the deterministic model is exactly half of that of the standard model with exponential service time. In other words, the performance of the queuing system can be improved quite dramatically by reducing the variance of the service time.

Example: Arrivals of customers at a single service desk follow a Poisson distribution, while the service time follows a general distribution. There is an average of $\lambda = 15$ arrivals, while the service time is 3 min on average with a standard deviation of 6 min. This means that $1/\mu = 1/20$ [hrs] and $\sigma^2 = 1/(10)^2 = 1/100$, so that the average number of customers waiting in line is then $L_q = 5.625$ and the average

15.1 Basic Queuing Models

waiting time is $W_q = 0.375$ hr $= 22.5$ min. If the service time were exponential, we obtain the standard M/M/1 system and the performance measures $L_q = 2.25$ and $W_q = 9$ min, while the deterministic model has $L_q = 1.125$ and $W_q = 4.5$ min.

Another well-known result involves the so-called *Kingman equation*. It was described by the British mathematician Sir John Kingman in 1961, and it is an approximation for the expected waiting time in the very general G/G/1 queue. Defining CV_λ and CV_μ as the coefficient of variation of the interarrival times and service times, respectively (recall that the coefficient of variation is the standard deviation divided by the mean), we can write the expected waiting time as

$$W_q \approx \frac{\rho}{1-\rho} \frac{CV_\lambda^2 + CV_\mu^2}{2} \frac{1}{\mu}.$$

If we were to apply this general formula to the special case of exponential interarrival and service time distributions, we obtain means of $1/\lambda$ and $1/\mu$, respectively, as well as standard deviations of $1/\lambda$ and $1/\mu$, respectively, so that $CV_\lambda = CV_\mu = 1$. Inserting these parameters into Kingman's equation, we obtain $W_q \approx \frac{\rho}{\mu-\lambda}$, which is exactly the formula for the M/M/1 case. As an illustration, consider again the above example with $\lambda = 15$, $\mu = 20$, and $\sigma^2 = 1/100$, so that $CV_\mu = \frac{1}{2}$. Assuming that $CV_\lambda = 2$, we obtain $W_q = 0.1875$ hr or $11\frac{1}{4}$ min.

Consider now the case of multiple service stations. Here, we have to make the distinction between one multi-station service center and a number of parallel single-service centers. The general rule is that each service center has only a single waiting line. As an example, consider what we may refer to as a "bank system" and a "supermarket system." In a bank system, there is a single queue and thus a single multi-server system. In contrast, a supermarket features multiple waiting lines, hence we deal with multiple single-server systems.

Let us deal with multiple single-server systems first, as they are a straightforward extension of the concepts discussed earlier. The usual assumptions include no balking (i.e., customers in need of service will join the queue regardless of its length) and no jockeying (i.e., customers do not change queues if they perceive that another queue may result in shorter waiting times). As an example, suppose again that arrivals follow a Poisson distribution with a mean arrival rate of $\lambda = 45$ customers per hour, while service time is exponentially distributed with an average service time of $1/\mu = 2$ min (or, equivalently, a service rate of $\mu = 30$ customers per hour). Before performing any computations, feasibility requires that $\rho = \lambda/\mu = 45/30 = 1\frac{1}{2}$ does not exceed the number of service stations c, requiring at least two service stations in this instance. Suppose now that $c = 3$ service stations are available. As a result, we now deal with three separate single-service systems, or, in terms of the taxonomy, $3 \times$ M/M/1 systems. Consider now the customers. While an average of $\lambda = 45$ customers are in need of service, each of the three systems receives only about one-third of this number, as customers may be assumed to randomly choose the system they want to be served by. Hence, for each of the three systems, we have an individual arrival rate of $\lambda' = 15$. The value λ' will then replace λ in all of the

Table 15.2 Formulas for M/M/c systems

$P_0 = \dfrac{1}{\left[\sum_{i=0}^{c-1}\dfrac{\rho^i}{i!}\right]+\dfrac{\rho^c}{c!(1-\frac{\rho}{c})}}$	$P_n = \begin{cases} \dfrac{\rho^n}{n!}P_0, \text{ if } 0 \leq n \leq c \\ \left(\dfrac{\rho^n}{c!c^{n-c}}\right)P_0, \text{ if } n > c \end{cases}$
$L_q = \dfrac{\rho^{c+1}}{(c-1)!(c-\rho)^2}P_0$	$L_s = L_q + \rho$

formulas. In our example, feasibility is guaranteed as $\rho' = \lambda/\mu = 15/30 = \frac{1}{2} < 1$ for each of the systems. We can then apply the usual steady-state formulas for M/M/1 models shown in Table 15.1. For instance, the average time a customer spends in the system is $W_s = \frac{1}{\mu-\lambda'} = 1/(30-15) = 1/15$ hrs $= 4$ minutes. On the other hand, the average number of customers waiting is $L_q = \frac{\rho^2}{1-\rho} = (\frac{1}{2})^2/(1-\frac{1}{2}) = \frac{1}{2}$, meaning that on average half a customer is waiting in line in each of the three subsystems. In other words, on average in all of our three service systems combined, 1.5 customers will be waiting for service.

Next, consider a single M/M/3 system such as the one we encounter in a bank with three tellers. Some of the relevant formulas are summarized in Table 15.2, where the values of the waiting times W. can be computed by using Little's formula (15.2).

In our example, first compute the probability that there are no customers in the system. Here, $P_0 = \frac{1}{\left[1+\frac{3}{2}+\frac{9}{8}\right]+\frac{(3/2)^3}{3!\left(1-\frac{3/2}{3}\right)}} = \frac{4}{19}$. This allows us to compute $L_q = 9/38 \cong .2368$, $L_s = 9/38 + 3/2 \cong 1.7368$ and, applying Little's formula (15.2), we obtain $W_q = L_q/\lambda = 0.3158$ min and $W_s = L_s/\lambda = 2.3158$ min.

The differences between the two models arise from the fact that in the former model with separate queues, it is possible that at least one customer is still waiting, while a service station is idle. This is based on the aforementioned assumption that jockeying is prohibited.

A simple extension is a model with self-service. It is easily derived from the standard M/M/c model by letting the number of service stations c tend to infinity. Such a model will not include any waiting time, so that $L_q = W_q = 0$ and $W_s = 1/\mu$, so that $L_s = \rho$. It can also be shown that in this case, $P_n = \frac{e^{-\rho}}{n!}\rho^n$ for any $n \geq 0$.

For more complex queuing models, it is useful to have either tables or figures at hand that provide the decision maker with a quick idea of what to expect without engaging in complex computations. Entire books with queuing graphs and queuing tables exist. Figure 15.2 shows a typical example of such queuing graphs. This particular graph depicts the situation of a $c \times$ M/M/1 graph, where the abscissa shows the utilization rate ρ, while the ordinate is related to the total number of customers in all systems cL_s. The steep solid line relates to the case of a single facility (i.e., $c = 1$), the broken, broken-and-dotted, broken-and-twice-dotted, and dotted lines are for $c = 2, 3, 5$, and 10 service stations, respectively.

Finally, we would like to have a look at multiple-server problems from a slightly different angle. Suppose that the decision maker specifies a certain service level and inquires how many service stations are needed in order to provide a prespecified level of service. Questions of this nature frequently arise in the public service, e.g.,

15.1 Basic Queuing Models

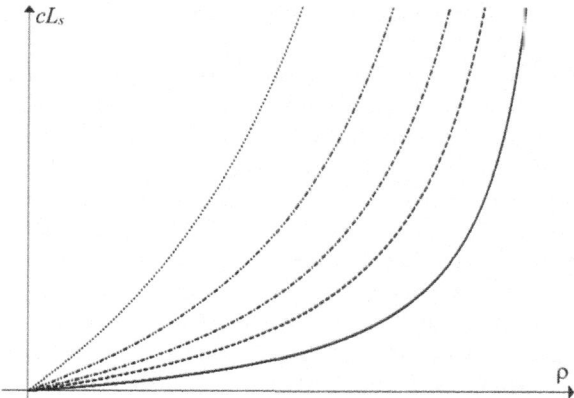

Fig. 15.2 Customers in the system as a function of the utilization rate, for various numbers of servers

how many police officers are needed so as to adequately protect an area. As an illustration, consider the following.

Example: Suppose that an average of 20 customers per hour randomly arrive at a hospital. The system uses a subsystem for each doctor on duty, and there will be an unknown number of c doctors available. Given that each doctor can deal with an average of ten patients per hour, hospital administrators want to ensure that the average waiting time for any patient does not exceed 10 min. What is the smallest number of doctors that allows this?

Solution: The system described above is a set of c parallel $M/M/1$ systems, and the individual arrival rate at each of these systems is then $\lambda' = \lambda/c$. Following the results in Table 15.1, the average waiting time is defined as $W_q = \frac{\rho}{\mu - \lambda}$, which, with $\rho' = \frac{\lambda'}{\mu} = \frac{\lambda}{\mu c} = \rho/c$, can be rewritten as $W_q = \frac{\rho/c}{\mu - \lambda/c} = \frac{2/c}{10 - 20/c}$. The condition is now that the average waiting time is no higher than 1/6 [hr], which can be written as

$$\frac{2/c}{10 - 20/c} \leq 1/6.$$

This inequality can be rewritten as $\frac{2}{10c - 20} \leq \frac{1}{6}$, so that $c \geq 3.2$. In other words, at least four doctors are needed to provide the desired service. It may also be interesting to note that with three doctors, the average waiting time is 12 min, while four doctors result in an average waiting time of only 6 min.

Finally, we would like to point out that some queuing calculators are readily available on the internet. One such example is the site https://www.supositorio.com/rcalc/rcalclite.htm that calculates waiting times, probabilities and queue lengths for $M/M/c$ models.

15.2 Optimization in Queuing

While queuing models are primarily designed to compute performance measures, they can also be applied in the context of optimization. As an example, consider a retail establishment. The owner of the store has to decide how many clerks to employ for the cash registers at the checkout counter. Clearly, increasing the number of clerks will increase the costs. However, at the same time, more clerks will result in less waiting time for customers, which, in turn, results in less ill will, lost sales, and other customer behavior detrimental to sales. One of the main problems applying these models is the quantification of the loss due to customer ill will.

The example of a tool crib is much easier to justify. A tool crib is a place in which expensive tools are kept that are not in constant use by the workers. Due to cost considerations, it would not be feasible to provide each worker with one tool, so that a service desk is established, where workers can sign out the tool whenever it is needed. The costs of the system include the costs of the tool crib clerks as well as the costs for the lost time of the workers. If c is the number of clerks and $\$_c$ and $\$_w$ denote the hourly wage of a clerk and a worker, respectively, and L'_s denoting the number of workers in each of the c service stations, then the costs can then be written as

$$C = (\text{cost of clerks}) + (\text{cost of worker's lost time}) = c(\$_c + \$_w L'_s) =$$
$$= c\left(\$_c + \$_w \frac{\lambda}{c\mu - \lambda}\right).$$

The idea is now to determine the optimal number of clerks so as to minimize the overall costs. As an illustration, consider the following

Example: The demand for a specialized tool occurs randomly at a rate of about 100 times per hour. Whenever the need arises, workers walk over to the tool crib, sign out the tool, use it, and then return it to the tool crib. All clerks are equally efficient with a service time of 3 min. For simplicity, we assume that the organization of the signing out follows $c \times M/M/1$ systems. Assume that the hourly wage of a clerk is $\$_c = 10$, while a worker's lost hour costs $\$_w = 25$.

Solution: With the given parameters of $\lambda = 100$ and $\mu = 20$ (and thus $\rho = 5$), we first note that due to the feasibility condition $\rho \leq c$, we will need at least six clerks. The optimal number of clerks can then be determined as follows. Consider the two cost curves that determined the total costs. The costs for the clerks increase linearly with the number of clerks. On the other hand, the cost for the workers' lost time decreases hyperbolically with an increasing number of clerks. This is shown in Fig. 15.3. Note that while only integer values of c are relevant, we display the costs for all real values of c so as to better show the shapes of the curves. Also note the similarity between this cost curve and those in inventory management (Fig. 15.3).

For the smallest possible number of clerks that still has a steady state (e.g., $c = 6$), waiting time for the workers will be very long, resulting in high costs. As the number of clerks increases, waiting times decrease and with them the costs. The reduction of workers' waiting times due to adding clerks to the system is very significant at first, but less and less so as the number of clerks decreases. Eventually, the benefit of an

15.2 Optimization in Queuing

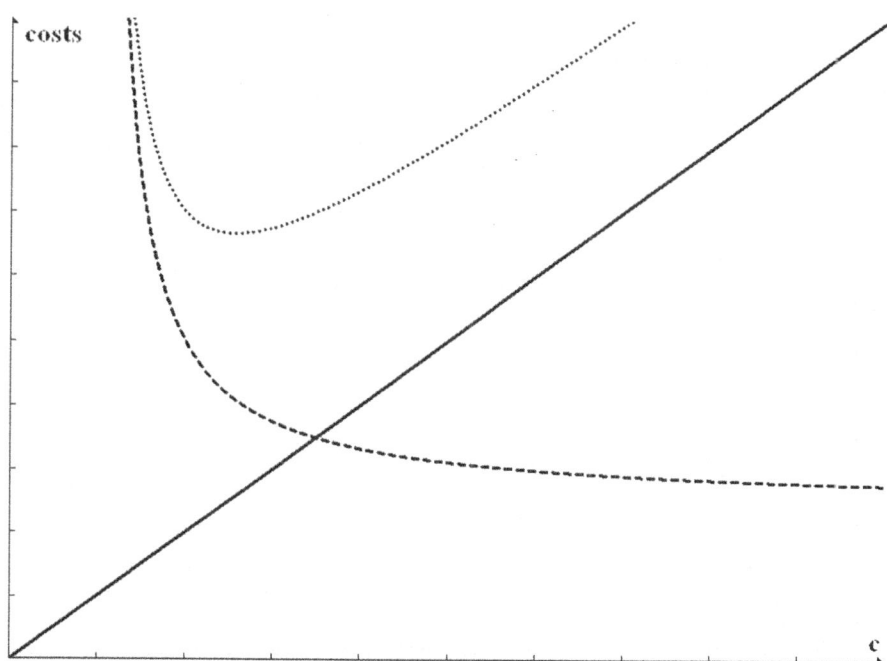

Fig. 15.3 Cost curves for the tool crib example

additional clerk is outweighed by his costs, so that the total costs start increasing again. This suggests a brute force search procedure, in which the total costs are computed for $c = 6, 7, 8, \ldots$ clerks until the costs that were initially decreasing, start increasing again. At that point, the optimal number of clerks has been found. Detailed computations are shown below.

$c = 6$: $\lambda = 100/6 = 16.6667$, $\rho = 16.6667/20 = 0.8333$, $L_s = 5$ in each system,
Cost $= 6(10) + 5(6)(25) = 810$.

$c = 7$: $\lambda = 100/7 = 14.2857$, $\rho = 0.7143$, $L_s = 2.5$ in each system,
Cost $= 7(10) + 2.5(7)(25) = 507.5$.

$c = 8$: $\lambda = 100/8 = 12.5$, $\rho = 0.6250$, $L_s = 1.6667$ in each system,
Cost $= 8(10) + 1.6667(8)(25) = 413.34$.

$c = 9$: $\lambda = 100/9 = 11.1111$, $\rho = 0.5556$, $L_s = 1.25$ in each system,
Cost $= 9(10) + 1.25(9)(25) = 371.25$.

$c = 10$: $\lambda = 100/10 = 10$, $\rho = 0.5$, $L_s = 1$ in each system,
Cost $= 10(10) + 1(10)(25) = 350$.

$c = 11$: $\lambda = 100/11 = 9.0909$, $\rho = 0.4545$, $L_s = 0.8333$ in each system,
Cost $= 11(10) + (0.8333)7(25) = 339.16$.

$c = 12$: $\lambda = 100/12 = 8.3333$, $\rho = 0.4167$, $L_s = 0.7143$ in each system,
Cost $= 12(10) + (0.7143)12(25) = 334.29$.

$c = 13$: $\lambda = 100/13 = 7.6923$, $\rho = 0.3846$, $L_s = 0.625$ in each system,
Cost $= 13(10) + 0.625(13)(25) = 333.13$.

$c = 14$: $\lambda = 100/14 = 7.1429$, $\rho = 0.3571$, $L_s = 0.5556$ in each system, Cost $= 14(10) + 0.5556(14)25 = 334.43$.

At this point, the costs start increasing again, so that it is optimal to have $c = 13$ parallel service stations.

An alternative procedure for the determination of the optimal number of clerks is this. For now, assume that the variable c is continuous. We can then set up the general cost function

$$\text{Cost} = \$_c c + \$_w L_s c \text{ with } L_s = \frac{\lambda}{c\mu - \lambda}.$$

Evaluating the derivative $\frac{d\text{Cost}}{dc}$ and setting it equal to zero results in the optimality condition

$$c^* = -\rho \pm \rho\sqrt{\frac{\$_w}{\$_c}}.$$

Since c^* is in all likelihood not integer, we need to examine the costs of its two neighboring integer values and choose the better one of the two. In our example, $c^* = 2.9$ or 12.9. Since $c \geq 6$, we examine $c = 12$ and $c = 13$, and find again $c = 13$ as the optimal number of clerks.

If we were to repeat the problem with a single waiting line and $M/M/c$ systems, we will start again with $c = 6$ clerks, which has 7.9376 customers in the system on average, costing us $6(10) + (25)(7.9376) = \$258.44$, dramatically less than the multiple-queue system considered above. The system with $c = 7$ clerks costs \$215.26, while the system with $c = 8$ clerks will cost \$211.97. We then find that with $c = 9$ clerks, the costs are \$217.52, so that in this scenario, 8 clerks are optimal.

Another possibility to incorporate optimization in queuing systems occurs, when retraining of clerks is considered. The basic setting is similar to that of the tool crib above (with the firm paying for service as well as wasted time). The retraining time for the clerks includes the actual (recurrent) retraining as well as costs for the time that the clerk is absent during training, at which time the position must be staffed by other clerks. It is typical that the costs to increase a clerk's service rate increase at an increasing rate. A numerical illustration is provided in the following.

Example: Customers arrive at a system at a rate of $\lambda = 30$ per hour. Keeping them waiting is estimated to cost \$20 per hour. At present, the service rate is $\mu = 40$, but with some additional training, this rate can be increased up to 60. Training to achieve a service rate of $\mu \in [40, 60]$ costs $2(\mu - 40)^2$. For simplicity, we assume that an $M/M/1$ system is used. To what service rate should the clerk be trained so as to minimize the training and the service costs?

Solution: The cost function under consideration includes again two components. They are the training costs of the clerk and the cost for waiting customers. Following the results in Table 15.1, the average waiting costs in an $M/M/1$ system are $L_q = \frac{\rho\lambda}{\mu-\lambda} = \frac{\lambda^2}{\mu(\mu-\lambda)}$.

Then the cost function is then

15.2 Optimization in Queuing

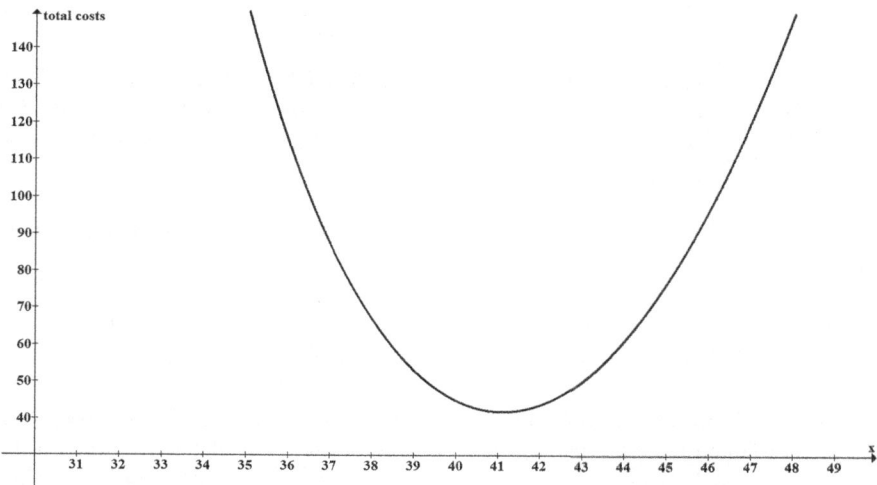

Fig. 15.4 Cost for the clerk training example

Table 15.3 Queuing costs for differently trained clerks

μ	40	41	42	50	60
C	45	41.9113	43.7143	218	810

$$C = \text{(retraining costs)} + \text{(waiting costs)} = 2(\mu - 40)^2 + 20\frac{900}{\mu(\mu - 30)}.$$

The graph in Fig. 15.4 shows the total costs in this example. In particular, the function reaches a minimum at $\mu = 41.12358$ with costs $C = \$41.8742$.

Total costs for a number of other service rates have also been computed. They are shown in Table 15.3.

In other words, leaving the clerk essentially un(re-)trained will cost $45, less than 10% off optimum, whereas training the clerk up to capacity will cost 18 times in total as much as leaving him untrained.

Exercises

Problem 1 (optimization of the number of channels and the service rate): Customers arrive at a retail outlet a rate of 12 per hour. The total time that customers spend in the store contributes to their dissatisfaction. A wasted customer hour has been estimated to cost $20. A clerk at the checkout counter typically earns $8 and can serve up to 10 customers per hour.

(a) What is the cost-minimal number of checkout counters?
(b) Suppose now that an alternative to the system under (a) is to employ two clerks who have been retrained. Their retraining enables them to serve up to 15 customers per hour, and they will earn $10 per hour. Is it worth considering this option?

Solution:

(a) $c = 2, \lambda' = 6$ each, $\rho = 6/10 = 0.6, L_s = 1.5$, so that $TC = 20(2)(1.5) + 2(8) = 76$
$c = 3, \lambda' = 4$ each, $\rho = 4/10 = 0.4, L_s = 2/3$, so that $TC = 20(3)(2/3) + 3(8) = 64$
$c = 4, \lambda' = 5$ each, $\rho = 3/10 = 0.3, L_s = 0.4286$, so that $TC = 20(4)(0.4286) + 4(8) = 66.29$

This implies that the optimal solution is to have 3 clerks. At optimum, the system will cost $64 per hour.

(b) With $\mu = 15$, we obtain $\rho = 0.4$ and $L_s = 2/3$. Then $TC = 20(2)(2/3) + 2(10) = \$46.67$, which is cheaper than the 3-clerk option in (a).

Problem 2 (optimization of the number of channels and sensitivity analysis): Joe plans to open his own gas station "Joe's Place." He has planned to open from 7 a.m. to 11 p.m. He estimates that 15 customers will arrive each hour during the day to fill up their tanks. Doing so takes typically 4 min plus 1 min for paying the bill. Joe now has to decide how many pumps to install. He has read in the industry magazine "Full of Gas" that each hour that a customer waits in line costs $15 in terms of loss of goodwill (i.e., patronizing a different gas station in the future, buying smokes and other emergency items elsewhere, etc.). Also, he has determined that installing a pump costs $100 per day.

(a) Determine the optimal number of pumps Joe should install.
(b) Joe has also heard that there may be a possible gasoline shortage—or at least the perception of one—in the near future. Joe read that in the past, this meant that customers do not really change their driving habits, but fill up their tanks twice as often. Would that change his plans?

Solution:

(a) Arrival rate per hour $\lambda = 15$, service time $1/\mu = 4 + 1 = 5$ min, or $\mu = 12$ customers per hour. Thus, we need at least $c = 2$ pumps for a steady state to exist. The daily (16 hrs) costs are then:

$c = 2: \lambda' = 7.5$ each, $\rho = 0.625, L_q = 1.0417$, so that $TC(c = 2) = 2(100) + 16(2)15(1.0417) = 700$,
$c = 3: \lambda' = 5$ each, $\rho = 0.4167, L_q = 0.2976$, so that $TC(c = 3) = 3(100) + 16(3)15(0.2976) = 514$,
$c = 4: \lambda' = 3.75$ each, $\rho = 0.3125, L_q = 0.1420$, so that $TC(c = 4) = 4(100) + 16(4)15(0.1420) = 536$,
so that it is optimal to install $c = 3$ pumps.

(b) Arrival rate per hour $\lambda = 30$, service time $1/\mu = 2 + 1 = 3$ min (as the fill up time is now only 2 min, since the customers fill up when the tank is half full), or $\mu = 20$ customers per hour. Again, at least $c = 2$ pumps are needed.

15.2 Optimization in Queuing

$c = 2$: $\lambda' = 15$ each, $\rho = 0.75$, $L_q = 2.25$,
so that $TC(c = 2) = 2(100) + 16(2)15(2.25) = 1280$,
$c = 3$: $\lambda' = 10$ each, $\rho = 0.5$, $L_q = 0.5$,
so that $TC(c = 3) = 3(100) + 16(3)15(0.5) = 660$,
$c = 4$: $\lambda' = 7.5$ each, $\rho = 0.375$, $L_q = 0.225$,
so that $TC(c = 4) = 4(100) + 16(4)15(0.225) = 562$,
$c = 5$: $\lambda' = 6$ each, $\rho = 0.3$, $L_q = 0.1286$,
so that $TC(c = 5) = 5(100) + 16(5)15(0.1286) = 592.59$.

Under these circumstances, it would be best for Joe to have $c = 4$ pumps. This represents a 9.34% cost increase over the case without the perception of a shortage.

Problem 3 (comparing queuing systems with fast and slow service): Customers arrive at a retail outlet at a rate of 30 customers per hour. The total time that customers spend in the store contributes to their dissatisfaction. A wasted customer hour has been estimated to cost $10. Management now has two options: either employ one fully trained fast clerk who is able to serve up to 50 customers per hour, or two less trained slower clerks, who can handle up to 30 customers per hour each. Each of the two clerks would have his own waiting line (the supermarket system). Each of the slow clerks earns $6 per hour, while the fast clerk is fully aware of his capability and asks for $16 per hour.

(a) Should we hire the two slower clerks or the one fast clerk?
(b) A new applicant for the job offers his services. The company tried him out and it turned out that he is able to handle no less than 75 customers per hour. Based on the result under (a), what is the maximal amount that we would pay him?

Solution:

(a) The arrival rate is $\lambda = 30$. The fast clerk offers $\mu = 50$, so that $\rho = 30/50 = 0.6$ and $L_s = \lambda/(\mu - \lambda) = 30/20 = 1.5$. The hourly costs are then (cost for clerk) + (costs for customers) = $16 + 1.5(10) = \$31$.

In case of the two clerks, there are two $M/M/1$ systems, each with an individual arrival rate of $\lambda' = 15$. With a service rate of $\mu = 30$ each, we obtain $\rho = 15/30 = 0.5$ each, so that $L_s = 15/(30 - 15) = 1$ each. The hourly costs are then (costs for two clerks) + 2(costs for customers in each system) = $2(6) + 2(1)(10) = 32$. As a result, we should hire the fast clerk, even though he charges more than the two slow clerks together and can handle fewer customers than the two slower clerks combined.

(b) Given a service rate of $\mu = 75$, we obtain $\rho = 0.4$ and $L_s = 2/3$. With an unknown wage w, this results in costs of $w + 2/3(10) = 6\tfrac{2}{3} + w$. This amount should not exceed the costs of the best-known solution (a single fast clerk with hourly costs of $31), so that the bound on the superfast clerk's wage is $6\tfrac{2}{3} + w \leq 31$ or $w \leq \$24.33$.

Simulation

Simulation is one of the major devices in an operations researcher's toolkit, and there is little doubt that it is among the most flexible and commonly used techniques. In the words of Budnick et al. (1988),

> Simulation is primarily concerned with experimentally predicting the behavior of a real system for the purpose of designing the system or modifying behavior.

In other words, simulation is a tool that builds a model of a real operation that is to be investigated, and then feeds the system with externally generated data. We generally distinguish between *deterministic* and *stochastic simulation*. The difference is that the data that are fed into the system are either deterministic or stochastic. As an example, a deterministic simulation may attempt to evaluate the effectiveness of different advertising campaigns. Starting with present sales, assume that we know the effects of different strategies as well as their costs. We can then simulate the outcomes of different strategies, with different assumptions, etc., all assuming that we know with certainty what happens. This chapter deals only with stochastic simulation, which, in the above example, includes uncertain events, e.g., how customers react to different types of advertising and how repeated ads affect purchasing behavior. Stochastic simulation is sometimes also referred to as *Monte Carlo simulation* in reference to the Monte Carlo Casinos and the (hopefully) random outcome of their games of chance.

Another distinction is between *continuous* and *discrete-event simulation*. Continuous simulation deals with processes that are continuous and that are modeled as continuous. Typical examples include the growth of plants, movement of vehicles, and temperatures. In contrast, discrete-event simulation (the only kind of simulation discussed in this chapter) has a finite number of points of time, during which events occur. This could be the demand for a product during a specific day, the number of times a website is visited, or the number of incidents of a specific disease at a regional hospital. Discrete-event simulation can, however, also discretize events, i.e., consider the continuous growth of plants, which are observed, say, every 2 hrs.

16.1 Introduction to Simulation

The main reason for a researcher to resort to simulation is twofold. First of all, simulation is probably the most flexible tool imaginable. Take queuing as an example. While it is very difficult to incorporate reneging, jumping queues, and other types of customer behavior in the usual analytical models (see, e.g., Chap. 15 of this volume), this presents no problem for simulation. Similarly, recall that the queuing formulas that have been derived refer to steady-state solutions. A system may have to run for a very long time to reach a steady state, assuming that one exists. As a result, a modeler may be more interested in transient states, which are easily available in a simulation.

The second reason is that simulation is very cheap. Building a model that simulates the opening of a new restaurant will most certainly be a lot less expensive than trying it out. Even if costs are no subject, the time frame can be compressed in a simulation. For instance, if we were to observe the demand structure of a product, a long time would be required, so that results would probably be available when the product has become technologically obsolete anyway.

The main steps of a discrete-event simulation include

1. Building the model
2. Assigning numbers to uncertain events according to their likelihoods
3. Generating uncertain events
4. Applying predetermined policies
5. Evaluating the results including verifying the model.

The generation of random numbers will be explained in some detail in the next section. The generation of uncertain events and the application of policies to them use an accounting procedure that is demonstrated on a queuing example and an inventory system in Sect. 16.3.

Before starting to discuss the generation of random numbers, we would like to discuss their assignment to random events. This is best explained by way of an example. Suppose that the owner of a store has observed the demand for a specific item and has determined that there is a 10% chance that the demand is 20, a 30% chance that the demand is 35, a 50% chance that the demand is 50, and a 10% chance that the demand is 60. The task is now to assign random numbers to these random events, so that the likelihood of choosing a random number that is assigned to an event equals the observed probability of the event. In our example, we could use single-digit random numbers. If we generate uniformly distributed random numbers, then all digits are equally likely to come up; i.e., the probability of each digit's appearance is 0.1. We could then assign the digit 3 to the demand of 20, the digits 0, 5, and 8 to a demand of 35, the digits 1, 2, 6, 7, and 9 to a demand of 50, and the digit 2 to the demand of 60. Since each digit has a 10% chance of appearing, randomly generated events will have the different demands come up with the observed probabilities. Alternatively, we could make the assignments of double-digit random numbers (in a somewhat more orderly fashion) as shown in

Table 16.1 Assignment of random numbers to demands

Demand	20	35	50	60
Probability	.1	.3	.5	.1
Numbers assigned to event	01–10	11–40	41–90	91–00

Table 16.1. Again, the numbers assigned to the discrete events reflect the observed probabilities.

For the assignment shown in Table 16.1, if the random numbers 15, 27, and 81 are drawn, they refer to demands of 35, 35, and 50, respectively. It is apparent that assigning different random numbers to random events—even while preserving their probabilities—will result in different demands being generated. In order to overcome the effects that are due to the specific assignment, the process should be repeated very often. For example, if the decision maker is interested in the demands for a product during a 12-month period, we would not generate demands for 1 year, but for thousands of years, so that differences due to different random number assignments will vanish.

Given the size (more so than the complexity) of the task, it is little surprise that all simulations are computer-based. While it is possible to write simulations in any all-purpose programming language such as Python, C/C++, Java, or R, special simulation languages have been around since the 1960s. Among them are commercial simulation languages such as AnyLogic, Arena, or FlexSim, and open-source programs such as Simula, SimPy, and others.

Whenever a simulation has been performed, the validation of the results is mandatory. Often this means checking the computer code and performing statistical tests. However, the validation of some of the behavioral assumptions and the structure of the model are at least as important. For instance, if we assume that customers make a special trip to a gas station, this assumption has to be validated. While this is a task that is typically performed before the simulation takes place, it is sometimes necessary to validate an assumption after the fact. As an example, consider a fast-food chain that attempts to locate a new branch. In addition to behavioral studies before the simulation, it may be very useful to apply the model with all of its assumption to an already existing branch and see whether or not it recreates a known situation. If it does not, the discrepancies will allow the modeler to pinpoint the aspects of the model, in which erroneous assumptions may have been made.

16.2 Random Numbers and Their Generation

Random numbers have been around for a long time. Among the earlier systematic efforts to generate random numbers is the work by statistician L.H.C. Tippet, who produced the first random number tables that were based on census numbers. In the mid-1950s, the Rand Corporation published a tome *A Million Random Numbers*. Today, we distinguish between *true random numbers* and *pseudo-random numbers*.

Roughly speaking, true random numbers are generated by way of a random process, while pseudo-random numbers are machine generated by means of a deterministic process. An obvious way to generate true random numbers is to roll dice. Assuming that we have a usual six-sided die which is not loaded or skewed, each side has a chance of 1/6 of coming up; i.e., the probability of each number is $16\frac{2}{3}\%$. It is not difficult to devise differently shaped dice that have ten sides, once for each possible digit. Note that for the time being we only deal with uniformly distributed random numbers, i.e., those in which all possible numbers have the same chance of appearing. If two-digit random numbers are sought, use multiple dice, roll them all, and add their numbers. Note, however, that care must be taken: taking, for instance, two standard six-sided dice, rolling them and adding up their numbers, will not result in uniformly distributed results. As an example, the outcome of "2" is only possible, if both dice show a "1," which has a probability of 1/36. On the other hand, an outcome of "8" has a probability of 5/36, as it can be realized from 2 and 6, 3 and 5, 4 and 4, 5 and 3, and 6 and 2.

However, changing the numbers on the faces enables us to use the same process to generate random numbers that have an equal probability of appearing. If the first die has the numbers 0, 1, 2, 3, 4 ,and 5 on its side, the second has 0, 6, 12, 18, 24, 30, and 36 on its sides, the third has 0, 36, 72, 108, 144, and 180, and the fourth die has numbers 0, 216, 432, 648, 864, and 1080, then the number that results from adding the face values of the four dice is a uniformly distributed random number between 0 and 1295.

Generating random numbers by way of rolling dice may be fun, but it certainly is not a viable method for industrial applications. This is when we resort to machine-generated sequences of random numbers. The basic idea of all of these generators is the same: given a *seed*, i.e., an initial user-determined value, we put this number into a "black box," which uses our input and its internal parameters to generate another number, which is used twofold, as the first random number, and also as the next input into the black box, which uses it to generate the next random number.

Among the first random number generators is the *Midsquare Method*, which is said to date back to Nobel Prize laureate John von Neumann. The idea is to start with a seed, square it, retain the middle digits as random number and next input, and continue in this fashion. The main reasoning behind choosing the center part of a number and delete its first last parts is this. The last digit(s) of a number is/are not necessarily random. For instance, if the last digit is 0, 1, 5, or 6, the square of the number, regardless of what the other digits of the number are, will have a last digit of 0, 1, 5, or 6 as well. Similarly, if the last digit is an even number, then the square of the number will also have an even last digit. As far as the leading digit is concerned, there is much less of a chance of getting an 8 or a 9 than getting a smaller first digit.

As an example of the Midsquare Method, consider the seed $x_0 = 107364$ and assume that we are interested in five-digit pseudo-random numbers. Squaring this number results in 115 27028 496, so that our first random number is $x_1 = 27028$, the center of the number shown by the appropriate spacing. Using x_1 as input and squaring it results in 73 05127 84, so that $x_2 = 05127$. Squaring x_2 results in

16.2 Random Numbers and Their Generation

262861 29 and $x_3 = 62861$. The process continues with $x_4 = 51505$, $x_5 = 52765$, and so forth.

The Midsquare Method is plagued by a multitude of problems, though. Take, for instance, the seed $x_0 = 41$ and generate a sequence of random numbers by deleting the first and last digit after squaring. This process results in the sequence $x_1 = 68$, $x_2 = 62$, $x_3 = 84$, $x_4 = 05$, $x_5 = 02$, and $x_6 = 00$, at which time the series has degenerated and will never generate anything but zeroes.

A much better choice is one of the so-called *Linear Congruence Methods*. They work with the function

$$x_i = (a + bx_{i-1}) \bmod c,$$

where a, b, and c are integer parameters, while x_i is the ith random number as usual. The "mod" function returns the remainder as the result of the division. As an example, consider a number of examples. In case 17 mod 5, we divide 17 by 5, which equals 3 and a remainder of 2, thus 17 mod 5 $\equiv$ 2. Similarly, consider 31 mod 9. Dividing 31 by 9 equals 3 and a remainder of 4, so 31 mod 9 $\equiv$ 4.

Suppose now that we use the parameters $a = 17$, $b = 3$, $c = 101$, and $x_0 = 53$. We can then compute

$$x_1 = [17 + 3(53)] \bmod 101 = 75,$$
$$x_2 = [17 + 3(75)] \bmod 101 = 40,$$
$$x_3 = [17 + 3(40)] \bmod 101 = 36,$$
$$x_4 = [17 + 3(36)] \bmod 101 = 24,$$
$$x_5 = [17 + 3(24)] \bmod 101 = 89,$$

and so forth. It is apparent that the largest number that can be generated in this example will be $c - 1 = 100$. This means that after at most 100 generated numbers, the sequence generated with our parameters will reach a number that has been generated before. And, since the parameters have not changed, the same sequence will be generated over and over again. The number of different random numbers that can be generated before the sequence repeats itself is called the *cycle length* or the *period* of the generator. Typically, the idea is to choose the parameters, so that the period is as long as possible. However, that is not the only criterion for a good set of random numbers. Consider a generator with $a = b = 1$, $c = 10$, and a seed $x_0 = 0$. The generator determines $x_1 = 1$, $x_2 = 2$, $x_3 = 3$, and so forth, until we obtain $x_9 = 9$, and $x_{10} = x_0 = 0$. Thus, the cycle length equals 10, but the sequence looks anything but random.

Another obvious criterion a sequence of uniformly distributed random numbers has to satisfy is that each digit will come up about 10% of the time. The above sequence 0, 1, 2, ..., 9, 0, ... does exactly that. However, the conditional probability of, say, a 5 coming up directly following a 2 is zero, while the probability of a 3 directly following a 2 is 1. This is an obvious test that this particular sequence fails. And that makes it a pseudo-random number. In contrast, remember the roll of a single die. Suppose that a 2 came up on one roll, and a 5 on the next. We keep on

Table 16.2 Cumulative distribution values for a Poisson distribution with $\lambda = 2.5$

x	0	1	2	3	4	5
$F(x)$	.0821	.2873	.5438	.7576	.8912	.9580
x	6	7	8	9	10	
$F(x)$	.9858	.9958	.9989	.9997	.9999	

rolling, until another 2 comes up. What is then the probability that the next roll will show a 5? With a perfect die, it will be 1/6. With any pseudo-random number, it will be 1, as whenever the same number comes up again, we are in a cycle, which repeats itself.

There is a variety of other tests that random number generators have to pass in order to be reasonable. The parameter c is particularly critical, and it is often chosen as a large prime number. Overall, each sequence of numbers generated in this way has a finite period.

Part of the importance of random numbers is not so much that they are used for simulations, but they are also crucial for Internet security by way of encryptions, and Internet gambling. Given the amount of money involved in these ventures, much is at stake. And the general idea is that random numbers can only truly be generated by a process that involves random elements. Some fairly unusual methods have been included, among others seeds that depend on the number of particles emitted from radioactive elements, atmospheric noise from resistors, and a patented method, a seed based on random movements observed in lava lamps. Another simple technique is to use digits of the value of the irrational number π, which have been computed to billions of digits.

So far, we have discussed random numbers that are uniformly distributed. This is not always desirable. However, fairly simple procedures can be employed to transform uniformly distributed random numbers into random numbers that follow other distributions. The easiest case is to determine random variables that are uniformly distributed on [0, 1[. If numbers with k digits to the right of the decimal point are sought, then each k-digit random number needs to be divided by the largest k-digit number that can be generated, viz., $10^k - 1$. Random numbers that are uniformly distributed on [0, 1[are typically the basis for the generation of random numbers that follow other distributions.

Consider now Poisson-distributed random numbers. The cumulative density function $F(x)$ of a random variable x that follows a (discrete) Poisson distribution with parameter λ can be found in many works with mathematical tables, e.g., Råde and Westergren (2004). Table 16.2 shows the cumulative functional values for the Poisson distribution with parameter $\lambda = 2.5$.

We can now generate Poisson-distributed random numbers by starting with uniformly distributed random numbers. If such a random number falls into an interval $F(x_1)$ and $F(x_2)$, then the Poisson-distributed random number is x_2. As a numerical example, consider the following uniformly distributed random numbers:

0.0537 0.7406 0.5926 0.8807 0.6603 0.7126 0.8016 0.7973 0.9584 0.6570 0.8457

The first random number is between 0 and $F(x=0)$, so that a random number of 0 results. The next random number 0.7406 is in the interval [.5438, .7576], so that the next random number equals the x-value of the upper end of the interval, viz., 3. Similarly, the next uniformly distributed random number .5926 is also located in the interval [.5438, .7576], so that again the x-value of the upper bound $x=3$ results as the next Poisson-distributed random number. Continuing in a similar fashion results in the ten random numbers 0, 3, 3, 4, 3, 3, 4, 4, 6, 3, and 4.

Consider now random numbers that follow an exponential distribution with parameter λ. Denoting again uniformly distributed random numbers in [0, 1[by u_i, we can then compute exponentially distributed random numbers x_i by using the formula $x_i = -\ln u_i/\lambda$. Given again the above 11 uniformly distributed random numbers, we can compute exponentially distributed random numbers with parameter $\lambda = 2.5$ as

1.1697, 1.2012, 0.2093, 0.0508, 0.1660, 0.1355, 0.0885, 0.0906, 0.0170, 0.1680, and 0.0670.

On the other hand, random variables that follow a standard normal distribution can be generated from uniformly distributed numbers in the [0, 1[interval in pairs. Letting u_i and u_{i+1} denote the ith pair of uniformly distributed random variables, we can obtain a pair of related standard normally distributed random variables x_i, x_{i+1} by using the relations

$$x_i = \sqrt{-2\ln u_i}\sin(2\pi u_{i+1}) \text{ and } x_{i+1} = \sqrt{-2\ln u_i}\cos(2\pi u_{i+1}).$$

The sequence of standard normally distributed random numbers derived from the first ten uniformly distributed random numbers is

0.1962, 2.4104, 0.0986, 1.0278, 0.0711, 0.9083, 0.0581, 0.6625, 0.0298, and 0.2908.

Random numbers that follow other distributions can be computed as well. In the next section, we will demonstrate how these random numbers can be used to model practical situations.

16.3 Examples of Simulations

This section presents two numerical examples of simple simulations. While they focus only on very specific aspects, they should be able to convey some of the fundamental ideas used in real-world simulations.

16.3.1 Simulation of a Waiting Line System

The main goal of simulation is to evaluate existing solutions and policies. In other words, we start with a policy (or a solution), and test how this policy or solution will fare in an uncertain environment. And this uncertain environment is recreated by generating scenarios, given the states of nature and their probabilities that have been observed.

This may best be explained by a few examples. First consider a waiting line system. For simplicity, we will work with a single-channel system. Customers arrive at the service station, so that the interarrival times are uniformly distributed on the integers between 4 and 9. The service times are also uniformly distributed, but on the integers between 5 and 7 (i.e., a service time of 5 min has a probability of 1/3, the same as a service time of 6 min and one of 7 min. The purpose is now to evaluate the performance of the system. The criteria used for that purpose can be manifold. On the customers' side, we could use the probability that a customer will have to wait, the average waiting time, and the average number of customers in the system (really a proxy for the congestion of the system). On the server's side, we could be interested in the average idle time during a workday, and, of course, the cost of the system.

First, we will have to assign events to random numbers. For simplicity, we will use single digits for the interarrival times. A random digit of 4 means an interarrival time of 4 min, a random number of 5 means an interarrival time of 5 min, and so forth. In case a random digit 1, 2, 3, or 0 comes up, we will reject the digit and move on to the next random number.

A similar procedure will be used for service times. Here, the random numbers 5, 6, and 7 denote the actual service times in minutes, all other random numbers will be rejected. A set of uniformly distributed random numbers is shown in Table 16.3. Tables of random numbers can be read from left to right and top to bottom, or right to left bottom up, or in any other more or less organized way. Here, we scan them row by row from left to right, starting with the first row.

The scenarios for the first 15 customers are generated in Table 16.4. The first column of the Table 16.4 is the customer number. Column 2 lists the interarrival times that are generated with the random numbers from Table 16.3. The first two digits are 2 and 0, and neither of them are assigned to actual interarrival times (only digits between 4 and 9 are), so that these digits are rejected. The next two digits are 4 and 9. They are assigned to interarrival times of 4 and 9, which are now the interarrival times of the first two customers. The next four digits are 9, 1, 3, and 5. Here, 1 and 3 are unassigned and are rejected, so that only the digits 9 and 5 are usable; they are the interarrival times of the next two customers. This process

Table 16.3 Random digits

2049	9135	6601	5112	5266	6728	2188	3846	3734	4017
7087	2825	8667	8831	1617	7239	9622	1622	0409	5822
6187	0189	5748	0380	8820	3606	7316	4297	2160	8973

16.3 Examples of Simulations

Table 16.4 Simulation of a queuing system

Customer #	Interarrival time (random number)	Arrival time	Service finished/ customer leaves	Customer wait time	Station idle time	Start service at	Service time (random number)	Cumulative wait time	Cumulative idle time
1	4	9:04	9:11	0	4	9:04	7	0	4
2	9	9:13	9:20	0	2	9:13	7	0	6
3	9	9:22	9:27	0	2	9:22	5	0	8
4	5	9:27	9:33	0	0	9:27	6	0	8
5	6	9:33	9:39	0	0	9:33	6	0	8
6	6	9:39	9:46	0	0	9:39	7	0	8
7	5	9:44	9:52	2	0	9:46	6	2	8
8	5	9:49	9:59	3	0	9:52	7	5	8
9	6	9:35	10:06	4	0	9:59	7	9	8
10	6	10:01	10:12	3	0	10:06	6	14	8
11	6	10:07	10:18	3	0	10:12	6	19	8
12	7	10:14	10:23	4	0	10:18	5	23	8
13	8	10:22	10:29	1	0	10:23	6	24	8
14	8	10:30	10:37	0	1	10:30	7	24	9
15	8	10:38	10:43	0	1	10:38	5	24	10

continues, until we have assigned interarrival times for the first 15 customers. So far, we have used random numbers of the first seven of the ten blocks of four digits each in the first row of Table 16.3.

We now use the same process to generate the service times for the first 15 customers. Starting with the second row of random numbers in Table 16.3 (alternatively, we could have simply continued where we left off with the interarrival times and started with the eighth block in the first row), we use only random numbers between 5 and 7 as service times, while rejecting all other digits. In the first block, we keep the first and fourth digit (both 7 sec), while rejecting the second and third digit (0 and 8), as they are unassigned. In the second block, the first three digits are unassigned, leaving on the fourth digit (with a value of 5). The third block has the first digit unassigned, but the remaining three digits (6, 6, and 7) are usable. The fourth block has all digits unassigned, and so forth. The service times for the first 15 customers are displayed in column 8 of Table 16.4.

We are now ready to perform the simulation. We start with customer 1, meaning the first row of Table 16.4. Given that the system starts operating at, say, 9 a.m., and the interarrival time is 4 min, the first customer will arrive at 9:04 (column 3). Since no other customer is presently in the system, service for customer 1 will start immediately (column 7). This means that there was no waiting time (column 5), while there was a 4 minute idle time before this customer's arrival (column 6). Given the service time of 7 min (column 8), service on customer 1 will be finished at 9:11 (column 4). For simplicity, we compute the aggregate waiting times (column 9) and the aggregate idle times (column 10) for each customer, so that they are easily available for our evaluations later on.

The computations for customer 2 are similar. Given the interarrival time of 9 min, this customer will arrive 9 min after customer 1 arrived, meaning at 9:13. Since customer 1 has left the service station at 9:11, there was a 2-minute idle time (column 6), but no waiting time for customer 2 (column 5). As a result, service begins immediately at 9:13 (column 7). Given that the service takes 7 min (column 8), customer 2 is done and leaves at 9:20.

This process continues in the same fashion until customer 6 leaves. Consider now customer 7. When the customer arrives at 9:44, customer 6 is still in the system, so that customer 7 will have to wait. More specifically, since customer is finished at 9:46, customer 7 will wait for 2 min (which is recorded in column 5). Service for customer 7 starts at 9:46 (reported in column 7). Also note that there is no idle time for the system, which is shown in column 6. Customer 7 will leave the system after 6 minute service time (column 8) at 9:52 (column 4). The process continues in a similar fashion for the remaining eight customers.

This is also the point, at which other behavioral aspects could easily be incorporated. Take, for instance, the possibility that a customer reneges, i.e., considers the lineup too long (as a proxy for the expected waiting time) and decides not to wait but go elsewhere instead or return at some other time. While such a behavior would be rather difficult to incorporate in an analytical method, it is easy to do so in a simulation. Here, if customer 7 will only enter the system, if there is no other customer in the system, we would record at this time that one potential customer was not served and continue with the next customer. The number of customers who leave the system unserved is another criterion in the evaluation of the performance of the system.

Suppose this very small simulation is to be evaluated. We first observe that there is some idle time in the beginning, while later on the system is very busy. This is hardly surprising: after all, the average interarrival time is 6.5 min, while the average service time is 6 min, leading to an average traffic intensity of $\rho = .9231$, a clear sign of a very busy system.

We also observe that the total waiting time for all the 15 customers that we have observed (column 9) is 24 min. In other words, there is an average waiting time of $24/15 = 1.6$ min per customer. This information will have to be evaluated by the decision maker, who will have to decide whether or not this time is too long. Note again that the waiting times are not even distributed throughout the day, but get longer later in the day.

Another issue is the idle time of the system. The system has been operating from its opening at 9:00 until the last customer left at 10:43, i.e., for a total of 103 min. The total idle time (row 15 of column 10) is 10 min. This means that the total idle time is $10/103 \approx 10\%$ of the time. Again, this time is clustered during the early stages of the operation.

These, and potentially other, criteria can then be used by the decision maker to evaluate and, if deemed necessary, improve the performance of the system by using appropriate measures.

16.3.2 Simulation of an Inventory System

Another popular area, in which Monte Carlo simulation is applied, deals with inventory management. Again, we first have to observe and quantify all system parameters, and then decide on a policy, meaning assign values to the variables under our jurisdiction. In this example, we distribute a single product, for which there is a daily demand that is uniformly distributed between 0 and 99 units. The lead time is assumed to be 3 days. Note that in more realistic applications, the lead time will also be a random variable. Furthermore, we assume that we have an opening inventory of 200 units. The unit costs are \$100 to place an order and receive the shipment, the daily holding costs are 10¢ per unit that is held overnight, and the shortage costs are 30¢ for each unit that is not available and for which a customer has to wait overnight. Unit holding costs and shortage costs are linear in time, meaning that if a customer has to wait for 5 days to obtain the desired good, the shortage costs will be $5(0.3) = \$1.50$. The present inventory policy does not place a new order, if another order is still outstanding, even if the inventory level is below the reorder point. Also, in case of shortages, we assume that units are backordered and there are no lost sales. Other than that, management has formulated the following.

Policy 1: Place an order of size $Q = 300$, whenever the inventory level observed at the end of a day falls below the reorder point $R = 150$.

Table 16.5 Random numbers

| 89 | 91 | 72 | 26 | 10 | 83 | 90 | 30 | 76 | 40 |

Table 16.6 Simulation for Inventory Policy 1

Day #	Inventory level before opening	Demand	Inventory level after closing	Costs: C_o, C_h, C_s
1	200	89	111 → order	100, 11.10, 0
2	111	91	20	0, 2.00, 0
3	20	72	−52	0, 0, 15.60
4	−52 + 300 = 248	26	222	0, 22.20, 0
5	222	10	212	0, 21.20, 0
6	212	83	129 → order	100, 12.90, 0
7	129	90	39	0, 3.90, 0
8	39	30	9	0, 0.90, 0
9	9 + 300 = 309	76	233	0, 23.30, 0
10	233	40	193	0, 19.30, 0

In order to perform the simulation, we first need random numbers. Table 16.5 provides the needed numbers, and we will assign a double-digit random number to a demand of the same magnitude; i.e., a random number 67 will symbolize a demand of 67.

Table 16.6 shows the simulation for 10 days. Some details of the computations will be elaborated upon below.

Before opening on Day 1, our inventory is 200 units as stated in the assumptions. The demand on this day (the first two random digits from the list) is 89, so that by the end of the day, our inventory level has decreased to 111. Comparing this value to our reorder point, we realize that the present inventory level has fallen below the reorder point of $R = 150$, so that we place an order. Given the (deterministic) lead time of 3 days, the shipment that relates to this order will arrive on Day 4 before we open the store. The costs on that day are $100 for placing an order, $111(0.10) = \$11.10$ in terms of inventory holding costs (the 111 units to be carried over to Day 2 multiplied by the unit holding costs), and zero shortage costs, as no shortages were encountered.

Day 2 is dealt with similarly. Note that even though by the end of the day, only 20 units remain in stock which is much less than the reorder point, a new order will not be placed according to the policy that prohibits placing a new order if another order is still outstanding. During Day 3, we encounter a shortage. Similar to carrying costs, shortages are assessed for the number of units we are short by the end of the day (here: 52 units short at 30¢ each for $15.60).

This brings us to the morning of Day 4, at which time a shipment with 300 new units comes in. This shipment belongs to the order placed on the evening of Day 1. From this shipment, the demand of all customers whose items were on backorder will be satisfied, before the regular demand takes over.

16.3 Examples of Simulations

Table 16.7 Simulation for Inventory Policy 2

Day #	Inventory level before opening	Demand	Inventory level after closing	Costs: C_o, C_h, C_s
1	200	89	111 → order	100, 11.10, 0
2	111	91	20	0, 2.00, 0
3	20	72	−52	0, 0, 15.60
4	−52 + 600 = 548	26	522	0, 52.20, 0
5	522	10	512	0, 51.20, 0
6	512	83	429	0, 42.90, 0
7	429	90	339	0, 33.90, 0
8	339	30	309	0, 30.90, 0
9	309	76	233	0, 23.30, 0
10	233	40	193	0, 19.30, 0

That way, we simulate the process for 10 days. At this point, we can evaluate Policy 1. We note that the ordering costs are $200, the carrying costs are $116.80, and the shortage costs are $15.60, for total inventory costs of $332.40. Other characteristics of the system can also be evaluated, for instance the service level, which may be expressed as the proportion of the demand that can be satisfied immediately rather than having to be backordered. For simplicity, we will concentrate on cost considerations. In the simulation of Policy 1, we note that the ordering costs dominate, while shortage costs are very low. This may lead to a revised policy, in which we place larger orders. In particular, we formulate:

Policy 2: Place an order of size $Q = 600$, whenever the inventory level observed at the end of a day falls below the reorder point $R = 150$.

Using the same random numbers (and thus the same demand throughout the 10 days), the workings of this policy are shown in Table 16.7. Since the calculations are very similar to those in Table 16.6, we just produce the results without further comments.

The individual total costs associated with Policy 2 are $C_o = 100$, $C_h = 266.80$, and $C_s = 15.60$ for total inventory costs of $382.40, a 15% increase over Policy 1. We note that the holding costs are now very high, while shortage costs are still at their previous low level. A (hopefully improved) new policy may be defined as having a reorder point of, say, $R = 100$, and an order quantity of $Q = 450$. We leave further experimentation to the reader.

In general, we would like to emphasize that the decision rule that is to be evaluated with simulation does not have to be a fixed rule that is determined once in the beginning and then left unchanged throughout the process. Instead, the rule can include periodic updates. For instance, a diet planner could optimize the food plan for a senior citizen's home and, once one or more of the parameters change, reoptimize and implement the new plan. Similarly, we could periodically update the reorder point and order quantity in an inventory system.

In our example, we could formulate the following policy that features dynamic readjustments of order quantity and reorder point:

Policy 3: (1) Starting with an order quantity of $Q = 300$, readjust the order quantity whenever an order is placed, so that it is 3 times the average daily demand since the last time an order was placed (i.e., 3 times the average daily demand in the last cycle).
(2) Starting with a reorder point of $R = 150$, update the reorder point whenever an order comes in, so that $R: = R \pm \frac{1}{2}$ (shortage/inventory level just before the new order comes in). In other words, if there is a shortage just before the arrival of the new order, the new reorder point equals the previous value of R plus half the shortage. If, on the other hand, there is still some inventory left, half of that amount is subtracted from the previous reorder point to obtain the new value of R.

This somewhat more elaborate policy requires some additional explanations. Again, we will use the same random numbers and thus the same demand as in the previous policies. Initially, we have a reorder point of $R = 150$ and an order quantity of $Q = 300$ as in Policy 1. As in Policy 1, the demand on Day 1 equals 89, so that our inventory level has fallen to 111 by the end of the day. The order quantity is now recalculated as three times the average daily demand since the last order was placed. Since this is the first order that we place, the average is computed for the time between the beginning of the simulation and the end of Day 1. Since only one daily demand has occurred, the order quantity is computed as $Q = 3(89) = 267$. Again, the shipment that relates to this order will arrive in the morning of Day 4.

The computations are the same as for Policy 1 until the morning of Day 4 when the shipment arrives. After deducting the backordered demand, we recomputed the reorder point. Since we had a shortage of 52 units before the shipment arrived, the new reorder point is $R = 150 + \frac{1}{2}(52) = 176$.

The process continues again until Day 6. In the evening of Day 6, the inventory level has fallen below the new reorder point of $R = 176$, and we place another order, which will arrive in the morning of Day 9. The order quantity is now computed on the basis of the average demand since the last order was placed. Here, the previous order was placed on Day 1, and the demand since then was 91 units on Day 2, 72 units on Day 3, 26 units on Day 4, 10 units on Day 5, and 83 units on Day 6 for an average of $282/5 = 56.4$ units. According to the policy, the order quantity is three times this amount, i.e., 169.2 units, which we round to the nearest integer, so that $Q = 169$.

From here, the inventory system continues without interruption until Day 9, when the shipment arrives that was ordered at the end of Day 6. After satisfying the demand with the backordered items, we still have 145 units in stock. Since there was a shortage of 24 units just before opening on Day 9, the new reorder point is calculated as the previous reorder point of 176 plus half of the latest shortage for $R = 176 + \frac{1}{2}(24) = 188$. Note that even though the opening inventory on Day 9 is below the reorder point, the policy allows orders to be placed only by the end of the day, which is done here at the end of Day 9. Since the last order on Day 6, the daily demand has been 90, 30, and 76 for an average of $196/3 = 65\frac{1}{3}$, so that the new

16.3 Examples of Simulations

Table 16.8 Simulation for Inventory Policy 3

Day #	Inventory level before opening	Demand	Inventory level after closing	Costs: C_o, C_h, C_s
1	200 ($R = 150$)	89	111 → order $Q = 267$	100, 11.10, 0
2	111	91	20	0, 2.00, 0
3	20	72	−52	0, 0,, 15.60
4	−52 + 267 = 215 ($R = 176$)	26	189	0, 18.90, 0
5	189	10	179	0, 17.90, 0
6	179	83	96 → order $Q = 169$	100, 9.60, 0
7	96	90	6	0, 0.60, 0
8	6	30	−24	0, 0, 7.20
9	−24 + 169 = 145 ($R = 188$)	76	69 → order $Q = 196$	100, 6.90, 0
10	69	40	29	0, 2.90, 0

Table 16.9 Conditional survival probabilities and associated random digits

	$P(Y|0)$	$P(Y|1)$	$P(Y|2)$	$P(Y|3)$	$P(Y| \geq 4)$
Probability	0.9	0.7	0.5	0.3	0.2
Random numbers	1–9	1–7	1–5	1–3	1–2

order quantity is computed as $Q = 3\left(65\frac{1}{3}\right) = 196$. The first 10 days of this simulation are shown in Table 16.8.

In this policy, the total ordering costs are $300, the carrying costs are $69.90, and the shortage costs are $22.80 for a grand total of 392.70. Since these costs are about 18% higher than those for Policy 1, further refinements are needed. However, it is worth pointing out that the time frame of 10 days is far too short to make any real recommendations. It was chosen here merely for illustrative purposes.

The above two examples were chosen for this book, as they deal with subject matter that was introduced in earlier chapters, and because they are very intuitive. Even these simple scenarios could be extended in a variety of directions so as to become quite involved. Still the basic ideas remain the same regardless of the complexity of the model.

Exercises

Problem 1 (simulation of a replacement problem): The lighting director of a theater is worried about the maintenance and replacement of five floodlights. They fail according to the number of weeks they have been installed and used. The probability that a bulb still functions after it has been used for t weeks (i.e., it is presently in its $(t + 1)$-st week of use) is denoted by $P(Y|t)$. The numerical values are shown in Table 16.9. In addition, the single-digit random numbers associated with the survival events are shown in the last row of the table.

Whenever a bulb fails, it has to be replaced immediately, as "the show must go on." Changing a bulb individually is expensive, as scaffolding must be put up. The entire process costs $350. On the other hand, changing all five bulbs once regardless of whether they still work or not costs $800.

The director ponders two replacement policies: either replace the bulbs only when they actually fail, or, alternatively, in addition to failures during the week which have to be attended to immediately, change all bulbs every 3 weeks regardless of whether they still work or not. It should be pointed out that even if multiple bulbs fail during the same week, they may do so at different times, so that this case has to be treated and paid for as individual failures.

Solution: Use the following random numbers to generate specific instances of survival and failure:

83638 51597 70322 35984 03933 30948 36142 72865 63348 28024

Tables 16.10 and 16.11 then display for each light its age in a given week, the random number, and an indication, if a bulb works during any given week (W), or if it fails (F). Consider, for instance, Light 4. In week 1, its age is 0 as the bulb is new. According to Table 16.9, the random digits associated with not failing are 1–9, and since the random digit is 3, it will symbolize proper functioning. This means that in the beginning of week 2, Light 4 is of age 2. The next random digit is 9, which, for a bulb in week 2, means failure. This means that during week 2, bulb 4 will be replaced and its age in week 3 will again be 0. The process continues in this fashion for all bulbs. The result of this replacement policy is that 16 bulb replacements are necessary for a total cost of $5600.

The simulation for the second policy is shown in Table 16.11. Here, we use the same random digits as before. As an explanation of the numbers in the table, consider Light 1. It works during weeks 1 and 2, but fails during week 3, when it has to be replaced during the week. At the beginning of week 4 (indicated by a "*" in the leftmost column), a group replacement is made, at what time the bulb of Light 1 is replaced again, even though it was just replaced individually during the previous week.

It turns out that this policy requires only 12 individual replacements for a total of $4200, plus three total replacements for $2400 for a grand total of $6600. This is more than the costs of individual replacement alone, making the former strategy preferable.

Problem 2 (evaluation of investment strategies via simulation of stock prices): The manager of an investment company manages a specific fund. There are $1,000,000 available for investment and three stocks, currently priced at $17, $59, and $103 per share, respectively, are considered for that fund. Planning is made on a weekly basis and any amounts that are not invested will be kept in a short-term money market account that pays 0.01% per week.

The manager considers three investment strategies. The first strategy would be to keep all of the money in the short-term account. This is the benchmark strategy. The second strategy will invest 50% of the available money at the end of any week that

16.3 Examples of Simulations

Table 16.10 Simulation given only individual replacements

Week	Light 1 Age	#	W/F	Light 2 Age	#	W/F	Light 3 Age	#	W/F	Light 4 Age	#	W/F	Light 5 Age	#	W/F
1	0	8	W	0	3	W	0	6	W	0	3	W	0	8	W
2	1	5	W	1	1	W	1	5	W	1	9	F	1	7	W
3	2	7	F	2	0	F	2	3	W	0	2	W	2	2	W
4	0	3	W	0	5	W	3	9	F	1	8	F	3	4	F
5	1	0	F	1	3	W	0	9	W	0	3	W	0	3	W
6	0	3	W	2	0	F	1	9	F	1	4	W	1	8	F
7	1	3	W	0	6	W	0	1	W	2	4	W	0	2	W
8	2	7	F	1	2	W	1	8	F	3	6	F	1	5	W
9	0	6	W	2	3	W	0	3	W	0	4	W	2	8	F
10	1	2	W	3	8	F	1	0	F	1	2	W	0	4	W
11	2			0			0			2			1		

Table 16.11 Simulation given individual replacements in addition to triweekly group replacements

Week	Light 1			Light 2			Light 3			Light 4			Light 5		
	Age	#	W/F	Age	#	W/F	Age	#	W/F	Age	#	W/F	Age	#	W/F
1	0	8	W	0	3	W	0	6	W	0	3	W	0	8	W
2	1	5	W	1	1	W	1	5	W	1	9	F	1	7	W
3	2	7	F	2	0	F	2	3	W	0	2	W	2	2	W
4*	0	3	W	0	5	W	0	9	W	0	8	W	0	4	W
5	1	0	F	1	3	W	1	9	F	1	3	W	1	3	W
6	0	3	W	2	0	F	0	9	W	2	4	W	2	8	F
7*	0	3	W	0	6	W	0	1	W	0	4	W	0	2	W
8	1	7	W	1	2	W	1	8	F	2	6	F	1	5	W
9	2	6	F	2	3	W	0	3	W	1	4	W	2	8	F
10*	0	2	W	0	8	W	0	0	F	0	2	W	0	4	W
11	1			1			0			1			1		

16.3 Examples of Simulations

Table 16.12 Probabilities $P(\Delta)$ and random numbers #

Δ	+2%	+1%	±0	−1%	−2%
$P(\Delta)$	0.05	0.15	0.60	0.15	0.05
#	01–05	06–20	21–80	81–95	96–00

Table 16.13 Probabilities $P(\Delta_1)$ and random numbers #

Δ_1	+3%	+1%	±0	−1%	−3%
$P(\Delta)$	0.03	0.15	0.70	0.10	0.02
#	01–02	03–12	13–82	83–97	98–00

Table 16.14 Probabilities $P(\Delta_2)$ and $P(\Delta_3)$ and random numbers #

Δ_2, Δ_3	+3%	+1%	±0	−1%	−4%
$P(\Delta_2), P\Delta_3)$	0.04	0.12	0.60	0.21	0.03
#	01–04	05 – 16	17–76	77–97	98–00

Table 16.15 Random numbers

6959915528	3270108890	4539882224	7576293709
1302727211	7576371930	9295108634	6867423785

has seen at least a 2% increase in value, and will sell at the end of any week that has seen a decline in stock price, provided that a gain can be realized. Otherwise, the stock is held until a gain can be made. The third strategy is to invest 50% of the available money in a stock whenever its price declines, and it will be sold as soon as a gain can be realized. The manager will not purchase new shares of a stock that is still held.

Stock prices are thought to follow two overlapping trends. On the one hand, their value will be determined by the "state of the economy" (measured by the relative value of the currency, unemployment figures, manufacturers' receipts, and similar factors) and, on the other hand, by stock-specific factors. The change of the state of the economy is denoted by Δ, while the changes of the standings of the three industries are Δ_1, Δ_2, and Δ_3, respectively. The stock prices are then thought to be the sum of Δ and Δ_j for stock j, $j = 1, 2, 3$. The probabilities of these changes are denoted by $P(\Delta)$, $P(\Delta_1)$, $P(\Delta_2)$, and $P(\Delta_3)$. These probabilities have been observed and are displayed in Table 16.12 for changes in the overall economy, Table 16.13 for the specific case of the first industry, and Table 16.14 for the second and third industry. In each table, double-digit random numbers # are listed that are associated with the individual changes Δ and Δ_j. In addition, we will use the random numbers shown in Table 16.15. They are read row by row from left to right. The task is to evaluate the different investment strategies of the investment manager.

Solution: We first use the random numbers to simulate the states of the economy Δ, followed by the simulation of the states of the three different industries Δ_1, Δ_2, and Δ_3. This allows us to compute the stock prices. All of these computations are shown in Table 16.16.

Strategy 1: Leaving the entire amount of $1,000,000 in the short-term account for 10 weeks will net us $1,001,000.45, or a 0.1% gain.

Strategy 2: In week 1, none of the stocks have increased by at least 2%, so that we keep our entire amount in the short-term account. By the end of week 2, we have $1,000,200.01. Since Stock 2 increases by 3% in week 2, we purchase it with half of the available money, i.e., $500,100. At the price of $61.38 a share, we obtain 8147.6051 shares. The first time we can realize a gain is at the end of week 8, at which point we sell the shares at $61.99 a share for a total of $505,070.04. This money is kept in the short-term account for 2 weeks, resulting in a payoff at the end of week 10 in $505,171.06. The remaining $500,100 that were not invested in week 2 will remain in the short-term account, resulting in $504,114.83 for a total of $1,009,285.89 or an increase of 0.929%.

Strategy 3: By the end of week 1, Stock 3 has declined in value, which leads the investor to invest half of the available money in that stock. The $1,000,000 has appreciated due to its investment in the short-term account for 1 week, so that 1,000,100 are available, half of which ($500,050) are invested in Stock 3. Each share costs $101.97, so that 4903.8933 shares are purchased. Since the shares will never exceed that value again during the 10 weeks, they will not be sold.

The remaining $500,050 are left in the short-term account for 2 weeks until the end of Week 3, when they have appreciated to $500,150.02. As Stocks 1 and 2 decreased in value in Week 3, half of the available amount is invested in each. (Note, by the way, that Stock 3 decreased in Week 2, but since we still hold shares of that stock, we do not invest in it again.) The sum of $250,075.01 is invested in Stock 1, which costs $16.66 per share, so that we obtain 15,010.5048 shares. We hold them until the end of Week 8, when their price increases to $16.83, which gives us $252,626.80.

Back to the end of Week 3, we invested $250,075.01 in Stock 2 at $60.77 a share, so that we obtain 4115.1063 shares. We sell these shares at the end of Week 5 for $61.38 each, resulting in $252,585.22. We hold this money in the short-term account until the end of Week 8, when the investment in Stock 1 is liquidated. By that time, we have $505,212.02. Since none of the stocks declined in Week 8, we hold the amount in the short-term account for a week, resulting in $505,262.54. We are now at the end of Week 9. During Week 9, we observed all stocks declining. As we still hold Stock 3, we cannot invest in it, so that we invest the entire remaining money in Stocks 1 and 2 in equal parts. For the 252,631.27 invested in Stock 1, we obtain 15,163.9418 shares, while for the same amount invested in Stock 2, we obtain 4116.5271 shares of Stock 2.

Since none of the stock prices increase during Week 10, the account by the end of the planning period consists of 15,163.9418 shares of Stock 1, 4116.5271 shares of Stock 2, and 4903.8933 shares of Stock 3. The total value of the portfolio is thus $982,989.70, for a loss of 1.7%.

Comparing the three strategies, it appears that the second investment strategy is best.

16.3 Examples of Simulations

Table 16.16 Simulation of stock prices

Week	State of economy		Stock 1				Stock 2					Stock 3			
	#	Δ	#	Δ_1	Total	Price	#	Δ_2	Total	Price		#	Δ_3	Total	Price
0						17.00				59.00					103.00
1	69	±0	45	±0	±0	17.00	13	+1%	±1%	59.59		92	−1%	−1%	101.97
2	39	±0	39	±0	±0	17.00	02	+3%	±3%	61.38		95	−1%	−1%	100.95
3	91	−1%	88	−1%	−2%	16.66	72	±0	−1%	60.77		10	+1%	±0	100.95
4	55	±0	22	±0	±0	16.66	72	±0	±0	60.77		86	−1%	−1%	99.94
5	28	±0	24	±0	±0	16.66	11	+1%	±1%	61.38		34	±0	±0	99.94
6	32	±0	75	±0	±0	16.66	75	±0	±0	61.38		68	±0	±0	99.94
7	70	±0	76	±0	±0	16.66	76	±0	±0	61.38		67	±0	±0	99.94
8	10	+1%	29	±0	+1%	16.83	37	±0	+1%	61.99		42	±0	+1%	100.94
9	88	−1%	37	±0	−1%	16.66	19	±0	−1%	61.37		37	±0	−1%	99.93
10	90	−1%	09	+1%	±0	16.66	30	±0	−1%	60.76		85	−1%	−2%	97.93

References

Budnick FS, McLeavey D, Mojena R (1988) Principles of operations research for management, 2nd edn. Richard D. Irwin, Homewood, IL

Råde L, Westergren B (2004) Beta: mathematics handbook, 5th edn. Springer, Berlin

Heuristic Methods

17

In this book, you will read about or even directly encounter a number of solution algorithms. All of these algorithms fall into two broad categories: *exact algorithms* (sometimes also somewhat misleadingly referred to as *optimal algorithms*) and *heuristic methods* usually simply called *heuristics*. Exact algorithms have the obvious advantage of providing the best possible solution there is, given the user-defined constraints, whereas heuristics do not. Some heuristics do have error bounds, some actually proven, while others are empirical, i.e., they state that a certain heuristic usually (typically on average) finds solutions that have a certain quality. On the other hand, there is computing speed. Some models are such that it takes an exact algorithm exceedingly long to find the optimal solution. Is this relevant? Well, it depends. If the task at hand is to, say, locate a landfill for millions of dollars, you will not care if it takes a laptop 2 or 3 weeks to run, so that it can find a solution that may potentially save hundreds of thousands of dollars. There are limits to this argument, of course: if it takes years or even longer to find a solution, most problems have either solved themselves or have become irrelevant by that time. So, this is not acceptable.

In order to make the case for heuristics, consider the situation of automated guided vehicles (*AGV*s). Suppose you have a number of individual workstations on the shop floor, each of which processes a given piece from a semi-finished product to the finished good. For that purpose, it will be necessary to move pieces, on which work has been finished at one station to the next station. Note that it is not necessarily the case that all goods are running through the same sequence of jobs, not all tasks have to be performed at all workstations, and different levels of customization are possible. The movement of goods may be accomplished by automated guided vehicles that receive a message from a workstation whenever a piece is ready for pickup, and the machine knows where the piece has to go next. However, temporarily there may be more pieces to be transported than the machine can handle, so that it will put the tasks in a list. Whenever it is ready, it will work on that list, depending on where it is at any point in time, how far it is to the destination for that particular transportation job, how many workstations will be idle if they have to wait for the next job, and many other considerations. It is apparent that solutions to that problem have to be found in real time, i.e., immediately. This is where heuristic algorithms come in.

Heuristic algorithms typically have two phases. The first phase is the *construction phase*, in which a solution is established. This phase starts with nothing, and by the end of the phase, we have a solution. Phase 1 should be followed by Phase 2, which is an *improvement phase*, in which the method uses simple modifications of the present solution that improve the quality of the solution. The best-known heuristics are the *Greedy Method* (a construction method) and the *Swap* (or *pairwise exchange*) method, which is an improvement method.

In order to illustrate the Greedy technique, consider the following example. Suppose that a hiker is lost in the woods. In order to be visible from the air for a rescue mission, he decides to climb as high as possible. Since he has neither a map, nor an idea of where he really is (and it is very foggy, so that visibility is very limited, making his vision myopic), he can only determine the shape of the land in close vicinity. At any point, he can examine the terrain to his North, Northwest, West, Southwest, South, Southeast, East, and Northeast. If there are higher points in any of these directions, he will go into that direction that features the highest nearby point (i.e., the best possible improvement, hence the name "Greedy"). If the points in all eight directions are lower than where he is right now, he will conclude that he is standing on top of a hill and stay there, assuming that he has arrived.

As an example, consider the situation shown in Fig. 17.1. In order to simplify matters, we have determined the altitudes of the points shown as big black dots and have them displayed again in Fig. 17.2.

Suppose now that the hiker is presently at the Northeasterly point in Fig. 17.2 at an altitude of 2160 ft. There are only three surrounding points, those to the West, the Southwest, and the South, with altitudes of 2210, 2230, and 2050, respectively. The highest of these is the point to the Southwest, which is then where the hiker walks to. This point now has eight neighbors, the highest of which is located directly to the West of the hiker's present position at an altitude of 2340. Consequently, the hiker relocates to this point. Repeating the procedure, the hiker notices that all points in the vicinity of his present location are lower. So, he concludes that he is standing on top of a hill. From the topographic map, we know that he is, but we also know that this is not the highest hill in the area of interest.

Formally, a point all of whose neighbors are lower (higher) is called a *local maximum* (*local minimum*). If a local maximum (local minimum) is also the highest (lowest) point overall, it is referred to as a *global maximum* (*global minimum*). As an illustration, consider the function $y = \sin x + 0.05x^2$. For values of x between -10 and $+10$, the function is shown in Fig. 17.3.

The function has four minima on the domain shown, they are at $x = -7.0689$ (with $y = 1.7911$), at $x = -1.4755$ (with $y = -0.8879$), at $x = 4.2711$ (with $y = 0.0079$), and at $x = 9.6788$ (with $y = 4.4327$). (Myopic) heuristics (as well as derivatives) will readily find local optima, but not necessarily global optima. The reason for this is apparent: if any myopic method arrives at, say, the rightmost of the three local minima in Fig. 17.3, how is the method to know whether or not there are

17 Heuristic Methods

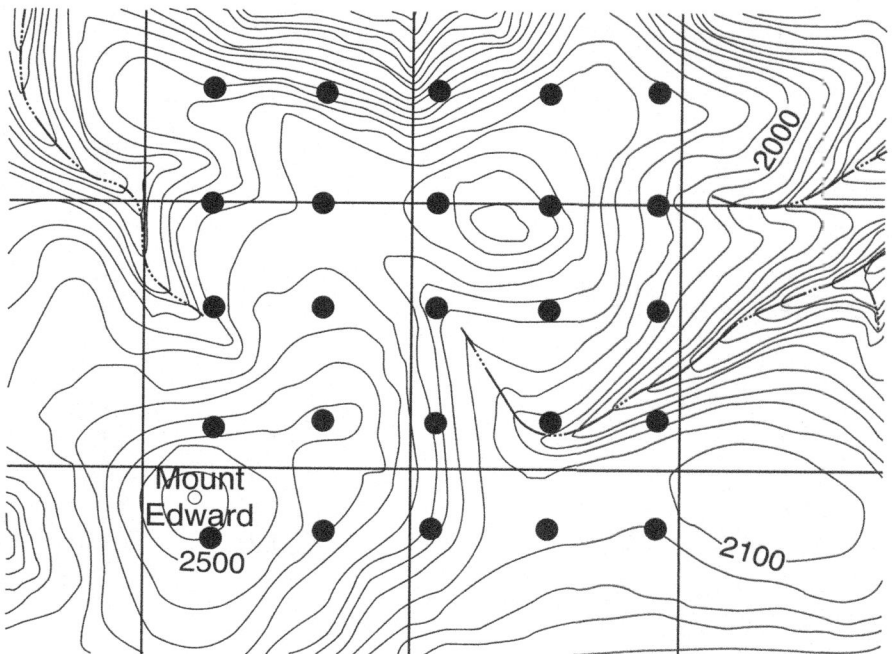

Fig. 17.1 Map for the lost hiker example © Department of Natural Resources Canada. All rights reserved

better points located to the left? If a graph such as this were available, it is easy to see, but typically it is not, as almost all problems are multidimensional, and thus cannot be graphed in three or fewer dimensions.

Consider now again the possible movements of the hiker from all possible starting points. Figure 17.4 shows the best move from each of the grid points, given that, in contrast to our previous discussion, the hiker can move only parallel to the axes. The figure shows that the hiker will end up at one of three local maxima, shown in the figure in red. In particular, starting anywhere in the northeast (above the red line), the hiker will finish at the local maximum with altitude 2340 ft. Starting anywhere in the middle (the red box that includes only four grid points), the hiker will terminate his walk at the local maximum at altitude 2460 ft. Only if the hiker starts his walk in the south or southwest will he end the hike at the global maximum point with altitude 2550 ft. The areas bordered by the red lines are also referred to as the *catchment areas* of the local maxima.

A similar situation applies to the function in Fig. 17.3. Starting a Greedy minimization method at any point to the left of $x = -5.2671$ (a local maximum), we will end up at the local minimum at $x = -7.0689$. Starting at any point between $x = -5.2671$ and $x = 1.7463$ will end at the global minimum at $x = -1.4276$, and starting anywhere to the right of $x = 1.7463$ will lead to the local minimum at $x = 4.2711$.

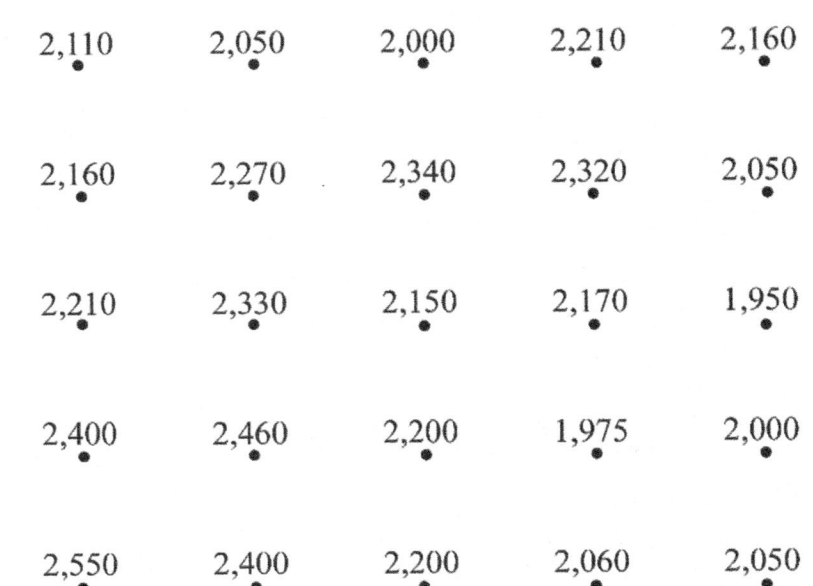

Fig. 17.2 Altitudes in feet at grid points in our example

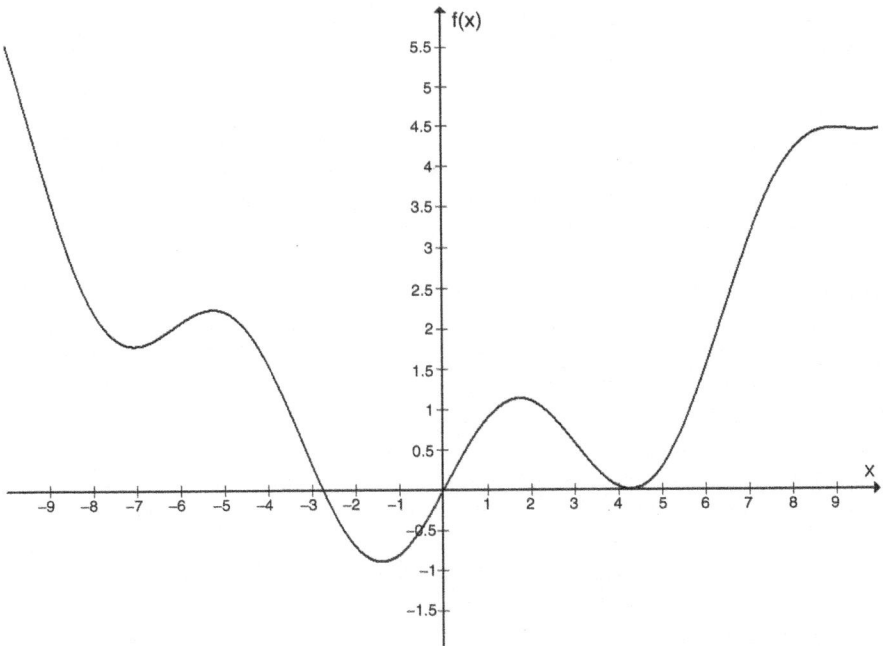

Fig. 17.3 Function with three local minima

17 Heuristic Methods

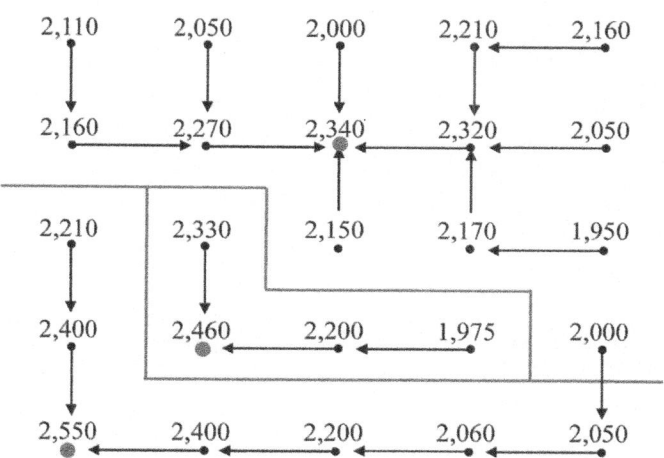

Fig. 17.4 Movements of hiker

This immediately suggests a technique that is called a *multistart method*. The idea is simply to apply a Greedy technique starting at a number of different points, compare the results, and choose the best (highest or lowest, depending on whether a minimum or a maximum is sought).

The hiker's plight described above might now be considered contrived and of little general interest, if it were not for the fact that each point on the map may represent a course of action (determined by the values of its coordinates), and the contour lines may represent their profit. So, the search for the highest hill has now become a search for the point of maximal profit, which is of considerable general interest.

Next, we will discuss an improvement method. The Swap method is easy to describe. Given a solution that has been obtained "somehow," it takes two components and exchanges them. Depending on the specific application, this may mean exchange their sequence, their inclusion/exclusion status, or whatever the problem commands. The method then computes the change of the value of the objective function. If the value has improved (i.e., has increased in case of a maximization problem or decreased in case of a minimization problem), the modified solution becomes the new starting point, and the previous solution is discarded. This step is repeated until no further Swap step can improve the solution any further.

As an example, consider the map in Fig. 17.5 and assume that the task at hand is to determine the shortest route from Memphis, Tennessee, to Reno, Nevada The traveler has outlined the tour in layers, so that a drive from a city in one layer to a city in the next is about a day's drive.

The distances are as follows: From Memphis to Omaha, Wichita, Oklahoma City, and Dallas, we have 650, 555, 460, and 460 miles, respectively. From Omaha to Denver and Albuquerque, there are 540 and 860 miles. The distances from Wichita

Fig. 17.5 Map of the US for the shortest route example

to Denver and Albuquerque are 510 and 585 miles, from Oklahoma City to Denver and Albuquerque are 630 and 550 miles, and from Dallas to Denver and Albuquerque are 780 and 640 miles, respectively. The distances from Denver to Salt Lake City and Phoenix are 505 and 835 miles, while from Albuquerque, there are 610 and 460 miles to Salt Lake City and Phoenix, respectively. Finally, the distance from Salt Lake City to Reno is 525 miles and from Phoenix to Reno there are 735 miles.

In order to obtain some solution, we use the Greedy algorithm, starting at the origin of our trip in Memphis. From here, Greedy will choose the nearest neighbor, which is either Dallas or Oklahoma City, both 460 miles away. We arbitrarily choose Dallas. The nearest neighbor of Dallas is Albuquerque, which is 640 miles away (as opposed to Denver, which is 780 miles from Dallas). From Albuquerque, we take the closest connection to Phoenix (460 miles), and from there we have no choice but take the last long trip to Reno (735 miles). The trip leads us on the route Memphis–Dallas–Albuquerque–Phoenix–Reno, and its total length is 2295 miles.

At this point, we start to swap. One possibility is to swap Phoenix and Salt Lake City. This means that we have to add the connections from Albuquerque to Salt Lake City (610 miles) and from Salt Lake City to Reno (525 miles) and subtract the connections from Albuquerque to Phoenix (460 miles) and from Phoenix to Reno (735 miles), for net savings of 60 miles. This is an improvement, and so our new route is Memphis–Dallas–Albuquerque–Salt Lake City–Reno, and its total length is 2235 miles.

We can now use the new solution and try other Swap moves. For instance, we could attempt to swap Albuquerque and Denver. The net change of such a swap move is +35 miles, so we will not make this change. Another possibility is to swap Dallas and Oklahoma City. The net change is −90 miles, so we make the change and obtain the route Memphis–Oklahoma City–Albuquerque–Salt Lake City–Reno, whose length is 2145 miles.

At this time, we may examine again the pair Albuquerque and Denver. This swap did not improve the solution the last time we tried it, but since then the solution has changed. In fact, swapping the two cities at this point results in a net change of +630 + 505 − 550 − 610 = −25, so that the change is made. This results in the new route Memphis–Oklahoma City–Denver–Salt Lake City–Reno, which is 2120 miles long.

We may now try to exchange Oklahoma City and Wichita, which leads to a net change of +555 + 510 − 460 − 630 = −25, for another reduction in terms of the total distance, which is now 2095 miles. The route leads from Memphis–Oklahoma City–Denver–Salt Lake City–Reno. At this point, we may try to further reduce the length of the tour, which is no longer possible with swap moves. As a matter of fact (unbeknownst to us when we are just using heuristics), the tour is actually optimal.

Our final example of a heuristic method deals with a much-studied field called *bin packing*. We have an unspecified number of bins, all of which are of the same prespecified length. We also have a number of rods that are to be placed into the bins. The problem is one-dimensional, in that the bins and the rods have the same height and width, so that only the length of the bins and the rods that are placed into them will decide whether or not they actually fit. For instance, if the bin is 20 ft and there are one 6 ft rod, one 3 ft rod, and one 5 ft rod, then these three rods will occupy 14 ft of the bin and leave 6 ft unoccupied. The task at hand is now to put the existing rods into the smallest number of bins possible.

Despite its apparent simplicity, the problem has been proven to be very difficult from a computational point of view. A Greedy-like heuristic is the so-called *First Fit (FF) Algorithm*. In order to implement the method, we first assume that a sufficiently large number of bins is available. These bins are numbered 1, 2, ... The First Fit Algorithm can be described as follows:

FF Algorithm: Put the next rod into the bin with the smallest number into which it will fit.

As an example, suppose that all bins are 19 ft long and we have six 11 ft rods, six 6 ft rods, and twelve 4 ft rods. In this type of situation, we can actually compute a very simple bound for the number of bins that will be needed. Here, the total length of the rods is $6(11) + 6(6) + 12(4) = 150$ ft. Given that each bin is 19 ft long, we will need at least $150/19 \cong 7.89$ bins. Since the number of bins must be integer, we will need at least 8 bins. This also means that if we were to find a solution to the problem that requires 8 bins, this solution must be optimal.

Apply now the First Fit Algorithm. Assigning the rods in order of their lengths (i.e., the 11 ft rods first, then the 6 ft rods, and finally the 4 ft rods), we notice that only a single 11 ft rod fits into each bin. This means that we have to put each of the 11 ft rods into one bin each, so that we now have dealt with all 11 ft rods and have used parts of six bins. Next, we assign the six 6 ft rods. Since each of them fits into one of the already partially used bins, we now have six bins with one 11 ft and one 6 ft rod each, leaving 2 ft of free space in each of the six bins. This is not sufficient for any of the remaining 4 ft bins, so that we have to use additional bins. We can place four 4 ft rods in each bin, leaving an empty (and unusable for us) space of 3 ft each, which requires another three bins. We now have assigned all rods to the bins. This solution requires a total of nine bins. There is no apparent pairwise exchange (swap) step able to improve the solution.

On the other hand, if we were to put one 11 ft and two 4 ft rods into each of six bins, this would leave no empty space at all. The remaining six 6 ft rods can be put into two bins. Having again assigned all rods, this solution requires only eight bins and, given the bound computed above, must be optimal.

While the bin packing example is somewhat artificial, the mathematical structure has important applications. Among them are employees (the bins with a capacity of 40 hrs a week) and tasks of different lengths (the rods), so that the bin packing problem is then to minimize the number of employees necessary to perform a given set of tasks.

A variety of other heuristics exists for this problem. An excellent (albeit difficult) pertinent reference is the book by Garey and Johnson (1979). An interesting extension of the problem makes available the different rods over time. This is reminiscent of ready times in machine scheduling. The solution obtained in such a case will be no better than the one found in the case in which all rods are available at the beginning of the process (simply because the problem that makes rods only available over time is more restrictive than the problem discussed here.).

The efficiency of the heuristics and their performance may be evaluated in different ways. An obvious evaluation is to use simulation (see Chap. 13 of this book). This will enable users to specify an average or expected *error bound* of the algorithm. For instance, in the above example, the heuristic method uses 9 instead of the optimal 8 bins, i.e., 12.5% more than optimal.

However, in some cases it is possible to determine theoretical error bounds, i.e., bounds that cannot be violated. For instance, it has been shown that the solution found by the First Fit Heuristic cannot be worse than about 70% higher than the optimal solution. One problem associated with the theoretical bounds is that while they represent a reliable, provable property, they tend to be very high.

Many other heuristics have been presented in the literature. Many improvement algorithms are *neighborhood searches*, whose main distinguishing feature is that they start with a given solution and search for better solutions in the neighborhood of the present solution. The Swap Method described above belongs to this class. Another very successful heuristic in this class is *tabu search*. The idea of this method is to get out of a local optimum by temporarily allowing the current solution to deteriorate. In order to avoid cycling between solutions that are better and those that are worse, a list of prohibited moves (a tabu list) is set up that is updated as the algorithm progresses. This procedure allows to "get over the hump" from the present point to other solutions that are hopefully better than the best solution known at this point. For example, in Fig. 17.3, if the best-known minimum is $x = -7.0689$, we may allow worse solutions (i.e., those with higher functional values) in our move to the left. This may allow us to find the global minimum at $x = -1.4755$.

Other techniques are based on observations made in the technical or the natural world. Examples are *simulated annealing*, a technique modeled after the way molten metal cools. Similar to tabu search, it allows moves to solutions worse than the best presently known solution. Such moves are allowed with a certain probability that decreases during the course of the algorithm. The formulas ensure that the probability to accept a move to a very bad solution is very small. Other methods follow some behavioral patterns of ant colonies or bees.

Exercise
Problem (Bin packing problem with Greedy and SWAP): A firm has ten tasks that need to be completed. Their durations are 10, 15, 7, 18, 22, 16, 19, 13, 23, and 11 hrs, respectively. The jobs cannot be split, i.e., the worker who started the job has to finish it. A large number of potential workers is available, all receive the same wage. Their weekly capacity is 40 hrs.

(a) Determine lower and upper bounds for the number of workers needed to perform all tasks.
(b) Use the *First Fit Algorithm* to allocate the jobs to employees, so that the number of employees needed to perform all tasks is minimized.
(c) Starting with the solution obtained under (a), use the *SWAP* technique to improve the solution.
(d) Repeat (a) and (b), given that the tasks are renumbered, so that tasks with a longer duration are scheduled first.

Solution

(a) Since the total number of hours required to perform all tasks is 154 and each worker has a capacity of 40 hrs, we would require a minimum of 4 workers (sufficient if it were possible to split jobs). This is a lower bound. A simple upper bound follows the consideration that each worker performs exactly one of the tasks, which requires ten workers, which provides an upper bound.
(b) Given the order to the tasks shown above, the first three tasks are allocated to Worker 1 (who is then occupied for $10 + 15 + 7 = 32$ hrs), Tasks 4 and 5 are assigned to Worker 2 (who now works $18 + 22 = 40$ hrs), Tasks 6 and 7 are assigned to Worker 3 (who works $16 + 19 = 35$ hrs), Tasks 8 and 9 are assigned to Worker 4 (who works $13 + 23 = 36$ hrs), and Task 10 is assigned to Worker 5, who works only 11 hrs.
(c) and (d) In the *SWAP* step, we may attempt to reassign jobs, so as to reduce one worker's load to no more than 29 hrs (i.e., 11 hrs of idle time), so that he can take over the light load of Worker 5, who will then no longer be needed. For instance, Worker 1's longest task may be exchanged for the shortest task of one of the

other workers, ensuring that the step results in feasible schedule for both workers. In our demonstration, we will not touch the workload of Worker 2, which is exactly 40 hrs, which cannot be improved upon. First attempt: Swap jobs between Workers 1 and 3. Since the shortest task of Worker 3 is longer than the longest task of Worker 1, no exchange can result in a shorter workload of Worker 1. Second attempt: Workers 1 and 4 swap tasks 2 and 8, so that Worker 1 now works on Tasks 1, 3, and 8 (for 30 hrs), while Worker 4 now performs Tasks 2 and 9 for a total of 38 hrs. A further reduction by simple *SWAP* steps is not possible. Note that an optimal solution allocates Tasks 2, 3, and 4 to Worker 1 (for $15 + 7 + 18 = 40$ hrs), Tasks 1, 7, and 10 to Worker 2 (for $10 + 19 + 11 = 40$ hrs), Tasks 6 and 9 to Worker 3 (for $16 + 23 = 39$ hrs), and Tasks 5 and 8 to Worker 4 (for $22 + 13 = 35$ hrs). If the tasks had been rearranged in the order of Tasks 2, 3, 4, 1, 7, 10, 6, 9, 5, 8, this exact solution would have been obtained. This suggests a multistart procedure with different permutations of the sequence of tasks.

Reference

Garey MR, Johnson DS (1979) Computers and intractability: a guide to the theory of NP-completeness. W.H. Freeman & Company, San Francisco, CA

Appendices

Appendix A: Vectors and Matrices

This appendix is intended to provide the reader with some basic refresher regarding some basic operations that involve matrices and vectors and the solution of systems of simultaneous linear equations. It is not designed to replace a text, but as a mere quick reference for some material used in this book.

Definition A.1 An $[m \times n]$-dimensional matrix

$$\mathbf{A} = (a_{ij}) = \begin{bmatrix} a_{11} & a_{12} & \cdots & a_{1n} \\ a_{21} & a_{22} & \cdots & a_{2n} \\ \vdots & \vdots & \ddots & \vdots \\ a_{m1} & a_{m2} & \cdots & a_{mn} \end{bmatrix}$$

is a two-dimensional array of elements a_{ij} arranged in m horizontal rows and n vertical columns, so that the element a_{ij} is positioned in row i and column j. If $m = n$, the matrix is said to be *square*, if $m = 1$, it is called a *row vector*, if $n = 1$, it is a *column vector*, and if $m = n = 1$, it is a *scalar*.

It is common practice to denote scalars by italicized letters, vectors by lowercase bold letters, and matrices by capitalized letters in boldface.

Definition A.2 Given an $[m \times n]$-dimensional matrix $\mathbf{A}$ and an $[n \times p]$-dimensional matrix $\mathbf{B}$, the product $\mathbf{C} = \mathbf{AB}$ is an $[m \times p]$-dimensional matrix $\mathbf{C} = (c_{ij})$, such that

$$c_{ij} = a_{i1}b_{1j} + a_{i2}b_{2j} + \ldots + a_{in}b_{nj}, \text{ for } i = 1, \ldots, m \text{ and } j = 1, \ldots, n.$$

Example A.1: With $\mathbf{a} = \begin{bmatrix} 3, \sqrt{2}, -6 \end{bmatrix}$ and

$$\mathbf{B} = \begin{bmatrix} 0 & 5 & \pi \\ 7 & -6 & 2 \\ \sqrt{3} & -11 & 1 \end{bmatrix},$$

we obtain $\mathbf{aB} = \begin{bmatrix} 7\sqrt{2} - 6\sqrt{3}, & 81 - 6\sqrt{2}, & 3\pi + 2\sqrt{2} - 6 \end{bmatrix}$, and

$$\mathbf{B}^2 = \mathbf{BB} = \begin{bmatrix} 35 + \pi\sqrt{3} & -11\pi - 30 & 10 + \pi \\ -42 + 2\sqrt{3} & 49 & -10 + 7\pi \\ -77 + \sqrt{3} & 55 + 5\sqrt{3} & -21 + \pi\sqrt{3} \end{bmatrix}.$$

Definition A.3 Given an $[m \times n]$-dimensional matrix $\mathbf{A}$, the *transpose* $\mathbf{A}^T = \left(a_{ij}^T\right)$ is the $[n \times m]$-dimensional matrix with $a_{ij}^T = a_{ji}$ for $i = 1, \ldots, m$ and $j = 1, \ldots, n$.

Example A.2: Use the vector $\mathbf{a}$ and the matrix $\mathbf{B}$ of Example A.1, we then obtain

$$\mathbf{a}^T = \begin{bmatrix} 3 \\ \sqrt{2} \\ -6 \end{bmatrix} \text{ and } \mathbf{B}^T = \begin{bmatrix} 0 & 7 & \sqrt{3} \\ 5 & -6 & -11 \\ \pi & 2 & 1 \end{bmatrix}.$$

Appendix B: Systems of Simultaneous Linear Equations

Definition B.1 A mathematical relation is written as

LHS R RHS,

where *LHS* denotes the left-hand side, *R* is the relation, and *RHS* is the right-hand side of the relation.

Typically (but not necessarily), *LHS* is a function $f(x_1, x_2, \ldots, x_n)$ of n variables $x_1, x_2, \ldots, x_n$, R is a relation of type $<, \leq, = \geq, >$, or $\neq$, and *RHS* is a scalar.

Definition B.2 A relation $f(x_1, x_2, \ldots, x_n) R b$ is said to be *linear*, if the function f can be written as $f(x_1, x_2, \ldots, x_n) = a_1 x_1 + a_2 x_2 + \ldots + a_n x_n$. We will refer to $f(x_1, x_2, \ldots, x_n)$ as *LHS*, while b is the *RHS*.

Example B.1: The relation $2x_1 - 0.7x_2 + \sqrt{11}x_3 \leq 59$ is linear, whereas the relations $2\sqrt{x_1} - 0.7x_2 + \sqrt{11}x_3 \leq 59$, $2x_1 - 0.7x_1 x_2 + \sqrt{11}x_3 \leq 59$, and $2x_1 - 0.7x_2 + \sqrt{11}x_3^3 \leq 59$ are not, due to the appearance of the square root $\sqrt{x_1}$, the product $x_1 x_2$, and the cubic x_3^3, respectively.

We will first deal with the case when the relation R is an equation. Assume now that we have a system of m linear equations with n unknowns, and that we want to find a solution, i.e., an assignment of values to the variables, that satisfies all equations. It is now possible to prove the following

Appendices

Theorem B.1 Consider a system of m simultaneous linear equations in n variables:

$$
\begin{aligned}
a_{11}x_1 + a_{12}x_2 + \ldots + a_{1n}x_n &= b_1 \\
a_{21}x_1 + a_{22}x_2 + \ldots + a_{2n}x_n &= b_2 \\
&\vdots \\
a_{m1}x_1 + a_{m2}x_2 + \ldots + a_{mn}x_n &= b_m
\end{aligned}
$$

The system has either no solution, exactly one solution, or an infinite number of solutions.

Example B.2: Consider the system

$$
\begin{aligned}
2x_1 + 3x_2 &= 7 \\
4x_1 + 6x_2 &= 10.
\end{aligned}
$$

It is known that if we simultaneously multiply right-hand side and left-hand side of an equation, we do not change its content. Multiplying the first equation by 2, we obtain $4x_1 + 6x_2 = 14$. Now the left-hand side of this equation and that of the second equation are equal, but its right-hand side differs, indicating that there is an inherent contradiction in the system. Thus, it is no surprise that the system has no solution.

On the other hand, consider the system

$$
\begin{aligned}
2x_1 + 3x_2 &= 7 \\
4x_1 + 6x_2 &= 14.
\end{aligned}
$$

Multiplying both sides of the first equation by 2 results in the second equation. In other words, the two relations have exactly the same informational content. This means that we really have only a single equation, whereas we need one equation to specify the value of each unknown. Hence this system has an infinite number of solutions x_1 and $x_2 = \frac{1}{3}(7 - 2x_1)$.

Consider now the case that has exactly one solution. Here, we are not concerned with conditions in which a system has exactly one solution, but our focus is on how to actually obtain such a solution, given that it exists. There are many different versions of the *Gaussian elimination technique* (named after the German mathematician Carl Friedrich Gauss, 1777–1855). In order to illustrate the technique, consider the system of simultaneous linear equations

$$2x_1 + 3x_2 - 5x_3 = 1 \qquad \text{(I)}$$

$$x_1 - 2x_2 + 4x_3 = -3 \qquad \text{(II)}$$

$$4x_1 + x_2 + 6x_3 = 2. \qquad \text{(III)}$$

The idea is to eliminate one variable, say x_3, from all equations but one. The system has then one equation in all (here: three) variables, whereas two equations include only the remaining variables (here: x_1 and x_2). Among these remaining variables, we now choose another variable to be eliminated (here: x_2), and the procedure is repeated. In the end, we have a single equation in just one variable, which is then replaced by its value in all of the remaining equations, resulting in another system, in which again one of the equations is just a function of a single variable, which is replaced by its value everywhere, and so forth, until the system is solved.

Applying this idea to our example, we first eliminate the variable x_3 from Eq. (II) by multiplying (I) by 4 and multiplying (II) by 5, adding them, and then replacing Eq. (II) by $4 \times$ (I) $+ 5 \times$ (II). The revised system can then be written as

$$2x_1 + 3x_2 - 5x_3 = 1 \quad (I)$$

$$13x_1 + 2x_2 = -11 \quad (II') = 4 \times (I) + 5 \times (II)$$

$$4x_1 + x_2 + 6x_3 = 2. \quad (III)$$

Next, we eliminate x_3 from Eq. (III) and replace Eq. (III) by $6 \times$ (I) $+ 5 \times$ (III). This results in the system

$$2x_1 + 3x_2 - 5x_3 = 1 \quad (I)$$

$$13x_1 + 2x_2 = -11 \quad (II')$$

$$32x_1 + 23x_2 = 16 \quad (III') = 6 \times (I) + 5 \times (III)$$

Since the Eqs. (II') and (III') now contain only the two variables x_1 and x_2, we can eliminate x_2 from Eq. (III') by replacing Eq. (III') by $23 \times$ (II') $- 2 \times$ (III'). This process results in

$$2x_1 + 3x_2 - 5x_3 = 1 \quad (I)$$

$$13x_1 + 2x_2 = -11 \quad (II')$$

$$235x_1 = -285. \quad (III'') = 23 \times (II') - 2 \times (III')$$

We say that the system is now in triangular form, due to the pattern of coefficients on the left-hand side. This allows us to obtain the values of the variables in a recursive procedure. First we can determine the value of x_1 from Eq. (III''). Clearly, $x_1 = -\frac{285}{235} = -\frac{57}{47}$. Inserting the value of x_1 into Eq. (II') allows us to solve for the variable x_2. In particular, we have $13\left(-\frac{57}{47}\right) + 2x_2 = -11$, which results in $x_2 = \frac{112}{47}$. Finally, inserting the values of x_1 and x_2 into Eq. (I), we can solve for x_3. The relation reads $2\left(-\frac{57}{47}\right) + 3\left(\frac{112}{47}\right) - 5x_3 = 1$, which results in $x_3 = \frac{35}{47}$. The system has now been completely solved. The solution is $[x_1, x_2, x_3] = \left[-\frac{57}{47}, \frac{112}{47}, \frac{35}{47}\right] \cong [-1.2128, 2.3830, 0.7447]$. Inserting the values of the unknowns into the original Eqs. (I), (II), and (III), we can verify that the solution is indeed correct. In analogous fashion, we can determine the solution of any system of linear equations. For further details, see any pertinent introductory text on linear algebra or the short summaries in Eiselt and Sandblom (2007) or (2004).

Appendix C: Probability and Statistics

While most chapters in this book deal with deterministic models, some include probabilistic models, in which concepts of probability are needed. This appendix is intended to briefly cover those probabilistic concepts needed in this book. It is by no means intended to replace a thorough knowledge of statistics.

Definition C.1 The *probability* of an event (or outcome) of a random experiment is a number p between 0 and 1, which measures the likelihood that the event will occur.

A value of p close to 0 indicates that the event is unlikely to happen, while a value of p close to 1 means that the event is very likely. We can interpret the probability as the proportion of times that the event or outcome will occur, if the experiment is repeated a large number of times.

Suppose now that the events are *mutually exclusive* (i.e., the events are distinct and do not overlap) and *collectively exhaustive* (meaning that exactly one of them will occur). We then obtain the following result

Theorem C.1 If $p_1, p_2, \ldots, p_n$ denote the probabilities of the mutually exclusive and collectively exhaustive outcomes of an experiment, then $p_1 + p_2 + \cdots + p_n = 1$.

Example C.1: Consider an experiment that involves tossing a fair coin three times. suppose that each outcome is either head or tail, standing on edge does not occur. Denote H for "head" and T for tail, the outcome "first tail, then head, then tail" is written as *THT*. Then there are eight possible outcomes: *HHH, HHT, HTH, HTT, THH, THT, TTH,* and *TTT*. Given a fair coin, all outcomes are equally likely, and since $p_1 + p_2 + \cdots + p_8 = 1$, we obtain the result that each outcome has a probability of $P(HHH) = P(HHT) = \cdots = P(TTT) = \frac{1}{8}$.

The probability of a *composite event* that consists of several outcomes is the sum of probabilities of the outcome of the event. For instance, the event "obtain exactly one tail in three flips of a fair coin" refers to the event $\{HHT, HTH, THH\}$, so that the probability of such an event is $P(\{HHT, HTH, THH\}) = \frac{1}{8} + \frac{1}{8} + \frac{1}{8} = \frac{3}{8}$. The *event space* is defined as the set of all possible events of an experiment.

Definition C.2 A *random variable* X is a function defined on the event space of an experiment.

Example C.2: Given the above coin tossing experiment, let X denote the number of heads that come up in the three tosses. There are four possibilities: $X = 0$ (which occurs only in the event $\{TTT\}$, so that the probability of this event is $P(X = 0) = \frac{1}{8}$, $X = 1$ (which happens if one of the events $\{HTT, THT, TTH\}$ occurs, so that the probability $P(X = 1) = \frac{3}{8}$), $X = 2$ (an event that occurs if one of $\{HHT, HTH, THH\}$ occurs, so that $P(X = 2) = \frac{3}{8}$), and finally $X = 3$ (which occurs only if $\{HHH\}$ happens, so that $P(X = 3) = \frac{1}{8}$). Again, the probabilities of all possible events add up to 1.

In general, we distinguish between two different types of random variables, *discrete* and *continuous random variables*.

Table C.1 Probability distribution for the coin toss example

X	P	F
0	$\frac{1}{8}$	$\frac{1}{8}$
1	$\frac{3}{8}$	$\frac{4}{8}$
2	$\frac{3}{8}$	$\frac{7}{8}$
3	$\frac{1}{8}$	$\frac{8}{8}$

Definition C.3 A random variable X is called *discrete*, if it can assume only one of (countably many) values $a_1, a_2, \ldots$. The function $P(a_j) = P(X = a_j) = p(a_j)$ is called the *discrete probability distribution* (function) of X. The function $F(a_j) = P(X \leq a_j)$ is called the *cumulative probability distribution* (function) of X.

Definition C.4 A discrete random variable X with a probability distribution function $p(x) = P(X = x) = \binom{n}{x} p^x (1-p)^{n-x} = \frac{n!}{x!(n-x)!} p^x (1-p)^{n-x}, x = 0, 1, \ldots, n$ is said to be *binomially distributed* with parameters n (integer) and p, where $0 \leq p \leq 1$. We write $X \sim Bin(n, p)$.

Example C.3: Let X be the random variable that denotes the number of heads in three tosses of a fair coin. Then $X \sim Bin(0.5, 3)$ and Table C.1 displays the probability function as well as the cumulative distribution functions.

We note that $F(a_j)$ is an increasing function of a_j that eventually reaches the value of 1 for the largest value of a_j (in case of finitely many outcomes, and that converges towards 1 for infinitely many outcomes a_j. Since P and F are probabilities, we must have $0 \leq P(a_j) \leq 1$ and $0 \leq F(a_j) \leq 1$ for all events a_j.

Definition C.5 A discrete random variable X with a probability distribution function $p(x) = P(X = x) = \frac{\lambda^x}{x!} e^{-\lambda}$, $x = 0, 1, 2, \ldots$ is called *Poisson-distributed* with parameter λ. In case a random variable follows this distribution, we will write $X \sim Po(\lambda)$, where Po stands for Poisson.

The Poisson distribution is named in honor of Siméon Denis Poisson, a French mathematician, 1781–1840. This distribution function is of major importance in queuing (Chap. 15 of this book) and will be extensively used in that context.

Example C.4: If the random variable $X \sim P_o(\lambda = 1.3)$, then $P(X = 2) = p(2) = \frac{(1.3)^2}{2!} e^{-1.3} = 0.2303$, and $F(X \leq 1) = p(0) + p(1) = 0.6268$.

Definition C.6 A random variable X is called *continuous*, if there exists a function $f(x)$, such that the cumulative distribution function $F(x)$ for X can be written as $F(x) = \int_{-\infty}^{t} f(t)dt$. The function $f(x)$ is called the *(probability) density function* of X.

Definition C.7 A continuous random variable X with cumulative distribution function $F(x) = 1 - e^{-(\lambda x)^\alpha}$ for $x > 0$, where α and λ are given positive parameters is said to have a *Weibull distribution*.

For $\alpha = 1$, the Weibull distribution reduces to the *exponential distribution* $Exp(\lambda)$.

Definition C.8 The *gamma function* $\Gamma(x)$ is given by

$$\Gamma(x) = \int_0^\infty e^{-t} t^{x-1} dt \text{ for } x > 0.$$

One can show that $\Gamma(1) = 1$ and $x\Gamma(x) = \Gamma(x+1)$, so that $\Gamma(n) = (n-1)!$, $n = 1, 2, \ldots$ and also that $\Gamma(\frac{1}{2}) = \sqrt{\pi}$.

Definition C.9 A continuous random variable X with density function $f(x) = \frac{\lambda}{\Gamma(\alpha)}(\lambda x)^{\alpha-1} e^{-\lambda x}$ for $x > 0$, where α and λ are given positive parameters, is said to follow a *gamma distribution*, and we write $X \sim \Gamma(\alpha, \lambda)$.

For $\alpha = n$ integer, we obtain the so-called *Erlang distribution* $\Gamma(n, \lambda)$, and if additionally $\alpha = 1$, we obtain the exponential distribution $Exp(\lambda) = \Gamma(1, \lambda)$.

Definition C.10 A continuous random variable X with density function $f(x) = \frac{1}{\sigma\sqrt{2\pi}} e^{-\frac{(x-\mu)^2}{2\sigma^2}}$ is said to follow a *normal distribution* (or, alternatively, *Gaussian distribution*) with parameters μ and $\sigma > 0$. In such a case, we write $X \sim N(\mu, \sigma)$. The distribution $N(0, 1)$, i.e., the normal distribution with $\mu = 0$ and $\sigma = 1$ is called the *standard normal distribution* and is usually denoted by Z. Its density function is called the *normal curve* (or *bell curve*), given by the function $f(x) = \frac{1}{\sqrt{2\pi}} e^{-\frac{1}{2}x^2}$.

Definition C.11 The *expected value* (or *mean* or *expectation*) $E(X)$ or μ of a discrete random variable X is given by $E(X) = \mu = a_1 P(X = a_1) + a_2 P(X = a_2) + \cdots$, meaning that we multiply each outcome by its associated probability and add them up. For a continuous random variable X, the expected value is given by $E(X) = \mu = \int_{-\infty}^{\infty} t f(t) dt$.

There are discrete and continuous random variables for which there exists no expected value.

Example C.5: For the random variable X of Examples $C2$ and $C3$, where X denotes the number of heads in three tosses of a fair coin, we obtain the expected value $E(X) = 0\left(\frac{1}{8}\right) + 1\left(\frac{3}{8}\right) + 2\left(\frac{3}{8}\right) + 3\left(\frac{1}{8}\right) = 1\frac{1}{2}$.

Theorem C.2 If $X \sim Bin(n, p)$, then $E(X) = np$, if $X \sim Po(\lambda)$, then $E(X) = \lambda$, if X has a Weibull distribution, then $E(X) = \frac{1}{\lambda}\Gamma\left(1 + \frac{1}{\alpha}\right)$, if $X \sim Exp(\lambda)$, then $E(X) = \frac{1}{\lambda}$, if $X \sim \Gamma(\alpha, \lambda)$, then $E(X) = \frac{\alpha}{\lambda}$, and if $X \sim N(\mu, \sigma)$, then $E(X) = \mu$.

Definition C.12 The *variance* $V(X) = \sigma^2$ of a random variable X with mean μ is defined as $E((X - \mu)^2)$. We call $\sigma = \sqrt{V(X)}$ the *standard deviation* of X. The expression $\sigma_\mu = \sigma/\mu$ is referred to as the *coefficient of variation* of X.

There are discrete and continuous variables for which the mean exists, but the variance does not. One can show that $V(X) = E(X^2) - \mu^2$ and that $V(X) \geq 0$ for all random variables X, as long as $E(X^2)$, $E(X)$, and $V(X)$ exist.

Example C.6: For the random variable X of Examples C2 and C3, where X denotes the number of heads in three tosses of a fair coin, we obtain the variance $V(X) = E(X^2) - \mu^2 = 3 - 2.25 = 0.75$. The standard deviation $\sigma = \sqrt{V(X)} = \sqrt{.75} \approx .8660$. The coefficient of variation $\sigma_\mu = \sqrt{.75}/1.5 \approx 0.5774$.

Theorem C.3 If the random variable $X \sim \text{Bin}(n, p)$, then $V(X) = np(1-p)$, if $X \sim Po(\lambda)$, then $V(X) = \lambda$, if $X \sim Exp(\lambda)$, then $V(X) = \frac{1}{\lambda^2}$, if $X \sim \Gamma(\alpha, \lambda)$, then $V(X) = \frac{\alpha}{\lambda^2}$, and if $X \sim N(\mu, \sigma)$, then $V(X) = \sigma^2$.

Theorem C.4 If the random variable $X \sim N(\mu, \sigma)$, then $Z = [(X - \mu)/\sigma] \sim N(0, 1)$.

Theorem C.5 For any random variable X and for any given numbers $a < b$, we have $P(a < X \leq b) = F(b) - F(a)$. For any continuous random variable X, $P(a \leq X \leq b) = P(a \leq X < b) = P(a < X < b) = F(b) - F(a)$ as well as $P(X = a) = P(X = b) = 0$.

Example C.7: Let the random variable $X \sim N(2.3, 0.9)$ and compute the probability $P(1.7 \leq X \leq 2.6)$. We find that $P(1.7 \leq X \leq 2.6) = P\left(\frac{1.7-2.3}{0.9} \leq \frac{X-2.3}{0.9} \leq \frac{2.6-2.3}{0.90}\right)$
$= P(-0.6667 \leq Z \leq 0.3333) = F(0.3333) - F(-0.6667)$. The function $F(x) = \int_{-\infty}^{x} \frac{1}{\sqrt{2\pi}} e^{-\frac{1}{2}t^2} dt$ is called area under the normal curve and it is tabulated at the end of this Appendix for values $x \geq 0$. By virtue of symmetry of the normal curve, we find that $F(-x) = 1 - F(x)$, therefore $F(0.3333) - F(-0.6667) = F(0.3333) - 1 + F(0.6667) = 0.6306 + 1 - 0.7475 = 0.3781$ by reading the table and interpolating as necessary.

It has been observed in practice that empirical data that are bell-shaped and symmetric share some common properties. In particular, the *empirical rule* holds, according to which about 68% of all observations lie within one standard deviation about the mean, about 95% of the observations are within two standard deviations about the mean, and virtually all observations are within three standard deviations about the mean. It is worth mentioning that this is not a provable property, but an observed rule that often occurs in practice for distributions of the aforementioned shape.

Making now stronger assumptions about the distribution, in particular that it is not only bell-shaped and symmetric but normal, we are able to confirm the assertion of the empirical rule. In particular, for a standard normal variable $Z \sim N(0, 1)$ we have $F(0) = 0.5$, $F(1) = 0.8413$, $F(2) = 0.9773$, and $F(3) = 0.9987$. Then

$P(-1 \leq Z \leq 1) = 2(0.8413 - 0.5) = 0.6826$, $P(-2 \leq Z \leq 2) = 2(0.9773 - 0.5) = 0.9546$, and $P(-3 \leq Z \leq 3) = 2(0.9987 - 0.5) = 0.9974$.

We will now consider events A, B, C, ..., which are sets of outcomes of experiments.

Definition C.13 Given the events A and B, the *union* $A \cup B$ is the event consisting of outcomes that are in A or in B or both. In contrast, the *intersection* $A \cap B$ is the event that consists of outcomes that are in both A and B.

Theorem C.6 (The *addition law for probabilities*): For any events A and B, we have $P(A \cup B) = P(A) + P(B) - P(A \cap B)$.

Example C.8: Recall Example C4, in which $X \sim Po(\lambda = 1.3)$. Furthermore, let $A = \{X = 0 \text{ or } 1\}$ and $B = \{1 \text{ or } 2\}$. Then $A \cup B = \{X = 0, 1, \text{or } 2\}$, while $A \cap B = \{X = 1\}$. We then find $P(A) = p(0) + p(1) = 0.6268$, $P(B) = p(1) + p(2) = 0.5846$, $P(A \cup B) p(0) + p(1) + p(2) = 0.8571$, $P(A \cap B) = p(1) = 0.3543$. In accordance with the theorem, we find that $P(A) + P(B) - P(A \cap B) = 0.6268 + 0.5846 - 0.8571 = P(A \cup B)$.

Definition C.14 The sets A and B are said to be *collectively exhaustive*, if their union $A \cup B$ includes all possible outcomes of an experiment. The two sets are called *mutually exclusive*, if their intersection is empty. The *complement* $\overline{A}$ of a set A (sometimes also written as $\neg A$) is the set of all possible outcomes not in A.

As far as probabilities are concerned, $P(\overline{A}) = 1 - P(A)$.

The addition law for probabilities can be generalized. For instance, for mutually exclusive sets A_1, A_2, ... A_m, we obtain $P(A_1 \cup A_2 \cup \ldots \cup A_m) = P(A_1) + P(A_2) + \ldots + P(A_m)$. To establish a multiplication law for probabilities, we need the following

Definition C.15 For any events A and B with $P(B) \neq 0$, the *conditional probability of A given B* is

$$P(A|B) = \frac{P(A \cap B)}{P(B)}.$$

Example C.9: Consider an experiment, in which the random variable X denotes that at most two heads come up in three tosses of a fair coin, i.e., $X \leq 2$, and define B as an event that sees at least one head in three tosses of a fair coin, i.e., $X \geq 1$. Then $A \cap B = \{X = 1 \text{ or } 2\}$, so that $P(A \cap B) = \frac{3}{8} + \frac{3}{8} = \frac{3}{4}$ and $P(B) = \frac{3}{8} + \frac{3}{8} + \frac{1}{4} = \frac{7}{8}$. Therefore, $P(A|B) = \frac{3/4}{7/8} = \frac{6}{7} \approx 0.8571$.

Theorem C.7 (The *multiplication law for probabilities*): For any events A and B, $P(A \cap B) = P(A|B)P(B)$.

Note that if $P(B) = 0$, then $P(A|B)$ is not defined, but in this case, the right-hand side is interpreted as being zero.

Definition C.16 The events A and B are said to be *statistically independent*, if $P(A \cap B) = P(A)P(B)$.

Theorem C.8 (*Bayes's theorem*): Let the events $A_1, A_2, \ldots A_m$ be mutually exclusive and collectively exhaustive. Then for any event B with $P(B) \neq 0$, we have
$$P(A_i|B) = \frac{P(B|A_i)P(A_i)}{P(B|A_1)P(A_1) + P(B|A_2)P(A_2) + \ldots + P(B|A_m)P(A_m)}, \quad i = 1, \ldots, m$$
The theorem of Bayes (Thomas Bayes, English clergyman, 1702–1761) will be used in Chap. 10 of this book. In the context of Bayes's theorem, the unconditional probabilities $P(A_i)$ are called *prior probabilities*, while the conditional probabilities $P(A_i|B)$ are referred to as *posterior probabilities*.

Area Under the Normal Curve

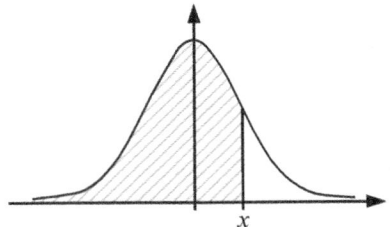

x	0.00	0.01	0.02	0.03	0.04	0.05	0.06	0.07	0.08	0.09
.00	.5000	.5040	.5080	.5120	.5160	.5199	.5239	.5279	.5319	.5359
.10	.5398	.5438	.5478	.5517	.5557	.5596	.5636	.5675	.5714	.5754
.20	.5793	.5832	.5871	.5910	.5948	.5987	.6026	.6064	.6103	.6141
.30	.6179	.6217	.6255	.6293	.6331	.6368	.6406	.6443	.6480	.6517
.40	.6554	.6591	.6628	.6664	.6700	.6736	.6772	.6808	.6844	.6879
.50	.6915	.6950	.6985	.7019	.7054	.7088	.7123	.7157	.7190	.7224
.60	.7258	.7291	.7324	.7357	.7389	.7422	.7454	.7486	.7518	.7549
.70	.7580	.7612	.7642	.7673	.7704	.7734	.7764	.7794	.7823	.7852
.80	.7881	.7910	.7939	.7967	.7996	.8023	.8051	.8079	.8106	.8133
.90	.8159	.8186	.8212	.8238	.8264	.8289	.8315	.8340	.8365	.8389
1.0	.8413	.8438	.8461	.8485	.8508	.8531	.8554	.8577	.8599	.8621
1.1	.8643	.8665	.8686	.8708	.8729	.8749	.8770	.8790	.8810	.8830
1.2	.8849	.8869	.8888	.8907	.8925	.8944	.8962	.8980	.8997	.9015
1.3	.9032	.9049	.9066	.9082	.9099	.9115	.9131	.9147	.9162	.9177
1.4	.9192	.9207	.9222	.9236	.9251	.9265	.9279	.9292	.9306	.9319
1.5	.9332	.9345	.9357	.9370	.9382	.9394	.9406	.9418	.9430	.9441
1.6	.9452	.9463	.9474	.9485	.9495	.9505	.9515	.9525	.9535	.9545
1.7	.9554	.9564	.9573	.9582	.9591	.9599	.9608	.9616	.9625	.9633
1.8	.9641	.9649	.9656	.9664	.9671	.9678	.9686	.9693	.9700	.9706
1.9	.9713	.9719	.9726	.9732	.9738	.9744	.9750	.9756	.9762	.9767
2.0	.9773	.9778	.9783	.9788	.9793	.9798	.9803	.9808	.9812	.9817
2.1	.9821	.9826	.9830	.9834	.9838	.9842	.9846	.9850	.9854	.9857
2.2	.9861	.9865	.9868	.9871	.9875	.9878	.9881	.9884	.9887	.9890
2.3	.9893	.9896	.9898	.9901	.9904	.9906	.9909	.9911	.9913	.9916
2.4	.9918	.9920	.9922	.9925	.9927	.9929	.9931	.9932	.9934	.9936

(continued)

(continued)

2.5	.9938	.9940	.9941	.9943	.9945	.9946	.9948	.9949	.9951	.9952
2.6	.9953	.9955	.9956	.9957	.9959	.9960	.9961	.9962	.9963	.9964
2.7	.9965	.9966	.9967	.9968	.9969	.9970	.9971	.9972	.9973	.9974
2.8	.9974	.9975	.9976	.9977	.9977	.9978	.9978	.9980	.9980	.9981
2.9	.9981	.9982	.9983	.9983	.9984	.9984	.9985	.9985	.9986	.9986
3.0	.9987	.9987	.9987	.9988	.9988	.9989	.9989	.9989	.9990	.9990
3.5	.999767									
4.0	.9999683									
4.5	.99999660									
5.0	.999999713									

References

Eiselt HA, Sandblom C-L (2004) Decision analysis, location models, and scheduling problems. Springer, Berlin

Eiselt HA, Sandblom C-L (2007) Linear programming and its applications. Springer, Berlin

Index

A
ABC classification, 401, 421
Activity
  critical, 304–318, 321, 324–327, 330
  duration, 311–314, 317, 320, 321, 325–327
Activity-on-arc (AOA) representation, 305
Activity-on-node (AON) representation, 305
Adjacency matrix, 216
AHP, *see* Analytic Hierarchy Process (AHP)
Algorithms, exact and heuristic, 172
All-integer linear programming (AILP), 162–166
Allocation problem, 28–34, 333
Analytic hierarchy process, 391–399
Anticipated payoffs, 359, 360, 376, 380
AOA representation, *see* Activity-on-arc (AOA) representation
AON representation, *see* Activity-on-node (AON) representation
Arc, 215–217, 221–223, 225–234, 239–250, 255, 256, 262, 263, 278, 286, 305, 356, 427
Arrival rate, 460–463, 465, 470, 471
Arrival time, 334, 481
Aspiration level, 134
Assignment problems, 15, 38–45, 67, 197–199, 211–213, 217, 250, 261, 293

B
Backorders, 149, 402, 404, 410–412, 417, 484
Backward pass, sweep, or recursion, 224, 226, 306–309, 311, 313, 321, 325, 372
Basis point, 99
Bayes's rule, 360, 367, 369, 372, 388
Bayes's theorem, 369, 514
Bin packing, 179, 501–503
Bisection search, 280, 282, 299

Blending problems, 34–38
Block diagram, 440, 441, 445
Boolean truth method, 444
Bottleneck, 28, 103, 107, 111, 227, 304, 309, 312
Branch-and-bound methods, 186, 188–194
Branching, 189, 191–193, 206, 207, 210
Breadth-first-search, 224
Break-even analysis, 9
Breakthrough, 161, 223–226, 254
Bridge (of a graph), 2, 215, 240, 242
Budget constraint, 4, 19, 31, 82, 133–135, 199, 203

C
Calling population, 458, 459
Capital budgeting, 171, 353
Catchment areas, 497
Center, absolute
  node, 278, 279
Center-of-gravity, 284, 285, 300
Center problems, 269, 277–281
Centrality, closeness
  betweenness, 218, 220
  degree, 218
  eigenvalue, 220
Certainty equivalent, 373, 374
Changes
  objective function coefficients, 94, 95, 97, 100, 108, 113
  right-hand side values, 82, 94, 97, 98, 100, 101, 103, 104, 108, 164
  structural and parameter, 91, 93, 94
Channel, 458, 462, 469, 470, 480
Chinese postman problem, 240–242, 246–248, 259, 262
Circuit, 216, 241–244, 451, 457

Coefficient of variation, 394, 395, 398, 399, 463, 512
Column
  dominated, 359
Competitive location, 292
Complementary slackness conditions, 119, 122
Completion time, 322, 334, 335, 344, 346
Congruence methods, 477
Conservation equations, 222, 228, 229, 231, 256
Consistency index, 394
Constraint method, 131, 133, 134, 140
Constraints
  addition and deletion of, 91
  binding, 64, 73, 74, 78, 86–88, 93, 100, 101, 103–105, 123, 176, 177
  complicating, 198, 199, 211–213
  conditional, 168, 172, 174, 182, 200, 203
  exclusion, 169, 173, 175
  linking, 173, 174, 176
Construction phase, 496
Contour lines, 75, 84, 499
Convex hull, 183, 184
Convexity, 79
Corner point
  theorem, 77, 80, 84, 163, 165, 183, 286
Cost, holding/carrying
  ordering, 149, 403–407, 410, 412, 415, 418–420, 424, 485, 487
  penalty, 375, 411, 417–420, 424
  purchasing, 54, 404–406, 412
  shortage, 404, 410, 411, 417, 421–423, 483–485, 487
Covering matrix, 271, 273, 294, 296
Covering problems, 270–277, 280, 281, 293
CPM, *see* Critical path method (CPM)
Crashing, 312
Crash time, 312, 314, 316, 317, 325, 326
Critical activity, 308–311
Critical path, 303–317, 321, 322, 324–328, 330, 332
Critical path method (CPM)
  project acceleration, 312–317
Cut, minimal, 227, 253, 254
Cutting plane method, 183–188, 204
Cutting stock problems, 52–70
Cycle length, 404, 406–408, 477

D
Dantzig cut, 184–186
Decision analysis, 353–383, 385, 388, 390, 425

Decision trees, 356, 358, 366–372, 375, 377, 379–383
Decomposition principle, 8
Degeneracy
  dual, 84–85, 88, 89, 185, 188
  primal, 85, 86, 89, 90, 99, 104, 185, 205
Degree (of a node), 218, 241, 245
Depth-first search, 224
Destination, 38–43, 113–115, 217, 220, 229, 230, 292, 495
Diet problem, 22–28, 109, 110, 125, 172–174, 176
Dijkstra method, 234, 256
Discrete event simulation, 473
Distribution
  binomial, 453, 510
  Erlang, 449, 511
  exponential, 446, 449, 459, 479, 511
  gamma, 449, 452, 511
  normal, 322, 479, 511
  Poisson, 459, 461–463, 478–479, 510
  Rayleigh, 449
  Weibull, 440, 449, 511
Divisibility, 16, 17, 143
Dominance, 357–359, 380, 382
Duality, 114–123, 152, 166, 167, 197–199, 211, 212
Dual problem
  relations to primal problem, 114–116, 119, 122
  setting up, 121
Dual variables, 114, 117–119, 143, 166, 168, 198, 199, 211–213
Due time, 334
Dynamic programming, 3, 223

E
Earliest due date (EDD) algorithm, 337, 349
Earliest possible finishing times (EF), 306, 307, 309, 311, 330
Earliest possible starting times (ES), 306–309, 311, 317
Economic order quantity (EOQ), 149, 404–416, 421, 422, 424
EDD algorithm, *see* Earliest due date (EDD) algorithm
Edge, 140, 215–219, 222, 238–246, 248, 261, 278, 280, 286, 509
EF, *see* Earliest possible finishing times (EF)
Efficiency, 38, 125, 180, 291, 292, 371, 380, 382, 383, 502
Either-or constraints, 168

Empirical rule, 321, 512
Employee scheduling, 49–52
EMV, see Expected monetary values (EMV)
EOQ, see Economic order quantity (EOQ)
EPII, see Expected payoff with imperfect information (EPII)
EPPI, see Expected payoff with perfect information (EPPI)
Error bound, 495, 502
ES, see Earliest possible starting times (ES)
Euclidean distance, 268, 279, 283–285, 300, 390, 396
Euler graph, 241
EVPI, see Expected value of perfect information (EVPI)
EVSI, see Expected value of imperfect (or sample) information (EVSI)
Excess variable, 90, 102–104, 106, 110, 111, 114, 117, 119, 122, 134, 185, 186, 205, 309, 332
Expected monetary values (EMV), 360, 363–367, 371, 372, 378, 379
Expected payoff with imperfect information (EPII), 371, 372
Expected payoff with perfect information (EPPI)
"equity" objectives, 291
Expected value, definition, 371
Expected value of imperfect (or sample) information (EVSI), 371, 372, 380, 383
Expected value of perfect information (EVPI), 366, 367, 371, 380, 382
Extreme points
number of, 125

F

Facilities, extensive
undesirable, 269, 290
Failure rate, decreasing, increasing, 449
Feasibility, 6–9, 79, 82, 130, 197, 463, 464, 466
Feasible
set, 28, 72, 73, 76–81, 84–86, 88, 89, 94, 98–100, 120, 130, 132, 133, 138, 140, 150, 154, 155, 159, 162–165, 167, 183–186, 188–191, 291
solution, nonexistence of, 80, 81, 100
Feasible direction methods, 79
First come, first served (FCFS), 459
First fit (FF) algorithm, 501
First passage time, 428, 429
Fixed charges, 175–178
Fleury's algorithm, 242, 244, 246, 248

Flow, pattern and value
balancing equations, 222
mean, 335, 336, 344, 347–350
time, 54, 334–336
Floyd–Warshall method, 235, 236, 256, 258
Forward pass, sweep, or recursion, 306
Frontier, nondominated, 130, 131, 138, 139, 141

G

Games against nature, 354
Gantt chart, 2, 317, 318, 320, 326–328, 330, 331, 337, 338, 340, 341, 346, 348–352
Gaussian elimination, 507
GERT, see Graphical evaluation and review technique (GERT)
Gini index, 291
Global optimum, 496
Goal programming, 126, 133–142, 181
Graph, augmented
bipartite, 217
connected, 216
directed, undirected, and mixed, 215
Graphical review and evaluation technique (GERT), 320
Graphical solution method, 18, 70–80, 83, 84, 86–88, 95, 96, 138
Greedy method, 194, 210, 289, 496
Guillotine cuts, 56

H

Halfplane, 72, 128
Halfspace, 71, 72, 81, 98, 127
Hazard function, 439, 440, 447–450, 454
Hazard rate, 448, 449, 455
Heuristic methods, 6, 168, 183, 194–213, 249, 252, 289, 318, 335, 340, 341, 495–504
Hub location, 292
100% rule, 97, 100, 101
Hyperplane, 71, 72, 74–76, 85, 86, 90, 98, 99

I

Implementability, 7
Improvement cone, 128, 130, 137, 138, 140, 141
Improvement phase, 496
Incremental technique, 79
Indegree, 241, 244, 247, 248, 262, 264
Influence diagrams, 355–357, 375, 376
Institute for Operations Research and Management Science (INFORMS), 1

Integer programming, 2, 17, 161–163, 165, 166, 170–179, 183–213, 217, 268, 293, 311, 319, 338, 339, 353
Integrality gap, absolute and relative, 166
Interarrival time, 460, 461, 463, 480–483
Interchange heuristic, 195
Interval constraints, 82
Inventory models, 43, 45–49, 59, 148, 401–424
Investment allocation, 30, 32
Iso-profit lines, 75–79, 127

J

Jackson's job shop algorithm, 347, 348
Jackson's rule, 337
Johnson's algorithm, 347
Just-in-time (JIT) inventory system, 402

K

Kanban systems, 402
Kendall's notation, 458
Kingman equation, 463
Kirchhoff node equations, 222
Knapsack problems, 122, 170–172, 194, 199
Königsberg bridge problem, 2, 215, 239, 240
K-out-of-n systems, 443
Kruskal's method, 239

L

Label correcting methods, 233
Labeling methods, 222
Label setting methods, 233
Lagrangean multipliers, 151, 167, 168, 197, 213
Lagrangean relaxation, 166, 167, 197–199, 211
Land use problem, 174, 175
LAPT algorithm, *see* Longest alternate processing time (LAPT) algorithm
Last in, first out (LIFO), 459
Lateness, 334, 335, 337, 338, 345, 346, 348, 349
Latest allowable finish time (LF), 307–309, 311
Latest allowable start time (LS), 307–309
Layout problems, 293
Lead time
  stochastic, 417–420
LF, *see* Latest allowable finish time (LF)
Linear congruence methods, 477
Linearity, 16, 17, 39
Linear programming

relaxation, 165, 166, 171, 183–189, 191, 192, 204
List scheduling methods, 340
Little's formula, 461, 464
Local optimum, 502
Location
  continuous, 268
  discrete, 268
Location-allocation heuristic, 288, 301
Location models, 230, 235, 267–302
Location set covering problem (LSCP), 271, 272
Logical variables, 161, 168, 171, 173
Longest alternate processing time (LAPT) algorithm, 345, 352
Longest processing time first (LPT) algorithm, 340–342, 348–350
Lorenz curve, 291
LPT algorithm, *see* Longest processing time first (LPT) algorithm
LS, *see* Latest allowable start time (LS)
LSCP, *see* Location set covering problem (LSCP)

M

Machine scheduling
  single, parallel, dedicate, 334, 336–352
Makespan, 335, 336, 340–342, 345–347, 349, 350
Management science, vii, 1, 3, 4
Manhattan distance, 268, 279, 282, 298, 299, 390, 391
Markov chains, 425–437
Markovian property, 426
Material requirements planning (MRP), 402
Matrix, 39, 40, 43, 65, 67, 211, 212, 216, 219, 220, 235–238, 245, 247, 251, 252, 258, 259, 264, 265, 271, 273, 276, 278, 286–289, 294–297, 301, 355, 357, 359–361, 363, 366, 368, 375, 376, 379, 380, 382, 385, 387–389, 391–396, 398, 399, 426–429, 431, 432, 434, 435, 437, 505, 506
Maximal covering location problem (MCLP), 274, 275
Maximal flow problem, 222, 227, 230
Maximal lateness, 335, 337, 338, 345, 346, 348
Maximax rule, 359
Maximin rule, 359
MCLP, *see* Maximal covering location problem (MCLP)
McNaughton's algorithm, 342

Mean recurrence time, 428, 431, 434, 436
Mean time between failures (MTBF), 447
Mean time to failure (MTTF), 447, 449, 451
Median problems, 269, 281–290, 299
Midsquare method, 476, 477
Min cost feasible flow problem, 227, 228, 231
Min-cut max flow theorem, 253
Minimax, 359
Mixed-integer linear programming (MILP), 162
Model, deterministic and stochastic, 473
Modeling process, 1, 10–13, 28, 91, 92, 136
MOdified DIstribution (MODI) method, 40
Multicriteria decision making (MCDM) problems, 126
Multiobjective programming, 125–142
Multistart procedure, 252, 504

N
Nash equilibria, 293
Nearest neighbor method, 251, 252, 500
Neighborhood search, 502
Network
 flows, 215, 220–230
Newspaper boy problem, 360
NIMBY, 269
Nodes
 active (in branch-and-bound), 188
 reachable, 216
Nonbreakthrough, 223, 226, 254
Nondominated frontier, 130, 131, 138, 139, 141
Normal duration, 312, 324, 325
Normalization technique, 392, 398

O
Objective
 function, gradient of, 75–77, 83, 84, 89, 94–96, 98, 127, 138
Operational life testing, 451
Operations research
 definition, 1
 elements, 4–10
 journals, 3
Opportunity costs, 101, 102, 104, 110, 117–119, 360, 361
Optimality, 3, 6, 7, 9, 41, 125, 126, 149–151, 166, 204, 284, 341, 344, 352, 366, 368
Optimality gap, 166
Optimal solution
 alternative, 84, 88, 89, 92, 96, 104, 206, 248, 251, 263, 349
 economic analysis of, 22

 unbounded, 83–84, 92–95, 104, 119, 167
 unique, 78, 95, 96, 122, 149, 155–159, 198, 211–213, 261
Optimist's rule, 359
Order point (s, S) policy, 420
Origin, 38–43, 72, 75, 76, 84, 99, 113–115, 147, 217, 220, 229, 267, 292, 500
Outdegree, 241, 244, 247, 248, 262, 264
Overshipments, 42

P
Page rank, 220
Pairwise comparisons, 358, 391, 392
Pairwise exchange, 196, 252, 265, 496
Parallel-series system, 441
Parallel structure, 441
Parameters, 5–7, 9, 16–18, 24, 39, 46, 59, 90, 91, 93, 94, 144, 149, 199, 268, 269, 274, 333, 334, 401–404, 406, 407, 410, 411, 414, 416, 417, 446, 448, 449, 451, 452, 454, 459, 461, 463, 466, 476–479, 483, 485, 510, 511
Path, 79, 84, 191, 215–220, 224–226, 230–238, 240, 244–248, 253, 254, 256–264, 268, 271, 280, 286, 303–317, 321, 322, 324–328, 330, 332, 427, 428, 430, 441, 442, 445
Payoff table, 380
Penalty term, 152, 198
Performance ratio, 341, 342, 348, 350
Periodic review (R, S) policy, 421
PERT, *see* Program evaluation and review technique (PERT)
Pessimistic rule, 359
Phase, construction and improvement, 496
Poisson process, 459
Political districting, 181–183
Pollaczek–Khintchine formula, 462
Polytope, 73
Postoptimality analyses, 90–114, 119
Precedence relations, 304–306
Preemption, 336, 342, 344, 350
Probabilistic, 16, 269, 304, 320, 333, 439, 509
Probability
 conditional, 367, 369, 372, 379, 381, 426, 429, 448, 477, 513, 514
 definition, 16
 distribution, 304, 320, 354, 355, 366, 417, 419, 420, 429, 440, 510
 indicator, 368, 369, 371, 372
 posterior, 369, 371, 372, 514
 prior, 369, 372, 380, 514

Processing time, 5, 20–22, 93, 176, 179, 334, 336–351
Process, stationary, 427
Production-inventory models, 43, 45–49, 59
Production-lot size model, 414–417
Production planning, 6, 9, 20–22, 56, 60, 105, 106
Program evaluation and review technique (PERT), 303, 304, 320–332
Project networks, 303–332, 335, 340

Q
Quantity discounts, 404, 412–414, 423

R
Random numbers, true and pseudo, 475–477
Random variables, 320, 417, 418, 426, 445, 447, 448, 478, 479, 483, 509–513
Reachable (in a graph), 216
Ready time, 334, 345, 349, 502
Rectilinear distance, 282, 284
Reduced costs, 102
Redundancy, 85, 86, 440, 450–451
Regret criterion, 359, 375, 382
Regret matrix, 359, 360, 376
Regularity condition, 415, 416, 422, 423, 460
Release time, 334, 336, 337
Reliability testing, 451
Reorder point, 409, 417–424, 483–486
Reshipments, 42
Resource requirement graph, 317–320, 327, 328, 330, 331
Resources, scarce, 1, 28, 30–33, 58, 277, 304
Review systems, continuous and periodic, 420
Risk
  aversion, 4, 373, 374
  neutral, 373, 375
  seeking, 373
Robust optimization, 359
Root (of a tree), 189, 192, 193, 207, 210, 367, 368
Routing, arc and node, 239–266
Row, dominated, 359

S
Satisficing, 134
Schedule length, 335, 340–343, 345–348, 350, 352
Seed, 476–478
Selection problem, 353–355

Sensitivity analyses
  graphical, 90–101
Series-parallel system, 443
Service
  rate, 149, 460–463, 468, 469, 471
  station, 149, 458–461, 463, 464, 466, 468, 480, 482
  time, 337, 459, 461–463, 466, 470, 480–483
Service in random order (SIRO), 459
Shadow prices, 102–104, 106, 108, 110–114, 117, 120, 332
Shop, open, flow, job, 334
Shortest path problems, 230–238
Shortest processing time (SPT) algorithm, 336, 349
Simplex method, 3, 15, 16, 79, 84, 86, 131
Simulated annealing, 503
Simulation, deterministic and stochastic, 473
Simultaneous linear equations, 73, 74, 77–80, 152, 430, 505–508
Sink, 220, 222–231, 234, 255, 264, 305–307, 311, 315, 316
Smith's ratio rule, 336
Solution
  definition, 6
  efficient, 130, 131
  nondominated, 130–133, 140
  noninferior, 130, 141
  pareto-optimal, 126, 130
  tree, 188–194, 206–208, 210, 433
Source, 10, 220, 222–231, 233–235, 254, 255, 264, 305–308, 311, 315, 316, 475
Spanning tree
  minimal, 239, 240, 259
Spider plot, 288, 289
SPT algorithm, see Shortest processing time (SPT) algorithm
Standard deviation, 321, 322, 324, 326, 327, 394, 450, 453, 462, 463, 512
Standby systems, 450
State, absorbing
  steady, 460
  transient, 460
State space, 425
Statistical independence, 514
Steady state solution, 430–431, 474
Stochastic processes, 425–437, 439, 445
Subtour elimination constraints, 250, 251
Supply chain management, 402
Swap heuristic, 195
System, k-out-of-n
  lifetime, 445, 446
  parallel, 443

Index 523

parallel-series, 442
series, 443
series-parallel, 443
standby, 450

T
Tabu search, 502, 503
Tardiness, 334, 335, 338, 339, 347
Target value, 134, 361–363, 375, 377
Testing
 (highly) accelerated life, 451
 operational life, 451
 reliability, 451
Time-state graph, 427
Time-to-failure (TTF), 445, 447
Tool crib, 466–468
TOPSIS, 390–391, 396
Total float, 308, 309, 311, 326, 327
Tradeoffs, 125, 126, 135, 277
Traffic intensity, 460, 483
Transient states, 460, 474
Transition probabilities, 426–428, 431, 435
Transportation problem, 38, 39, 41, 43, 44, 49, 65, 113, 183, 248, 263, 264
 (un-)balanced, 39
Traveling salesman problem, 45, 241, 249, 264
Tree (graph), 216–217
Truncation, 451
Two-bin system, 409, 420

U
Uncertainty, 354, 355, 357–364, 375, 417

Undesirable facilities, 269, 290, 291
Utility theory, 353, 373–383
Utilization rate, *see* Traffic intensity

V
Variables
 addition and deletion of, 91
 deviational, 134–136, 181
 slack, 19, 102, 103, 119, 122, 134, 186, 188
 surplus, 102
Variance, 180, 291, 321, 322, 326, 327, 394, 459, 462, 512
Varignon frame, 285, 286
Vector, 75, 126–133, 137, 139, 140, 275, 286–289, 297, 301, 359, 392, 394, 398, 429, 430, 432, 505–506
Vector optimization, 126–133, 137, 139, 140
Vertex (graph), 81, 130, 290, 297, 299–301
Vertex substitution method, 300–301
von Stackelberg solutions, 293

W
Wald's rule, 359, 360, 374
Weighted shortest processing time (WSPT) algorithm, 336, 337
Weighting method, 131–133, 135, 140
Weiszfeld method, 284, 300
Workload balancing, 179–181
Wrap-Around Rule, 342, 343, 350, 351
WSPT algorithm, *see* Weighted shortest processing time (WSPT) algorithm

The manufacturer's authorised representative in the EU is Springer Nature Customer Service Centre GmbH, Europaplatz 3, 69115 Heidelberg, Germany. If you have any concerns regarding our products, please contact ProductSafety@springernature.com

Printed and bound by CPI Group (UK) Ltd, Croydon, CR0 4YY
25/03/2026
02078174-0015